5th International Conference on Bioinformatics, Biotechnology, and Biomedical Engineering (BioMIC 2023)

BIO Web of Conferences Volume 75 (2023)

Yogyakarta, Indonesia
6-7 September 2023

Editors:

S.A. Kristina
M.-H. Hsu
A. Syahrom
H. Ismail
T.Q. Vo

ISBN: 978-1-7138-8329-6

Printed from e-media with permission by:

Curran Associates, Inc.
57 Morehouse Lane
Red Hook, NY 12571

Some format issues inherent in the e-media version may also appear in this print version.

For additional information, please contact EDP Sciences – Web of Conferences at the address below.

EDP Sciences – Web of Conferences
17, Avenue du Hoggar
Parc d'Activité de Courtabœuf
BP 112
F-91944 Les Ulis Cedex A
France

Phone: +33 (0) 1 69 18 75 75
Fax: +33 (0) 1 69 28 84 91

contact-edps@webofconferences.org

Additional copies of this publication are available from:

Curran Associates, Inc.
57 Morehouse Lane
Red Hook, NY 12571 USA
Phone: 845-758-0400
Fax: 845-758-2633
Email: curran@proceedings.com
Web: www.proceedings.com

TABLE OF CONTENTS

BIOINFORMATICS AND DATA MINING

BIOMEDICAL SCIENCES AND ENGINEERING

https://doi.org/10.1051/bioconf/20237500001

Preface

BioMIC 2023 - The 5th International Conference on Bioinformatics, Biotechnology, and Biomedical Engineering is an annual meeting organized by Directorate of Research Universitas Gadjah Mada (UGM), Yogyakarta, Indonesia. This year's conference was held on 6–7 September 2023. The scientific symposium program consisted of plenary sessions and parallel sessions and included the following topics: Big Data for Public Health Policy; Bioinformatics and Data Mining; Biomedical Sciences and Engineering; Biomolecular and Biotechnology; and Drug Development and Nutraceutical.

We are grateful for the support of the faculties in the Universitas Gadjah Mada which scientifically supported the conference from the reviewers until the editing of the papers published in the BIO Web on Conferences. The BioMIC 2023 had about 165 attendees including the presenters and participants. Among those, the conference received 108 submitted papers and all papers were subjected to at least 2–3 reviews by experts on the related topics. For this, only 40 papers were selected for publication from six countries: Indonesia, Morocco, Vietnam, United Stated, Thailand, and Ireland.

We believed that during the two days conference, we had enabled to facilitate plenty of networking opportunities and provided the participants with the opportunity to meet and interact with the leading scientists and researchers, friends as well as colleagues. We hope that this proceeding will be beneficial for the development of life science in general and biological sciences and public health in particular.

Editorial boards

Welcoming Remarks from the Chair

Ladies and gentlemen, esteemed guests, and distinguished speakers and researchers, on behalf of the organizing committee, I extend a warm welcome to each and every one of you gathered here, representing various corners of the world, to this remarkable scientific gathering, the 5th International Conference on Bioinformatics, Biotechnology, and Biomedical Engineering (BioMIC 2023).

Held as part of the Universitas Gadjah Mada Annual Scientific Conferences (UASC 2023) series, the 5th International Conference on Bioinformatics, Biotechnology, and Biomedical Engineering (BioMIC 2023) provides an ideal academic platform for researchers to present the latest research findings and describe emerging technologies and directions in bioinformatics, biotechnology, and biomedical engineering.

This year, the conference will take the theme "Accelerating translational healthcare by leveraging big data and machine learning" with five symposia: Big Data for Public Health Policy symposium, Bioinformatics and Data Mining symposium; Biomedical Sciences and Engineering symposium; Biomolecular and Biotechnology symposium; and Drug Development and Nutraceutical symposium.

I am happy to announce that we have received 108 submissions in the beginning of the call for paper period. After a thorough review, 40 papers are selected to participate and be presented in this conference. Our participants are various, coming from different parts of the world, Indonesia, Morocco, Vietnam, United Stated, Thailand, and Ireland. Importantly, we are very grateful for our speakers, Prof. Teruna Siahaan from Kansas University, US; Prof. Hideki Enomoto from Kobe University, Japan; Prof. Dr. Min-Huei Hsu from Taipei Medical University, Taiwan; Assoc.Prof. Dr. Ardiyansyah Syahrom from University of Technology Malaysia, Malaysia; and Dr. apt. Hilda Ismail from Universitas Gadjah Mada, Indonesia, who are willing to share their expertise and knowledge in BioMIC 2023.

Translational healthcare research has changed over time as technologies have advanced. Translational research enhanced the process of leveraging healthcare data from all stakeholders in order to facilitate both the discovery and translation, as well as the health outcomes associated with new health technologies and therapies.

With the long-term goal of enhancing public health, translational research promotes the transdisciplinary integration of basic research, patient-oriented research, and population-based research. The following stage of evolution will use Big Data and ML (Machine Learning) to extract meaningful insight from Real World Evidence and, in the process, enable strategies that close crucial knowledge and practice gaps and assist the government in estimating the true value of treatments in terms of improving health and lowering costs.

The integration of knowledge across health system borders and the use of big data, analytics, and procedures to promote collaboration and innovation are key components of accelerating medical and health research. Utilizing technology that simplifies the information acquisition, representation, inference, and explanation processes can improve innovation. Large-scale access to clinical, biological, and healthcare data offers advancements in disease detection and treatment as well as improvements in patient health.

Interdisciplinary approaches to problem solving and collaboration are becoming increasingly important in facilitating knowledge discovery and integration. Big Data and ML technologies promise to have a profound impact – enabling reproducibility, aiding in discovery, and accelerating and transforming medical and healthcare research across the healthcare ecosystem.

Together, let us seize this remarkable opportunity to connect, collaborate, and inspire one another, as we embark on an extraordinary voyage of knowledge and discovery at BioMIC 2023. Thank you for joining us on this remarkable journey, and I wish you all a rewarding and memorable conference experience.

Yogyakarta, September 6, 2023

Chair of BioMIC 2023
Prof. Dr. apt. Susi Ari Kristina, S.Farm., M.Kes.

 https://doi.org/10.1051/bioconf/20237500002

Welcoming Remarks
from the Rector of Universitas Gadjah Mada

Dear Distinguished Keynote Speakers,
Distinguished Invited Speakers and Participants,
Conference's Committee,
Ladies and Gentlemen.

On behalf of Universitas Gadjah Mada, it is my pleasure and privilege to welcome you all to the fifth International Conference on Bioinformatics, Biotechnology, and Biomedical Engineering (BioMIC 2023), hosted by Universitas Gadjah Mada (UGM). First of all, I would like to extend my gratitude to distinguished speakers. Please allow me to express my sincere appreciation for the honorable speakers: Prof. Teruna Siahaan from Kansas University, US; Prof. Hideki Enomoto from Kobe University, Japan; Prof. Dr. Min-Huei Hsu from Taipei Medical University, Taiwan; Assoc. Prof. Dr. Ardiyansyah Syahrom from University of Technology Malaysia, Malaysia; and Dr. apt. Hilda Ismail from Universitas Gadjah Mada, Indonesia.

As a pioneering university, the history of UGM's education has opened the boundaries between academics and professionals across the world, to discover critically scientific inventions as the precious roots of knowledge for the benefit of humankind. This conference also has a role to answer the critical global health issues through the bioinformatics, biotechnology, and biomedical engineering views.

The conference will take the theme "Accelerating translational healthcare by leveraging big data and machine learning". The significance of accelerating translational healthcare through the utilization of big data and machine learning cannot be overstated. It has sparked a transformative wave in the healthcare industry, fundamentally changing our understanding, diagnosis, and treatment of diseases. With the ever-increasing volume and complexity of healthcare data, innovative methods have become imperative to unlock its immense potential. By leveraging big data analytics and machine learning algorithms, healthcare professionals now have unprecedented access to valuable insights, patterns, and correlations that were previously concealed. This empowers them to make informed decisions and deliver personalized care to patients based on concrete evidence. The ability to rapidly analyze vast amounts of data has drastically expedited the translation of scientific breakthroughs into practical applications, resulting in advancements in disease prevention, early detection, and treatment optimization. Additionally, the integration of machine learning algorithms has further enhanced the accuracy and efficiency of medical diagnostics, enabling the prediction of outcomes and identification of new therapeutic targets. Embracing the capabilities of big data and machine learning propels translational healthcare forward, leading us into an era of precision medicine, improved patient outcomes, and a healthier future for all.

The conference is bridging the gap among disciplines to bring and share their innovation, research, and ideas about our scientific issues today. UGM is proud to be leading the way in facilitating the interdisciplinary research dissemination of cutting-edge sharing information in diverse subjects. Let us think widely to identify research supports in the biomedical engineering field and bioinformatics to facilitate contact with biomolecular, drug development, and big data for public health.

After 5 years, BioMIC 2023 as a part of the Annual Scientific Conference Series, holding annual gatherings for the brilliant ideas from Indonesia and overseas to share the latest findings in their fields. It proves UGM's consistency to preserve international academic relations. This series has been an enormous success to bring collaboration with our international partners, shaping the scientific networks, increasing Indonesia author's greatness in the global publication's scopes, and with a global readership, and underscoring UGM's place as a standard-bearer of scientific development.

We are honored and humbled to many experts who have attended this year's conference. We thank the speakers for the expertise and knowledge that will bring to great discussion during. Special thanks are also extended to the organizing committee members in the BioMIC 2023 preparation, for their hard work, as well as the entire staff of Directorate of Research. I also would like to thank all the conference participants who contribute to making this truly the most memorable BioMIC 2023. I wish you all to enjoy this conference, and above all a successful BioMIC 2023.

Thank you.

Rector of Universitas Gadjah Mada,
Prof. dr. Ova Emilia, M.Med.Ed., Sp.OG(K)., Ph.D.

UNIVERSITAS GADJAH MADA | UGM BioMIC

The 5th International Conference ON BIOINFORMATICS, BIOTECHNOLOGY, AND BIOMEDICAL ENGINEERING

Conference Day : **6-7 September 2023** | Yogyakarta, Indonesia

Keynote Speakers

Prof. **HIDEKI ENOMOTO**
Graduate School of Medicine, Kobe University, Japan

Assoc.Prof. Dr. **ARDIYANSYAH BIN SYAHROM**
University of Technology Malaysia, Malaysia

Prof. **MIN-HUEI HSU**
Taipei Medical University, Taiwan

Prof. **TERUNA J. SIAHAAN**
School of Pharmacy, Kansas University, US

Symposia

- Big Data for Public Health Policy
- Bioinformatics and Data Mining
- Biomedical Sciences and Engineering
- Biomolecular and Biotechnology
- Drug Development and Nutraceutical

Theme

Accelerating Translational Healthcare by Leveraging Big Data and Machine Learning

Registration Link

https://biomic.ugm.ac.id/participate

Important Date

31 July 2023 — Full paper submission deadline

11 Aug 2023 — Last Notification of acceptance

14 Aug 2023 — Early bird payment deadline

18 Aug 2023 — Registration deadline

24 Aug 2023 — Camera-ready and presentation submission deadline

6-7 SEP 2023 — THE CONFERENCE

The accepted and presented papers will be submitted for proceedings/journals that are indexed by Scopus

biomic.ugm.ac.id

biomicugm | biomic@ugm.ac.id | s.id/biomicsupport | fb.me/biomicugm edp sciences — Web of Conferences

Statement of Peer review

In submitting conference proceedings to *Web of Conferences*, the editors of the proceedings certify to the Publisher that

1. They adhere to its **Policy on Publishing Integrity** in order to safeguard good scientific practice in publishing.
2. All articles have been subjected to peer review administered by the proceedings editors.
3. Reviews have been conducted by expert referees, who have been requested to provide unbiased and constructive comments aimed, whenever possible, at improving the work.
4. Proceedings editors have taken all reasonable steps to ensure the quality of the materials they publish and their decision to accept or reject a paper for publication has been based only on the merits of the work and the relevance to the journal.

Title, date and place of the conference

BioMIC 2023 - 5th International Conference on Bioinformatics, Biotechnology, and Biomedical Engineering

Yogyakarta, Indonesia. 6-7 September 2023

Proceedings editor(s):

1. Prof. Dr. Susi Ari Kristina, S.Farm., M.Kes., Apt.; Universitas Gadjah Mada, Indonesia (susiari_k@ugm.ac.id)
2. Prof. Min-Huei Hsu; Taipei Medical University, Taiwan (701056@tmu.edu.tw)
3. Assoc.Prof. Dr. Ardiyansyah bin Syahrom; University of Technology Malaysia, Malaysia (ardi@utm.my)
4. Dr. apt. Hilda Ismail, M.Si.; Universitas Gadjah Mada, Indonesia (hilda_fa@ugm.ac.id)
5. Assoc Prof. Vo Quang Trung; Pham Ngoc Thach University of Medicine, Vietnam (trungvq@pnt.edu.vn)

Date and editor's signature

October 31th, 2023

Prof. Dr. apt. Susi Ari Kristina, S.Farm., M.Kes.

Bibliometric Analysis of *Basella* ssp. as an Antioxidant

Dewa Ayu Swastini[1,2], *Ronny* Martien[4], *Jajah* Fachiroh[5], and *Agung* Endro Nugroho[3*]

[1]Doctoral Program, Faculty of Pharmacy, Universitas Gadjah Mada, Yogyakarta 55281, Indonesia
[2]Pharmacy Study Programme, Faculty of Mathematics and Natural Sciences, Universitas Udayana, Badung 80361, Bali
[3]Departement of Pharmacology and Clinical Pharmacy, Faculty of Pharmacy, Universitas Gadjah Mada, Indonesia
[4]Departement of Pharmaceutics, Faculty of Pharmacy, Universitas Gadjah Mada, Yogyakarta 55281, Indonesia
[5]Departement of Histology and Cell Biology, Faculty of Medicine, Universitas Gadjah Mada, Yogyakarta 55281, Indonesia

Abstract. The last ten years have seen the discovery of free radicals and their damaging impacts. Increasing exogenous antioxidant intake could reduce the damage caused by oxidative stress. Several plants have been shown to have antioxidant activity, and one such plant is *Basellal.* It is high in phytochemicals which can act as antioxidants, and its consumption may help fight free radicals generated by the body. In particular, this plant is essential for stimulating normal wound healing response. To the best of our knowledge, no bibliometric analysis of published data on Basella as an antioxidant has been done. The goal of this study is to conduct a bibliometric analysis of the research on *Basella*'s antioxidant properties in the Scopus database using the VOSviewer and RStudio tools. There were 56 articles on Basella as an antioxidant according to the bibliometric analysis. The countries with the highest research output was India (27 documents), and the most productive institution was Chiang Mai University (15 documents). The most productive source was the International Journal of Pharmacy and Pharmaceutical Sciences. P. Giridhar had significant significant impact on papers on Basella as an antioxidant (H-index of 5). The most common keywords were "antioxidant" (859 occurrences with 1,340 total link strength) and "Basella alba" (606 occurrences with 1,048 total link strength). Findings from this data suggest the novelties of Basella as an antioxidant.

Keywords : *Basella, antioxidant, bibliometric, VOSviewer, RStudio*

1 Introduction

The last ten years have seen the discovery of free radicals and their damaging impacts. These are toxic substances that the body produces during its normal metabolic process, along with poisons and wastes [1]. By neutralizing free radicals and damaging by-products of normal cell metabolism, antioxidants have protective effects. By contrast, disrupting this balance in humans might result in serious medical conditions. Normally, the body's antioxidant system can remove free radicals, maintaining the proper balance between oxidation and antioxidation [2]. However, excessive creation of free radicals could hinder the procedure from being done successfully, and its effectiveness also decreases with advancing age. Increasing antioxidant intake can minimize health issues and prevent diseases. Therefore, increasing exogenous antioxidant intake could reduce the damage caused by oxidative stress [2]. The majority of exogenous antioxidants come from foods and medicinal plants, such as *Basella*, native to tropical Southern Asia and likely to came from Indonesia or India [3].

Basella (*alba* and *rubra*) is a perennial vine of the Basellaceae family that can withstand intense heat and is also known as Malabar spinach, Indian spinach, Ceylon spinach, and Vine spinach [4]. The antioxidant capabilities of *Basella* are enhanced by its richness in phenols and other secondary metabolites. They can act as organic defenses against free radicals that their metabolisms release [5]. The mechanism of action of preventative antioxidants that decrease the rate of oxidation by a variety of means.Chain-breaking antioxidants *scavenge free radicals, inhibit the initiation stage or stop the propagation step of lipid oxidation [6]. Members of the *Basella* genus are high in phytochemicals, and their consumption may help provide antioxidants that fight free radicals generated by the body [5]. Free radicals are strong oxidizing agents that cause cell damage, but they are also helpful, in particular, they are essential for stimulating normal wound-healing response [7].

When normal anatomical structure and function of skin tissue are disrupted, a complicated and multifaceted process known as wound healing occurs. [8]. Among natural antioxidants, mounting evidence points to polyphenols as potential treatment options for oxidative stress-induced impared wound healing [9]. In this situation, an ideal therapeutic method to accelerate wound healing is the use of safe and efficient antioxidants in the wound bed to combat excessive reactive oxygen species (ROS). Therefore, these *Basella* phytochemicals can control one or more stages of the wound-healing process [7]. Many scientific studies have been conducted on *Basella* in the past few decades. Numerous biomaterials

*Corresponding author: nugroho_ae@ugm.ac.id

have been designed and tested in response to an increased interest in employing antioxidant compounds for wound therapy [7]. We belive that no bibliometric analysis of published data on *Basella* as an antioxidant has been done. The goal of this study is to conduct a bibliometric analysis of the research on *Basella*'s antioxidant properties in the Scopus database using the VOSviewer and RStudio tools. An increase in publications, nation, institution, source contributions, authorship analysis; paper and keyword occurrences on *Basella* as an antioxidant are specifically mentioned. This review can serve as a useful starting point for future studies on *Basella* as an antioxidant.

2 Methodology

2.1 Study technique and search strategy

Research articles included in this study were taken from the Scopus database (https://www.scopus.com) on July 2, 2023. The keywords "*Basella*" AND "antioxidant" were given particular attention. All articles, reviews, and conference proceedings in English from the Scopus database were eligible for inclusion criteria. Articles that are not in the Scopus database and are not in English are excluded. Following a thorough data cleaning process, our search confirmed that the collected papers covered *Basella* and antioxidants, and then we checked for data duplication. The articles from Scopus were imported into Microsoft Excel and saved as Comma Separated Values (CSV) files.

2.2 Data Analysis

The CSV files were imported into VOSviewer 1.6.19 from the Center for Science and Technology at Leiden University in the Netherlands to perform the bibliometric analysis. This program examines organizations, sources, writers, works, and co-occurrences of keywords. To prevent data duplication, data cleansing used the thesaurus in Excel. Then, the bibliometric data were analyzed using the RStudio software 2023.03.0-386 via the bibliometrics from the Department of Economics and Statistics, University of Naples Federico II, Italy. This program examines publication trends and the contributors' sources, nations, and writing styles.

3 Result and discussion

3.1 Data searches

Searches with the terms "Basella" AND "antioxidant" from the Scopus database returned 56 articles. A bibliometric analysis of Basella and antioxidant using the Scopus database was applied to characterize and map knowledge concepts related to the expansion of research on *Basella* as an antioxidant. The research criteria, study questions, and analytical approach selection steps were used to construct the bibliometric analysis. Performance analysis and scientific mapping are methodologies used for bibliometric analysis [10]. Performance analysis considers the contributions of academics from different countries, institutions, sources, and authors that could increase the productivity of the papers generated [11]. On the other hand, scientific mapping analysis based on bibliographic networks could be used to extract knowledge from the intellectual, social, or conceptual structures of a study topic [10].

3.2 Publication trend

The average number of citations per publication and trends in publishing data are displayed in Table 1. Only 56 documents discussing *Basella*'s antioxidant activity are listed in the Scopus database. The first journal was published in 2004, resulting in an annual production rate of 2.8 documents. The year with the highest number of publications was 2015 (7 articles), followed by 2012, 2018, 2021, and 2022 with 6 articles each. In terms of citations, the year with the highest citation rate is 2004 (8.65 citations per year), followed by 2010 (6.57 citations per year) and 2021 (5.17 citations per year). A paper with a significant number of citations is likely to have an impact on other researchers' use of the knowledge that reflects intellectual influence [12]. Citations are increasingly used in research policy and the research system as performance measurements. Citations are typically seen as indicating the importance or standard of the research [13].

Table 1. Publication data trend by year using Rstudio application

Year	Number of Articles	Mean Total of Citation per Year	Mean Total of Citation per Articles	Citable Year
2004	1	8.65	173	20
2010	2	6.57	92	14
2011	1	0.38	5	13
2012	6	2.79	33.5	12
2013	2	4.27	47	11
2014	2	3.5	35	10
2015	7	1.89	17	9
2016	4	0.97	7.75	8
2017	2	0.86	6	7
2018	6	1.75	10.5	6
2019	1	4.2	21	5
2020	5	2.05	8.2	4
2021	6	5.17	15.5	3
2022	6	1.92	3.83	2
2023	5	0	0	1
Average	2.8	2.248	23.76	

3.3 Analysis of contributing country

To determine which country contributed the most to the studies on *Basella's* antioxidant activity, analysis of contributing country was carried out using RStudio. The heatmap of all the nations included in *Basella's* research is displayed in Figure 1. Countries in darker blue had more articles published. Countries producing the most publications were India (27 documents), Nigeria (4 documents), and China (3 documents). A pink line shows partnerships with other countries. Most collaborative articles originated in Saudi Arabia and India (2 collaborations).

In addition to the quantityof documents created, each country's citation rate was examined. Figure 2 reveals that India (554 citations), the United States (173 citations), and the United Kingdom (140 citations) were the three most cited countries. This suggests that these countries may have the most influence on research on Basella's antioxidant activity.

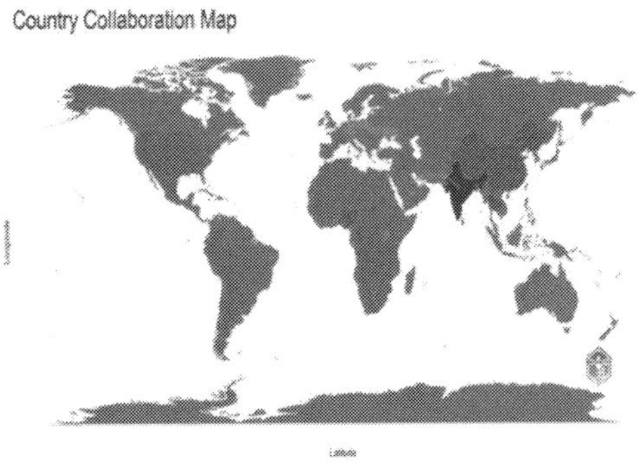

Fig. 1. Country production and colaboration heatmap of articles discussing *Basella's* antioxidant activity

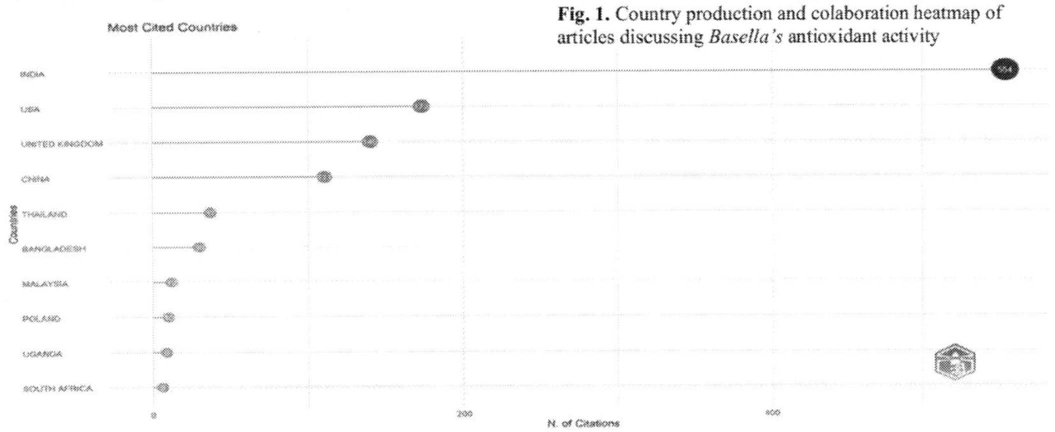

Fig. 2. Most-cited Country of articles discussing Basella's antioxidant activity

3.4 Analysis of contributing institution

Identifying the most prolific institutions was one of the most important tasks in bibliometric analysis [14]. The Scopus database was used for the analysis. According to the database, institutions with the highest number of publication of articles on *Basella* as an antioxidant were Chiang Mai University (15 documents), Icar-Indian Institute of Vegetable Research (12 documents), and three other universities each with 7 articles: Dayananda Sagar University, National Chiayi University, and the University of Tahjshahi (7 documents each) as shown in Table 2.

Table 2. The institution with the highest productivity using the RStudio application

Institution	Country	Documents
Chiang Mai University	Thailand	15
Icar-Indian Institute of Vegetable Research	India	12
Dayananda Sagar University	India	7
National Chiayi University	Taiwan	7
University of Rajshahi	India	7
Chinese Culture University	China	6

3.5 Analysis of contributing source

A total of 56 documents from 48 sources worldwide are listed in the Scopus database as research on Basella's antioxidant activity. The most productive sources, as indicated in Table 3, were the International Journal of Pharmacy and Pharmaceutical Sciences (3 articles), the Journal of Agricultural and Food Chemistry, and the Journal of Ethnopharmacology with 2 articles each. The American Journal of Clinical Nutrition, the Journal of Ethnopharmacology (149 citations), and Chemosphere were three journal with the highest number of citations relative to the number of papers published (139 citations).

Table 3. The most productive source using the RStudio application

Sources (Abbreviation)	Docum ents	Citations	Average
Int. J. Pharm. Pharm. Sci.	3	42	14
J. Agric. Food Chem.	2	54	27
J. Ethnopharmacol.	2	149	74.5
Am. J. Clin. Nutr.	1	173	173
Chemosphere	1	139	139
Environ. Sci. Pollut. Res.	1	67	67

The association between the author's country (AU_CO), sources/ journals (SO), and affiliation (AU_UN) is shown in Figure 3. The top 20 sources, top 20 authors, and top 20 keywords used in published works were the focus of this study Gray lines connect the three regions. The length of the rectangle indicates the number of related items in each box. The length of the rectangle and the number of items in each box become increasingly apparent as the rectangle elongates. According to the inflow analysis, the source with the greatest correlation was the Latin American Journal of Pharmacy, which had papers published by writers with two separate affiliations. The Latin American Journal of Pharmacy outflow analysis revealed that the journal published articles by writers from three of the seventeen countries that had published research about Basella's antioxidant activities

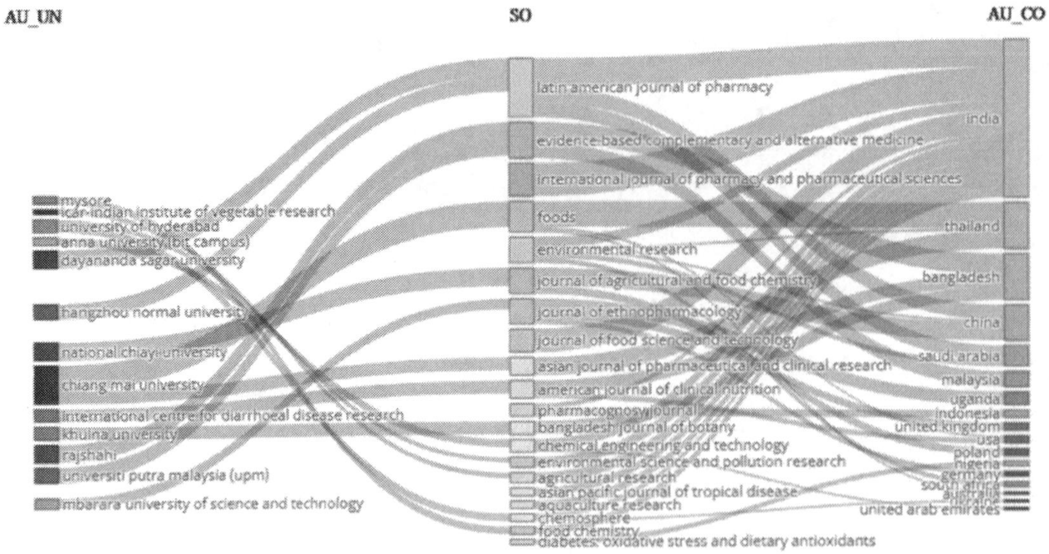

Fig. 3. Three-fields plot between author affiliations (AU_UN), Source (SO), and author country (AU_CO) using the RStudio application

3.6 Analysis of contributing author

The RStudio software was used to analyze the contributing author. The contribution of the associated author and author impact were examined using the RStudio software as well [15]. Table 4 shows the top authors who published the most articles that linked with other authors who wrote articles about *Basella* as an antioxidant. P. Giridhar published the most articles (5), followed by S.S Kumar, from the Central Food Technological Research Institute published 3 articles with an average of 28.6 citations per article. P. Manoj who had the same affiliation with P. Giridhar and S.S. Kumar published 3 articles with an average of 30.6 citations per article. D.S. Arkoyo from the Cape Peninsula University of Technology published 2 articles with an average of 3 citations per article, while V. Sagar from the Indian Institute of Vegetable Research published 2 articles with an average of 3.5 citations per article..

The author's impact is shown by the value of the H-index in Table 4. P. Giridhar's papers on Basella as an antioxidant had a significant impact, with the highest H-index of 5, followed by S.S. Kumar and P. Manoj with H-index of 3. F. Giampieri and S. Kolayli had the same H-index of 9, and V. Nanda had an H-index of 8. D.S. Arkoyo and V. Sagar had the same H-index of 1. Table 4 also shows the three most recent papers. This information is highly useful for authors, especially researchers, in planning and discussing the significance of study findings.

3.7 Analysis of contributing paper

An article's citation count shows the study that has made the most significant contribution. The higher the citation counts, the greater its impact on advancing research on Basella as an antioxidant. Fifty-six papers contributed to research on Basella as an antioxidant. Table 5 lists the top 5 most-cited articles. The most influential article, titled "Daily consumption of Indian spinach (Basella alba) or sweet potatoes has a positive effect on total-body vitamin A stores in Bangladeshi men, was written by Haskell et al. (2004) with 173 citations. The effect of 60 days of daily supplementation with 750 g retinol equivalents (RE) of cooked and puréed sweet potatoes, cooked and puréed Indian spinach (*Basella alba*), or synthetic sources of vitamin A on total-body vitamin A stores in Bangladeshi men was investigated in this study. This study found that eating cooked, puréed green leafy vegetables or sweet potatoes daily boosts vitamin A stores in populations at risk of vitamin A deficiency [16].

Table 4. Top 5 most productive authors using RStudio

No	Authors	3 Latest Tittle of Documents (Year)
1.	Giridhar P - Total Documents: 5 - Citations: 133 - Average Citation per Document: 26.6 - H index: 5 - Affiliation: Central Food Technological Research Institute	1. Nanoliposomal encapsulation mediated enhancement of betalain stability: characterization, storage stability and antioxidant activity of *Basella rubra* L. Fruits for its applications in vegan gummy candies (2020) 2. Influence of photoperiod on growth, bioactive compounds and antioxidant activity in callus cultures of *Basella rubra* L. (2020) 3. Fruit extracts of *Basella rubra* that are rich in bioactives and betalains exhibit antioxidant activity and cytotoxicity against human cervical carcinoma cells (2015)
2.	Kumar SS - Total Documents: 3 - Citations: 86 - Average Citation per Document: 28.6 - H index: 3 Affiliation: Central Food Technological Research Institute	1. Influence of photoperiod on growth, bioactive compounds and antioxidant activity in callus cultures of *Basella rubra* L. (2020) 2. Fruit extracts of *Basella rubra* that are rich in bioactives and betalains exhibit antioxidant activity and cytotoxicity against human cervical carcinoma cells (2015) 3. Nutrition facts and functional attributes of foliage of *Basella* spp. (2015)
3.	Manoj P - Total Documents: 3 - Citations: 92 - Average Citation per Document: 30.6 - H index: 3 Affiliation: Central Food Technological Research Institute	1. Fruit extracts of *Basella rubra* that are rich in bioactives and betalains exhibit antioxidant activity and cytotoxicity against human cervical carcinoma cells (2015) 2. Nutrition facts and functional attributes of foliage of *Basella* spp. (2015) 3. A method for red-violet pigments extraction from fruits of malabar spinach (*Basella rubra*) with enhanced antioxidant potential under fermentation
4.	Arokoyo DS - Total Documents: 2 - Citations: 6 - Average Citation per Document: 3 - H index: 1 Affiliation: Cape Peninsula University of Technology	1. *Basella alba*, oxidative stress, and diabetes (2020) 2. Antioxidant activities of *Basella alba* aqueous leave extract in blood, pancreas, and gonadal tissues of diabetic male wistar rats (2018)
5.	Sagar V - Total Documents: 2 - Citations: 7 - Average Citation per Document: 3.5 - H index: 1 Affiliation: Indian Institute of Vegetable Research	1. The inheritance of betalain pigmentation in *Basella alba* L. 2. Seed priming with ZnO and Fe_3O_4 nanoparticles alleviate the lead toxicity in *Basella alba* L. through reduced lead uptake and regulation of ROS

An article entitled "Traditionally used Thai medicinal plants: in vitro anti-inflammatory, anticancer and antioxidant activities" written by Siriwatanametanon et al. (2010) was the second most influential article with 140 citations. This study assessed the traditional Thai claims about the therapeutic potential of medicinal plants. It selected plants for future phytochemical research; nine plant species with anti-inflammatory properties, such as Basella alba L. and Basella rubra L. (Basellaceae), were selected from Thai textbooks. This research also investigated their in vitro anti-inflammatory, antiproliferative, and antioxidant activities. This study provides in vitro evidence of the use of Thai plants. Ethyl acetate extract of Basella alba has an antioxidant activity in inhibiting DPPH with IC50 of 5.32 µg/mL, while that of Basella rubra has a value of 34.58 µg/mL. Ethyl acetate extract of Basella alba also shows an NF-kB inhibitory activity with an IC50 value of 83.28 µg/mL, and that of Basella rubra shows a value of 162.83 µg/mL [17].

The third most influential article was "Antioxidant Activity in Extracts of 27 Indigenous Taiwanese Vegetables" written by Chao et al. (2014) with 64 citations. The objective of this study was to identify the antioxidants and antioxidant activity in 27 Taiwan's indigenous vegetables, one of them being Basella alba. This study found that Basella alba contains various antioxidant compounds in acid hydrolysates, such as polyphenol (7.12 ± 1.40 mg GAE/g DW), flavonoids

(42.71 ± 3.06 mg QUE/g DW), and flavonols (7.73 ± 2.19 mg QUE/g DW). The IC_{50} of the DPPH scavenging activity of *Basella alba* was 427.78 ± 0.48 µg/mL [18].The article "Studies on the spectrometric analysis of metallic silver nanoparticles (AgNPs) using Basella alba leaf for the antibacterial activities" written by Mani et al. (2021) was the fourth most influential article with 51 citations. In this study, AgNPs were synthesized using aqueous Basella alba leaves extract. The antioxidant studies revealed significant scavenging activity ranging from 13.71 percent to 67.88 percent. Green synthesized AgNPs have well-organized biological activities in terms of antioxidant and antibacterial activities, which can be used in various biological applications [19].

The fifth most influential article was "Fruit extracts of *Basella rubra* that are rich in bioactive and betalains exhibit antioxidant activity and cytotoxicity against human cervical carcinoma cells", written by Kumar et al. (2015) with 48 citations. Fruit extracts of *Basella rubra*, which are high in bioactive phenolics, flavonoids, and betalains, were tested for antioxidant and anticancer activities against human cervical carcinoma (SiHa) cells. Fruit extracts in water and aqueous methanol showed significant free radical scavenging and ferric-reducing antioxidant power. Fruit extracts at 50 mg/mL demonstrated strong (81%) cytotoxic activity against human cervical carcinoma cells. Thus, fruit extracts may have applications in cancer treatment and nutraceutical or dietary supplements [20].

Table 5. Top 5 most cited articles using RStudio

No	Author	Title	Source	Total Citation
1	[16]	Daily consumption of Indian spinach (*Basella alba*) or sweet potatoes has a positive effect on total-body vitamin A stores in Bangladeshi men	The American Journal of Clinical Nutrition	173
2	[17]	Traditionally used Thai medicinal plants: in vitro anti-inflammatory, anticancer and antioxidant activities	Journal of Ethnopharmacology	140
3	[18]	Antioxidant Activity in Extracts of 27 Indigenous Taiwanese Vegetables	Nutrients	64
4	[19]	Studies on the spectrometric analysis of metallic silver nanoparticles (AgNPs) using *Basella alba* leaf for the antibacterial activities	Environmental Research	51
5	[20]	Fruit extracts of *Basella rubra* that are rich in bioactives and betalains exhibit antioxidant activity and cytotoxicity against human cervical carcinoma cells	Journal of Functional Food	48

3.8 Analysis of keyword co-occurrence

The following analysis used the VOSviewer to examine keyword co-occurrence across all terms. Based on the publication's content, this analysis can map existing or future research issues on *Basella* as an antioxidant [11]. The number of documents that contain a specific keyword is displayed by occurrence. Because the calculation method was full counting, the number of keyword occurrences obtained in the study represents the total number of times certain keywords appeared in all documents. The "All keyword" category includes author keywords (article titles, abstracts, and full texts) as well as indexed keywords [21]. Network visualization of the topic by VOSviewer can be seen in Figure 4.

Twelve clusters were generated through the VOSviewer, and every cluster indicates how one subject links to the others. This software can display bibliometric mapping (Figure 4). The keywords are denoted by colored circles, and the size of the circles shows how frequently they appear in titles and abstracts. As a result, the size of the letters and circles depended on how frequently they occur. The more frequent a keyword appears, the more frequently the letters and circles exist. Based on the data gathered from the articles containing *Basella* as an antioxidant, 1,510 keywords were found from 765 articles. The clusters in each of the examined issue areas are shown in Figure 4. For example, the phrases "antioxidant", "*Basella alba*", "oxidative stress", and "lipid peroxidation" share the same circle color, indicating that they have a close link and are grouped. The most common keywords were "antioxidant" (859 occurrences with 1,340 total link strength) and "B*asella alba*" (606 occurrences with 1,048 total link strength). This data may suggest the novelties of Basella as an antioxidant.

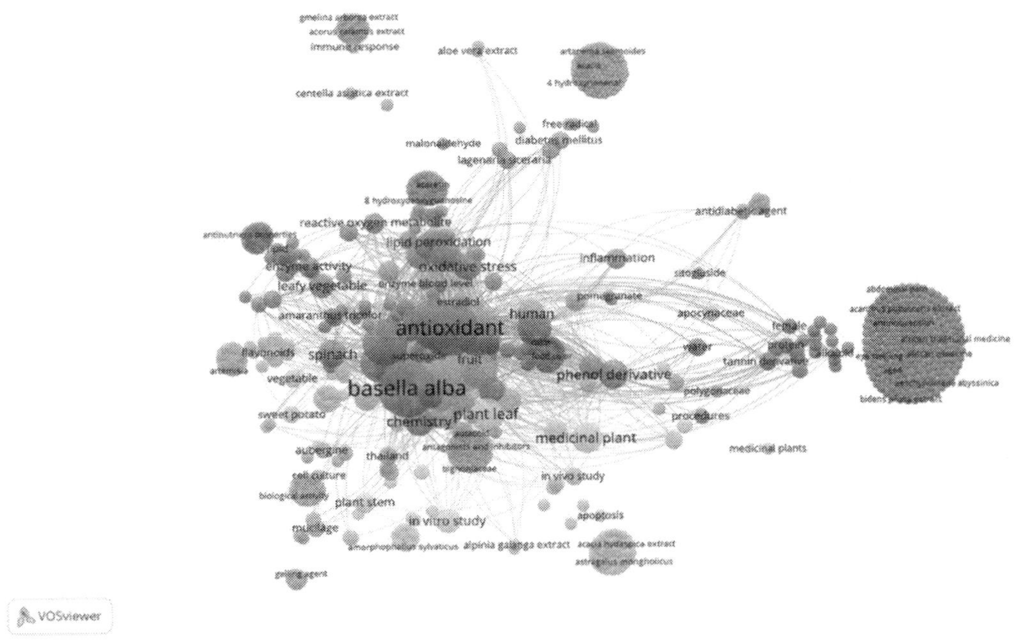

Fig 4. Network visualization of the topic by VOSviewer

4 Conclusion

Reasearch on Basella as an antioxidant has been widely undertaken in a number of countries, including Thailand which published the most (15 publications). According to this bibliometric study, P. Giridhar authored the most articles about Basella as an antioxidant and had a large impact on articles on this topic, with the highest H-index of 5. Haskell et al.'s (2004) article "Daily consumption of Indian spinach (Basella alba) or sweet potatoes has a positive effect on total-body vitamin A stores in Bangladeshi men1-3" received the most citations (173). "Antioxidant" (859 occurrences with 1,340 total link strength) and "Basella alba" (606 occurrences with 1,048 total link strength) were the most common keywords. These results provide insights to stimulate pharmaceutical research collaborations and reveal open issues about Basella as an antioxidant. Nevertheless, this study had many limitations, such as our knowledge of the literature influencing keyword selection, which could affect the amount and diversity of articles included in our analysis.

5 Acknowledgments

We gratefully acknowledge the Ministry of Research, Technology, and Higher Education Indonesia for financing under the PPS-PDD Programs with contract number 3085/UN1/DITLIT/Dit-Lit/PT.01.03/2023

References

1. Y. Shebis, D. Iluz, Y. Kinel-Tahan, Z. Dubinsky, and Y. Yehoshua, *Food Nutr Sci* 04, 643 (2013).
2. D. P. Xu, Y. Li, X. Meng, T. Zhou, Y. Zhou, J. Zheng, J. J. Zhang, and H. Bin Li, *Int J Mol Sci* 18, 1 (2017).
3. S. A. Deshmukh and D. K. Gaikwad, *J Appl Pharm Sci* 4, 153 (2014).
4. S. Kumar, A. K. Prasad, S. V. Iyer, and S. K. Vaidya, *Journal of Pharmacognosy and Phytotherapy* 5, 53 (2013).
5. T. A. Adenegan-Alakinde and F. M. Ojo, *International Journal of Vegetable Science* 25, 431 (2019).
6. S. K. Reshmi, K. M. Aravinthan, and P. Suganya, *Int J Pharmtech Res* 4, 900 (2012).
7. I. M. Comino-Sanz, M. D. López-Franco, B. Castro, and P. L. Pancorbo-Hidalgo, *J Clin Med* 10, 1 (2021).
8. N.' Izzah Ibrahim, S. K. Wong, I. N. Mohamed, N. Mohamed, K. Y. Chin, S. Ima-Nirwana, and A. N. Shuid, *Int J Environ Res Public Health* 15, 1 (2018).
9. A. Casado-Diaz, J. M. Moreno-Rojas, J. Verdú-Soriano, J. L. Lázaro-Martínez, L. Rodríguez-Mañas, I. Tunez, M. La Torre, M. B. Pérez, F.

Priego-Capote, and G. Pereira-Caro, *Pharmaceutics* 14, 1 (2022).
10. I. Sajovic and B. Boh Podgornik, *Sage Open* 12, 1 (2022).
11. N. Donthu, S. Kumar, D. Mukherjee, N. Pandey, and W. M. Lim, *J Bus Res* 133, 285 (2021).
12. S. Kraus, M. Filser, M. O'Dwyer, and E. Shaw, *Review of Managerial Science* 8, 275 (2014).
13. D. W. Aksnes, L. Langfeldt, and P. Wouters, *Sage Open* 9, 1 (2019).
14. P. Gao, F. Meng, M. N. Mata, J. M. Martins, S. Iqbal, A. B. Correia, R. M. Dantas, A. Waheed, J. Xavier Rita, and M. Farrukh, *Journal of Theoretical and Applied Electronic Commerce Research* 16, 1667 (2021).
15. S. F. Fatimah, E. Lukitaningsih, R. Martien, and A. K. Nugroho, *Pharmacia* 69, 467 (2022).
16. M. J. Haskell, K. M. Jamil, F. Hassan, J. M. Peerson, I. Hossain, G. J. Fuchs, and K. H. Brown, *Am J Clin Nutr* 80, 705 (2004).
17. N. Siriwatanametanon, B. L. Fiebich, T. Efferth, J. M. Prieto, and M. Heinrich Michael, *J Ethnopharmacol* 130, 196 (2010).
18. P. Y. Chao, S. Y. Lin, K. H. Lin, Y. F. Liu, J. I. Hsu, C. M. Yang, and J. Y. Lai, *Nutrients* 6, 2115 (2014).
19. M. Mani, S. Pavithra, K. Mohanraj, S. Kumaresan, S. S. Alotaibi, M. M. Eraqi, A. D. Gandhi, R. Babujanarthanam, M. Maaza, and K. Kaviyarasu, *Environ Res* 199 (2021).
20. S. S. Kumar, P. Manoj, P. Giridhar, R. Shrivastava, and M. Bharadwaj, *J Funct Foods* 15, 509 (2015).
21. F. H. Arifah, A. E. Nugroho, A. Rohman, and W. Sujarwo, *South African Journal of Botany* 151, 128 (2022).

Citrus Anticancer Research: A Bibliometric Mapping of Emerging Topics

Febri Wulandari[1*], *Asti Arum Sari*[1], *Mila Hanifa*[2], *Muhammad Haqqi Hidayatullah*[1]

[1]Faculty of Pharmacy, Universitas Muhammadiyah Surakarta, Indonesia
[2]Cancer Chemoprevention Research Center, Faculty of Pharmacy, Universitas Gadjah Mada, Yogyakarta 55281, Indonesia

Abstract. Research on the potential anticancer effects of citrus has been widely published in scientific journals. Still, a bibliometric analysis concerning this topic has not been executed. This study employed bibliometric mapping to analyze articles related to citrus anticancer from the Scopus Database and visualized the results using the VOSviewer. In this review, 442 papers published between 1995 and 2023 were selected. Jeju National University in South Korea is recognized as a top contributor. According to the analysis, apoptosis and anticancer are the two specific keywords in the field with the highest co-occurrence. The other keywords in the selected papers were hesperidin, naringenin, nobiletin, apoptosis, and flavonoids. We also found the following steps in this research area: formulation, synthesis, and *in vivo* preclinical studies. Research trends have shifted from the crude extract to practical applications of specific flavonoid compounds with structure modification to improve their anticancer properties. Still, clinical trials in humans are lacking in this research area and should be further investigated to embrace citrus flavonoids as an anticancer candidate. This analysis and mapping provide a comprehensive understanding of research on the potential anticancer effect of citrus.
Keywords: citrus, flavonoid, anticancer, bibliometric, VOSviewer

1 Introduction

Cancer is an intricate environment comprised of tumor cells that proliferate uncontrollably in the body, triggering immune system dysfunction and possibly leading to mortality [1]. It is a multistep process in which the genetic composition of individual cells is altered, transforming normal cells into aberrant cells, and driving the change towards malignancy [2]. Chemotherapy is a common technique for treating cancer in the initial stages [3], although it may have negative effects on normal cells and occasionally fails to manage the malignant type [4]. Given the adverse effects and lack of selectivity of synthetic medications, the development of innovative and natural chemotherapeutic agents is in great demand. In recent decades, there has been a rise in research focused on producing anticancer candidates from nature. The ideal option is to utilize natural products, which are rich in bioactive chemicals that provide significant health effects [5]. Citrus flavonoids and their structural derivates have been explored as possible anticancer drugs because of their high chemical variety [6, 7]. Several promising lead compounds have recently been found [8]; however, only a handful are in clinical trials [9].

Citrus is the world's fourth-most-produced crop and contains numerous nutrients that are beneficial to humans [8]. Individual citrus components, such as vitamins, hesperidin, naringin, limonene, quercetin, tangeretin, nobiliten, etc., all contribute to various pharmacological effects [8, 10–12], including anticancer properties. Citrus in its entirety, including peel [6], seed [13], and juice [14], is an invaluable source of secondary metabolites and bioactive compounds. The anticancer properties of these compounds include cytotoxicity, inhibition of proliferation [15], induction of apoptosis, retardation of metastasis [16], suppression of angiogenesis [17], and enhancement of chemotherapy [18]. Multiple studies, including those on breast cancer, colon cancer, and pancreatic cancer, have demonstrated that hesperidin [19] and naringenin [20] inhibit the proliferation of cancer cells. These flavonoids have tremendous potential as an alternative cancer prevention and treatment strategy. In addition, the effects of citrus consumption have been studied using a variety of experimental models, and an estimate of citrus' toxicity has been established [21]. Citrus-derived bioactive chemicals have also been utilized in numerous toxicological studies, which have demonstrated that their safety profile is comparable to that of other bioactive compounds. The potential for citrus to be developed as a novel anticancer agent provides an essential opportunity. However, only a couple of studies have described the next phase of the preclinical evaluation of citrus flavonoids as a potential anticancer agent.

*Corresponding author: fw548@ums.ac.id

Despite extensive research and several review articles on citrus anticancer research, an absence of bibliometric studies came to light. With a recent surge in the number of citrus anticancer research, researchers have found it difficult to locate pertinent information. As a result, bibliometric analysis may be utilized to deal with this massive volume of data [22]. Bibliometric analysis may be used to discover research trends, assess the effect of individual publications, assess the productivity of academics and institutions, and find possible partners [23]. Furthermore, it could be used to provide information on how far research has progressed as well as to guide financing, policy, and strategy choices. Bibliometric tools, such as the Visualization of Similarities viewer (VOSviewer) [24], and scientific databases, such as Scopus, can be used to study and evaluate large volumes of scientific data. Using the Web of Science database, recent research presented a scientometric investigation of citrus flavonoids as an anticancer and neuroprotective drug. In the other investigations, a bibliometric analysis was successful in determining worldwide trends in triple-negative breast cancer nanomedicine research. This study employs a bibliometric technique to examine the citrus anticancer research trend using the Scopus Database to answer several questions on significant themes, most productive authors, leading institutions and journals, leading countries, co-occurrence of authors, organization, and keywords in the citrus anticancer research. From this study, we hope the findings can elucidate the research gap and the prospects for future research.

2 Methodology

2.1.Data Source and Search Query

Publications on anticancer research in citrus were retrieved on May 31, 2023, from the Scopus database using the following search query: TITLE-ABS-KEY ("citrus") AND TITLE-ABS-KEY ("anti-cancer") OR TITLE-ABS-KEY ("anticancer"). Search strategies were limited to: (Document type, "Article") AND (Source type, "Journal") AND (LANGUAGE, "English"). No year restriction was applied. Manual filtering by assessing abstract eligibility was used to improve the quality of the collected data. The bibliographic data of selected papers was exported in CSV format and continued to be analyzed using the VOSviewer software. The flowchart of the bibliometric analysis is summarized in Fig. 1.

2.2.Bibliometric Mapping and Data Analysis

VOSviewer [24] was used to search for bibliographical and author keywords in selected articles. The network of interest items in the selected articles, such as contributing countries, authorships, keyword occurrences, and citations, were also built by VOSviewer using full-counting methods; the total of an analysis is equal to the number of links obtained as a result of the analysis. A trend of publications by year was illustrated by a graph. The top 10 of most countries, institutions, authors, journals, papers, and keywords were displayed in a table to assess the influence and impact of the research conducted.

The terms (countries, authors, and keywords) are represented in a bubble, and the size of the bubble corresponds to the number of documents or occurrences of each term. Lines are used to connect each of the bubbles to the others. The distance that exists between two bubbles is a representation of the degree to which the words are associated in terms of co-authorship, co-occurrence, or bibliographic coupling. The category that the words are categorized into may be inferred from the color of the bubble.

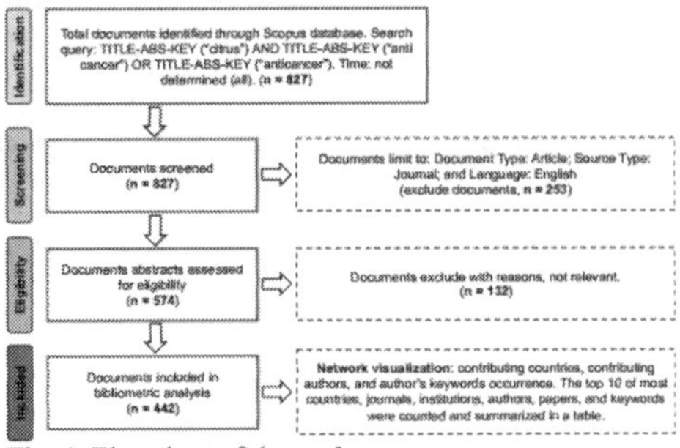

Fig. 1. Flowchart of the study.

3 Results

3.1 Trends of Publications by Year

For 28 years (1995–2003), a total of 442 research articles on citrus as an anticancer agent were published (Fig. 2). As of 1995, the first publication by the *British Journal of Cancer* revealed the apoptosis induction of the citrus flavone tangeretin against HL-60 leukemia cell lines [25]. The latest paper published by Amalina et al. [26] in the *Journal of the Egyptian National Cancer Institute* demonstrates the in vitro synergistic effect of citrus flavonoid hesperidin to enhance the efficacy of doxorubicin against 4T1 metastatic breast cancer cell lines.

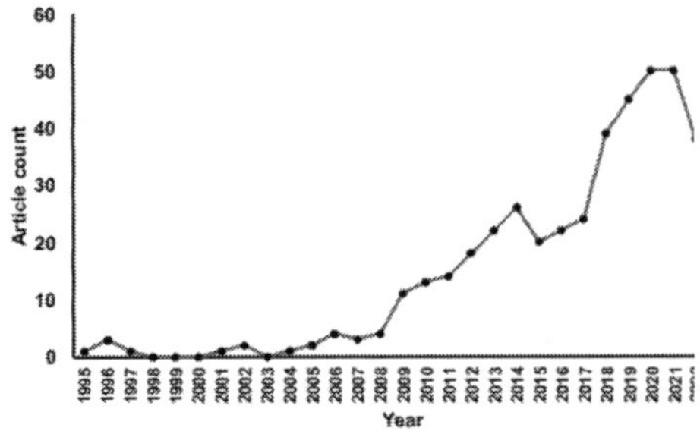

Fig. 2. Publication trends by year of citrus in the field of anticancer activities (n = 442).

The trend of publications by year has been relatively increasing over time, though from 1995 to 2008, related research on citrus anticancer was noted in under five publications per year. In 2009, the number of publications increased to eleven from four in 2008. The annual growth rate (AGR) climbed by 90% in 2009 and more than doubled between 2013 and 2014. Our graph showed that the overall number of publications decreased from 2015 to 2017, fluctuating around 20 publications, but showed a rebound in 2018 to double the previous and reach its peak in 2021 (n = 50). More than 50% of the publications published after 2018 (n = 250). In addition, the vast majority of articles published in the Biochemistry, Genetics, and Molecular Biology; Pharmacology, Toxicology, Pharmaceuticals; and Medicine subject areas We believed that the annual publications of citrus anticancer still became a challenging topic to explore and may continue to gain traction among researchers in the near future.

3.2 Assessment of Contributing Journals

The top 10 most productive journals are shown in Table 1. A total of 442 selected papers were published by 160 different journals, and the majority published their papers within the past five years. The top 10 journals contributed 5% of the total number of selected papers. The most productive journal in publishing citrus anticancer research is *Molecules* (n = 14), with an average number of citations (AC) of 33.36%. In second and third place are the *Journal of Agricultural and Food Chemistry* and *Nutrition and Cancer*, which come out with 11 and 10 publications, with AC values of 59.18% and 79.90%, respectively. Although in third place, *Nutrition and Cancer* has the most cited paper with 799 citations. This data could help academics and researchers find the most credible journal both for literature searching and submitting their research on those specific topics. In brief, the *Molecules* could potentially be selected as a popular journal destination for publishing research on citrus anticancer. On the other hand, based on the latest articles published by the top 10 journals, studies on citrus anticancer have progressed to the anticancer exploration of specific flavonoid compounds from *Citrus sp.*, such as tangeretin, hesperetin, hesperidin, nobiletin, and sinensetin. The research being carried out is no longer directed towards the use of crude extracts but rather

Table 1. Top 10 Journals with the most papers

No.	Journal	Number of papers	Number of Citations	AC[1] per paper	The title of the latest paper (year of publication)
1	Molecules	14	467	33,36	Metabolomic Analysis of Phytochemical Compounds from Ethanolic Extract of Lime (*Citrus aurantifolia*) Peel and Its Anti-Cancer Effects against Human Hepatocellular Carcinoma Cells (2023)
2	Journal of Agricultural and Food Chemistry	11	651	59,18	Identification and Quantification of Both Methylation and Demethylation Biotransformation Metabolites of 5-Demethylsinensetin in Rats (2022)
3	Nutrition and Cancer	10	799	79,90	Evaluation of the Effects of Nobiletin on Toll-Like Receptor 3 Signaling Pathways in Prostate Cancer In Vitro (2021)
4	International Journal of Biological Macromolecules	9	286	31,78	Construction and antitumor activity of selenium nanoparticles decorated with the polysaccharide extracted from *Citrus limon* (L.) Burm. f. (Rutaceae) (2021)
5	International Journal of Molecular Sciences	9	138	15,33	Acetylation Enhances the Anticancer Activity and Oral Bioavailability of 5-Demethyltangeretin (2022)
6	Phytomedicine	8	168	21,00	Nobiletin inhibits breast cancer cell migration and invasion by suppressing the IL-6-induced ERK-STAT and JNK-c-JUN pathways (2023)
7	Plos One	8	345	43,13	Experimental, molecular docking and molecular dynamic studies of natural products targeting overexpressed receptors in breast cancer (2022)
8	Bioorganic and Medicinal Chemistry Letters	7	330	47,14	Isolation and biological activity of agrostophillinol from kaffir lime (*Citrus hystrix*) leaves (2020)
9	Food and Function	7	289	41,29	Hesperetin ameliorates hepatic oxidative stress and inflammation: Via the PI3K/AKT-Nrf2-ARE pathway in oleic acid-induced HepG2 cells and a rat model of high-fat diet-induced NAFLD (2021)
10	International Journal of Oncology	7	231	33,00	Hesperetin induces apoptosis in A549 cells via the Hsp70-mediated activation of Bax (2022)

towards single compounds and their formulations. The latest papers published by *Molecules* (2023) reported the metabolomic analysis of citrus peel extract and its cancer activity towards hepatocellular carcinoma [27]. The superior anticancer effect of citrus peel extract likely results from hesperidin and limonin, the major compounds in *Citrus aurantifolia* peel. Another recent paper published by Phytomedicine (2023) revealed an inhibitory effect on breast cancer metastasis of nobiletin, one of the interesting flavonoid compounds contained in *citrus sp.*, through *in vitro* and *in vivo* elucidation [28]. The results verified the safety and efficacy of metastasis inhibition by downregulating the ERK-STAT and JNK-c-JUN pathways, indicating the potential of nobiletin as an anticancer agent. From this analysis, it can be summarized that the research on citrus anticancer is increasingly directed toward exploring in depth specific compounds from citrus on their molecular pathways.

3.3 Analysis of Contributing Countries and Their Collaborations

The present study revealed that the top 10 countries with the most papers contributed to the growth of citrus anticancer research globally (Table 2). The selected articles (n = 442) came from 55 different countries. About 38% of publications were contributed by China and India, indicating that these two countries are key players in the progress of citrus anticancer research. With 91 papers, China was the most productive country, followed by India (n = 80) and South Korea (n = 68). Although China was the most productive country based on paper count, South Korea had a greater impact in terms of AC per paper (48.48), followed by Italy with an AC value of 32.53. The other 8 countries contribute equally, with an AC value of ± 20. We then analyse the collaboration between countries with at least one publication in collaboration (Fig. 3a). China, India, the United States, and South Korea displayed a big bubble compared to other countries. The size of the bubble represents the number of publications, and the thickness of the line between countries indicates the strength of collaboration. From the visualization, China showed strong collaboration with the United States, and the United States showed equal collaboration with Taiwan. India was the top country with the most collaboration networks (19 countries), followed by the United States (15 countries). China and South Korea showed collaboration with a similar linkage of 12 countries. In the overlay visualization (Fig. 3b), the green bubbles displayed that China (average publication year 2018) was the most up-to-date country compared with South Korea (average publication year 2015), the United

Table 2. Top 10 countries with the most papers.

No.	Country	Number of papers	Number of citation	AC[1] per paper
1	China	91	2263	24,87
2	India	80	1788	22,35
3	South Korea	68	1929	28,37
4	United States	61	2957	48,48
5	Japan	28	747	26,68
6	Iran	27	652	24,15
7	Egypt	26	691	26,58
8	Taiwan	22	607	27,59
9	Italy	17	553	32,53
10	Saudi Arabia	14	290	20,71

[1]AC: average number of citations

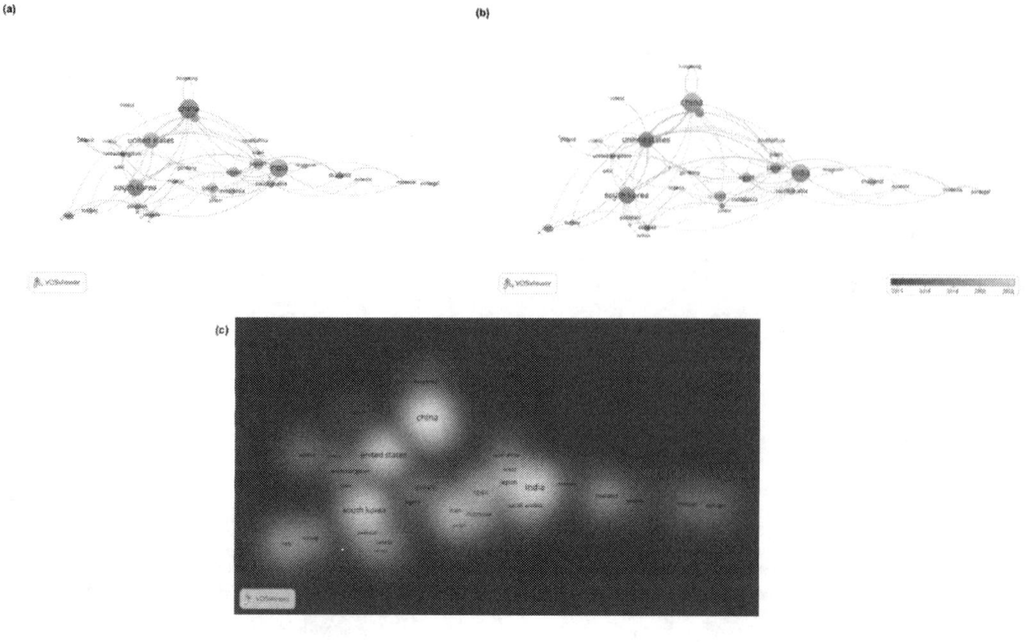

Fig. 3. Network visualization (a), overlay visualization (b), and item density visualization (c) of contributing countries.

States (average publication year 2014), and India (average publication year 2017) in publishing citrus anticancer research. These results are in line with item density visualization that showed China, India, the United States, and South Korea were the most productive countries with a large number of publications on this topic, indicated by a stronger yellow color intensity (Fig. 3c).

3.4 Institution Partnerships

A total of 442 selected papers were contributed by 160 institutions, which contain 1362 different organizations. A country can have several institutions, and an institution can consist of more than one organization. We summarized the top 10 institutions with the most papers and citations (Table 3). The most productive institution worldwide in citrus anticancer research was Jeju National University, which ranked as the top institution with 22 publications. This data correlated to the fact that Jeju Island was famous for its special citrus, the Jeju Tangerine and Jeju Mandarin [29]. Besides, Gyeongsang National University and Kyung Hee University from South Korea are in the third and fifth positions, respectively, and influenced 16 and 10 publications, respectively. In the second and fourth positions were Ruthers University-New Beunswick, United States, and the National Research Centre, Egypt, which contributed 19 and 12 publications, respectively. The other five institutions were China Medical University, the University of Madras, the University of Massachusetts Amherst, Texas A&M University, and Chiang Mai University, which contributed 10, 10, 9, 9, and 8 publications, respectively. Interestingly, tough contributed 9 publications, Texas A&M University ranked first with an AC value 64.78, indicates that their publication topics more influential and became a reference than other institutions. The publication of citrus anticancer research in 2014 from Texas A&M University demonstrated that obacunone, a unique compound from lemon seeds, inhibits estrogen-responsive breast cancer by activating apoptosis, aromatase enzymes, and inflammatory pathways [30]. Whereas Jeju National University, as the most productive institution, contributed their last papers in 2022 about the involvement of eriodictyol, a plant-derived flavonoid found in citrus fruits, in regulating phosphorylation of JNK, ERK, and FAK/AKT in SNU213 and Panc-1 pancreatic cell lines [31].

Table 3. Top 10 institutions with the most papers

No.	Institutions	Country	Number of Papers	Number of Citations	AC[1] per paper	The title of the latest paper (year of publication)
1	Jeju National University	South Korea	22	417	18,95	Eriodictyol induces apoptosis via regulating phosphorylation of JNK, ERK, and FAK/AKT in pancreatic cancer cells (2022)
2	Ruthers University-New Beunswick	United States	19	437	23,00	Acetylation Enhances the Anticancer Activity and Oral Bioavailability of 5-Demethyltangeretin (2022)
3	Gyeongsang National University	South Korea	16	590	36,88	Pectolinarigenin induced cell cycle arrest, autophagy, and apoptosis in gastric cancer cell via PI3K/AKT/mTOR signaling pathway (2018)
4	National Research Centre	Egypt	12	206	17,17	Synthesis, characterization and cytotoxic activity of naturally isolated naringin-metal complexes (2019)
5	Kyung Hee University	South Korea	10	323	32,30	Hesperidin ameliorates benign prostatic hyperplasia by attenuating cell proliferation, inflammatory response, and epithelial-mesenchymal transition via the TGF-β1/Smad signaling pathway (2023)
6	China Medical University	Taiwan	10	379	37,90	Systematic analysis of the mechanism of aged citrus peel (Chenpi) in oral squamous cell carcinoma treatment via network pharmacology, molecular docking and experimental validation (2022)
7	University of Madras	India	10	444	44,40	Cytotoxic and apoptotic effect of citrus flavonoid naringin in treating pa-1 ovarian cancer cells (2020)
8	University of Massachusetts Amherst	United States	9	375	41,67	Identification of Xanthomicrol as a Major Metabolite of 5-Demethyltangeretin in Mouse Gastrointestinal Tract and Its Inhibitory Effects on Colon Cancer Cells (2020)
9	Texas A&M University	United States	9	583	64,78	Obacunone exhibits anti-proliferative and anti-aromatase activity in vitro by inhibiting the p38 MAPK signaling pathway in MCF-7 human breast adenocarcinoma cells (2014)
10	Chiang Mai University	Thailand	8	221	27,63	Sesame extract promotes chemopreventive effect of hesperidin on early phase of diethylnitrosamine-initiated hepatocarcinogenesis in rats (2021)

[1]AC: average number of citations

According to the title of the most recent paper from the top ten institutions, hesperidin is a popular citrus flavonoid that has been explored for its anticancer activity. Hesperidin has been reported to attenuate benign prostate hyperplasia through the TGF-B1/Smad signaling pathway [32]. Nowadays, the study of anticancer drugs has reached the molecular mechanisms that are specifically targeted by compounds.

3.5 Assessment of Influential Authors

The 442 selected papers were contributed by 2125 authors from several organizations, institutions, or countries. The number of papers produced by an author reflects the author's contribution and engagement in this topic, as well as the contribution of the institution and country. Table 4 outlines the most prolific citrus anticancer authors from three countries, including the United States (1 author), South Korea (8 authors), and India (1 author). Interestingly, though China was crowned the most productive country with the most papers produced, China's authors were not included in the top ten productive authors. This can be explained by the possibility that, indeed, many authors from China published research on anticancer citrus, but they did not dedicate themselves to researching this topic totally and continually. The most productive author was Ho c.-t. from the United States, who contributed 15 papers. In second until ninth position, authors came from South Korea, named Kim g.s., Hong g.e., Kim g.e., Lee w.s., Lee h.j., Cho s.k., Ngappan a., and Choi y.h., each with 10, 9, 9, 9, 9, 9, 8, and 7 papers, respectively. The last author in the top ten list was Yumnam s. from India, who contributed seven papers. However, their latest papers were published before 2020, except for Ho c.-t., who published his latest paper in 2022. As the most productive author with the most papers produced, Ho c.-t. is also known to have the highest H-index value of 103. A high H-index value can indicate that the author was productive in producing papers and that the papers were highly cited by other papers.

The collaboration among the authors is visualized by the co-authorship analysis visualization generated from VOSviewer. The minimum number of documents for an author was one paper. Among 2125 authors, only 691 constituted the largest set of linked authors in 31 clusters (Fig. 4a). From the overall (Fig. 4b) and density (Fig. 4c) visualizations, we noted that most of the authors published their work between 2015 and 2022,

Table 4. Top 10 authors with the most papers

No.	Author name	Country	H-index	Number of Papers	Number of Citations	AC[1] per paper	The title of the latest paper (year of publication)
1	Ho c. -t.	United States	103	15	333	22,20	Acetylation Enhances the Anticancer Activity and Oral Bioavailability of 5-Demethyltangeretin (2022)
2	Kim g.s.	South Korea	32	10	378	37,80	Pectolinarigenin induced cell cycle arrest, autophagy, and apoptosis in gastric cancer cell via PI3K/AKT/mTOR signaling pathway (2018)
3	Hong g.e.	South Korea	16	9	293	32,56	Korean Byungkyul -*Citrus platymamma* Hort.et Tanaka flavonoids induces cell cycle arrest and apoptosis, regulating MMP protein expression in Hep3B hepatocellular carcinoma cells (2017)
4	Kim e.h.	South Korea	19	9	362	40,22	Pectolinarigenin induced cell cycle arrest, autophagy, and apoptosis in gastric cancer cell via PI3K/AKT/mTOR signaling pathway (2018)
5	Lee w.s.	South Korea	32	9	309	34,33	Pectolinarigenin induced cell cycle arrest, autophagy, and apoptosis in gastric cancer cell via PI3K/AKT/mTOR signaling pathway (2018)
6	Lee h.j.	South Korea	9	9	326	36,22	Pectolinarigenin induced cell cycle arrest, autophagy, and apoptosis in gastric cancer cell via PI3K/AKT/mTOR signaling pathway (2018)
7	Cho s.k.	South Korea	33	9	217	24,11	Supercritical Fluid Extraction of *Citrus iyo* Hort. ex Tanaka Pericarp Inhibits Growth and Induces Apoptosis Through Abrogation of STAT3 Regulated Gene Products in Human Prostate Cancer Xenograft Mouse Model (2017)
8	Nagappan a.	South Korea	21	8	274	34,25	Proteomic analysis of selective cytotoxic anticancer properties of flavonoids isolated from *Citrus platymamma* on A549 human lung cancer cells (2016)
9	Choi y.h.	South Korea	70	7	104	14,86	The reactive oxygen species/AMP-activated protein kinase signaling pathway's role in the apoptotic induction of MCF-7 human breast cancer cells caused by the ethanol extract of *Citrus unshiu* peel (2018)
10	Yumnam s.	India	17	7	235	33.57	Korean Byungkyul -*Citrus platymamma* Hort.et Tanaka flavonoids induces cell cycle arrest and apoptosis, regulating MMP protein expression in Hep3B hepatocellular carcinoma cells (2017)

[1]AC: average number of citations

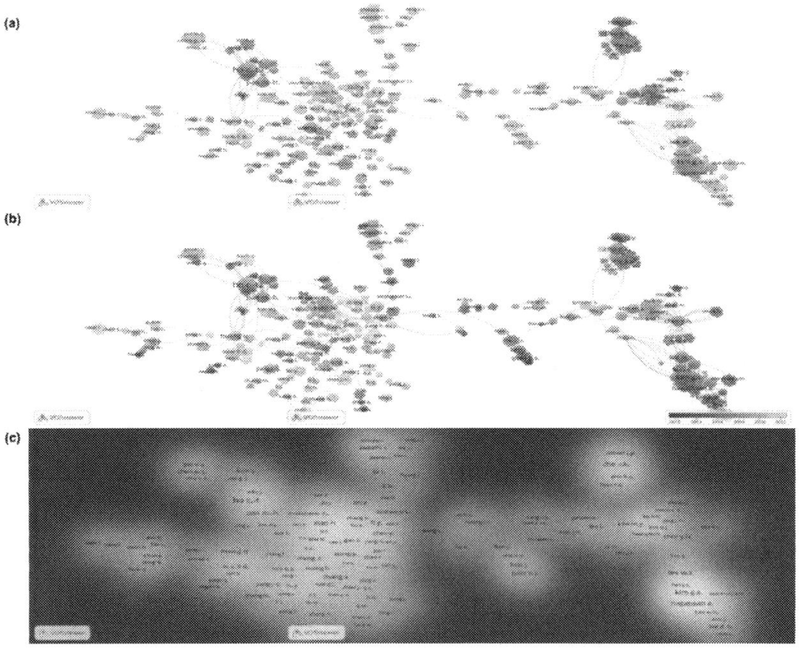

Fig. 4. Network visualization (a), overlay visualization (b), and item density visualization (c) of contributing authors.

and they had similar color density, which means that overall authors produced a similar number of articles. The most collaborative partner was Ho c.-t., displayed by the highest link strength of 64. The link strength reflects the number of papers co-authored by two authors. The latest publication from Ho c.-t. reported the effect of acetylation on 5-Demethyltangeretin's (5-DTAN) anticancer activity and oral bioavailability [33]. The acetylation contributed to G2/M phase arrest stimulation, cancer cell migration suppression, and a higher maximum concentration (Cmax) and area under the curve (AUC) in plasma. This research goes beyond the anticancer activity examination of natural flavonoids from citrus but goes more deeply into how structural modifications can be made to increase the anticancer activity. The latest papers from the top ten authors showed overall research in citrus anticancer had been done on various kinds of flavonoids, but none had led to formulation or clinical trials. This data showed a research gap: there was no research bridging citrus to be developed as a dosage form and directed to clinical trials.

3.6 Keywords Co-occurrence Analysis

The analysis of keywords from selected papers was conducted to find the common terms and guide researchers in identifying the most popular subjects. Only keywords that appear at least five times are used in this visualization (Fig. 5a–c). Out of a total of 1239 keywords, only 49 meet the threshold. The selected forty-nine keywords were grouped into six clusters, with 250 links and a total link strength of 450. "Apoptosis" appeared as the most frequent keyword co-occurrence, displayed as the largest bubble size, the stronger intensity of the bubble yellow color (Fig. 5c), and the highest link value of 36. As the keywords used as queries in this bibliometric analysis, "anticancer", "anticancer activity", and "citrus" occupy the second, eighth, and ninth positions in the top ten keywords (Table 5). This is possible because many publications directly mention flavonoid compounds, for example "hesperidin", "naringenin', nobiletin", "naringin", "tangeretin", and "hesperetin", rather than mentioning "citrus". From the visualization, it is also noted that the anticancer activity from citrus flavonoids has been done against "colon cancer", "gastric cancer", "lung cancer", "colorectal cancer", "breast cancer", and "prostate cancer". Anticancer parameters that have been widely studied include "apoptosis", "anti-inflammation", "antiproliferative", "cytotoxicity", "oxidative stress", -

Table 5. Top 10 authors' keywords with the most co-occurrences.

No.	Keywords	Co-occurrences
1	Apoptosis	74
2	Anticancer	31
3	Flavonoids	28
4	Cytotoxicity	22
5	Hesperidin	21
6	Naringenin	20
7	Nobiletin	20
8	Anticancer activity	20
9	Citrus	19
10	Antioxidant	16

Table 6. Top 10 most cited papers.

No.	Authors	Number of Citations	Title	Brief of studies
1	So f.v. et al. (1996)	449	Inhibition of human breast cancer cell proliferation and delay of mammary tumorigenesis by flavonoids and citrus juices	Aims: To assess two citrus flavonoids, hesperetin and naringenin for their effects on proliferation and growth of a human breast carcinoma cell line, MDA-MB-435. Results: Citrus flavonoids are effective inhibitors of human breast cancer cell proliferation in vitro, especially when paired with quercetin, which is widely distributed in other foods.
2	Mathey j.a. et al. (2002)	331	Antiproliferative activities of citrus flavonoids against six human cancer cell lines	Aims: To demonstrate the antiproliferative property of synthetic methoxylated flavones against additional human cancer cell lines. Results: The strong antiproliferative activities of the polymethoxylated flavones suggest that they may have use as anticancer agents in humans.
3	Morley k.l. et al. (2007)	201	Tangeretin and nobiletin induce G1 cell cycle arrest but not apoptosis in human breast and colon cancer cells	Aims: To evaluate the antiproliferative activity of tangeretin and nobiletin against human breast cancer cell lines MDA-MB-435 and MCF-7 and human colon cancer line HT-29. Results: Tangeretin and nobiletin could be effective cytostatic anticancer agents. Inhibition of proliferation of human cancers without inducing cell death may be advantageous in treating tumors as it would restrict proliferation in a manner less likely to induce cytotoxicity and death in normal, non-tumor tissues.
4	Hirano t. et al. (1995)	191	Citrus flavone tangeretin inhibits leukaemic HL-60 cell growth partially through induction of apoptosis with less cytotoxicity on normal lymphocytes	Aims: To observe the apoptosis induction of tangeretin. Results: Tangeretin inhibits growth of HL-60 cells in vitro, partially through induction of apoptosis, without causing serious side-effects on immune cells.
5	Zygmunt k. et al. (2010)	182	Naringenin, a citrus flavonoid, increases muscle cell glucose uptake via AMPK	Aims: To examine the direct effects of naringenin on skeletal muscle glucose uptake and investigated the mechanism involved. Results: Naringenin increases glucose uptake by skeletal muscle cells in an AMPK-dependent manner.
6	Poulose s.m. et al. (2005)	161	Citrus limonoids induce apoptosis in human neuroblastoma cells and have radical scavenging activity	Aims: To evaluate the apoptosis induction of citrus lumonoids and its radical scavenging activity. Results: Citrus limonoid glucosides are toxic to SH-SY5Y cancer cells through apoptosis by an as yet unknown mechanism of induction. Individual limonoid glucosides differ in efficacy as anticancer agents, and this difference may reside in structural variations in the A ring of the limonoid molecule.
7	Dudai n. et al. (2005)	145	Citral is a new inducer of caspase-3 in tumor cell lines	Aims: To investigate the anti-cancer potential of citral and its mode of action. Results: Citral induced apoptosis, accompanied by DNA fragmentation and caspase-3 catalytic activity induction. The apoptotic effect of citral depended on the α,β-unsaturated aldehyde group.
8	Chen j. et al. (1997)	143	Two New Polymethoxylated Flavones, a Class of Compounds with Potential Anticancer Activity, Isolated from Cold Pressed Dancy Tangerine Peel Oil Solids	Aims: To isolate citrus compounds and determined their biological activity. Results: Compounds II and VII are novel natural products; compounds IV, V, and VIII have been reported with significant activity against various strains of carcinoma cells; and compounds I and IV decrease erythrocyte aggregation and sedimentation in vitro.
9	El-Readi m.z. et al. (2010)	128	Inhibition of P-glycoprotein activity by limonin and other secondary metabolites from Citrus species in human colon and leukaemia cell lines	Aims: To investigate the effects of nine naturally occurring compounds isolated from *Citrus jambhiri* Lush and *Citrus pyriformis Hassk* (Rutaceae) for their potential to modulate the activity of P-gp in the multidrug-resistant human leukaemia cell line CEM/ADR5000. Results: The isolated Citrus compounds could be considered as good candidates for the development of novel P-gp/MDR1 reversal agents which may enhance the accumulation and efficacy of chemotherapy agents.
10	Celia c. et al. (2013)	126	Anticancer activity of liposomal bergamot essential oil (BEO) on human neuroblastoma cells	Aims: To evaluated the BEO liposomes on their anticancer activity in vitro against human SH-SY5Y neuroblastoma cells. Results: The results warrant further investigation of BEO liposomes for in vivo applications.

[1]AC: average number of citations

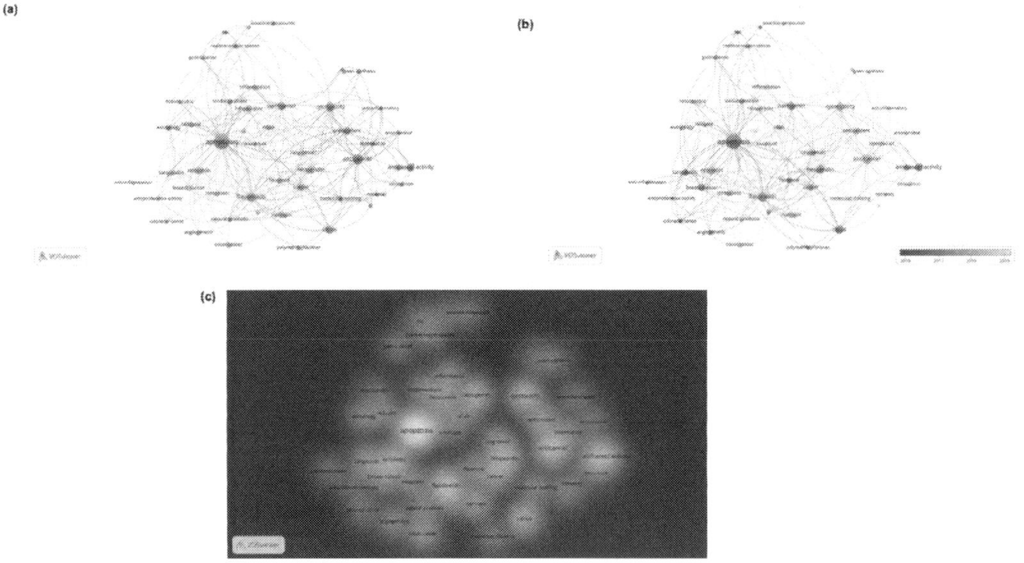

Fig. 5. Network visualization (a), overlay visualization (b), and item density visualization of authors' keywords co-occurrence

"autophagy", "cell cycle", "antioxidant", "metastasis", and "angiogenesis", as well as exploration using "molecular docking". All the selected author's keywords in visualization (Fig. 5b) tend to start being used around 2015. Topics marked in blue bubbles were popular at the start of around 2015, and topics marked in green bubbles were more popular between 2017 and 2018. In the year around 2019, topics in yellow bubbles dominated.

The terms "nanoparticle", "ultrasound", and "green synthesis", indicate that the citrus anticancer research also led to single compound development that focused on drug-delivery systems and advanced biotechnological approaches. From these findings, exploration of drug-delivery strategies may have future research promise in terms of enhancing molecular bioavailability[34, 35]. *In vivo* preclinical and clinical investigations can follow the drug-delivery strategy to elaborate on this specific issue.

3.7 Most Cited Papers

Among 442 selected citrus anticancer papers, we figured out the top ten most cited papers. The top 10 most-cited papers received 126 to 447 citations, with an average of 205.7 (Table 6). Only three papers from the top ten list were published after 2010. The most cited paper by So f.v. et al., [36] with 447 citations, was the oldest paper among the top 10. The research investigated the breast cancer suppression of orange juice and citrus flavonoids using *in vitro* and *in vivo* experiments. The results provide evidence of the anticancer properties of common fruits and orange juice, which contain several flavonoids that are effective inhibitors of cancer proliferation. The latest papers from the top 10 list were published by Celia c. et al., [37] and reported the anticancer properties of liposomal bergamot essential oil against human neuroblastoma cells. This finding showed that the formulation of citrus flavonoids or extracts to overcome physical and chemical limitations, such as poor water solubility, stability, and limited bioavailability, had been increasingly explored.

3.8 Top 10 citrus compounds

Table 7. Top 10 citrus compounds explored

No.	Compounds
1	Hesperidin
2	Naringenin
3	Nobiletin
4	Naringin
5	Tangeretin
6	Hesperetin
7	Polymethoxyflavones
8	Pectin
9	Didymin
10	Diosmin

We figured out the top ten explored citrus compounds to elucidate what the most promising compound is to develop as an anticancer agent [38]. Table 7 displayed that hesperidin, naringenin, and nobiletin were compounds widely explored as anticancer agents. The latest publication reported that hesperidin ameliorates benign prostatic hyperplasia through the TGF-β1/Smad signaling pathway [32]. Indeed, numerous studies have also highlighted the potency of hesperidin as an anti-breast cancer candidate, especially as a co-chemotherapeutic agent [26]. On the other side, naringenin and nobiletin, as an aglycon flavonoid from hesperidin, are also used as a co-adjuvant strategy in breast cancer [39]. Though not as deeply studied as the three compounds previously mentioned, the other citrus compounds have received greater attention in isolation and have been extensively explored for pharmacological activities besides anticancer.

3.9 Top 10 citrus species

Table 8. Top 10 citrus species explored

No.	Species
1	*Citrus aurantifolia*
2	*Citrus limon*
3	*Citrus hystrix*
4	*Citrus reticula*
5	*Citrus sinensis*
6	*Citrus maxima*
7	*Citrus unshiu*
8	*Citrus limetta*
9	*Citrus bergamia*
10	*Citrus platymamma*

Over 1000 species of citrus fruits grew all over the world, enriched with numerous kinds of hybridization results [21]. Despite the wide variety of citrus cultivars that are commercially accessible, the facilitation of resource allocation would be enhanced. However, it is essential to address this issue with particular attention due to the inherent diversity of their chemical constituents. We found out the top ten citrus species that have been extensively investigated for their pharmacological activities, including anticancer. Citrus aurantifolia, or key lime, is reported to be the most favorite species for anticancer research. The phytochemical composition of many different species would exhibit an appealingly consistent form, yet the proportions would vary. It serves to point out the importance of making precise determinations in research as a way to ensure the comprehensiveness and accuracy of the database [40].

4 Discussion

A bibliometric method was utilized in this study to examine the available literature on citrus anticancer, find trends in cancer research, and embark on a research agenda for the next few decades. Furthermore, this method analyzes authorship, article keywords, citations, and co-citations, which is a critical subject. The citation trend analysis enables us to comprehend how the subject of interest is related to other study areas [41]. Citations have long been used to assess the usefulness of a research work. Citation reflects a paper's academic value and the extent to which it is acknowledged by other academics. A bibliometric research essay differs from a review article in that it focuses on the most recent progress, challenges, and likely future directions of a certain field [42]. This goal has the ability to increase awareness as well as progress in the field of research. This study provides an in-depth examination of the anticancer properties of citrus and appeared as a research article in journals included in the Scopus database. Despite the fact that there are several review papers on the issue, there are no instances of bibliometric analysis in the literature. According to the findings, published research on citrus anticancer countermeasures began in 1995 and will continue through 2023. This study has been ongoing since 1995. The number of publications has steadily increased over the last five years, although this trend has not yet reached its pinnacle. In 2022, there was a slight decrease in the number of publications, which might be attributed to the COVID-19 pandemic [43] and the fact that the bulk of research is focused on immunotherapy [44]. If there is no compromise in investigating the mechanism of action, then the lack of research related to citrus anticancer may be attributed to the difficulty in developing anticancer drugs from natural components owing to the diversity of chemicals contained in citrus. Obtaining pure single-use chemicals from natural sources requires tremendous work and incurs major financial expense. In reality, in order for these substances to be transformed into pharmaceuticals, they must be produced in large quantities. As a consequence, efforts will be necessary in the future, from synthesis through formulation.

In a trend-topic scan, initial research was focused on crude extract [45, 46], total flavonoids [47], and cytotoxicity [25], but now it is focused more on single compounds, deeper mechanisms, and prospective action for overcoming physico-chemical limitations and improving anticancer properties, such as biosynthesized **[48–50]**, micronized [51, 52], and nanoformulation **[53–55]**. Research on citrus anticancer will continue to gain traction in the coming years because there are still many opportunities to reach as a result of this growth. Still, the research on citrus anticancer still stacks on the targeted mechanism exploration, and no compound continues to be clinically tested in humans. From these findings, we noted that there are no authors, organizations, or countries that are constantly focused on research on citrus anticancer. The most persistent author was Ho c.-t. from the United States, who focused on tangeretin's development as an anticancer candidate. The latest publication demonstrated the structure modification of tangeretin with acetylation to enhance its bioavailability *in vivo* as well as increase its anticancer properties [33]. Structure modification can be done to improve the physical and chemical limitations of natural compounds. It is different in formulation but has the same objectives. Structure modification would facilitate the dosage formulation and design of synthesis so that efficacy can be achieved with minimal cost.

In a bibliometric mapping, keywords may give immediate details on topics, emerging topics, and the core subject of research. Word analysis may reveal the topics on which a given research project focuses [56]. Apoptosis, anticancer, flavonoid, citrus, hesperidin, naringenin, tangeretin, and nobiletin were the most frequently used terms in our investigation during the previous decade. The bulk of these keywords were found in the center of a word network and were significant as core terms. Apoptosis has been recognized as an essential term since the majority of citrus flavonoid mechanisms as anticancer agents induce apoptosis. Then, we may use the general search phrase "citrus flavonoid" since the most often used terms in the recent decade may be selected by researchers to obtain citrus anticancer studies in this field. In our study, we evaluated the course of time to identify the keywords with the most citation bursts in the word network. Year-by-year trends in citrus anticancer research are revealed by keywords with high citation counts. Between 2016 and 2017, in the case of citrus flavonoid, the word hesperidin was used the most. Positive developments in citrus anticancer research over the past five years indicate the direction of the trend for citrus flavonoids in the field of structural modification. Further research can be directed toward in-depth exploration related to selectivity and efforts to improve compound limitations to be developed in dosage formulation, followed by in vivo preclinical and clinical studies.

At the end, citrus showed promising potential as anticancer agent, especially hesperidin, naringenin, and nobiletin. The challenges of the development of citrus as an anticancer candidate is also related to the many species that exist in the world, but it will give big advantages that people are presented with a powerful, high-value resource to be consumed daily. Indeed, for anticancer treatment, it needs a single compound to be isolated, purified, and established its mechanism of action with selective, effective, and targeted. From this bibliometric mapping, future direction of the research on citrus anticancer would-be developing dosage formulation to increase the chemical and biological effectivity. This study's limitations should be considered when evaluating results. We used Scopus database for relevant papers and did not examine items in non-Scopus journals. Another important factor is selecting research articles over conference papers, book chapters, reviews, etc. Despite these limitations, these studies contribute to citrus anticancer research mapping and provide basic understanding for the research gap and future direction of anticancer exploration.

5 Conclusion

This study gives significant data on the overall number of publications from 1995 to 2023 in the field of citrus anticancer research. There has been a rising tide of publications on the subject, especially after 2017. The examination of recent trends in the field of citrus anticancer demonstrates that scientists have been paying greater attention to the role of a certain flavonoid chemical in cytotoxicity, cell cycle, apoptosis, autophagy, and metastasis. The most fruitful countries are China and India, and the journal "Molecules" has the most articles on citrus's anticancer properties. The investigation has also uncovered the top institutions and emerging trends in the field.

Overall, the synthesis and formulation of citrus flavonoids (specific citrus compound), as well as *in vitro* preclinical trials, have advanced citrus anticancer research and became the future perspective of citrus anticancer. More clinical trials are required to determine the optimal dose of citrus flavonoids for treating cancer as well as address any safety issues that may arise from using these compounds. As an anticancer drug, citrus flavonoid still warrants investigation into its formulation technique, followed by *in vivo* preclinical and clinical studies.

At the end, research collaboration between leading countries and top journals in the publication of scientific articles on citrus anticancer research is urgently needed to speed up the research progress. Through publication on a relevant academic impact and thus establish an important channel to promote studies for the development of citrus anticancer research.

6 Acknowledgement

This study received no specific grant from any funding agency.

7 Conflict of Interest

The authors have no conflict interest to disclose.

8 Ethical Approvals

This study does not involve experiments on animal or human subjects.

9 Author Contribution

All authors made substantial contributions to the conception and design, acquisition of data, or analysis and interpretation of data; took part in drafting the article; gave final approval of the version to be published; and agree to be accountable for all aspects of the work.

10 References

1. Ferlay J, Colombet M, Soerjomataram I, et al (2019) Estimating the global cancer incidence and mortality in 2018: GLOBOCAN sources and methods. Int J Cancer 144:1941–1953. https://doi.org/10.1002/ijc.31937
2. Abbas T, Keaton MA, Dutta A (2013) Genomic Instability in Cancer. Cold Spring Harbor Perspectives in Biology 5:a012914–a012914. https://doi.org/10.1101/cshperspect.a012914
3. Zhou H, Wang Y, Zhang Z, et al (2023) A novel prognostic gene set for colon adenocarcinoma relative to the tumor microenvironment, chemotherapy, and immune therapy. Front Genet 13:975404. https://doi.org/10.3389/fgene.2022.975404
4. Karagiannis GS, Condeelis JS, Oktay MH (2019) Chemotherapy-Induced Metastasis: Molecular Mechanisms, Clinical Manifestations, Therapeutic Interventions. Cancer Res 79:4567–4576. https://doi.org/10.1158/0008-5472.CAN-19-1147
5. Wong SC, Kamarudin MNA, Naidu R (2023) Anticancer Mechanism of Flavonoids on High-Grade Adult-Type Diffuse Gliomas. Nutrients 15:797. https://doi.org/10.3390/nu15040797

6. Koolaji N, Shammugasamy B, Schindeler A, et al (2020) Citrus Peel Flavonoids as Potential Cancer Prevention Agents. Current Developments in Nutrition 4:nzaa025. https://doi.org/10.1093/cdn/nzaa025
7. Rawson NE, Ho C-T, Li S (2014) Efficacious anti-cancer property of flavonoids from citrus peels. Food Science and Human Wellness 3:104–109. https://doi.org/10.1016/j.fshw.2014.11.001
8. Qiu M, Wei W, Zhang J, et al (2023) A Scientometric Study to a Critical Review on Promising Anticancer and Neuroprotective Compounds: Citrus Flavonoids. Antioxidants 12:669. https://doi.org/10.3390/antiox12030669
9. Bisol Â, Campos PS, Lamers ML (2020) Flavonoids as anticancer therapies: A systematic review of clinical trials. Phytotherapy Research 34:568–582. https://doi.org/10.1002/ptr.6551
10. Harishkumar M, Masatoshi Y, Hiroshi S, et al (2013) Revealing the Mechanism of *In Vitro* Wound Healing Properties of *Citrus tamurana* Extract. BioMed Research International 2013:1–8. https://doi.org/10.1155/2013/963457
11. Pamungkas Putri DD, Maran GG, Kusumastuti Y, et al (2022) Acute toxicity evaluation and immunomodulatory potential of hydrodynamic cavitation extract of citrus peels. J App Pharm Sci 136–145. https://doi.org/10.7324/JAPS.2022.120415
12. Kim MY, Choi EO, HwangBo H, et al (2018) Reactive oxygen species-dependent apoptosis induction by water extract of *Citrus unshiu* peel in MDA-MB-231 human breast carcinoma cells. Nutr Res Pract 12:129. https://doi.org/10.4162/nrp.2018.12.2.129
13. Banjerdpongchai R, Wudtiwai B, Khaw-on P, et al (2016) Hesperidin from Citrus seed induces human hepatocellular carcinoma HepG2 cell apoptosis via both mitochondrial and death receptor pathways. Tumor Biol 37:227–237. https://doi.org/10.1007/s13277-015-3774-7
14. Cirmi S, Maugeri A, Ferlazzo N, et al (2017) Anticancer Potential of Citrus Juices and Their Extracts: A Systematic Review of Both Preclinical and Clinical Studies. Front Pharmacol 8:420. https://doi.org/10.3389/fphar.2017.00420
15. Manthey JA, Guthrie N (2002) Antiproliferative Activities of Citrus Flavonoids against Six Human Cancer Cell Lines. J Agric Food Chem 50:5837–5843. https://doi.org/10.1021/jf020121d
16. Do Prado SBR, Shiga TM, Harazono Y, et al (2019) Migration and proliferation of cancer cells in culture are differentially affected by molecular size of modified citrus pectin. Carbohydrate Polymers 211:141–151. https://doi.org/10.1016/j.carbpol.2019.02.010
17. Wang L, Wang J, Fang L, et al (2014) Anticancer Activities of Citrus Peel Polymethoxyflavones Related to Angiogenesis and Others. BioMed Research International 2014:1–10. https://doi.org/10.1155/2014/453972
18. Arafa E-SA, Zhu Q, Barakat BM, et al (2009) Tangeretin Sensitizes Cisplatin-Resistant Human Ovarian Cancer Cells through Downregulation of Phosphoinositide 3-Kinase/Akt Signaling Pathway. Cancer Research 69:8910–8917. https://doi.org/10.1158/0008-5472.CAN-09-1543
19. Aggarwal V, Tuli HS, Thakral F, et al (2020) Molecular mechanisms of action of hesperidin in cancer: Recent trends and advancements. Exp Biol Med (Maywood) 245:486–497. https://doi.org/10.1177/1535370220903671
20. Stabrauskiene J, Kopustinskiene DM, Lazauskas R, Bernatoniene J (2022) Naringin and Naringenin: Their Mechanisms of Action and the Potential Anticancer Activities. Biomedicines 10:1686. https://doi.org/10.3390/biomedicines10071686
21. Adokoh CK, Asante D-B, Acheampong DO, et al (2019) Chemical profile and in vivo toxicity evaluation of unripe Citrus aurantifolia essential oil. Toxicology Reports 6:692–702. https://doi.org/10.1016/j.toxrep.2019.06.020
22. Donthu N, Kumar S, Mukherjee D, et al (2021) How to conduct a bibliometric analysis: An overview and guidelines. Journal of Business Research 133:285–296. https://doi.org/10.1016/j.jbusres.2021.04.070
23. Mejia C, Wu M, Zhang Y, Kajikawa Y (2021) Exploring Topics in Bibliometric Research Through Citation Networks and Semantic Analysis. Front Res Metr Anal 6:742311. https://doi.org/10.3389/frma.2021.742311
24. Perianes-Rodriguez A, Waltman L, Van Eck NJ (2016) Constructing bibliometric networks: A comparison between full and fractional counting. Journal of Informetrics 10:1178–1195. https://doi.org/10.1016/j.joi.2016.10.006
25. Hirano T, Abe K, Gotoh M, Oka K (1995) Citrus flavone tangeretin inhibits leukaemic HL-60 cell growth partially through induction of apoptosis with less cytotoxicity on normal lymphocytes. Br J Cancer 72:1380–1388. https://doi.org/10.1038/bjc.1995.518
26. Amalina ND, Salsabila IA, Zulfin UM, et al (2023) In vitro synergistic effect of hesperidin and doxorubicin downregulates epithelial-mesenchymal transition in highly metastatic breast cancer cells. J Egypt Natl Canc Inst 35:6. https://doi.org/10.1186/s43046-023-00166-3
27. Phucharoenrak P, Muangnoi C, Trachootham D (2023) Metabolomic Analysis of Phytochemical Compounds from Ethanolic Extract of Lime (Citrus aurantifolia) Peel and Its Anti-Cancer Effects against Human Hepatocellular Carcinoma Cells. Molecules 28:2965. https://doi.org/10.3390/molecules28072965
28. Wu Y, Li Q, Lv L, et al (2023) Nobiletin inhibits breast cancer cell migration and invasion by suppressing the IL-6-induced ERK-STAT and JNK-c-JUN pathways. Phytomedicine 110:154610. https://doi.org/10.1016/j.phymed.2022.154610
29. Yun Y, Park S-H, Kim I (2019) Antioxidant effect of Kimchi supplemented with Jeju citrus

https://doi.org/10.1051/bioconf/20237501002

concentrate and its antiobesity effect on 3T3-L1 adipocytes. Food Sci Nutr 7:2740–2746. https://doi.org/10.1002/fsn3.1138
30. Kim J, Jayaprakasha GK, Patil BS (2014) Obacunone exhibits anti-proliferative and anti-aromatase activity in vitro by inhibiting the p38 MAPK signaling pathway in MCF-7 human breast adenocarcinoma cells. Biochimie 105:36–44. https://doi.org/10.1016/j.biochi.2014.06.002
31. Oh UH, Kim D-H, Lee J, et al (2022) Eriodictyol induces apoptosis *via* regulating phosphorylation of JNK, ERK, and FAK/AKT in pancreatic cancer cells. JABC 65:83–88. https://doi.org/10.3839/jabc.2022.011
32. Kim H-J, Jin B-R, An H-J (2023) Hesperidin ameliorates benign prostatic hyperplasia by attenuating cell proliferation, inflammatory response, and epithelial-mesenchymal transition via the TGF-β1/Smad signaling pathway. Biomedicine & Pharmacotherapy 160:114389. https://doi.org/10.1016/j.biopha.2023.114389
33. Tsai H-Y, Yang J-F, Chen Y-B, et al (2022) Acetylation Enhances the Anticancer Activity and Oral Bioavailability of 5-Demethyltangeretin. IJMS 23:13284. https://doi.org/10.3390/ijms232113284
34. Nurlaila SR, Rachmadani AD, Harismah K (2022) Formulation and Evaluation of Physical Stability Natural Acne Gel Based on Aloevera Gel with Essential Oil Blend. UJAS 2:34–42. https://doi.org/10.53017/ujas.163
35. Sukmawati A, Nafarin A, Yuliani R, et al (2018) IN VITRO CYTOTOXIC EVALUATION OF DOXORUBICIN AND CURCUMIN ANALOGUE LOADED MODIFIED CHITOSAN NANOPARTICLES. RJC 11:1657–1662. https://doi.org/10.31788/RJC.2018.1144084
36. So FV, Guthrie N, Chambers AF, et al (1996) Inhibition of human breast cancer cell proliferation and delay of mammary tumorigenesis by flavonoids and citrus juices. Nutrition and Cancer 26:167–181. https://doi.org/10.1080/01635589609514473
37. Celia C, Trapasso E, Locatelli M, et al (2013) Anticancer activity of liposomal bergamot essential oil (BEO) on human neuroblastoma cells. Colloids and Surfaces B: Biointerfaces 112:548–553. https://doi.org/10.1016/j.colsurfb.2013.09.017
38. Monteiro SS, De Oliveira VM, Pasquali MADB (2022) Probiotics in Citrus Fruits Products: Health Benefits and Future Trends for the Production of Functional Foods—A Bibliometric Review. Foods 11:1299. https://doi.org/10.3390/foods11091299
39. Kisacam MA (2023) Nobiletin is capable of regulating certain anti-cancer pathways in a colon cancer cell line. Naunyn-Schmiedeberg's Arch Pharmacol 396:547–555. https://doi.org/10.1007/s00210-022-02354-9
40. Sardari S, Shokrgozar MA, Ghavami G (2009) Cheminformatics based selection and cytotoxic effects of herbal extracts. Toxicol Vitro 23:1412–1421. https://doi.org/10.1016/j.tiv.2009.07.011
41. Lee M, Song M (2020) Incorporating citation impact into analysis of research trends. Scientometrics 124:1191–1224. https://doi.org/10.1007/s11192-020-03508-3
42. Skute I, Zalewska-Kurek K, Hatak I, De Weerd-Nederhof P (2019) Mapping the field: a bibliometric analysis of the literature on university–industry collaborations. J Technol Transf 44:916–947. https://doi.org/10.1007/s10961-017-9637-1
43. Ali S, Alam M, Khatoon F, et al (2022) Natural products can be used in therapeutic management of COVID-19: Probable mechanistic insights. Biomedicine & Pharmacotherapy 147:112658. https://doi.org/10.1016/j.biopha.2022.112658
44. Alberca RW, Teixeira FME, Beserra DR, et al (2020) Perspective: The Potential Effects of Naringenin in COVID-19. Front Immunol 11:570919. https://doi.org/10.3389/fimmu.2020.570919
45. So FV, Guthrie N, Chambers AF, et al (1996) Inhibition of human breast cancer cell proliferation and delay of mammary tumorigenesis by flavonoids and citrus juices. NUTR CANCER 26:167–181. https://doi.org/10.1080/01635589609514473
46. Breinholt VM, Nielsen SE, Knuthsen P, et al (2003) Effects of commonly consumed fruit juices and carbohydrates on redox status and anticancer biomarkers in female rats. Nutr Cancer 45:46–52. https://doi.org/10.1207/S15327914NC4501_6
47. Kim H, Moon JY, Mosaddik A, Cho SK (2010) Induction of apoptosis in human cervical carcinoma HeLa cells by polymethoxylated flavone-rich Citrus grandis Osbeck (Dangyuja) leaf extract. Food Chem Toxicol 48:2435–2442. https://doi.org/10.1016/j.fct.2010.06.006
48. Ahmed S, Kaur G, Sharma P, et al (2018) Fruit waste (peel) as bio-reductant to synthesize silver nanoparticles with antimicrobial, antioxidant and cytotoxic activities. J App Biomed 16:221–231. https://doi.org/10.1016/j.jab.2018.02.002
49. Almessiere MA, Khan FA, Auwal IA, et al (2022) Green synthesis, characterization and anti-cancer capability of Co0.5Ni0.5Nd0.02Fe1.98O4 nanocomposites. Arab J Chem 15:. https://doi.org/10.1016/j.arabjc.2021.103564
50. Atta EM, Hegab KH, Abdelgawad AAM, Youssef AA (2019) Synthesis, characterization and cytotoxic activity of naturally isolated naringin-metal complexes. Saudi Pharm J 27:584–592. https://doi.org/10.1016/j.jsps.2019.02.006
51. Salehi H, Karimi M, Raofie F (2021) Micronization and coating of bioflavonoids extracted from Citrus sinensis L. peels to preparation of sustained release pellets using supercritical technique. J Iran Chem Soc 18:3235–3248. https://doi.org/10.1007/s13738-021-02262-4

52. Wu J-J, Shen C-T, Jong T-T, et al (2009) Supercritical carbon dioxide anti-solvent process for purification of micronized propolis particulates and associated anti-cancer activity. Sep Purif Technol 70:190–198. https://doi.org/10.1016/j.seppur.2009.09.015
53. Liu B, Li C, Han J, et al (2023) Biosynthesized gold nanoparticles using leaf extract of Citrus medica inhibit hepatocellular carcinoma through regulation of the Wnt/β-catenin signaling pathway. Arab J Chem 16:. https://doi.org/10.1016/j.arabjc.2023.104800
54. Abdelsattar AS, Kamel AG, El-Shibiny A (2023) The green production of eco-friendly silver with cobalt ferrite nanocomposite using Citrus limon extract. Results Chem 5:. https://doi.org/10.1016/j.rechem.2022.100687
55. Alipanah H, Farjam M, Zarenezhad E, et al (2021) Chitosan nanoparticles containing limonene and limonene-rich essential oils: potential phytotherapy agents for the treatment of melanoma and breast cancers. BMC Compl MedTherapies 21:. https://doi.org/10.1186/s12906-021-03362-7
56. Chen G, Xiao L (2016) Selecting publication keywords for domain analysis in bibliometrics: A comparison of three methods. Journal of Informetrics 10:212–223. https://doi.org/10.1016/j.joi.2016.01.006

Depression, anxiety, and stress disorders detection in students during the Covid-19 pandemic using Naïve Bayes algorithm

Annisa Rahmadani[1*], *Casi* Setianingsih[1], *Fussy* Mentari Dirgantara[1], *Ayub* Rosihan Ambarita[1], *Hafid* Ikhsan Arifin[1], *Indratama* Pangasian Manalu[1], and *Muhammat* Lio Pratama[1]

[1]School of Electrical Engineering, Telkom University, Bandung, Indonesia

Abstract. During the Covid-19 pandemic, students in Indonesia carried out bold learning from home as a social effort during the pandemic. This bold learning process is considered to be still less effective and efficient and has resulted in some students, especially students having homework during the brave learning period. This has an impact on the psychology of students such as the emergence of depression, anxiety, and stress. Sources of psychological disorders not only from academics but from within themselves also affect mental health. The results of a survey on mental health during the pandemic conducted by the Association of Indonesian Mental Medicine Specialists (PDSKJI) showed that 64.8% of respondents experienced psychological problems in the age group of 19-24 years and over 60 years. In this study the author will make a system of Tests for Depression, Anxiety and Stress Disorders in Students. The results of this test are the severity of each psychological disorder and treatment recommendations based on the test results. The psychological scale used in this study is the DASS-42 (Depression, Anxiety, and Stress Scale) which has 42 statements and 3 categories of disorders, namely depression, anxiety, and stress. Each category has 5 levels, namely normal, mild, moderate, severe, and very severe. The Test System for Depression, Anxiety, and Stress Disorders for Students uses the Naïve Bayes method with the accuracy of the dataset obtained by 86.44%, so it can be said that this system is running according to the purpose.
Keywords: Covid-19 Pandemic, Psychological Disorders, PDSKJI, DASS-42, Naïve Bayes

1 Introduction

The Covid-19 pandemic has had a major impact on all fields, one of which is education. The Indonesian government has issued an online/distance learning policy since March 2020. This was done to stop the spread of the Covid-19 outbreak. Online learning has advantages and disadvantages in its application. The Association of Indonesian Mental Medicine Specialists (PDSKJI) conducted a survey on mental health during the COVID-19 pandemic. The results of this self-examination showed that 64.8% of respondents experienced psychological problems with the proportion of 64.8% anxious, 61.5% depressed, and 74.8% traumatized. Most psychological problems are found in the age group of 17-29 years and above 60 years [1].

For students, this pandemic causes stress and anxiety because it is related to changes in the lecture process and daily life. Therefore, a test is needed to determine the level of depression, anxiety, and stress for students in the midst of this covid-19 period, so that they can take preventive measures from the start before going to a more severe level. Depression, Anxiety and Stress Scale (DASS) is a self-assessment scale used to measure a person's negative emotions. for DASS development, not only the factor structure but also the relative performance of each item was found to be nearly the same in clinical and non-clinical samples [2]. However, the main purpose of measurement with the DASS in this study was to determine the severity of symptoms of depression, anxiety and stress and recommend some treatment. In applying the Depression, Anxiety, and Stress Disorder Test to Students, the author uses the Naïve Bayes algorithm. With this application, students and the general public can carry out initial screening and find out the level of depression, anxiety and stress based on the symptoms experienced and recommend treatment to help users in the first treatment.

2 Related Work

Naïve Bayes algorithm have been used in much research in health problems. Triyanna Widiyaningtyas, Ilham A Zaeni, and Nadiratin Jamilah (2020) in their research developed a method to diagnose the symptoms of fever in both diseases. Their Naïve Bayes algorithm used to classify the diagnosing fever symptoms. Algorithm testing is done using k-fold cross-validation, with k equal to 10. The evaluation of the algorithm is measured by calculating the value of accuracy, precision, and recall from prediction results. The results showed that the average accuracy rate was 94%, precision was 90%, and recall was 92%. This shows that the Naïve Bayes algorithm has good performance in diagnosing fever in patients [3].

There are several studies on the prediction of anxiety and depression in elderly patients using machine learning technology. Arkaprabha Sau and Ishita Bhakta (2017) in their

* Corresponding author: annisarahmadani@student.telkomuniversity.ac.id

research developed a predictive model for automatic diagnosis of anxiety and depression in geriatric patients. The data used in this study are socio-demographic factors and patient health. Geriatric patients were also classified into two using the Hospital Anxiety and Depression (HADS) scale classification process using ten algorithms including Bayesian Networks, Logistics, Multiple Layered Perceptrons, Naïve Bayes, Random Forests, Random Trees, J48, Sequential Minimal Optimization, Space Sub Random, and K Star. Results from 10 machine learning classifications were evaluated and Random forFst (RF) got a prediction accuracy of 91% and false positives only 10%. this accuracy is tested by 10-fold cross-validation [4].

Another research from Setiyo Budiyanto and Harry Candra Sihombing (2019) explained about measuring the tendency of depression and anxiety through social media using the closed loop method with Facebook text mining posts. with pre-processing stages including text extraction using the Naïve Bayes model for text classification and symptoms of depression and anxiety were measured using Depression, Anxiety, Stress Scale (DASS)-21. Facebook post data used as training data is 22,934 and the result is an analysis of user social demographic mapping which is usually a trigger for depression, and anxiety, such as sadness, illness, household affairs, children's education and others are available [5].

Apart from predicting anxiety and depression through social media, there is a study on the assessment of anxiety, depression, and stress using machine learning by Prince Kumara and Shruti Garg (2020) where the tools used are also DASS-21 and DASS-42. The difference in the five severity levels of anxiety, stress and depression were predicted using eight machine learning algorithms. The method used is divided into four different categories: Bayes, neural network, lazy tree, the K-star hybrid technique and the random forest method. All methods were applied to two different databases, DASS-42 and DASS-21. the results showed that the Radial Basis Function Network (RBFN) performed the best for depression in both datasets. random forest results are 100 percent for anxiety in the DASS-21 database [6].

In a study by Anu Priya and Shruti Garg (2019) the prediction of anxiety, depression and stress was made using a machine learning algorithm. The scale used is DASS-21 and is predicted based on five severity levels using five different machine learning algorithms. The algorithm used is Naive Bayes, Support Vector Machine (SVM), K-Nearest Neighbors (KNN), Random Forest and Decision Tree. Naive Bayes accuracy was found to be the highest with the percentages of depression, anxiety, and stress at 85%, 73% and 74%, respectively [7].

3 Research Method

This section discusses the methods used in this study.

3.1 Depression, Anxiety, Stress Scale (DASS-42)

Depression, Anxiety, Stress Scale (DASS) is a measuring tool developed by Lovibond and Lovibond in 1995. This DASS test consists of 42 statement items with three psychological scales namely depression, anxiety and stress. Each psychological scale consists of 14 items, which are further divided into sub-scales consisting of 2 to 5 items which are estimated to measure the same thing [8]. The severity of the DASS-42 is based on the mean population score obtained from a large and relatively heterogeneous sample. If a person has significant symptoms of depression, anxiety, and stress, they are still referred to a psychologist.

In the DASS-42 standard, the distribution of items/symptoms that affect certain disorders can be seen in Table 1 and when taking sample data, items using Indonesian [9].

Table 1. Question items.

No	Item
1	I found myself getting upset by quite trivial things
2	I was aware of dryness of my mouth
3	I couldn't seem to experience any positive feeling at all
4	I experienced breathing difficulty (eg, excessively rapid breathing, breathlessness in the absence of physical exertion)
5	I just couldn't seem to get going
6	I tended to over-react to situations.
7	I had a feeling of shakiness (eg, legs going to give way)
8	I found it difficult to relax
9	I found myself in situations that made me so anxious I was most relieved when they ended
10	I felt that I had nothing to look forward to
11	I found myself getting upset rather easily
12	I felt that I was using a lot of nervous energy
13	I felt sad and depressed
14	I found myself getting impatient when I was delayed in any way (eg, elevators, traffic lights, being kept waiting)
15	I had a feeling of faintness
16	I felt that I had lost interest in just about everything
17	I felt I wasn't worth much as a person
18	I felt that I was rather touchy
19	I perspired noticeably (eg, hands sweaty) in the absence of high temperatures or physical exertion
20	I felt scared without any good reason
21	I felt that life wasn't worthwhile
22	I found it hard to wind down
23	I had difficulty in swallowing
24	I couldn't seem to get any enjoyment out of the things I did
25	I was aware of the action of my heart in the absence of physical exertion (eg, sense of heart rate increase, heart missing a beat)
26	I felt down-hearted and blue
27	I found that I was very irritable
28	I felt I was close to panic
29	I found it hard to calm down after something upset me
30	I feared that I would be "thrown" by some trivial but unfamiliar task
31	I was unable to become enthusiastic about anything
32	I found it difficult to tolerate interruptions to what I was doing
33	I was in a state of nervous tension
34	I felt I was pretty worthless
35	I was intolerant of anything that kept me from getting on with what I was doing
36	I felt terrified
37	I could see nothing in the future to be hopeful about
38	I felt that life was meaningless
39	I found myself getting agitated
40	I was worried about situations in which I might panic and make a fool of myself
41	I experienced trembling (eg, in the hands)

Table 2. Item Number.

Psychological disorder	No Item
Depression	3,5,10,13,16,17,21,24,31,34,37,38,42
Anxiety	2,4,7,9,15,19,20,23,25,28,30,36,40,41
Stress	1,6,8,11,12,14,18,22,27,29,32,33,35,39

In each item there are four types of answers with different weights, namely never: 0, sometimes: 1, quite often: 2, very often: 3. In Table 2 is the division of each item for symptoms of depression, anxiety and stress consisting of 14 items. After the final score is calculated, it will be labeled according to the severity level, namely "Normal", "Mild", "Medium", "Severe", and "Very Severe". Table 3 is an assessment indicator of severity [10]:

Table 3. Rating Indicators.

Level	Depression	Anxiety	Stress
Normal	0-9	0-7	0-14
Mild	10-13	8-9	15-18
Medium	14-20	10-14	19-25
Severe	21-27	15-19	26-33
Very Severe	>28	>20	>33

3.2 Naïve Bayes

Bayes theorem is an approximation to uncertainty as measured by probability. Bayes theorem is used for classification purposes and to assume that classification is an independent predictor. It is assumed that the Naive Bayes classifier in the presence of certain features in the class is not related to other features. The Naive Bayes model is compatible for very large data sets to build on and beyond analysis. This model is a very simple and sophisticated classification method, and it well done even in complicated scenarios [11].

The Naive Bayes algorithm is suitable for classifying datasets of nominal and numeric types. If the dataset is of numeric type, the calculation of the gaussian distribution is used. The calculation of the gaussian distribution can be seenin equation 1 [12].

$$P(Xi = xi|Y = yj) = \frac{1}{\sqrt{2\pi\sigma}} e^{-\frac{(xi-\mu ij)^2}{2\sigma^2 ij}} \quad (1)$$

Description:
P: opportunity.
Xi: Attribute to *i*.
xi: Attribute value to *i*.
Y: The class you are looking for.
yj: subclass *Y* you are looking for.

Calculate the mean value according to equation 2:

$$\mu = \frac{\Sigma^n{}_i xi}{n} \quad (2)$$

Description:
μ: arithmetic mean (mean).
xi: sample value *i*.
n: number of samples.

Standard deviation value according to equation 3:

$$\sigma = \sqrt{\frac{\Sigma_i^n (xi-\mu)^2}{n-1}} \quad (3)$$

Description:
μ: arithmetic mean (mean).
xi: sample value *i*.
n: number of samples.
σ: standard deviation.

3.3 Synthetic Minority Oversampling Technique (SMOTE)

Synthetic Minority Oversampling Technique SMOTE is used to provide a solution to the problem of unbalanced data with a different processing base than the previously introduced oversampling. The basis of the SMOTE procedure is to interpolate between neighboring minority class instances. Thus, he was able to increase the number of minorities class examples by introducing new minority class examples in the environment, thereby helping classifiers to increase their generalizability capacity. The SMOTE preprocessing technique became a pioneer for the research community on imbalances initial classification. Since its release, many extensions and alternatives have been proposed to improve its performance under different scenarios. Due to its popularity and influence, SMOTE is considered as one of the most influential data preprocessing/sampling algorithms in machine learning and data mining [13].

4 System Design and Overview

4.1 System Overview

First, the user inputs the name first and then performs the Depression, Anxiety, Stress Scale (DASS)-42 test. The results of the DASS-42 test will be processed using the naive Bayes method based on the results of the severity of depression, anxiety, and stress. The output of the system will display the severity of depression, anxiety, and stress as well as treatment recommendations for users.

4.2 Treatment and Rules mapping

Table 4 is a treatment for the recommendations of this system. This treatment is collected by discussing with psychologists and what is recommended to users is a treatment that can be done alone without the need for a therapist. However, this treatment is only temporary because it does not eliminate the root of the problem from the user or only as a first treatment. The rules used for treatment recommendations obtained by discussing with psychologists are listed in Table 5.

Table 4. Treatment.

No	*Treatment*	Code
1	Take care of mental health	A
2	Food Nutrition	B
3	Sport	C
4	Enough sleep	D
5	Fun activities	E
6	journaling	F
7	yoga	G
8	meditation	H
9	art therapy	I
10	Keep building relationships/interpersonal therapy	J
11	Write down your worries	K
12	Music therapy	L
13	Using aromatherapy	M
14	Time management well	N
15	Healthy lifestyle	O
16	Breathing exercises	P

Table 5. Rules mapping.

No.	DASS test factor			*Treatment*
	D	A	S	
1	Normal	Normal	Normal	A
2	Normal	Normal	Mild	A D
3	Normal	Normal	Medium	A C
4	Normal	Normal	Severe	A O
5	Normal	Normal	Very Severe	A L
6	Mild	Normal	Normal	A E
7	Mild	Normal	Mild	A B
8	Mild	Normal	Medium	A C
9	Mild	Normal	Severe	A M
10	Mild	Normal	Very Severe	A G
11	Medium	Normal	Normal	A C
12	Medium	Normal	Mild	A D
13	Medium	Normal	Medium	A N
14	Medium	Normal	Severe	A F
15	Medium	Normal	Very Severe	AI
16	Severe	Normal	Normal	A O
17	Severe	Normal	Mild	A H
18	Severe	Normal	Medium	AN
19	Severe	Normal	Severe	A G
20	Severe	Normal	Very Severe	AG
21	Very Severe	Normal	Normal	AL
22	Very Severe	Normal	Mild	AF
23	Very Severe	Normal	Medium	AN
24	Very Severe	Normal	Severe	A L
25	Very Severe	Normal	Very Severe	A I
26	Normal	Mild	Normal	P A
27	Mild	Mild	Normal	P E
28	Medium	Mild	Normal	P C
29	Severe	Mild	Normal	P H
30	Very Severe	Mild	Normal	P F
31	Normal	Mild	Mild	P B
32	Mild	Mild	Mild	P D
33	Medium	Mild	Mild	P C
34	Severe	Mild	Mild	P O
35	Very Severe	Mild	Mild	P I
36	Normal	Mild	Medium	P G
37	Mild	Mild	Medium	P M
38	Medium	Mild	Medium	P C
39	Severe	Mild	Medium	P H
40	Very Severe	Mild	Medium	P I
41	Normal	Mild	Severe	PO
42	Mild	Mild	Severe	PO
43	Medium	Mild	Severe	PJ
44	Severe	Mild	Severe	PH
45	Very Severe	Mild	Severe	PI
46	Normal	Mild	Very Severe	PL
47	Mild	Mild	Very Severe	PI
48	Medium	Mild	Very Severe	PF
49	Severe	Mild	Very Severe	PO
50	Very Severe	Mild	Very Severe	PN
51	Normal	Medium	Normal	HA
52	Normal	Medium	Mild	HD
53	Normal	Medium	Medium	HC
54	Normal	Medium	Severe	HO
55	Normal	Medium	Very Severe	HL
56	Mild	Medium	Normal	HE
57	Mild	Medium	Mild	HB
58	Mild	Medium	Medium	HC
59	Mild	Medium	Severe	HM
60	Mild	Medium	Very Severe	HG
61	Medium	Medium	Normal	HC
62	Medium	Medium	Mild	HD
63	Medium	Medium	Medium	HN
64	Medium	Medium	Severe	HF
65	Medium	Medium	Very Severe	HI
66	Severe	Medium	Normal	H O
67	Severe	Medium	Mild	HM
68	Severe	Medium	Medium	HN
69	Severe	Medium	Severe	HG
70	Severe	Medium	Very Severe	HG
71	Very Severe	Medium	Normal	HL
72	Very Severe	Medium	Mild	HF

No.	DASS test factor			*Treatment*
	D	A	S	
73	Very Severe	Medium	Medium	HN
74	Very Severe	Medium	Severe	HL
75	Very Severe	Medium	Very Severe	HI
76	Normal	Severe	Normal	KA
77	Mild	Severe	Normal	KE
78	Medium	Severe	Normal	KC
79	Severe	Severe	Normal	KH
80	Very Severe	Severe	Normal	KF
81	Normal	Severe	Mild	KB
82	Mild	Severe	Mild	KD
83	Medium	Severe	Mild	KC
84	Severe	Severe	Mild	KO
85	Very Severe	Severe	Mild	KI
86	Normal	Severe	Medium	KG
87	Mild	Severe	Medium	KM
88	Medium	Severe	Medium	KC
89	Severe	Severe	Medium	KG
90	Very Severe	Severe	Medium	KI
91	Normal	Severe	Severe	KO
92	Mild	Severe	Severe	KO
93	Medium	Severe	Severe	KJ
94	Severe	Severe	Severe	KH
95	Very Severe	Severe	Severe	KI
96	Normal	Severe	Very Severe	KL
97	Mild	Severe	Very Severe	KI
98	Medium	Severe	Very Severe	KF
99	Severe	Severe	Very Severe	KO
100	Very Severe	Severe	Very Severe	KN
101	Normal	Very Severe	Normal	IA
102	Normal	Very Severe	Mild	ID
103	Normal	Very Severe	Medium	IC
104	Normal	Very Severe	Severe	IO
105	Normal	Very Severe	Very Severe	IL
106	Mild	Very Severe	Normal	IE
107	Mild	Very Severe	Mild	IB
108	Mild	Very Severe	Medium	I C
109	Mild	Very Severe	Severe	I M
110	Mild	Very Severe	Very Severe	I G
111	Medium	Very Severe	Normal	I C
112	Medium	Very Severe	Mild	I D
113	Medium	Very Severe	Medium	I N
114	Medium	Very Severe	Severe	I F
115	Medium	Very Severe	Very Severe	IL
116	Severe	Very Severe	Normal	I O
117	Severe	Very Severe	Mild	I M
118	Severe	Very Severe	Medium	IN
119	Severe	Very Severe	Severe	IG
120	Severe	Very Severe	Very Severe	IG
121	Very Severe	Very Severe	Normal	IL
122	Very Severe	Very Severe	Mild	IF
123	Very Severe	Very Severe	Medium	IN
124	Very Severe	Very Severe	Severe	IF
125	Very Severe	Very Severe	Very Severe	IL

4.3 Dataset

The dataset used in this study was obtained from openpsychometrics.org, this dataset is used in the process of building a Naïve Bayes model to map the severity of depression, anxiety, and stress with treatment recommendations. The depression anxiety stress scales dataset is in the form of answers to 42 item questions. The DASS test dataset was updated in 2018 and consists of 39,775 rows. Figure 1 is an example dataset from https://openpsychometrics.org/.

Q1A	Q1I	Q1E	Q2A	Q2I	Q2E	Q3A	Q3I	Q3E	Q4A	Q4I	Q4E	Q5A	Q5I	Q5E	Q6A	Q6I	Q6E
4	28	3890	4	25	2122	2	16	1944	4	8	2044	4	34	2153	4	33	2416
4	2	8118	1	36	2890	2	35	4777	3	28	3090	4	10	6078	4	40	2790
3	7	5784	1	33	4373	4	41	3242	1	13	6470	4	11	3827	3	9	3704
2	23	5081	3	11	6837	2	37	5521	1	27	4556	3	28	3269	3	26	3231
2	36	3215	2	13	7731	3	5	4156	4	10	2802	4	2	5628	2	9	6522

Fig. 1. Dataset.

4.4 Preprocessing

After the data is obtained from openpsychometrics.org, There are 172 columns and 39,975 rows in the dataset. Then the data cleaning process is carried out where the value of E in Figure 1 which represents the position of the questions from 42 questions and the value of I as the recording time is ignored or omitted. taken only the value of A as a collection of answers from respondents. Figure 2 is the result of data cleaning:

Q10A	Q11A	Q12A	Q13A	Q14A	Q15A	Q16A	Q17A	Q18A	Q19A	Q1A	Q20A	Q21A	Q22A	Q23A	Q24A	Q25A	Q26A
1	4	4	4	4	4	4	3	4	3	4	3	1	4	4	4	4	4
2	2	2	4	4	3	3	4	2	1	4	1	2	3	1	2	2	3
4	2	1	4	1	4	4	4	2	2	3	1	4	3	2	4	2	1
3	2	1	1	4	2	2	3	1	1	2	2	1	1	1	1	1	2
3	2	4	4	4	4	3	4	4	4	2	4	3	3	4	2	4	4
...	...	...	...	...	...	...	...	...	...	...	...	...	...	...	...	...	...

Fig. 2. Dataset after cleaning.

Figure 3 is the result of scoring symptoms of depression, anxiety, stress labeled score_D, score_A and Score_S. Then there is a category label for each symptom that contains the severity of psychological symptoms such as "Normal", "Ringan" means Mild, "Sedang" means Medium, "Parah" means Severe, and "Sangat Parah" means very severe. After that, the category will be mapped by treatment using the rules in Table 5.

score_D	score_A	score_S	category_depresi	category_anxiety	category_stress	treatment_label
27	34	40	parah	sangat parah	sangat parah	IG
24	17	27	parah	parah	parah	KH
39	12	17	sangat parah	sedang	ringan	HF
16	17	16	sedang	parah	ringan	KC
32	40	29	sangat parah	sangat parah	parah	IF
13	8	12	ringan	normal	normal	AE

Fig. 3. Dataset after scoring.

4.5 SMOTE

In this case, because the DASS-42 dataset experiences class imbalance, with the difference in each class being very much different in number, the SMOTE method is adopted. After oversampling with SMOTE, all classes have sizes of 4,683 each. Figure 4 is a class imbalance before applying SMOTE and Figure 5 is a class that is already balanced.

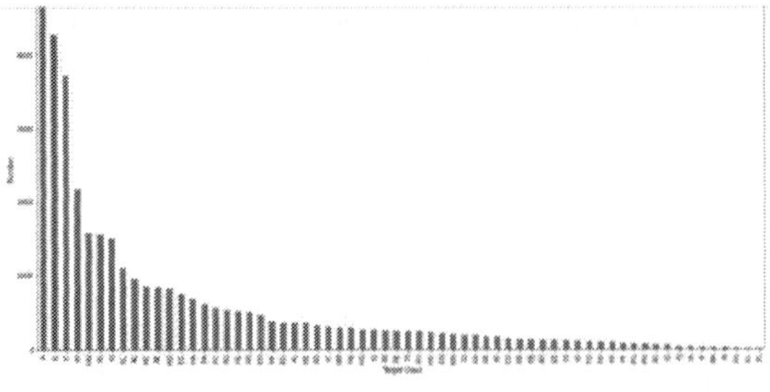

Fig. 4. Imbalanced class.

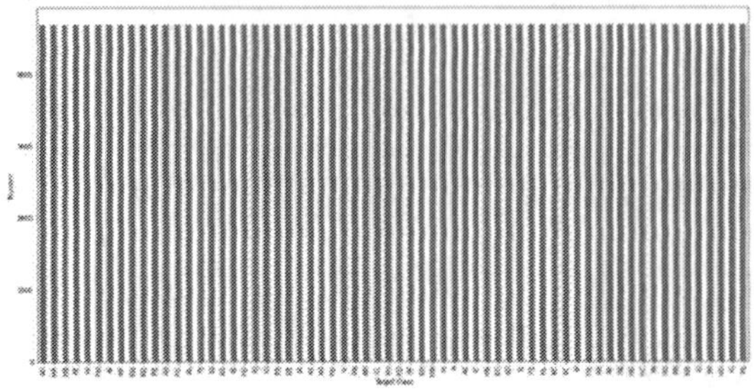

Fig. 5. Balanced class.

4.6 Naïve Bayes Algorithm Implementation

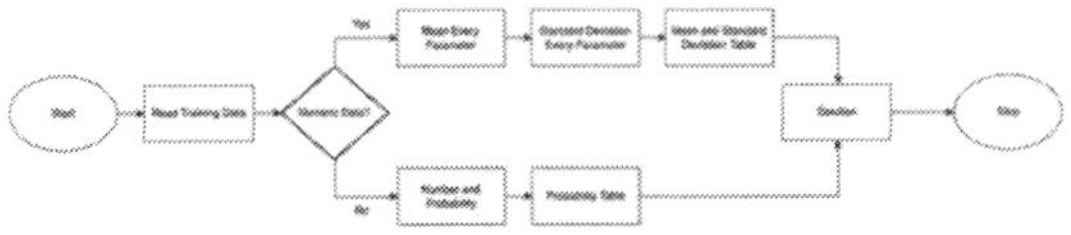

Fig. 6. Naive Bayes path.

After all classes are balanced, it is continued to build a naive Bayes model. in Figure 6 is a naive Bayes path that is carried out first, namely reading the training data, then if the data is numeric then look for the mean and standard deviation of each parameter which is numerical data. Find the probabilistic value by calculating the number of matching data from the same category divided by the number of data in that category [14].

Table 6 is an example of a case that will be calculated using manual calculations from the Nave Bayes algorithm. The training data taken are 8 data and 1 data for testing with two classes, namely "IF" and "IG".

Table 6. Sample Case.

No	D	A	S	*Treatment*
1	32	40	29	IF
2	19	20	27	IF
3	35	39	33	IF
4	27	34	40	IG
5	33	21	32	IF
6	14	22	28	IF
7	38	35	29	IF
8	24	22	27	IG
9	26	23	30	IG

Calculate the probability value of each class and calculate the mean and standard deviation of each attribute using equations 2 and 3. In Table 7 is the result of the mean value and Table 8 is the standard deviation.

Table 7. Mean.

	Mean		
Class	D	A	S
IF	27,8	27,4	29,8
IG	25,6	26,3	32,3

Table 8. Standard Deviation.

	Standard Deviation		
Class	D	A	S
IF	10,61	8,9	2,58
IG	1,52	6,65	6,8

Calculate the value of the Gaussian distribution with the testing data as follows in Table 9:

Table 9. Data Testing.

D	A	S	Treatment
32	40	29	?

From the test data above, then calculate the value of the Gaussian distribution with equation (1)

$$P\ (D = 32\ |\text{treatment} = IF) = \frac{1}{\sqrt{2\pi.10{,}61602562}} e^{-\frac{(32-27{,}8)^2}{2(10{,}61602562)^2}} = 0{,}113253318$$

$$P\ (A = 40|\ \text{treatment} = IF) = \frac{1}{\sqrt{2\pi.8{,}905055}} e^{-\frac{(40-27{,}4)^2}{2(8{,}905055)^2}} = 0{,}049143861$$

$$P\ (S = 29|\ \text{treatment} = IF) = \frac{1}{\sqrt{2\pi.2{,}588436}} e^{-\frac{(29-29{,}8)^2}{2(2{,}588436)^2}} = 0{,}236460745$$

$$P\ (D = 32|\ \text{treatment} = IG) = \frac{1}{\sqrt{2\pi.1{,}5275252}} e^{-\frac{(32-25{,}66666667)^2}{2(1{,}5275252)^2}} = 5{,}97257E-05$$

$$P\ (A = 40|\ \text{treatment} = IG) = \frac{1}{\sqrt{2\pi.6{,}658328}} e^{-\frac{(40-26{,}33333333)^2}{2(6{,}658328)^2}}$$

$$= 5{,}97257E - 05$$

$$P\ (S = 29|\ treatment = IG) = \frac{1}{\sqrt{2\pi.6{,}806859}} e^{-\frac{(29-3\ ,33333333)^2}{2(6{,}806859^2}} = 0{,}135666654$$

Multiply all the results according to the testing data that the class will look for.

P (x | treatment = IF) = 0,113253318 × 0,049143861 × 0,236460745 = 0,001316071

P (x | treatment = IG) = 5,97257E-05× 0,018814348× 0,135666654 = 1,52449E-07

P (treatment = IF|X) = 0,001316071 × 0,625 = 0,000822544

P (treatment = IG|X) = 1,52449E-07× 0,375 = 5,71683E-08

From the results above, it can be seen that the highest probability value is in the class (P|IF) so it can be concluded that the classification of the test data belongs to the "IF" class.

5 Testing and Result

This chapter describes the results and discussion of the tests that have been carried out to determine the success of the system. The tests carried out are testing the Naive Bayes algorithm and the results of the interface implementation.

5.1 Naïve Bayes Algorithm Testing

Then to test this Naive Bayes model using data partition by dividing the portion of the dataset into two parts, namely training data and test data. This test parameter also determines the random state value for the best data partition. data partitioning was carried out five times and the results of the data partitioning can be seen in Table 10 where the accuracy obtained in this system is 84.66% with data partitioning 90% training data that contain 526,838 data, 10% test data that contain 58,537 data and the random state is 34.

Table 10. Accuracy Algorithm.

Random State	Precision (%)	Recall (%)	Accuracy (%)
2	83.94	84.34	84.34
10	84.55	84.51	84.51
18	84.17	84.56	84.56
26	83.67	84.28	84.28
34	84.55	84.66	84.66
42	83.98	84.41	84.41
50	83.59	83.71	83.71

5.2 Interface Implementation

Fig. 7. DASS-42 test page.

Fig. 8. DASS-42 test results page/

Figure 7 is the Depression, Anxiety, Stress Scale (DASS)-42 test page which contains 5 question items on one page and the user can fill in the responses according to the circumstances experienced. Figure 8 shows the results page display. This page displays test results in the form of the severity of depression, anxiety, and stress as well as treatment recommendations for the first treatment for users.

6 Conclusion

The conclusion of this research is that the Test System for Depression, Anxiety and Stress Disorders in Students can run according to its purpose, namely knowing the level of user depression, anxiety, and stress disorders, and recommending treatment based on the results of the Depression, Anxiety, Stress Scale (DASS)-42 test. The method used is Naive Bayes with an accuracy obtained of 86.44% with 90% training data and 10% partition test data.

References

1. PDSKJI, *Infografik Masalah Psikologis Terkait Pandemi Covid-19 di Indonesia*, PDSKJI http://pdskji.org/home (accessed July 20, 2023).
2. P. F. Lovibond, S. H. Lovibond, *The structure of negative emotional states: Comparison of the Depression Anxiety Stress Scales (DASS) with the Beck Depression and Anxiety Inventories* in Behaviour Research and Therapy, **3**, 335-343 (1995).
3. T. Widiyaningtyas, I. A. E. Zaeni, N. Jamilah, *Diagnosis of Fever Symptoms Using Naive Bayes Algorithm* in Proceedings of the 5th International Conference on Sustainable Information Engineering and Technology (SIET '20). Association for Computing Machinery, New York, USA, 23–28 (2021).
4. A. Sau, I. Bhakta, *Predicting anxiety and depression in elderly patients using machine learning technology* in Healthcare Technology Letters, **4**, 238-243, (2017).
5. S. Budiyanto, H. C. Sihombing, and F. Rahayu, *Depression and anxiety detection through the Closed-Loop method using DASS-21* in Telkomnika (Telecommunication Computing Electronics and Control), **17** (2019).
6. P. Kumar, S. Garg, A. Garg, *Assessment of Anxiety, Depression and Stress using Machine Learning Models* in Procedia Computer Science, **171**, 1989-1998, (2020).
7. A. Priya, S. Garg, and N. P Tigga, *Predicting Anxiety, Depression and Stress in Modern Life using Machine Learning Algorithms* in Procedia Computer Science, **167**, 1258-1267, (2020).
8. S. S. Imam, *Depression Anxiety Stress Scales (DASS): Revisited* in The 4th International Postgraduate Research Colloquium, IPRC Proceedings, **3**, 184-196 (2008).
9. UNSW, *Depression Anxiety Stress Scale (DASS)*, UNSW, http://www2.psy.unsw.edu.au/groups/dass/ (accessed July 20, 2023).
10. R. C. B. Vignola, and A. MarcassaTucci, *Adaptation and validation of the depression, anxiety and stress scale (DASS) to Brazilian Portuguese* in Journal of Affective Disorders, **155**, 104-9, (2014).
11. V. jackins, S.Vimal, M. Kaliappan, and M. Y. lee, *AI-based smart prediction of clinical disease using random forest classifier and Naive Bayes* in The Journal of Supercomputing , **77**, 5198–5219, (2020).
12. N. Hasan, and D. Avianto, *Sistem Diagnosa Awal Gangguan Psikologis Pada Remaja Menggunakan Metode Naïve Bayes* (2019).
13. F. Alberto, G. Salvador, H. Francisco, and C. Nitesh, *SMOTE for Learning from Imbalanced Data: Progress and Challenges, Marking the 15-year Anniversary* in Journal of Artificial Intelligence Research, **61**, 863-905 (2018).
14. F. Harahap, A. Y. N. Harahap, E. Ekadiansyah, R. N. Sari, R. Adawiyah and C. B. Harahap, *Implementation of Naïve Bayes Classification Method for Predicting Purchase* in 6th International Conference on Cyber and IT Service Management, CITSM, 2018 (2018).

Exploring Reinforcement Learning Methods for Multiple Sequence Alignment: A Brief Review

Chaimaa Gaad[1*], *Mohamed-Amine Chadi*[2], *Mohamed Sraitih*[3], and Ahmed Aamouche[1]

[1] *LISA Laboratory, National School of Applied Sciences, University of Cadi Ayyad,* Marrakech, Morocco
[2] *LISI Laboratory, Computer science department, Faculty of Sciences Semlalia, University of Cadi Ayyad,* Marrakech, Morocco
[3] *MSC Laboratory, National school of applied sciences, University of Cadi Ayyad,* Marrakech, Morocco

Abstract. Multiple sequence alignment (MSA) plays a vital role in uncovering similarities among biological sequences such as DNA, RNA, or proteins, providing valuable information about their structural, functional, and evolutionary relationships. However, MSA is a computationally challenging problem, with complexity growing exponentially as the number and length of sequences increase. Currently, standard MSA tools like ClustalW, T-Coffee, and MAFFT, which are based on heuristic algorithms, are widely used but still face many challenges due to the combinatorial explosion. Recent advancements in MSA algorithms have employed reinforcement learning (RL), particularly deep reinforcement learning (DRL), and demonstrated optimized execution time and accuracy with promising results. This is because deep reinforcement learning algorithms update their search policies using gradient descent, instead of exploring the entire solution space making it significantly faster and efficient. In this article, we provide an overview of the recent historical advancements in MSA algorithms, highlighting RL models used to tackle the MSA problem and main challenges and opportunities in this regard.
Keywords: Multiple sequence alignment, Reinforcement learning, Computational complexity, Bioinformatics, Brief review.

1 Introduction

Multiple sequence alignment (MSA) is a crucial problem in bioinformatics that involves aligning multiple biological sequences, such as RNA, DNA, or protein sequences, to identify their evolutionary relation and functional similarity. In essence, MSA aims to organize sequences in a way that maximizes the similarity between the corresponding positions of the sequences. This is accomplished by including spaces (gaps) of various lengths within the sequences, aligning homologous positions in a manner similar to aligning beads of the same color in an abacus (Figure 1). Evolutionwise, these gaps symbolize indels (i.e., insertions and deletions) that are believed to have happened during the evolutionary process from a shared ancestor [1]. MSA is widely used in various applications in bioinformatics, including protein structure prediction, phylogenetic tree construction, and functional element identification in genomes. A recent paper [2] highlights the widespread use of multiple sequence alignment (MSA) in the field of biology. One of the most commonly used MSA methods, according to the study, is ClustalW [3], it is ranked as the 10th most cited scientific paper of all time. This demonstrates the significant impact that MSA has had on a wide range of in-silico analysis.

Since its inception, MSA has been tackled using a variety of methods, including probabilistic models [4], dynamic programming (DP) [5] such as the Needleman-Wunsch [6] and Smith-Waterman [7]. Dynamic programming is also used in the progressive alignment (PA) algorithm, which was first described by Hogeweg and Hesper [8]. Progressive alignment is one of the most used heuristics, A guide tree is constructed at the beginning of the algorithm to determine the order in which the input sequences will be incorporated into the final alignment. At each step, the algorithm performs a pairwise alignment between two sequences, profiles, or a combination of both, using a global dynamic programming algorithm. This combination of a tree based approach and a global pairwise alignment is the foundation of many popular MSA methods, including T-Coffee [9], and ProbCons [10]. These methods can be further improved by using iterative strategies [11], which involves repeating the process of estimating the guide tree and aligning sequences until they converge, as performed in methods such as MAFFT [12], MUSCLE [13], and Clustal Omega . These methods have been successful in providing accurate and fast solutions to the MSA problem, especially for small and moderate-sized datasets. However, as the size and complexity of biological datasets continue to increase, there is a growing need for more advanced and efficient MSA methods that can handle large-scale datasets and complex evolutionary relationships. The development of MSA as a valuable modeling tool has required

* Corresponding author: chaimaa.gaad@ced.uca.ma

overcoming a set of complex computational and biological challenges. Due to its NP- complete (non--

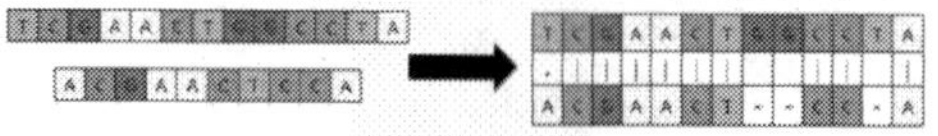

Fig. 1. Representation of 2 sequences of nucleotides (left), and the optimal alignment of these 2 sequences (right). Where (.) represents a substitution, (|) represents a match and ( ) represents a gap or an indel.

-deterministic polynomial time) nature [15], which means that finding an optimal solution for large-scale sequence datasets is difficult and often requires exponential time, the calculation of a precise MSA has been a long-standing issue in the field, resulting in the creation of over 100 distinct methods in the past thirty years [16].

Multiple sequence alignment (MSA) metrics are crucial in assessing the accuracy and quality of an alignment produced by a given method or software. The most commonly used metrics to evaluate the MSA quality are the sum-of-pairs (SP-score) and the column scores (CS). The SP score is a measure of the percentage of correctly aligned residue pairs in the alignment that indicates how well the concerned program is able to align at least some of the sequences [17]. On the other hand, the column score measures the percentage of the columns aligned correctly in the alignment, which assesses the program's ability to align all the sequences accurately. The calculation of column score (CS) involves taking the ratio between the number of columns that perfectly match (i.e., the exact match EM) across the entire alignment and the total length of the alignment (i.e., alignment's length AL) (CS = EM / AL). Both scores are valuable in determining the overall performance of the program in generating a reliable multiple sequence alignment [18].

$$SP = \sum_{i=1}^{N} \sum_{j=1}^{n-1} \sum_{k=j+1}^{n} s(c_i^j c_i^k) \quad (1)$$

Recently, reinforcement learning (RL), a branch of artificial intelligence (AI), that involves training an agent to make decisions in an environment by maximizing a reward signal, has been proposed as a new approach for solving the MSA problem. RL-based MSA methods aim to align sequences by learning a policy that guides the agent to make optimal alignments based on a reward signal that reflects the quality of the alignment. The key advantage of RL-based MSA methods is that they can effectively capture the complex relationships between sequences and handle different datasets by leveraging the generalization and representation learning capabilities of deep neural networks. The following article aims to provide a brief and comprehensive understanding of the application of RL multiple sequence alignment (MSA) problem. To achieve this goal, the article is structured into several sections. The first section is the introduction, which provides a comprehensive overview of MSA, the traditional tools to solve its problem and its relevance to bioinformatics. The next section will define RL, its components, as well as its types to give readers a decent understanding of the fundamental concepts related to the topic and will delve into the various types of RL used in

Table 1. Notation used in the SP-Score function.

Notation	Meaning
n	is the number of sequences in the alignment
N	is the number of columns in the alignment
c_i , $i \in \{1, \dots, N\}$	is the column i from the alignment
c_j , $j \in \{1, \dots, N\}$	is the jth character from the ith column
$s(c_i^j c_i^k)$	is a measure that indicates the comparing score between characters c_i^j and c_i^k

the state-of-the-art research on MSA, including Q-learning and A3C. The third section is a literature survey, which will provide an overview on the applications of RL based approaches for MSA problem, a comparison in term of performance for these applications and a discussion. Finally, the conclusion section will summarize the findings and provide insights into future directions of research in this field.

2 Reinforcement learning methods

2.1 Key concepts

Reinforcement Learning is a type of machine learning (ML), that focuses on teaching an agent to make decisions by taking actions in an environment to maximize a reward signal (Figure 2) [19]. Deep RL (DRL) when coupled with neural networks, we get. The goal of RL is to discover the best policy, or a mapping from states to actions, that maximizes the expected reward over time.

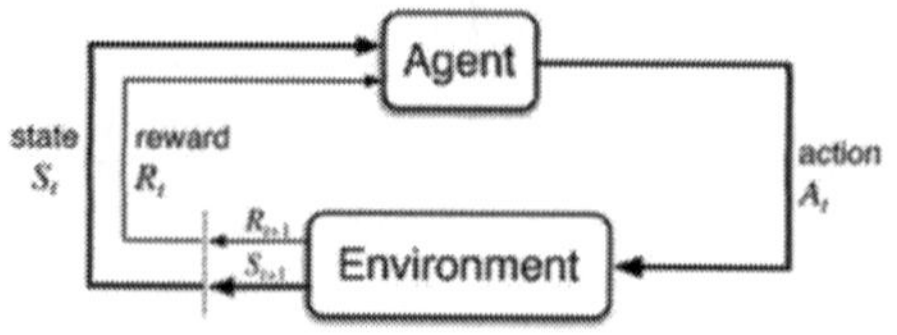

Fig. 2. The agent-environment interaction in RL.

Reinforcement Learning is based on several key concepts, including the agent, environment, state, action, reward, and policy (Table 2). There are several different types of RL algorithms, including value-based such as Q-learning algorithms, which was first introduced in 1992 by Watkins and Dayan [20], policy-based algorithms such as REINFORCE [21], and actor-critic algorithms that combines the two set of information treatment such as A3C (Asynchronous Advantage Actor-Critic) [22]. These algorithms differ in how they approach the problem of finding an optimal policy, but all rely on the same basic concepts. RL offers several advantages for MSA, such as (1) The ability to learn from experience: RL algorithms can learn from experience, through trials, which means that they can improve the performance over time without the need for explicit supervision. (2) Flexibility: RL algorithms are flexible and can be applied to a wide range of problems, including problems in robotics, control, and optimization (3) Adaptability: RL algorithms are able to adapt to changes in the environment [23], which is important in MSA where the data and the conditions of the environment can change over time, since we are dealing with a text-based data. (4) Robustness: RL algorithms are robust and can handle noisy and incomplete information, which is common in MSA. (5) Optimization: RL algorithms can be used to optimize the parameters of MSA algorithms, leading to improved performance.

2.2 RL algorithms used for MSA

Multiple sequence alignment (MSA) is a central problem in bioinformatics that has attracted a lot of attention in recent years, especially, with the challenge of aligning multiple sequences with different lengths.

In recent years, there has been limited research exploring the application of RL algorithms into MSA. RL-based MSA methods view the MSA problem as a sequential decision-making process, in which the goal is to maximize a reward function that evaluates the quality of the alignment. There are several RL-based MSA methods that have been proposed in recent years, each with its own approach to defining the environment, state, action, reward, and policy. Some of these methods have been based on the value-based algorithms, such as Q-learning and deep-Q-learning (DQN), while others are based on actor-critic algorithms, such as A3C.

2.2.1 Q-learning and DQN

Q-learning is among the very first models introduced in RL. It is a popular RL algorithm used to solve problems where the agent must make a sequence of decisions in an uncertain, dynamic environment to maximize a long-term reward. The key idea behind Q-learning is to learn an action-value function, also called a Q-function, which assigns a value to each action in a given state. The Q-function represents the expected cumulative reward that the agent will receive if it takes a particular action in a specific state and then follows the policy thereafter. During training, the agent uses the Q-function to choose its actions based on the current state and updates the Q-values of the action-state pairs based on the rewards obtained and the estimated value of the next state-action pair. This update rule is known as the Bellman equation [24], which is based on the principle of optimality stating that the optimal policy must satisfy the property where the value of the state is equal to the expected return for the best action. DQN (Deep Q-Network) is a RL algorithm that combines Q-learning with deep neural networks. Popularized in 2013 and has been used in various applications, including playing video games at superhuman level, robotic control [25], and natural language processing (NLP) [26].

Table 2. Key components of RL algorithm.

Element	Style
Agent	An agent is the decision-making entity in RL. It is responsible for taking actions in the environment and receiving rewards for those actions
Environm ent	The environment is the setting in which the agent operates. It can be a real-world environment, such as a stock market or a video game, or a simulated environment created in a laboratory. The environment provides information about the state of the world and gives the agent the opportunity to take actions that affect the world.
State	A state is a description of the environment at a given time. i.g, in a video game, the state might include information about the position of the player, the position of enemies, and the state of the game world
Action	An action is a decision that the agent makes in the environment. i.g, in a video game, an action might be to move left, right, up, or down
Reward	A reward is a signal that provides feedback to the agent about the quality of its actions, it can be a negative one (penalty). For instance, in a video game, the agent would possibly receive a reward for collecting coins or a penalty for losing a life. The agent uses the reward signal to update its policy and improve its performance over time
Policy	A policy is a mapping that goes from states to actions. It is the strategy that the agent learns through trial and error in the environment. The goal of RL is to discover the best policy that maximizes the expected reward over time

2.2.2 Asynchronous Advantage Actor-Critic

Asynchronous Advantage Actor-Critic (A3C) is a deep RL algorithm that is designed to achieve more efficient and effective learning. It combines the actor-critic approach with asynchronous parallelization. The actor-critic approach is a RL technique that consists of two components: the actor and the critic. The actor selects actions based on the current policy, while the critic evaluates the value of the current state. The actor uses feedback from the critic to improve its policy. In A3C, a neural network is used to represent both the actor and the critic. A3C also utilizes asynchronous parallelization, which enables multiple instances of the environment to run in parallel. Each instance updates the shared neural network parameters asynchronously, allowing the algorithm to take advantage of multiple cores or machines. This can greatly speed up the learning process and make it more efficient. By combining the actor-critic approach with asynchronous parallelization. A3C can improve learning efficiency and speed. This makes it a popular choice for applications in which a large amount of training data and computational power are needed.

3 RL applications: MSA insights

3.1 Applications from the literature

Tackling the MSA problem using RL was first introduced in 2016 [27]. In this paper, using Q-learning, the RL task is formulated as a search problem sequence and is represented as a set of multi-dimensional arrays, each array is a sequence of nucleotides {C, T, A, G} and (-) represents the gap, where the goal is to find the optimal path through a state space that leads to the optimal solution (i.e., the optimal alignment of the input sequences). The state space consists of all possible permutations of the input sequences, where each permutation represents a possible order for aligning the sequences. The RL agent's goal here is to find the permutation that leads to the optimal alignment of the input sequences. The agent's actions consist of selecting one of the input sequences at each step. The agent can only select each sequence once, and it must select all the sequences exactly once, in an order that leads to the optimal alignment. The reward function is defined as the score of the alignment associated with the selected permutation of the input sequences, the score is determined by comparing the aligned sequences to a reference alignment and summing the scores of the individual matches, mismatches, and gaps. These steps are done by using a matrix that is a representation of each nucleotide and gap locus.

In another paper [28], the A3C algorithm was employed to address the problem of speed in MSA. The scoring scheme used to optimize the SP score is the linear gap penalty approach, with a score of (-1, +1, -1) for (gap, match, mismatch). In this paper, authors defined the state as an $n \times b$ matrix (n is the number of sequences of maximum length L) called the game board, $b \geq L$, and each cell contains either a gap or a nucleotide. The agent's role is to push nucleotides to the left or to the right to alter the alignment. The reward function assigns a real value to each possible (state, action) pair based on the SP-score of the resulting alignment. The goal state is not known ahead of time, so the agent is allowed to perform nSteps (which is a fixed number of steps) before the episode is stopped. The state with the maximum SP-score is labeled the ad hoc goal state, and actions taken after visiting that state are discarded. The policy followed by the agent consists of selecting an action based on the current state. The value of a (state, action) pair corresponds to its long-term expected discounted reward. The study approximates the value for each of the possible actions.

Additionally, the work presented in [29] proposed another approach based on the Markov decision process (MDP) framework with a state space consisting of all possible states, and an action space consisting of n actions, one for each sequence. The reward function is defined based on the SP-score, and the agent receives a large negative reward ($-\infty$) if it chooses an action that is not distinct. The ultimate goal is to find an optimal policy (to maximize the expected reward). The model uses a neural network (NN) architecture based on Long Short-Term Memory (LSTM) to approximate the Q-values. The LSTM network has three hidden layers, each with 40 neurons, and a dropout layer to prevent overfitting. The input to the network is the state information needed to compute the SP-score, and the output is a list of chosen sequences. The model uses the experience replay method to prevent overfitting and improve stability, and the actor-critic algorithm for faster convergence, in this context, the actor refers to a policy that needs to select an action, while the critic is represented by the Q-function, which provides feedback to the actor regarding the optimality of its chosen action. The training loop of the model involves initializing the replay memory and the Q-network with random weights, selecting a random action with a small probability (to ensure sufficient exploration), observing the next reward and state, and storing the transition in the replay memory. The model then trains the network using mini-batch sampling and gradient descent, updating the weights with each iteration.

In a more recent study, the authors proposed an approach based on DQN [30], that is designed to guide progressive alignment using a combination of a new profile algorithm and negative feedback policy. Here the environment is the MSA task and the agent is responsible for computing the profile-sequence alignment, selecting the next sequence to align, and updating the state. The state in this RL environment is represented by a matrix T, which is padded with zeros (0) to a fixed size, the non-zero entries represent the sequences that have already been aligned, and the order of the entries must follow a specific rule to ensure that all non-zero numbers are adjacent. The action the agent takes is selecting the next sequence to align, and the reward is determined by the quality of the alignment resulting from the chosen action. The reward function also includes a negative reward for choosing a sequence that has already been aligned. The negative feedback policy role is to randomly select an action with a certain probability or uses the DQN model to predict the action value for each possible action. The DQN model is trained using a replay memory that saves all the invalid transitions that aligns the current profile with the new sequence selected by the agent and updates the current state. The transition function is a quad-tuple consisting of (the current state, the selected action, the resulting next state, and the associated reward).

3.2 Benchmarking

To conduct our benchmark, we used various well-known datasets that have been used in previous state-of-the-art studies. the first, contains DNA sequences from three distinct species: lemur, gorilla, and mouse. The average length of these sequences is 93 nucleotides. Similarly, the rat, lemur, and opossum dataset, which was used in the same study [31], consists of the third exon sequences of the beta globin gene and has an average sequence length of 129 nucleotides. We also included two datasets from the oxbench mdsa all [32] database: dataset 429, which has an average sequence length of 171 nucleotides, and dataset 469, which has an average sequence length of 332 nucleotides. Finally, we used two datasets from the European Molecular Biology Laboratory (EMBL) Nucleotide Sequence Database [33]: one dataset containing 10 sequences from the Hepatitis C virus (sutype 1b), with an average sequence

Table 3. Comprehensive comparison of the SP-Score metric.

Dataset	Number of sequences	Average length	Q-Learning [27]	DQN [29]	DQN [30]	ClustalW [27]	MAFFT [27]
Lemur, gorilla, mouse, dataset	3	93	345	345	**348**	345	345
429 from oxbench_mdsa_msa	12	171	8668	**10218**	8784	9575	10218
Papio Anubis	5	1093	18719	18860	**18968**	18827	18860
469 from oxbench_mdsa_all	3	332	565	565	**587**	464	549
Rat, lemur, opssum dataset	3	129	**486**	**486**	482	480	471
Hepatitis C virus	10	211.9	**18627**	**18627**	**18627**	**18627**	**18627**

length of 211 nucleotides, and another dataset containing 5 sequences from the Papio (Anubis olive baboon) organism, with an average sequence length of 1093 nucleotides. Note: In Table 3 below, we included all papers presented previously that used RL for solving the MSA problem. However, for [28], where authors used the A3C algorithm, the metric results were not mentioned, therefore, we did not include it in Table 3.

3.3 Discussion: challenges and opportunities

Multiple sequence alignment is a mandatory and fundamental technique in bioinformatics that aims to evaluate, distinguish, and identify similarities and differences between sequences. This knowledge is critical for scientists to understand the structure and function of biological molecules and can help with drug development, evolutionary biology, and comparative genomics studies.

Several approaches have been used such as probabilistic models [4], dynamic programming (DP), and progressive alignment. These classical techniques have been widely used for decades in the research area and have successfully provided accurate and fast solutions to perform the MSA, especially for limited datasets. However, these techniques are limited to a certain amount of data and a certain level of complexity, which elevates the need for more advanced methods suitable for dealing with large-scale data and intricate evolutionary interactions. In recent years, reinforcement learning emerged as a promising technique to elevate this issue. Nevertheless, as previously mentioned, only four studies suggested the RL for solving the MSA so far. This may appear insufficient for a full review, even though, reviewing these papers can provide insights into how these methods perform and if there is room for improvements. Each one of the reviewed papers used a RL model to perform the MSA by introducing a different formalization of the problem, yet to achieve the same goal.

As mentioned previously, RL is based on four core concepts (state, action, environment, reward). However, each author used these concepts differently by making the agent choose the sequence to use for the next round to extract the profile [29] and [30] or to find the path to the highest score linked to a certain alignment as in [27]. Or by taking actions on the sequence itself by moving the nucleotides or inserting gaps [28]. Therefore, the performances are quite different from one approach to the next for the same dataset, whereas, for others, all the approaches achieved the same results. Table 1 illustrates the performance of the SP score achieved by each approach evaluated on different datasets. The authors in this paper [30] achieved the highest SP score in terms of the majority of the experimented datasets by arising 0.57%, 0.87%, and 3.89% for Papio Anubis, Lemur, gorilla, mouse, and "629 from oxbench_mdsa"_all datasets, respectively. On the other hand, [28] and [29] achieved lower performance results for the same previous two datasets. Additionally, for the "429 from the oxbench_mdsa_msa" dataset, achieved the highest SP score by a rise of 16.32% in comparison with other approaches, keeping in mind that this is the largest dataset among the others. This could be because the authors implemented a strategy that makes the agent receive a large negative reward (-∞) if it chooses an action that is not distinct in finding the optimal policy (to maximize the expected reward). As well as a high performance, similar to [28] for the Rat, lemur, opssum dataset by a rise of 0.82%, which is the highest value it could be achieved specifically for this dataset as [30] didn't achieve better than that. Finally, for Hepatitis C virus dataset, all techniques achieved the same top SP score. This may be explained by the fact that for this specific dataset, all approaches were able to obtain the same ideal alignment, which reflects the greatest SP score. Furthermore, all these approaches have been compared to the classical methods, which illustrates that the novel approaches achieved better and equal performance in some cases and a lower SP score in others. However, this could be explained by the fact that the agent could find the optimal MSA by reaching the highest SP score. In other words, the RL approaches can explore the search space in an efficient reward-targeted way, which can provide a higher chance to find the optimal MSA for a given dataset.

Even with these performances, there are still several challenges that need to be addressed to boost the efficiency of reinforcement learning in MSA. One of the main challenges is having a high generalization ability, as the sequence size and number vary in different datasets. This issue can be investigated by using a diverse and representative training set with different model parameters. In addition to efficient reward functions. However, as the sequence scale is large, it is more complex, time and memory-consuming, which leads to the exponentially growing size of the state space. This needs to be solved by investigating transfer learning, where the agent is trained on a reasonable number of datasets to initialize networks for larger-scale sequences. Additionally, using the profiling algorithm, which reduces the computational complexity of the iterative process, improves speed without compromising accuracy.

4 Conclusion

In conclusion, the four articles presented different RL-based approaches for solving the MSA problem in bioinformatics. While each approach had its own strengths and limitations, they all showed promising results in improving the accuracy and efficiency of MSA. One key challenge for future research is to address the issue of generalization to longer and more numerous sequences, while using different RL algorithms, which is currently limited by the exponential growth of the state space. Transfer learning strategies and better use of space invariance and symmetries are potential solutions.

Additionally, the lack of appropriate objective function, designing models with high generalization ability and the complexity and resource requirements of large-scale sequence alignments are the challenges that need to be addressed in future research. To conclude, RL based approaches have the potential to significantly enhance the reliability and effectiveness of MSA in bioinformatics.

References

1 M. Chatzou, C. Magis, J. M. Chang, C. Kamena, G. Bussotti, I. Erb, C. Notredame, 'Multiple sequence alignment modeling: methods and applications', *Briefings in Bioinformatics*, vol. 17, no. 6, pp. 1009–1023, Nov. 2016, doi: 10.1093/bib/bbv099.

2 R. V. Noorden, B. Maher, and R. Nuzzo, 'Nature explores the most-cited research of all time.' *The top 100 papers*. Nature 2014;514:550–3.

3 J. D. Thompson, D. G. Higgins, and T. J. Gibson, 'CLUSTAL W: improving the sensitivity of progressive multiple sequence alignment through sequence weighting, position-specific gap penalties and weight matrix choice', *Nucleic Acids Res* 1994;22:4673–90.

4 S. R. Eddy, 'A Probabilistic Model of Local Sequence Alignment That Simplifies Statistical Significance Estimation', *PLoS Computational Biology*, vol. 4, no. 5, p. e1000069, May 2008, doi: 10.1371/journal.pcbi.1000069.

5 C. Lee, C. Grasso, and M. F. Sharlow, 'Multiple sequence alignment using partial order graphs', *Bioinformatics*, vol. 18, no. 3, pp. 452–464, Mar. 2002, doi: 10.1093/bioinformatics/18.3.452.

6 S. B. Needleman and C. D. Wunsch, 'A general method applicable to the search for similarities in the amino acid sequence of two proteins', *Journal of Molecular Biology*, vol. 48, no. 3, pp. 443–453, Mar. 1970, doi: 10.1016/0022-2836(70)90057-4.

7 T. F. Smith and M. S. Waterman, 'Identification of common molecular subsequences', *Journal of Molecular Biology*, vol. 147, no. 1, pp. 195–197, Mar. 1981, doi: 10.1016/0022-2836(81)90087-5.

8 P. Hogeweg and B. Hesper, 'The alignment of sets of sequences and the construction of phyletic trees: An integrated method', *Journal of Molecular Evolution* vol. 20, no. 2, pp. 175–186, Jun. 1984, doi: 10.1007/BF02257378.

9 C. Notredame, D. G. Higgins, and J. Heringa, 'T-coffee: a novel method for fast and accurate multiple sequence alignment 1 1Edited by J. Thornton', *Journal of Molecular Biology*, vol. 302, no. 1, pp. 205–217, Sep. 2000, doi: 10.1006/jmbi.2000.4042.

10 C. B. Do, M. S. P. Mahabhashyam, M. Brudno, and S. Batzoglou, 'ProbCons: Probabilistic consistency-based multiple sequence alignment', *Genome Research*, vol. 15, no. 2, pp. 330–340, Feb. 2005, doi: 10.1101/gr.2821705.

11 I. M. Wallace, O. Orla, and D. G. Higgins, 'Evaluation of iterative alignment algorithms for multiple alignment', *Bioinformatics*, vol. 21, no. 8, pp. 1408–1414, Apr. 2005, doi: 10.1093/bioinformatics/bti159.

12 K. Katoh, 'MAFFT: a novel method for rapid multiple sequence alignment based on fast Fourier transform', *Nucleic Acids Research*, vol. 30, no. 14, pp. 3059–3066, Jul. 2002, doi: 10.1093/nar/gkf436.

13 R. C. Edgar, 'MUSCLE: a multiple sequence alignment method with reduced time and space complexity', *BMC Bioinformatics*, vol. 5, no. 1, p. 113, 2004, doi: 10.1186/1471-2105-5-113.

14 F. Sievers, A. Wilm, D. Dineen, T. J. Gibson, K. Karplus, W. Li, R. Lopez, H. McWilliam, M. Remmert, J. Soding, J. D. Thompson and D. G. Higgins, 'Fast, scalable generation of high-quality protein multiple sequence alignments using Clustal Omega', *Molecular Systems Biology*, vol. 7, no. 1, p. 539, Jan. 2011, doi: 10.1038/msb.2011.75.

15 L. Wang and T. Jiang, 'On the Complexity of Multiple Sequence Alignment', *Journal of Computational Biology*, vol. 1, no. 4, pp. 337–348, Jan. 1994, doi: 10.1089/cmb.1994.1.337.

16 C. Kemena and C. Notredame, 'Upcoming challenges for multiple sequence alignment methods in the high-throughput era', *Bioinformatics*, vol. 25, no. 19, pp. 2455–2465, Oct. 2009, doi: 10.1093/bioinformatics/btp452.

17 D. J. Lipman, S. F. Altschul, and J. D. Kececioglu, 'A tool for multiple sequence alignment.', *Proceedings of the National Academy of Sciences. U.S.A.*, vol. 86, no. 12, pp. 4412–4415, Jun. 1989, doi: 10.1073/pnas.86.12.4412.

18 J. D. Thompson, P. Koehl, R. Ripp, and O. Poch, 'BAliBASE 3.0: Latest developments of the multiple sequence alignment benchmark', *Proteins*, vol. 61, no. 1, pp. 127–136, Jul. 2005, doi: 10.1002/prot.20527.

19 R. S. Sutton and A. G. Barto, Reinforcement learning: an introduction, Second edition. in *Adaptive computation and machine learning series*. Cambridge, Massachusetts: The MIT Press, 2018.

20 J. Clifton and E. Laber, 'Q-Learning: Theory and Applications', *Annual Review of Statistics and Its Application*, vol. 7, no. 1, pp. 279–301, Mar. 2020, doi: 10.1146/annurev-statistics-031219-041220.

21 R. J. Williams, 'Simple statistical gradient-following algorithms for connectionist reinforcement learning', *Machine Learning* 8(23)

22 V. Mnih , A. P. Badia, M. Mirza, A. Graves, T. Harley, T. P Lillicrap, D. Silver, K. Kavukcuoglu., 'Asynchronous Methods for Deep Reinforcement

Learning', *International Conference on Machine Learning* 2016.
23 F.-M. Luo, S. Jiang, Y. Yu, Z. Zhang, and Y.-F. Zhang, 'Adapt to Environment Sudden Changes by Learning a Context Sensitive Policy', *The Association for the Advancement of Artificial Intelligence*, vol. 36, no. 7, pp. 7637–7646, Jun. 2022, doi: 10.1609/ aaai. v36i7.20730.
24 R. Bellman, *Dynamic programming*. Princeton, NJ: Princeton Univ. Pr, 1984.
25 V. Mnih, K. Kavukcuoglu, D. Silver, A. Graves, I. Antonoglou, D. Wierstra, M. Riedmiller., 'Playing Atari with Deep Reinforcement Learning', *arXiv preprint* arXiv:1312.5602, 2013.
26 J. Luketina, N. Nardelli, G. Farquhar, J. Foerster, J. Andreas, E. Grefenstette, S. Whiteson, T. Rocktäschel., 'A Survey of Reinforcement Learning Informed by Natural Language'. *arXiv*, Jun. 10, 2019. Accessed: Jul. 25, 2023. Available: http://arxiv.org/abs/1906.03926
27 I.-G. Mircea, I. Bocicor, and G. Czibula, 'A Reinforcement Learning Based Approach to Multiple Sequence Alignment', in *Soft Computing Applications*, vol. 634, V. E. Balas, L. C. Jain, and M. M. Balas, Eds., in Advances in Intelligent Systems and Computing, vol. 634. , Cham: Springer International Publishing, 2018, pp. 54–70. doi: 10.1007/978-3-319-62524-9_6.
28 R. Kinattinkara Ramakrishnan, J. Singh, and M. Blanchette, 'RLALIGN: A Reinforcement Learning Approach for Multiple Sequence Alignment', in *2018 IEEE 18th International Conference on Bioinformatics and Bioengineering (BIBE)*, Taichung, Taiwan: IEEE, Oct. 2018, pp. 61–66. doi: 10.1109/BIBE.2018.00019.
29 R. Jafari, M. M. Javidi, and M. Kuchaki Rafsanjani, 'Using deep reinforcement learning approach for solving the multiple sequence alignment problem', *SN Applied Sciences* vol. 1, no. 6, p. 592, Jun. 2019, doi: 10.1007/s42452-019-0611-4.
30 Y. Zhang, Q. Zhang, Y. Liu, M. Lin, and C. Ding, 'Multiple Sequence Alignment based on deep Q network with negative feedback policy', *Computational Biology and Chemistry*, vol. 101, p. 107780, Dec. 2022, doi: 10.1016/j.compbiolchem.2022.107780.
31 X. Xiang, D. Zhang, J. Qin, and Y. Fu, 'Ant Colony with Genetic Algorithm Based on Planar Graph for Multiple Sequence Alignment', *Information Technology J.*, vol. 9, no. 2, pp. 274–281, Feb. 2010, doi: 10.3923/itj.2010.274.281.
32 H. Carroll, *W. Beckstead, T. O'Connor, M. Ebbert, M. Clement, Q. Snell, et D. McClellan.*, 'DNA reference alignment benchmarks based on tertiary structure of encoded proteins', *Bioinformatics*, vol. 23, no. 19, pp. 2648–2649, Oct. 2007, doi: 10.1093/bioinformatics/btm389.
33 C. Kanz, 'The EMBL Nucleotide Sequence Database', *Nucleic Acids Research*, vol. 33, no. Database issue, pp. D29–D33, Dec. 2004, doi: 10.1093/nar/gki098.

Genome Analysis of 10K SARS-COV-2 Sequences to Identify the Presence of Single-Nucleotide Polymorphisms

Husna Nugrahapraja[1*,2], *Nandrea* Hasna Syahira [1], and *Alidza* Fauzi[1]

[1]School of Life Sciences and Technology, Institut Teknologi Bandung, Bandung, 40132, West Java, Indonesia
[2]University Center of Excellence for Nutraceuticals, Bioscience and Biotechnology Research Center, Institut Teknologi Bandung, Bandung, 40132, West Java, Indonesia

Abstract. A new type of coronavirus was identified in Wuhan, China, in December 2019, which was named SARS-CoV-2 (Severe Acute Respiratory Syndrome Coronavirus-2). The high mutation rate of SARS-CoV-2 makes it challenging to develop effective vaccines for all variants. Substitution is the most common type of mutation that occurs in SARS-CoV-2. This research was conducted to identify the genetic variability of mutations in SNP of SARS-CoV-2 and analyse the impact. About 15,000 sequences of SARS-CoV-2 were downloaded from GISAID, which were isolated from 33 different countries around the world from February 2020 to July 2021. Sequence analysis was done using the MAFFT and the Nextclade. The results of this study are expected to help identify conserved regions in SARS-CoV-2 which can be used as probes for the virus identification process and can be used as target areas in vaccine development. Furthermore the results showed that the most common variants were variants 20B, 20A, and 20I (Alpha), with a population percentage of 32.12%, 23.95% and 17.39% of the total population, respectively. Furthermore, SNPs were called in the samples using the SNP-sites and extracted using Excel. Of the 10,107 sequences of SARS-CoV-2 studied, 154 SNPs were found with the highest number of SNPs in the spike, nsp3 and nucleocapsid genes. The ratio of the number of mutations to the most extensive sequence length was in the ORF8, ORF7a, and ORF7b genes with respective values of 0.537, 0.474, and 0.419.
Keywords: SARS-CoV-2, SNP, mutation, MAFFT, Nextclade

1 Introduction

A new type of coronavirus was identified in Wuhan, China, in December 2019. The virus has caused the COVID-19 outbreak which was declared a pandemic by WHO on March 11 2020. SARS-CoV-2 is the name given to the virus which comes from its taxonomic relationship with the coronavirus that caused the SARS outbreak [1]. Severe Acute Respiratory Syndrome Coronavirus-2 (SARS-CoV-2) is a virus that causes severe acute respiratory syndrome [2]. Furthermore symptoms of COVID-19 include fever, cough and shortness of breath [3]. In more severe cases, the infection can cause pneumonia, kidney failure and eventually death [4]. Every day thousands of new cases are revealed. According to the records of July 16 2021, GISAID recorded that globally more than 189 million people around the world were affected by this deadly virus, with more than 4 million deaths [5].

Coronavirus belongs to the order Nidovirales and the Coronaviridae family. Coronaviruses are classified into 4 genera namely, Alphacoronavirus, Betacoronavirus, Deltacoronavirus, and Gammacoronavirus. SARS-CoV-2 belongs to the Betacoronavirus genus and the Sarbecovirus sub-genus [6,1]. Based on genome analysis, SARS-CoV-2 belongs to an RNA virus. RNA viruses have a higher mutation rate than their hosts [7]. Mutations occur due to an error when the viral genetic material is replicated so that one or more of the nucleotide bases change. Mutations are so common that, in some cases, they do not cause significant changes in the organism. However, the high mutation rate in RNA viruses makes it more likely to create new variants more quickly. Genome analysis of SARS-CoV-2 over the last 1 year shows a nucleotide substitution rate in SARS-CoV-2 of ~1 × 10-3 substitutions per year [8].

This number is comparable to the substitution rate in Ebola virus, which is 1.42 × 10-3 [9]. Even so, this number is still slightly lower than the mutation rate in SARS-CoV, which is 0.8 – 2.38 × 10-3 [10]. Currently, most of the treatment therapy and diagnosis of COVID-19 are still based on the genome sequence isolated in Wuhan at the start of the pandemic. With the many mutations that have occurred so far, the detection of the SARS-CoV-2 virus and the efficiency of antiviral drugs can be affected by variations and changes in viral phenotypes [11,12]. Single-Nucleotide Polymorphism or SNP (pronounced snip) is a type of point mutation. Mutations that can be categorized as SNP are point mutations that have been found in more than 1% of the population [13]. Often, these SNPs do not have a significant impact on gene expression, but some SNPs on

*Corresponding author: nugrahapraja@sith.itb.ac.id

both coding and non-coding proteins can cause changes in protein structure, protein function, and regulation of protein expression [14]. In addition, SNPs can change the virulence, pathogenicity, and immunogenicity of viruses [15].

Several studies have shown a tendency to increase the transmissibility of each variant against the variants that existed before. The Alpha variant is estimated to have a higher transmissibility rate of 43-90% than the SARS-CoV-2 that appeared early in the pandemic [16]. The Beta variant is estimated to have a transmission rate of 1.5 times higher than the previous variant [17]. Research conducted in Brazil showed that the gamma variant has an increased transmissibility of 1.7-2.4 times compared to the previous variant [18]. Other studies have shown an increase in transmissibility in the Delta variant by 1.64 times that of the Alpha variant [19]. With the continued increase in the transmission rate of COVID-19, further studies are needed to continue to monitor the progress of this virus. Studies show that in a relatively short time, this virus is capable of mutating into various variants [11].
A study showed that mutations in the E (C26340T) and N (C29200T) genes affected the detection of target genes by 2 assay methods in 8 and 1 patient respectively [20]. Both mutations are transitions from C>T which is a common type of SNP and are related to the mechanism of mRNA editing in the host which is known as the apoliprotein B mRNA-editing enzyme [21, 22]. Another study found that the G>U transition at position 29140 affects the sensitivity of N gene-based detection [23].

Several countries have started promoting vaccine activities for their citizens, including Indonesia. There are various types of vaccines currently available. However, the emergence of SNPs in SARS-CoV-2 in large quantities can reduce the efficiency of the vaccine and can escape the current detection of COVID-19 [24]. Research conducted by Nasreen et al (2021), shows that there are differences in the effectiveness of the Pfizer, Moderna, and Astrazeneca vaccines against the COVID-19 Alpha, Beta, Gamma, and Delta variants. However, this still needs further research to find out the causes of differences in the effectiveness of these vaccines [25].

Research on SNP has been quite a lot done. However, SARS-CoV-2 still exists and continues to evolve. Therefore, studies on SNP need to be continued to monitor the development of this virus. Previous research was conducted by Yuan et al (2020), found 119 SNPs had occurred from 11,183 sequences of SARS-CoV-2. Among them, there are 74 non-synonymous mutations, 43 synonymous mutations, and 2 mutations in the intergenic section. The results of this study also showed a high frequency of mutations in the nsp2, nsp3, and Spike protein genes [26]. Another study was conducted with a sample of 714 SARS-CoV-2 sequences and found a total of 108 SNPs. The study also showed that the highest number of non-synonymous mutations occurred in the nsp3 gene, nucleocapsid, and spike protein. Of the 108 SNPs, 100 SNPs are found in the coding region and 35 of them are synonymous mutations [27]. Another study conducted on 10,664 SARS-CoV-2 sequences from 73 countries found 107 SNPs. Based on this study, 5 SNPs are predicted to have a harmful impact, namely mutations T85I, Q57H, R203M, F506L, and S507C [28]. In addition, virus detection and epitope-based synthetic vaccine design require large amounts of genome analysis to find the most conserved parts so that global RT-PCR methods can be developed and used worldwide [29].

Identification of the SARS-CoV-2 genome characters such as selection patterns is necessary because the virus continues to evolve. This knowledge is important in the diagnosis and control of disease [30, 31]. This research was conducted to identify the genetic variability of mutations in the form of Single-nucleotide polymorphism (SNP). The results of this study are expected to help deepen knowledge about the characteristics of SARS-CoV-2. In this study, SNP analysis was carried out using the short and accurate Multiple Sequence Alignment method using MAFFT [32], then SNP extraction was carried out using the SNP-sites program which is capable of performing SNP calling at high-speed multi-FASTA alignment [33].

2 Materials and Methods

2.1 Sample Data-Mining

The study was conducted in silico on a web server and local computer using the SARS-CoV-2 sequence data downloaded from GISAID with reference sequences taken from NCBI. The MAFFT web server version 7.481 is used with the –addfragments package [34]. SNP-sites software is also used [33] to perform SNP calling. In the search for variants, the Nextstrain web server was used with the Nextclade tools [35]. Data processing to extract SNP was carried out using Microsoft Excel. Excel is also used to visualize data in the form of graphs, tables and diagrams.

A total of 15,000 sequences of SARS-CoV-2 from 35 different countries around the world were downloaded from GISAID [5] on 12 July 2021. The time span for isolation of these sequences was from February 2020 to 12 July 2021. The selected sequences have high and full genome coverage, and is a complete genome with an average sequence length of 29.8 bp. In addition, sequences with low coverage are also excluded. As a reference, the complete sequence was taken from NCBI (National Center for Biotechnology Information) with accession number NC_045512.2 which was isolated from Wuhan, China. The sequences are also used to map samples. The selected sample has been isolated from the following countries, namely; Indonesia, India, Singapore, Japan, Saudi Arabia, China, Russia, Malaysia, United States, Canada, Costa Rica, Mexico, Belgium, Spain, France, Italy, Turkey, Germany, Australia, Colombia, Peru, Brazil, Kenya, Reunion, South Africa, Ghana, Madagascar, Malawi, Mali, Mauritius, Mayotte, Morocco and Mozambique.

2.2 Sample Processing and Alignment

Next, the samples were processed with BioEdit to be aggregated and sorted. Samples that have a sequence fragment of 'NNNNNN' in the sequence are deleted. In addition, sequences containing other than A, C, T, and G were deleted due to low quality. In this process, as many as 5000 sequences were deleted due to poor quality, leaving 10,107 sequences to go to the next stage.

Furthermore processed samples need to be aligned and mapped to the reference sequence. Multiple Sequence Alignment was performed to align the samples. This process is carried out using the MAFFT web-server (version 7.481) using default parameter [32]. The method used is progressive alignment using the –add fragments package contained in [34]. This method was chosen because the samples used came from the same species, so to speed up the alignment process each sequence was only aligned with the reference genome to form the entire MSA [34]. This process also simultaneously maps the sample against the reference sequence. After that, a consensus sequence is obtained to be able to extract point mutations in the sample. The output from the MSA will be in the FASTA format.

2.3 SNP-sites Calling

To be able to find mutation points, Single-nucleotide polymorphism was called on samples using SNP-sites with default parameter [16]. SNP-sites is a program that can call SNP on samples in the form of multi-fasta alignment. The resulting output can be a file with VCF or PHYLIP format. In this study, the selected output has a VCF format so that genome analysis can be carried out.

2.4 SARS-COV-2 Variant Identification

Furthermore, the sequences were processed with the Nextclade tool version 1.5.4 on the Nextstrain web server with default parameter [35] to identify the variants of each sequence. Nextclade is a tool for sequence analysis that is able to identify variants of SARS-CoV-2 based on the Nextstrain classification of sequences uploaded by users.

After obtaining the point of occurrence of substitution mutations, the VCF file is imported and opened using Microsoft Excel for further processing. Microsoft Excel is used to filter mutations to determine SNP, as well as to visualize research results. In this study, substitutions that occur in more than 1% of the sample are considered Single-nucleotide polymorphism, so these substitutions must occur in at least 101 samples.

3 Results and Discussion

Table 1. Profile of sample countries before and after sorting using BioEdit

Continent	Country	Number of Sequences	After Sorted	
Africa	South Africa	2604	1878	2358
	Ghana	214	22	
	Kenya	422	100	
	Madagascar	3	3	
	Malawi	9	0	
	Mali	1	1	
	Marocco	11	9	
	Mauritius	2	0	
	May otte	683	295	
	Mozambique	2	2	
Asia	China	486	699	4055
	Hongkong	340	121	
	India	464	368	
	Indonesia	492	368	
	Japan	569	406	
	Malaysia	652	608	
	Russia	736	572	
	Saudi Arabia	466	427	
	Singapore	529	408	
	Taiwan	96	82	
North America	USA	498	267	1277
	Canada	469	401	
	Costa Rica	398	370	
	Mexico	379	239	
South America	Brazil	387	267	720
	Peru	448	279	
	Colombia	470	174	
Europe	Belgium	540	325	1683
	Spain	497	334	
	France	666	573	
	Italia	594	125	
	Turkey	350	211	
	Germany	91	49	
	Switzerland	247	66	
Oceania	Australia	14	12	14
	New Zealand	2	0	
	Guam	2	2	
Total		15569	10107	

3.1 Datamine Collection

This study was conducted to find SNPs in 10,107 SARS-CoV-2 genome sequences. Samples taken from GISAID were isolated from February 2020 to July 2021. The samples selected are complete genomes and have high coverage. A total of 15,911 samples were downloaded. The samples were isolated from 35 different countries. Then, the samples were sorted using BioEdit to remove sequences containing letters other than A, C, T, and G. In this process, sequences containing the 'NNNN' fragment will also be deleted. In this process, more than 5000 sequences were deleted leaving 10,107 sequences. A list of all the countries in the 10,107 sequences can be found in Table 1. It can also be seen that there is an uneven distribution of the sample. There are several countries with very few samples, especially in Africa, namely Madagascar, Malawi. These countries do not upload many samples of SARS-CoV-2. Most of the samples representing the African continent were taken from South Africa. In addition, there are also a small number of sequences from the Oceania continent. And, only 3 countries can be taken from Oceania, namely Australia, New Zealand, and Guam. However, due to poor quality, samples from New Zealand were erased during processing with BioEdit. Few samples were taken from Oceania because there were relatively few cases in Australia and New Zealand, and other Oceania countries did not upload many samples to GISAID.

3.2 The Results of the Identification of Sample Variants

Sequence analysis using Nextclade was performed to determine the variance of each sequence. Figure 1. is the result of data processing obtained from Nextclade. Based on the figure, it can be seen that the highest number of variants is variant 20B with a percentage of 32.12% of the total population. Followed by the high number of 20A variants, namely 23.95% and 20I (Alpha) variants of 17.39% of the total population. These three variants alone dominate more than 50% of the SARS-CoV-2 variants in the world. Other variants that have the lowest prevalence percentage include 21G (Lambda), 21H, and 21D (Eta) with respective percentages of 0.07%, 0.03% and 0.02%. Even so, the data used in this study is the sequence of SARS-CoV-2 isolated from February 2020 to July 2021. According to data from Nextstrain in August 2021, the currently most common variant is variant 20I (Alpha) [35]. So, currently there is a possibility that there is a shift from the previously most common variant, namely the 20B variant, to the 20I (Alpha) variant. This indicates that SARS-CoV-2 has adapted and produced mutations that are beneficial to the SARS-CoV-2 variant 20I (Alpha), so that its number increases.

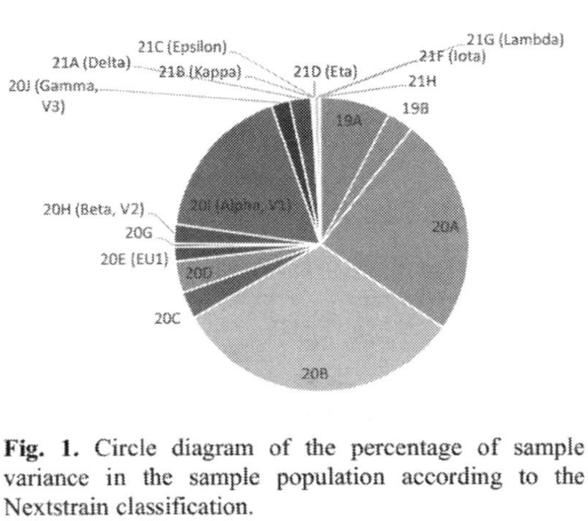

Fig. 1. Circle diagram of the percentage of sample variance in the sample population according to the Nextstrain classification.

3.3 Single-Nucleotide Polymorphism in the SARS-CoV-2 genome sequence

A total of 10,107 samples were processed with the SNP-Sites program to obtain their SNP positions. The regions analyzed were only the translated parts of the genome, namely at the nucleotide base positions 266-29676, while the remaining parts were 3'-UTR and 5'-UTR (untranslated region). From the results of data processing, a total of 154 SNPs were obtained. Based on Figure 2., it can be seen that mutations in the genome positions C14408T, C3037T, and A23403G occurred in more than 8000 samples. In addition, there is a high mutation rate, namely more than 5000 sample populations at sequential genome positions, namely G28881A, G28882A, and G28883C.

The C14408T mutation occurs in the nsp12 gene or it is also called RNA-dependent RNA polymerase (RdRp). Based on the research results in Figure 1., this mutation has the highest prevalence among other mutations. RdRp in SARS-CoV-2 plays an important role in the process of viral replication and transcription. The RdRp gene of SARS-CoV-2 has a high level of homology with SARS-CoV and is a conserved part, this shows that both of them can have the same mechanism [36, 37]. RdRp works with a complex mechanism with nsp7 and nsp8 in carrying out replication and transcription processes. RdRp also has a proofreading function, so mutations in RdRp can result in the emergence of new mutations or even an increase in the mutation rate [37]. Several studies have demonstrated high mutation rates following the C14408T mutation. This needs to be considered because currently there are various antiviral drugs that target RdRp. The C14408T mutation is among the conserved areas. Therefore, further studies are needed to see whether these mutations can affect the efficiency of these drugs [37]. According to Wang, et al (2021), the C14408T mutation is related to the A23403G mutation, which is a mutation with a high prevalence rate. The increasing number of C14408T mutations is also in line with the increasing number of COVID-19 patients which indicates a link between the C14408T mutation and the transmission of SARS-CoV-2 [38].

The A23403G mutation occurs in the spike protein. This mutation resulted in the D614G mutation which is located in a conserved part of this species [39]. Mutation A23403G is also one of the key mutations in variants 20A and 20I (Alpha), which are the 2 variants with the highest number in the sample population. The high frequency of the A23403G mutation indicates that this mutation has a beneficial effect on the SARS-CoV-2 virus. This is in line with research conducted by Wang et al (2021), this study states that there has been an increase in the number of A23403G mutations in the SARS-CoV-2 virus isolated in the United States over time. This increase in the number of mutations coincides with a sharp increase in COVID-19 cases in the United States [38]. Several studies have also carried out molecular docking of this mutation and have found that the A23403G mutation causes changes in the structure of the resulting protein resulting in a more infectious variant of SARS-CoV-2 [12, 40].

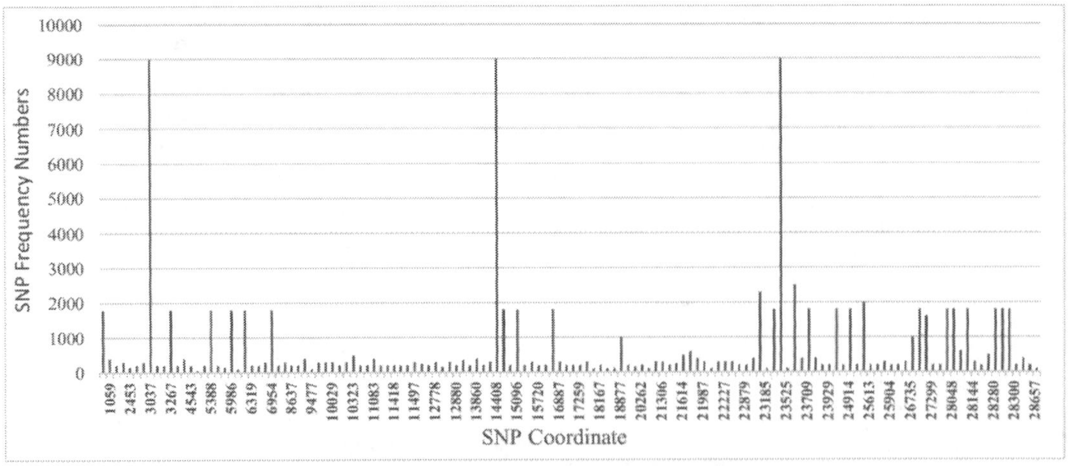

Fig. 2. Graph of SNP frequencies at each SNP coordinate

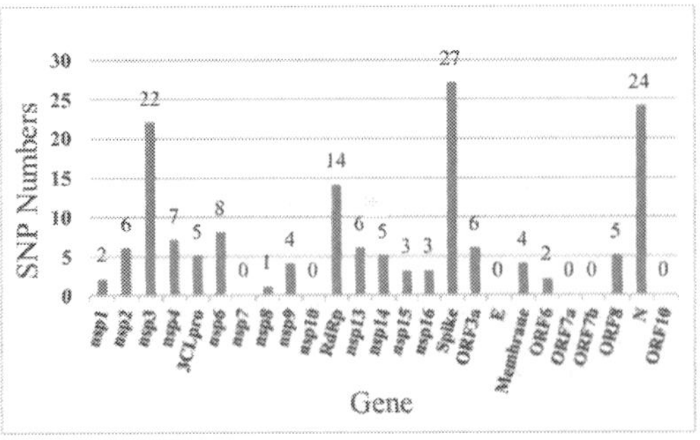

Fig. 3. Graph of the number of SNPs in each gene

The next highest mutations were 3 consecutive mutations in the gene encoding the nucleocapsid (N) protein, namely mutations G28881A, G28882A, and G28883C. The nucleocapsid is included in the structural protein in SARS-CoV-2 which plays an important role in RNA packaging, release of viral particles, and formation of the ribonucleoprotein core [41]. If these mutations occur simultaneously, it results in a change in the nucleotide base arrangement of GGG to AAC. This nucleotide change is quite significant. However, further research is needed to prove that this mutation can cause structural changes in the resulting protein.

The C3037T mutation is located in the nsp3 gene. The high population of variants 20A and 20I (Alpha) contributed to the large number of C3037T mutations. The C3037T mutation is one of the key mutations in the two variants. According to molecular analysis conducted by Yuan et al (2020), these mutations are synonymous which do not cause changes to the resulting amino acid sequence [26].

Figure 3 shows the number of SNPs in each gene. Based on the figure, the spike, nsp3, and nucleocapsid genes have the highest number of SNPs compared to the other genes.

3.4 Mutations in SARS-CoV-2

The highest number of SNPs were in the spike, nucleocapsid, and nsp3 genes. However, from these results it is not yet known whether these genes do have high variability or are influenced by other factors

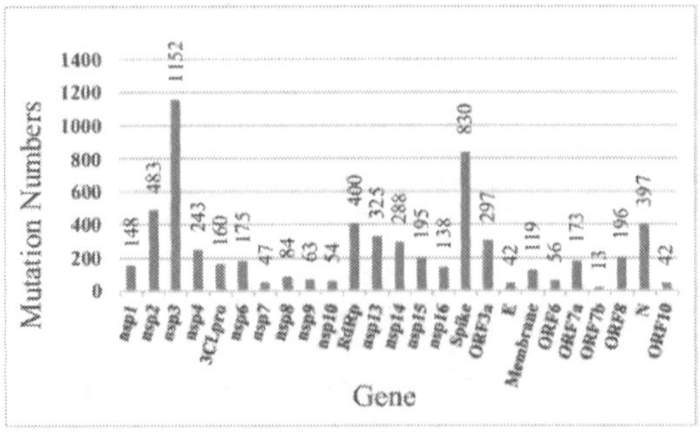

Fig. 4. Graph of the number of mutations in each gene

Figure 4. shows the number of mutations in each gene. In the figure it can be seen that the nsp3, spike, and nsp2 genes have the highest number of mutations. This is in accordance with previous studies which also found that the highest number of mutations occurred in these three genes [26]. However, these genes have different sizes. Longer genes tend to have a higher number of mutations than short genes. Thus, these results need to be normalized by the length of the sequence of each gene.

Based on Table 2, the ratio of the number of mutations to the relatively high sequence length occurred in the ORF8, ORF7a and ORF7b genes with respective values of 0.537, 0.474, and 0.419. This number is higher when compared to the ratio of the number of mutations to the length of the sequence in the nsp3 and nsp2 genes, which were only 0.197 and 0.252. An evolutionary study conducted by Pereira (2020) shows that among other accessory genes, ORF8 is the gene that has the highest level of variation. However, the function of ORF8 is still

https://doi.org/10.1051/bioconf/20237501005

not fully known, so it cannot be concluded whether this variation has an effect on SARS-CoV-2 [42].

Table 2. The ratio of the number of mutations to the length of the sequence of each gene

Gene	Number of mutations per gene	Sequence length	Mutation ratio to gene length
nsp1	148	539	0.275
nsp2	483	1914	0.252
nsp3	1152	5835	0.197
nsp4	243	1500	0.162
3CLpro	160	918	0,175
nsp6	175	870	0.201
nsp7	47	249	0.189
nsp8	84	594	0.141
nsp9	63	339	0.186
nsp10	54	417	0.129
RdRp	400	2795	0.143
nsp13	325	1808	0.180
nsp14	288	1581	0.182
nsp15	195	1038	0.188
nsp16	138	898	0.154
Spike	830	3822	0.217
ORF3a	297	828	0.359
E	42	228	0.185
Membrane	119	669	0.178
ORF6	56	186	0.303
ORF7a	173	366	0.474
ORF7b	13	32	0.419
ORF8	196	366	0.537
N	397	1260	0.315
ORF10	42	117	0.362

The parts with the lowest mutations were the nsp8 and RdRp genes with mutation ratios and sequence lengths of 0.141 and 0.143. This is related to the function of the RdRp gene itself which is important for the process of viral replication. Thus, the stability of these genes is necessary for viruses [37]. In addition, the M (membrane) and E (envelope) genes also have a relatively low ratio of the number of mutations to the length of the sequence among other structural proteins with a ratio of 0.178 and 0.184 respectively. The low level of variability of the M and E proteins indicates that these two genes tend to be more stable than the other genes, as well as the link between these two genes with housekeeping functions [43]. The following Table 2 shows the ratio of mutations to the length of the sequence.

3.5 Subtitutions in the SARS-CoV-2 genome

Of the 154 SNPs obtained in this study from Figure 5, 80 SNPs were associated with C>T substitutions which represented most of the transitions. Table 3 shows the C>T substitution that occurs in SARS-CoV-2. This substitution can be caused by several factors, namely the factor of RNA deamination and the cost of biosynthesis of the nucleotide base itself. In synonymous mutations, the C>T substitution may be caused by deamination of the host's RNA, resulting in a C>T or A>G substitution. This is because humans and various animal and plant species have adenosine-inosine and cytidine-uridine deamination mechanisms as RNA diversification steps in cells that cause mismatches in the viral replication process [44], [45]. Several studies have also demonstrated high levels of deaminated RNA in the sequences of SARS-CoV-2 [46]. This factor is one of the strong factors for the high C>T mutation. This is also supported by the number of A>G mutations which is the second most common mutation after C>T mutations. Even so, C>T mutations are still far more numerous than A>G mutations.

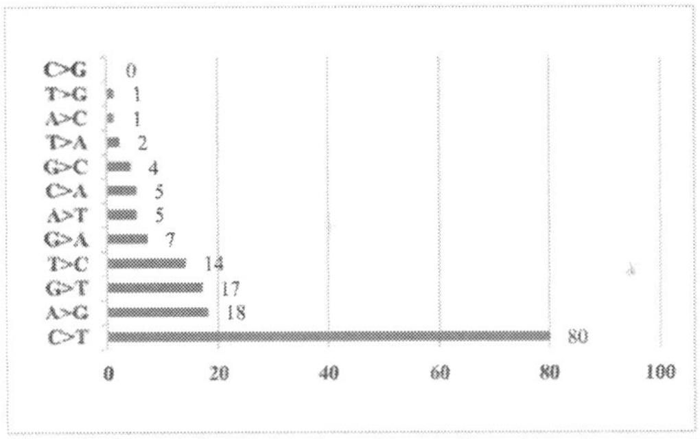

Fig. 5. Graph of the number of mutations in each type of substitution

Apart from these factors, another factor is the cost of nucleotide base biosynthesis. Thymine biosynthesis has a lower cost than cytosine, where thymine requires less ATP in its biosynthetic process. The biosynthesis of nucleotide bases requires a number of ATP molecules, the amount varies between nucleotide bases in the following order A > G > C > T. The biosynthesis of thymine is the lowest compared to other nucleotide bases [47]. Biosynthesis that requires less ATP is preferred in the process of natural selection [45]. Cost reduction in the biosynthetic process can be one of the causes of high C>T mutations.

This RNA deamination does not only occur in SARS-CoV-2, but also in various other viruses that attack animals and humans. This mutation pattern was also observed in other viruses namely Bat RaTG13 and other coronaviruses. Among other betacoronaviruses (SARS and MERS), SARS-CoV-2 experienced the most extreme RNA deamination [46, 48]. Additional details on the

https://doi.org/10.1051/bioconf/20237501005

containing mutations which are C>T mutations can be found in the following Table 3.

Table 3. Mutations with substitution type C>T

Position	Alteration	Frequency	Gene
683	C > T	169	nsp1
913	C > T	1761	nsp2
1059	C > T	648	nsp2
1191	C > T	123	nsp2
2110	C > T	278	nsp2
2453	C > T	110	nsp2
2749	C > T	208	nsp3
3037	C > T	9015	nsp3
3096	C > T	188	nsp3
3140	C > T	129	nsp3
3267	C > T	1762	nsp3
3828	C > T	215	nsp3
4002	C > T	362	nsp3
4543	C > T	402	nsp3
5944	C > T	149	nsp3
5986	C > T	1776	nsp3
6286	C > T	166	nsp3
6762	C > T	288	nsp3
8637	C > T	206	nsp4
8782	C > T	310	nsp4
8917	C > T	126	nsp4
9891	C > T	197	nsp4
10029	C > T	327	nsp4
10954	C > T	127	3CLpro
11195	C > T	158	nsp6
11497	C > T	280	nsp6
11514	C > T	116	nsp6
12778	C > T	211	nsp9
12789	C > T	113	nsp9
12815	C > T	104	nsp9
12880	C > T	162	nsp9
13536	C > T	369	RdRp
13730	C > T	436	RdRp
13860	C > T	224	RdRp
14120	C > T	361	RdRp
14408	C > T	9035	RdRp
14676	C > T	1769	RdRp
14805	C > T	174	RdRp
15279	C > T	1761	RdRp
15324	C > T	123	RdRp
15720	C > T	147	RdRp
16887	C > T	193	nsp13
17012	C > T	146	nsp13
17518	C > T	225	nsp13
18167	C > T	151	nsp14
18747	C > T	163	nsp14
18877	C > T	1352	nsp14
19524	C > T	406	nsp14
21306	C > T	147	nsp16
21516	C > T	214	nsp16
21575	C > T	193	Spike
21614	C > T	444	Spike
21638	C > T	229	Spike
22227	C > T	241	Spike
22444	C > T	340	Spike
23185	C > T	144	Spike
23525	C > T	249	Spike
23664	C > T	484	Spike
23709	C > T	1783	Spike
23731	C > T	366	Spike
23929	C > T	468	Spike
24642	C > T	278	Spike
25613	C > T	107	ORF3a
25844	C > T	123	ORF3a
25904	C > T	305	ORF3a
26681	C > T	251	Membrane
26735	C > T	1207	Membrane
26985	C > T	118	Membrane
27972	C > T	1772	ORF8
28311	C > T	470	Nucleocapsid
28657	C > T	132	Nucleocapsid
28854	C > T	680	Nucleocapsid
28863	C > T	118	Nucleocapsid
28887	C > T	608	Nucleocapsid
28932	C > T	158	Nucleocapsid
28977	C > T	1770	Nucleocapsid
29197	C > T	125	Nucleocapsid
29366	C > T	113	Nucleocapsid
29421	C > T	119	Nucleocapsid

4 Conclusions

It is concluded that out of 10,107 samples of SARS-CoV-2 studied, 154 SNPs were found. The genes with the highest number of SNPs were the spike, nsp3, and nucleocapsid genes. To determine the variability of each gene, the ratio of the number of mutations to the length of the sequence is used. The mutation ratios with the largest sequence lengths were in the ORF8, ORF7a, and ORF7b genes with respective values of 0.537, 0.474, and 0.419. Based on these results, the high number of mutations and SNPs in a gene does not necessarily reflect the level of variability of that gene. This can be seen from how the spike, nsp3, and spike genes have a high number of mutations, but when normalized by the sequence length, the number is relatively not too high compared to the accessory proteins. Therefore these results indicate that the high SNP in a gene cannot be used as a benchmark for whether or not the mutation rate of that gene is high.

Acknowledgements

The authors thank the School of Life Sciences and Technology for partially funding this research under the Biology Study Program.

Conflicts of interest statement

The authors declared that they have no conflicts of interest in this study.

References

1. Gorbalenya, A. E., Baker, S. C., Baric, R. S., de Groot, R. J., Drosten, C., Gulyaeva, A. A., Ziebuhr, J. (2020). The species Severe acute respiratory syndrome-related coronavirus: classifying 2019-nCoV and naming it SARS-CoV-2. *Nat. Microbiol*, **5**, 536-544.

2. Zhu, N., Zhang, D., Wang, W., Li, X., Yang, B., Song, J.,Tan, W. (2020). A novel coronavirus from patients with pneumonia in China, 2019. N. Engl. *J. Med*, **382**, 727–733. doi:https://doi.org/10.1056/NEJMoa2001017

3. Chen, N., Zhou, M., Dong, X., Qu, J., G. F., H. Y., Zhang, L. (2020). Epidemiological and clinical characteristics of 99 cases of 2019 novel coronavirus pneumonia in Wuhan, China: a descriptive study. *Lancet*, **395**, 507–513. doi:https://doi.org/10.1016/S0140-6736(20)30211-7.

4. Zhou, P., Yang, X., & Wang, X. (2020). A pneumonia outbreak associated with a new coronavirus of probable bat origin. *Nature*, 579, 270-273. doi:https://doi.org/10.1038/s41586-020-2012-7

5. Shu, Y., & McCauley, J. (2017). GISAID: Global initiative on sharing all influenza data—from vision to reality. *Eurosurveillance*, **22**, 30494. doi:https://doi.org/10.2807/1560-7917.ES.2017.22.13.30494

6. Fehr, A., & Perlman, S. (2015). Coronaviruses: An overview of their replication and pathogenesis. *Methods mol. biol.*, **1282**, 1–23. doi: https://doi.org/10.1007/978-1-4939-2438-7_1

7. Jenkins, G., Rambaut, A., Pybus, O., & Holmes, E. (2002). Rates of molecular evolution in rna viruses: a quantitative phylogenetic analysis. *J. Mol. Evol*, **54**, 156–165. doi:https://doi.org/10.1007/s00239-001-0064-3

8. Duchene, S., Featherstone, L., Haritopoulou-Sinanidou, M., Rambaut, A., Lemey, P., & Baele, G. (2020). Temporal signal and the phylodynamic threshold of SARS-CoV-2. *Virus Evol*, **6**(2). doi:https://doi.org/10.1093/ve/veaa061

9. Carroll, M., Matthews, D., Hiscox, J., Elmore, M., Pollakis, G., Rambaut, A., Becker. (2015). Temporal and spatial analysis of the 2014-2015 Ebola virus outbreak in West Africa. *Nature*, **524**, 97–101. doi:https://doi.org/10.1038/nature14594

10. Zhao, Z., Li, H., Wu, X., Zhong, Y., Zhang, K., Zhang, Y., Fu, Y. (2004). Moderate mutation rate in the SARS coronavirus genome and its implications. *BMC Evol Biol*, **4**, 21. doi:10.1186/1471-2148-4-21

11. Wang, C., Liu, Z., Chen, Z., Huang, X., Xu, M., He, T., & Zhang, Z. (2020). The establishment of reference sequence for SARS-CoV-2 and variation analysis. *J. Med. Virol.*, **92**(6), 667-674. doi: https://doi.org/10.1002/jmv.25762

12. Korber, B., Fischer, W., Gnanakaran, S., Yoon, H., Theiler, J., Abfalterer, W., Tang, H. (2020). Tracking changes in SARS-CoV-2 Spike: evidence that D614G increases infectivity of the COVID-19 virus. *Cell*, **182**(4), 812–827. doi:10.1016/j.cell.2020.06.043

13. Sherry, S., Ward, M., & Sirotkin, K. (1999). dbSNP—Database for Single Nucleotide Polymorphisms and Other Classes of Minor Genetic Variation. *Genome Res*, **9**(8), 677–679.

14. Khalid, Z., & Sezerman, O. (2020). A comprehensive study on identifying the structural and functional SNPs of human neuronal membrane glycoprotein M6A (GPM6A). *J. Biomol. Struct*, **39**(8), 2693-2701, 1-9.

15. André, N. M., Cossic, B., Davies, E., Miller, A. D., & Whittaker, G. R. (2019). Distinct mutation in the feline coronavirus spike protein cleavage activation site in a cat with feline infectious peritonitis-associated meningoencephalomyelitis. *J. Feline Med. Surg. Open Rep*, **5**(1), 2055116919856103.

16. Davies, N. G., Abbott, S., Barnard, R. C., Jarvis, C. I., Kucharski, A. J., Munday, J. D., Wong, K. L. (2021). Estimated transmissibility and impact of SARS-CoV-2 lineage B.1.1.7 in England. *Science*, **372**(6538).

17. Pearson, C. A., Russell, T. W., Davies, N. G., Kucharski, A. J., CMMID COVID-19 working group, Edmunds, W. J., & Eggo, R. M. (2021). Estimates of severity and transmissibility of novel South Africa SARS-CoV-2 variant 501Y. V2. *Preprint*, ***50***, 1-4.

18. Faria, N. R., Mellan, T. A., Whittaker, C., Claro, I. M., Candido, D. d., Mishra, S., Andrade, P. (2021). Genomics and epidemiology of the P.1 SARS-CoV-2 lineage in Manaus, Brazil. *Science*, **372**(6544).

19. Allen, H., Vusirikala, A., Flannagan, J., Twohig, K. A., Zaidi, A., Chudasama, D., ... & Kall, M. (2022). Household transmission of COVID-19 cases associated with SARS-CoV-2 delta variant (B. 1.617. 2): national case-control study. *Lancet Reg. Health - Eur.*, *12*.

20. Hasan, M. R., Sundararaju, S., Manickam, C., Mirza, F., Al-Hail, H., Lorenz, S., & Tang, P. (2021). A novel point mutation in the N gene of SARS-CoV-2 may affect the detection of the virus by reverse transcription-quantitative PCR. *J. Clin. Microbiol.*, **59**(4), 10-1128.

21. Artesi, M., Bontems, S., Gobbels, P., Franckh, M., Maes, P., Boreux, R., Durkin, K. (2020). A recurrent mutation at position 26340 of SARS-CoV-2 is associated with failure of the E gene quantitative reverse transcription-PCR utilize. *J. Clin. Microbiol.*, 58.

22. Ziegler, K., Steininger, P., Ziegler, R., Steinmann, J., Korn, K., & Ensser, A. (2020). SARS-CoV-2 samples may escape detection because of a single point mutation in the N gene. *Euro Surveillance*, **25**(39). doi:https://doi.org/10.2807/1560-7917.ES.2020.25.39.2001650

23. Vanaerschot, M., Mann, S., Webber, J., Kamm, J., Bell, S., Bell, J., Tato, C. (2020). Identification of a polymorphism in the N gene of SARS-CoV-2 that adversely impacts detection by reverse transcription-PCR. *J. Clin. Microbiol.*, 59.

24. Nguyen, T., Zhang, Y., & Pandolf, P. (2020). Virus against virus: a potential treatment for 2019-nCov (SARS-CoV-2) and other RNA viruses. *Cell Res*, **30**, 189–90.

25. Nasreen, S., Chung, H., He, S., Brown, K. A., Gubbay, J. B., Buchan, S. A., ... & Canadian Immunization Research Network (CIRN) Provincial Collaborative Network (PCN) Investigators. (2022). Effectiveness of COVID-19 vaccines against symptomatic SARS-CoV-2 infection and severe outcomes with variants of concern in Ontario. *Nat. Microbiol.*, **7**(3), 379-385.

26. Yuan, F., Wang, L., Fang, Y., & Wang, L. (2020). Global SNP analysis of 11,183 SARS-CoV-2 strains reveals high genetic diversity. *Transbound Emerg Dis*, 1-17. doi:https://doi.org/10.1111/tbed.13931

27. Das, A., Khurshid, S., Ferdausi, A., Nipu, E., Das, A., & Ahmed, F. (2021). Molecular insight into the genomic variation of SARS-CoV-2 strains from current outbreak. *Comput. Biol. Chem*, **93**, 107533. doi:https://doi.org/10.1016/j.compbiolchem.2021.107533

28. Ghosh, N., Saha, I., Sharma, N., Nandi, S., & Plewczynski, D. (2021). Genome-wide analysis of 10664 SARS-CoV-2 genomes to identify virus strains in 73 countries based on single nucleotide polymorphism. *Virus Res.*, 289. doi:https://doi.org/10.1016/j.virusres.2021.198401

29. Saha, I., Ghosh, N., Pradhan, A., Sharma, N., Maity, D., & Mitra, K. (2021). Whole genome analysis of more than 10 000 SARS-CoV-2 virus unveils global genetic diversity and target region of NSP6. *Bioinformatics*, **22**(2), 1106–1121. doi:10.1093/bib/bbab025

30. Lu, R., Zhao, X., Li, J., Niu, P., Yang, B., Wu, H., & Wang, W. (2020). Genomic characterisation and epidemiology of 2019 novel coronavirus: implications for virus origins and receptor binding. *Lancet*, **395**(10224), 565-574.

31. Ayres, J. (2020). Surviving COVID-19: a disease tolerance perspective. *Sci Adv*, **6**(18).

32. Katoh, K., Misawa, K., Kuma, K.-i., & Miyata, T. (2002). MAFFT: a novel method for rapid multiple sequence alignment based on fast Fourier transform. *Nucleic Acids Res*, **30**(14), 3059-3066. doi:https://doi.org/10.1093/nar/gkf436

33. Andrew J. Page, c. B., Delaney, A. J., Soares, J., Seemann, T., Keane, J. A., & Harris, S. R. (2016). SNP-sites: rapid efficient extraction of SNPs from multi-FASTA alignments. *Microbial Genomics*, **2**(4). doi:10.1099/mgen.0.000056

34. Katoh, K., & Frith, M. C. (2012). Adding unaligned sequences into an existing alignment using MAFFT and LAST. *Bioinformatics, 28*(23), 3144-3146. doi:https://doi.org/10.1093/bioinformatics/bts578

35. Hadfield, J., Megill, C., Bell, S., Huddleston, J., Potter, B., Callender, C., Neher, R. (2018). Nextstrain: Real-time tracking of pathogen evolution. *Bioinformatics*, **34**, 4121–4123. doi:https://doi.org/10.1093/bioinformatics/bty407

36. Kirchdoerfer, R. N., & Ward, A. B. (2019). Structure of the SARS-CoV nsp12 polymerase bound to nsp7 and nsp8 co-factors. *Nat. Commun.*, **10**(1), 2342.

37. Pachetti, M., Marini, B., Benedetti, F., Giudici, F., Mauro, E., Storici, P., Ippodrino, R. (2020). Emerging SARS-CoV-2 mutation hot spots include a novel RNAdependent-RNA polymerase variant. *J. Transl. Med.*, **18**, 1-9.

38. Wang, R., Chen, J., Gao, K., Hozumi, Y., Yin, C., & Wei, G.-W. (2021). Analysis of SARS-CoV-2 mutations in the United States suggests presence of four substrains and novel variant. *Commun. Biol.*, **4**, 228. doi: https://doi.org/10.1038/s42003-021-01754-6

39. Walls, A., Parl, Y., Tortorici, M., Wall, A., McGuire, A., & Veesler, D. (2020). Structure, function, and antigenicity of the SARS-CoV-2 spike glycoprotein. *Cell*, **181**, 281–292.

40. Yurkovetskiy, L., Wang, X., Pascal, K. E., Tomkins-Tinch, C., Nyalile, T. P., Wang, Y., . . . Veinotte, K. (2020). Structural and functional analysis of the D614G SARSCoV-2 spike protein variant. *Cell*, 739–751.

41. Zeng, W., Liu, G., Ma, H., Zhao, D., Yang, Y., Liu, M., Jin, T. (2020). Biochemical

characterization of SARS-CoV-2 nucleocapsid protein. *Biochem Biophys Res Commun,* **527**(3), 618-623. doi:10.1016/j.bbrc.2020.04.136

42. Pereira, F. (2020). Evolutionary dynamics of the SARS-CoV-2 ORF8 accessory gene. *Infect. Genet. Evol.,* 85. doi:10.1016/j.meegid.2020.104525

43. Laha, S., Chakraborty, J., Das, S., Manna, S., S, B., & Chatterjee, R. (2020). Characterizations of SARS-CoV-2 mutational profile, spike protein stability and viral transmission. *Infect. Genet. Evol., 85.* doi:https://doi.org/10.1016/j.meegid.2020.104445

44. Bass, B. (2002). RNA editing by adenosine deaminases that act on RNA. *Annu. Rev. Biochem,* **71**, 817–846.

45. Yu, Y., Li, Y., Dong, Y., Wang, X., Li, C., & Jiang, W. (2021). Energy efficiency trade-offs drive nucleotide usage in transcribed regions. *Future Virology,* 16. doi:https://doi.org/10.2217/fvl-2021-0078

46. Li, Y., Yang, X., Wang, N., Wang, H., Yin, B., Yang, X., & Jiang, W. (2020). Mutation profile of over 4500 SARS-CoV-2 isolations reveals prevalent cytosine-to-uridine deamination on viral RNAs. *Future Microbiol,* **15**(14), 1343–1352.

47. Chen, W., Lu, G., Bork, P., Hu, S., & Lercher, M. (2016). Energy efficiency trade-offs drive nucleotide usage in transcribed regions. *Nat. Commun,* 7.

48. Xia, X. (2020). Extreme Genomic CpG Deficiency in SARS-CoV-2 and Evasion of Host Antiviral Defense. *Mol. Biol. Evol,* **37**(9), 2699–2705. doi:https://doi.org/10.1093/molbev/msaa094

Influencing factors that improve mental conditions patients with complementary therapy at Nur Hidayah Hospital, Bantul, Yogyakarta

Santi Wulan Purnami[1*], *Kevina* Windy Arlianni[1], *Shofi* Andari[1], *Sagiran* Sagiran[2], *Estiana* Khoirunnisa[2], and *Wahyudi* Widada[3]

[1]Department of Statistics, Institut Teknologi Sepuluh Nopember, Surabaya, Indonesia
[2]Faculty of Medicine, Universitas Muhammadiyah Yogyakarta, Yogyakarta, Indonesia
[3]Program of Nursing, Universitas Muhammadiyah Jember, Jember, Indonesia

Abstract. Mental health is a concern of many parties. Complementary therapy is carried out in line with physical treatment with the aim of reducing the number of cases of mental health disorders, one of them at Nur Hidayah Hospital. This study used random forest and extended Cox regression to determine improvement in the mental condition of patients with complementary therapy. The research data used data from the medical records of 102 patients with Al-Quran Complementary Therapy. Random forest classifier does not give good accuracy only 0.58. Log-rank test showed that the variables attitude, problem Prayer, interaction history and treatment reactions had different survival curves between the categories of each variable. The interaction history variable does not meet the proportional hazard assumption so that the Cox model used is stratified and time-dependent Cox. The best model is the Cox stratified model with the interaction having the smallest AIC value of 286.84. Factors that have a significant effect on the time to improve mental conditions are gender and the main complaint in the patient strata who have a history of interaction and the age variable in the stratum of patients who have had a history of interaction.
Keywords: Al-Quran, Complementary Therapy, Extended Cox, Mental Conditions, Survival.

1 Introduction

Mental health is concerning issue around the globe. It is related to mental state and well-being enabling a person to cope with pressures in life, realize their abilities, study well and work well, and contribute to society. Hence, mental health plays an important role in supporting the productivity and quality of physical health. Up to 20% of the Indonesian population has the potential of mental disorders [1]. This number is unexpectedly high is due to the limited number of experts capable of carrying out prevention and control efforts and the lack of a recording and reporting system related to mental disorders.

Complementary and alternative medicine (CAM) is a treatment that includes a variety of approaches ranging from a mental approach through religious aspects (Al-Qur'an, prayer), spiritual which shows positive results on the patient's mental health condition [2]. This complementary medicine has been used in many countries [3-5]. The use of this therapy has also been developed in Indonesia, one of which is carried out by the Nur Hidayah Hospital, Bantul, with the practice of complementary therapy in the form of Al-Quran therapy in complementary units. This unit accepts patients based on the results of medical observations who have a low mental condition so that they require complementary therapy and can come from the wishes of the patient without medical complaints. Complementary therapy is expected to have a significant positive influence in the process of improving the patient's mental condition. Information about the patient's mental improvement time with complementary therapy is also very important to note in treatment efforts. Medical personnel will receive related information so that it becomes educational material for patients and as an evaluation for the development of complementary therapies.

Complementary therapy Al-Quran combine mental focus and spirituality to help relax the body and mind. Religion and spirituality are known as important roles in coping with illness. Previous studies examining the holistic nursing approach should include religion and spiritual interventions, since pain, negative experiences such as mental stress are frequently present in medical problems [6] and other studies use of CAM in Saudi older adults using multivariable Cox regression showed that the death hazard ratio in 2006 and 2015 showed no

* Corresponding author: santi_wp@statistika.its.ac.id

difference in mortality between groups of CAM users and nonusers [7]. Many previous studies have examined descriptively [2,4,7,8] while the analysis regarding the time to improve complementary therapy is still limited. Some of the factors that are suspected to be the affected improvement of mental health condition is gender. Information on the patient's mental improvement time with complementary therapy is also something needs to be considered as a treatment effort. Therefore, it is needed to see the relationship between time to event and the factors that are suspected to influence. The survival analysis is useful statistical methods to analyse the timing of event.

Therefore, the objectives of this study are to classify patient status based on predictor variables using random forest and discuss the time of mental improvement of patients with complementary therapy using extended Cox regression. This study was conducted to provide additional literature regarding the analysis of time for improving mental conditions and to find out the influencing factors, especially for cases of complementary therapy patients at Nur Hidayah Hospital using Cox regression survival analysis. This model aims to determine which factors that influence and how much its influence on the time of improving mental health in patients with complementary therapy Al-Quran.

Time of improving mental conditions is calculated from the patient's admission to the hospital until the mental condition improves (mental conditions improve with a marked decrease in attitude values, prayer problem and interaction history to zero). The value of attitude, prayer problem and interaction history are represented by 0, 1 and 2, respectively. Factors that are hypothesized to have an influence are attitude values, religious disturbances, history of interaction, gender, age, marital status, education, main complaint, chronic disease history and therapeutic reactions. The outcomes of this study are essential for healthcare providers and decision-makers to evaluate healthcare practices and to develop strategies that improve mental health.

2 Material and methods

2.1 Data source

The data used in this study is secondary data taken from the medical records of 102 patients at the Complementary Unit of Nur Hidayah Hospital, Bantul Regency from 1 January 2020 to 15 May 2022.

2.2 Complementary therapy Al-Quran

This Complementary unit accepts patients based on the results of medical observations who have a low mental condition so that they require complementary therapy and can come from the wishes of the patient without medical complaints. Complementary therapy practices carried out at the Complementary Unit of Nur Hidayah Hospital use Al-Qur'an therapy with a therapist who will read the Al-Qur'an in front of patient. Condition while therapy will be recorded on the assessment form and then a quick screening is carried out to assess the patient's initial condition through the SGR value then therapy is carried out. Patients are given a value of 0,1,2 for each SGR value, namely attitude (S), prayer problem (G) and interaction history (R).

The attitude value (S) describes the patient's psychological condition and is categorized into normal (0), anxious (1), unconscious (2). Prayer problem value (G) describes the history of prayer performed by the patient and is categorized as no distractions (0), lazy (1), not prayer (2). The value of interaction history (R) describes the history of supernatural in patients and families and is categorized into none (0), ever had a history of interaction (1), had a history of interaction (2).

After screening, patients are given complementary therapy using Al-Quran therapy by trained therapists. The implementation of therapy at Nur Hidayah Hospital can be done repeatedly according to the patient's mental condition. If the patient's mental condition is not normal, it is advisable to make a repeat visit. Based on information from officers and therapists at Nur Hidayah Hospital, this therapy is expected to help improve the patient's mental condition to normal. Normal conditions are characterized by all S, G and R values being zero.

2.3 Variables

This study used several variables consisting of response variables and predictor variables. The response variable in this study was the survival time of complementary therapy patients at Nur Hidayah Hospital. Survival time is the length of time (days) the patient enters the Complementary Unit of Nur Hidayah Hospital until an improvement in mental condition occurs. The observed event was an improvement in mental condition which was marked by all S, G and R values being zero. S, G, and R values each have a value of 0, 1, and 2. If there is only one S/G/R value that is not zero then the data is censored. An illustration of determining event data can be seen in Table 1.

Table 1. Determining event

First SGR Value			Final SGR Value			Censoring
S1	G1	R1	S2	G2	R2	
1	0	0	0	0	0	Event
1	1	0	0	1	0	Censored
2	2	1	1	2	0	Censored
2	2	1	0	0	0	Event
0	1	1	0	0	1	Censored
2	2	2	1	0	1	Censored
0	0	0	0	0	0	Event

The predictor variables attitude (S), prayer problem (G), supernatural interaction history (R), gender, age, marital status, education, main complaint, chronic disease history, and therapy reaction. Gender was classified into male and female. Age is the age of respondent in years. Marital status was classified into unmarried and married Education was classified into elementary school, junior high school, high school, and college.

Main complaint was classified into mentally and physically. The mentally category was addressed to respondents who has complaints about psychological mental conditions such as anxiety, confusion, and other mental disorders. The physical category was addressed to respondents who has complaints about physical conditions such as body aches, dizziness and other physical problems. Therapy reaction was classified into active and passive. The active category was addressed to respondents who had which reacts actively when treated. The passive category was addressed to respondents who do not reacts when treated.

2.4 Methods

The method used in this study is the classification of random forest and extended Cox. A random forest is a classifier consisting of a collection of tree-structured classifiers. The common element in all of these procedures is that for the *k*th tree, a random vector is generated, independent of the past random vectors but with the same distribution and a tree is grown using the training set and random vector. After a large number of trees is generated, they vote for the most popular class [9]. Label class in this study based on status patient. That are code by Improve explain who patient an improvement in mental condition and code by Not Improve explain that patient was not have an improvement in mental condition.

The stages of survival analysis are carried out by conducting Kaplan-Meier analysis is a curve that describes the relationship between the estimated survival function at time t and survival time and log-rank testing compared the Kaplan-Meier survival curves in each group [10]. Proportional hazard assumption testing for states that the hazard function of different individuals is proportional or the hazard ratio of different individuals is constant. If the proportional hazard assumption is violated, the Cox proportional hazard model cannot be carried out then we can continue proceed with extended Cox modelling including stratified and time-dependent Cox [10]. Comparison of accuracy measures of model is carried out using the AIC value. The best model has the smallest AIC value [11].

Step analysis in this study was carried out by

1. Describe the characteristics of complementary therapy patients at Nur Hidayah Hospital.
2. Performed classification analysis using random forest using 5-fold cross validation.
3. Create the confusion matrix and performance of the random forest model.
4. Obtain the survival curve using the Kaplan-Meier method and testing the differences using the log-rank test.
5. Checking the proportional hazard assumption on each predictor variable that is suspected of influencing the improvement of the patient's mental condition.
6. Establish a time model for improving the patient's condition using the extended Cox model.
   a. Forming strata from variables that do not meet the proportional hazard assumption.
   b. Performing model using stratified Cox models without interaction and stratified Cox models with interactions.
   c. Conduct interaction tests between stratified Cox models without interaction and stratified Cox models with interactions.
   d. Feature elimination using stepwise in each model to find the most influential.
   e. Modelling with the time-dependent Cox model. Form a model with a time function $g(t)=t$, $g(t)=ln(t)$, and heaviside function.
7. Selecting the best Cox model using the smallest AIC criterion. Calculate the hazard ratio and interpret the best Cox model. Draw conclusions from research results.

3 Results

3.1 Patient characteristics

The average age of complementary therapy patients at Nur Hidayah Hospital is 40.86 years, with the youngest being 2 years old and the oldest being 79 years old. Based on the event explanation used in this study, it was found that 54 patients have improvement in their mental condition (improved event) and 48 patients were censored data because they did not experience an improvement in their mental condition.

The Unconscious patients only 27% experienced mental improvement. There were 67% of prayer patients who experienced an improvement in their mental condition and there were only 31% of not prayer patients who experienced an improvement in their mental condition. Only 43% of patients who had a history of interaction experienced an improvement in their mental condition. Male patients have a greater percentage of experiencing an improvement in their mental condition compared to women, which is 61%. The percentage of patients for all marital status is >50% where unmarried patients experience a greater percentage of improving their mental condition than married patients.

All levels of education had a percentage that experienced an improvement in mental conditions >50%, but in junior high schools only 20% experienced improvement. Patients with major mental or physical complaints had the same percentage of improvement, namely 53%. The percentage of patients who experienced an improvement in their mental condition was greater for patients who did not have a history of chronic disease, namely 56%. Patients who reacted passively when therapy was carried out had a greater percentage of improving their condition than reacting passively, which was 74%.

3.2 Random forest classification analysis

Variable importance from random forest classification shown in Figure 1. The five most important variables for classifying patients using a random forest are age, education, interaction history, attitude, and prayer problem.

In the random forest classification of each decision tree which is grown based on data resampling will carries its own error in each decision tree. Calculation of the error obtained on each tree decisions that are built are the result of misclassification observations. Figure 2 shows us that the Out-of-Bag (OOB) error rate was the lowest with 100 trees. The optimal number of variables was set at eight.

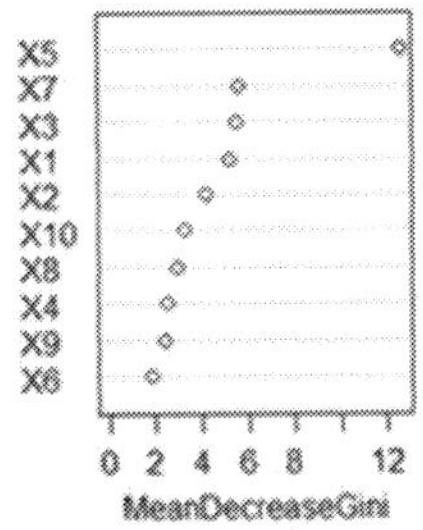

Fig. 1. Variables Importance

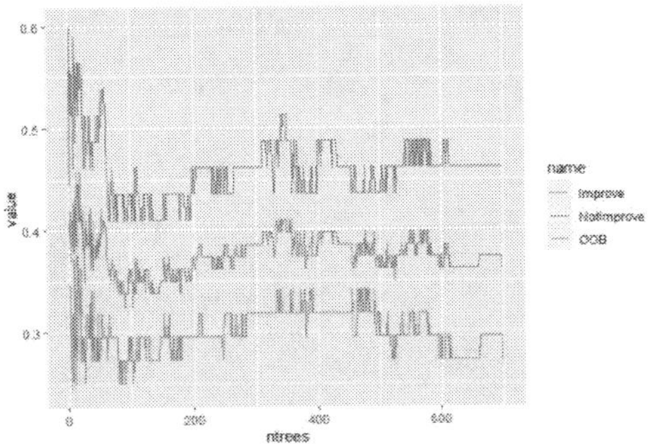

Fig. 2. Comparison error rates in increasing trees

The model using 5-fold cross validation give us result confusion matrix in Table 2 and performance random forest model in Table 3. Misclassification of patient status still occurs and this model has an accuracy of 0.58.

Table 2. Confusion matrix

	Actual	
Predicted	Improve	Not Improve
Improve	8	6
Not Improve	2	3

Random forest classification is used to predict patient status based on predictors, but this classification is not good enough to see whether an individual enters the improve class or not, so the analysis in this study is continued with survival analysis.

Table 3. Performance of the random forest model

	Random Forest Model
Accuracy	0.58
Sensitivity	0.80
Specificity	0.33
AUC	0.57

3.3 Survival analysis

Previous research examined only descriptively and compared CAM effectiveness with simple statistics so that this study will increase the complexity of the analysis by accommodating time to event. The median time that complementary therapy patients did not improve their mental condition was 30, meaning that patients had a 50% chance of not improving on the 30th day. It can be seen in Figure 3 that from day 0 to day 120 the survival curve drops slowly, this indicates that the longer the time, the greater the chance of improvement. On day 120 to day 370 it shows the lowest and constant line so the chance of not improving on day 120 to day 370 is the same and the chance of improving after day 120 is greater than before day 120.

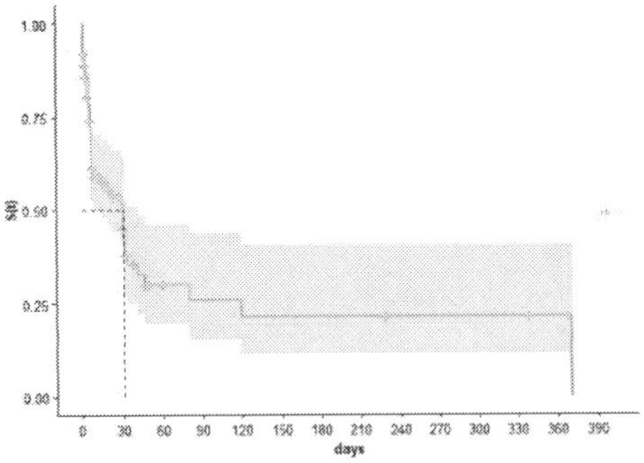

Fig. 3. Kaplan-Meier curve of time to improve mental conditions

The curve of patients with loss of consciousness is above the curve of patients with an anxious attitude, which means that the chance of not experiencing improvement in mental condition is higher in patients with loss of consciousness than patients with an anxious attitude. The curve of patients with non-prayer conditions is above the curve of patients with prayer conditions, which means that the chance of not experiencing improvement in their mental condition is higher in patients who do not prayer compared to patients who prayer.

The curve of patients who have a history of interaction is above the curve of patients who have and do not have a history of interaction, which means that the chances of not experiencing improvement in their mental condition are higher in patients who have a history of interaction than patients who do not have a history of interaction. The curve of patients who experience active therapy reactions is above passive reactions, which means that the chance of not experiencing improvement in mental condition is higher in patients with passive reactions than patients with active reactions.

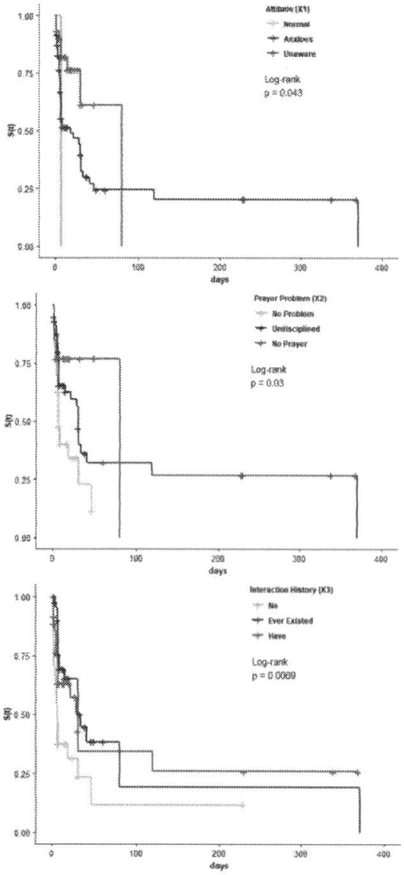

Fig. 4. Kaplan-Meier of attitude, prayer problem, and interaction history

Kaplan-Meier Curve Figure 4 attitude, prayer problem, interaction history and therapy reaction variables intersect between each category, meaning that patients with different categories have different chances of experiencing improvement in their mental condition. The Kaplan-Meier curves for women and men coincide with each other so that there is no difference in the chances of improving mental conditions between the sexes. The curves of marital status, education, main complaint and chronic disease history also coincide with each category so that patients with different categories have the same probability of experiencing an improvement in their mental condition.

The results of the log-rank test for each variable that is thought to affect the time to improve the mental condition of patients with complementary therapy at Nur Hidayah Hospital can be seen in Table 4. attitude, prayer problem, interaction history and therapy reaction variables have different survival curves between each category variable. Variables gender, marital status, education, main complaint and chronic disease history also coincide with each category so that patients with different categories have the same probability of experiencing an improvement in their mental condition. There is no difference in the chances of improving mental conditions between gender.

Table 4. Log-rank test

Variable	Chisq	*P*-value
Attitude (X_1)	6,30	0,040*
Prayer Problem (X_2)	7,00	0,030*
Interaction History (X_3)	9,90	0,007*
Gender (X_4)	0,80	0,400
Marital Status (X_6)	0,70	0,400
Education (X_7)	1,60	0,700
Main Complaint (X_8)	0,01	0,900
Chronic Disease History (X_9)	0,80	0,400
Therapy Reaction (X_{10})	6,20	0,010*

*Significant

3.3.1 Assumption proportional hazard

The Goodness of fit test is used to obtain a more objective decision. Table 5 shows that with a significance level of α of 5%, the predictor variable has a p-value greater than α except for the Interaction History variable. The Interaction History variable gives a decision to Reject H_0 which indicates that this variable does not meet the proportional hazard assumption.

Table 5. Goodness of fit

Variable	Statistic	*P*-value
Attitude (X_1)	4,634	0,0986
Prayer Problem (X_2)	0,685	0,7101
Interaction History (X_3)	8,175	0,0168*
Gender (X_4)	0,149	0,6995
Marital Status (X_6)	0,116	0,7336
Education (X_7)	0,235	0,6280
Main Complaint (X_8)	1,240	0,7434
Chronic Disease History (X_9)	0,316	0,5742
Therapy Reaction (X_{10})	0,669	0,4133

*Significant

There is one variable that does not meet the proportional hazard assumption so that the strata that are formed are based on the number of categories of the interaction history variable, namely three strata are formed. Stratum 1 is not having a history of interaction, stratum 2 ever had a history of interaction and stratum 3 is having a history of interaction.

3.3.2 Extended Cox Modelling

Extended Cox modelling including stratified and time-dependent Cox. Stratum for the modelling used from previous section. There are three stratum so the model was carried out using stratified Cox without interaction and with interaction. Stratified Cox model with interactions using variables that interact with the stratification variables are prayer problem, gender, main complaint, age and treatment reactions while the other variables are included in the model but do not interact with the stratified variables.

Interaction testing is carried out to determine whether there is interaction between the stratified variables and the variables included in the model. The results of the interaction test on the Cox stratified model show that with a degree of freedom of 12, the test statistic value is 31.63 and the p-value is 0.00158. By using a significance level of $\alpha = 0.05$, the p-value $< \alpha$. The decision that can be taken is to reject H_0, there is an interaction between the Interaction History variable and the predictor variable. The conclusion that can be drawn is that the stratified Cox model with interaction is better than the model without interaction.

Stepwise elimination was performed on the Cox stratified model with interactions to obtain the most significant variables from the model. The smallest AIC value of all stepwise stages is in the 3rd stage of 286.84. Variables formed in the stratified Cox model with the smallest AIC are attitude, chronic disease history and interaction variables, namely the interaction history variable with prayer problem, the interaction history variable with gender, the interaction history with age variable, the interaction history with main complaint variable, and the variables interactions history with therapeutic reactions.

Time-dependent variables are formed from variables that do not meet the proportional hazard assumptions that depend on time so that new variables are formed. The model with the time function $g(t)=t$ has a new variable namely Time × Interaction History. The model with the time function $g(t)=ln(t)$ has a new variable, namely ln(Time) × Interaction History. The heaviside function is used to overcome the difference in hazard ratio at different time intervals. The heaviside function used with cut of in 100th day.

Table 6. Model comparison

Model	AIC
Cox Proportional Hazard	414.272
Stratified Cox without Interaction	298.925
Stratified Cox with Interaction	286.839
Time-dependent Time Function	409.332
Time-dependent Heaviside Function	404.856

There are four Cox models formed, so a comparison of accuracy measures is carried out using the AIC value. The best model that show in Table 6 has the smallest AIC value. The smallest AIC value of all models is 286.839, namely in the stratified Cox model with interactions. Based on this, it was found that the best model that can be used to model the improvement of the mental condition of patients with complementary therapy at Nur Hidayah Hospital, Bantul, is the stratified Cox model with Interaction.

3.3.3 Model interpretation

Parameter estimation of the model stratified Cox with interaction shown in Table 7. Hazard ratio of significant variables shown in Table 8. Male patients who do not have a history of interaction have a 21.277 times tendency to improve their mental condition compared to female patients who have a history of interaction. Male patients who have had and still have a history of interaction and female patients who have and do not have a history of interaction have the same tendency as male patients who do not have a history of interaction in experiencing an improvement in their mental condition.

Table 7. Parameter estimation

Variable	Parameter Estimates	*P*-value
Attitude (1)	0,075	0,948
Attitude (2)	-1,636	0,199
Prayer problem (1)	0,154	0,792
Prayer problem (2)	1,057	0,459
Gender (1)	0,014	0,982
Age	0,020	0,384
Main Complaint (1)	0,554	0,482
Chronic Disease History (1)	-0,835	0,131
Therapy Reaction (1)	0,489	0,459
Interaction History (1) × Prayer problem (1)	-1,200	0,152
Interaction History (2) × Prayer problem (1)	-0,053	0,961
Interaction History (1) × Prayer problem (2)	-2,473	0,161
Interaction History (2) × Prayer problem (2)	-0,765	0,675
Interaction History (1)× Gender (1)	0,012	0,989
Interaction History (2)× Gender (1)	-3,056	**0,003**
Interaction History (1) × Age	-0,075	**0,019**
Interaction History (2) × Age	-0,059	0,108
Interaction History (1) × Main Complaint (1)	-0,654	0,518
Interaction History (2) × Main Complaint (1)	-3,699	**0,011**
Interaction History (1) × Therapy Reaction (1)	-1,094	0,246
Interaction History (2) × Therapy Reaction (1)	-1,902	0,100

Patients with mental complaints who do not have a history of interaction have a 40 times tendency to improve their mental condition compared to patients with physical complaints who have a history of interaction. Patients with mental complaints who have had and still have a history of interaction and patients with physical complaints who have had and do not have a history of interaction have the same tendency towards patients with mental complaints who do not have a history of interaction in experiencing improvement in their mental condition.

Table 8. Hazard ratio estimates

Variable	Hazard Ratio Estimates (95% CI)
Interaction History (2) × Gender (1)	0,047(0,006 ;0,358)
Interaction History (1) × Age	0,928(0,872 ;0,988)
Interaction History (2) × Main Complaint (1)	0,025(0,001 ;0,431)

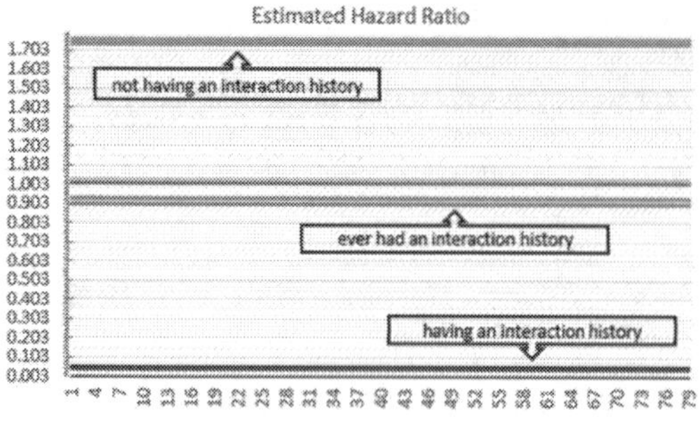

Fig. 5. Estimated hazard ratio for each category in interaction history variable

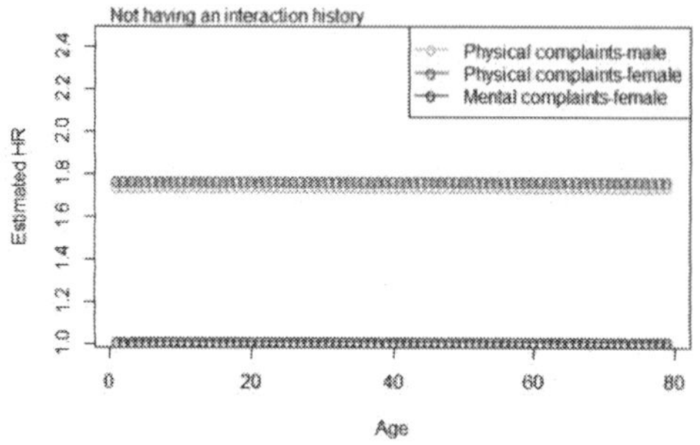

Fig. 6. Estimated hazard ratio for not having an interaction history

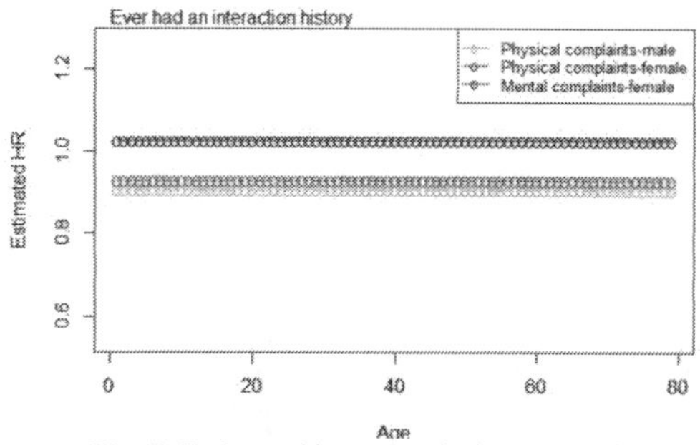

Fig. 7. Estimated hazard ratio for ever had an interaction history

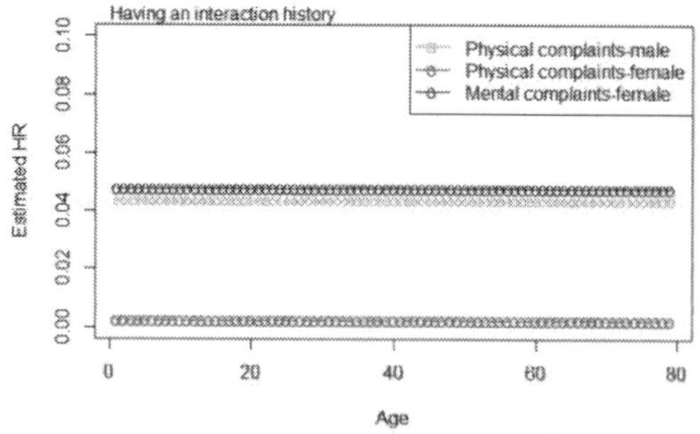

Fig. 8. Estimated hazard ratio for having an interaction history

Based on Figure 5, the increasing tendency of patients to experience improvement in mental condition is indicated by the lower the value of the interaction history variable where the hazard ratio value at stratum 1 (not having an interaction history) is the highest. The red line in Figure 6, Figure 7, Figure 8 shows that female patients with a main physical complaint have a tendency to improve their mental condition which varies according to interaction history variable and experience an increasing tendency to improve when the interaction history value is lower, the value is lowest when the patient has an Interaction History.

This best model already considering the time needed for the patient to improve mental health condition so that it can explain properly the factors that affect the patient's time mental improvement sooner or later. Factors that did not have a significant effect were attitude, prayer problem, chronic disease history, and therapy reactions. Based on these conclusions obtained some suggestions as follows. Future research can add medical variables where these variables have definite measurements. The hospital can record complementary therapy medical records in a more integrated manner with other medical records so that patient monitoring can be more effective. The therapist can pay attention to the factors that influence the time to improve mental conditions so that they can provide education to patients as an effort to make improvements better.

4 Conclusion

Attitude, prayer problem, interaction history and therapy reaction variables have different affect the time to improve the mental condition. The best model for modelling the improvement of the mental condition of patients with complementary therapy at Nur Hidayah Hospital is using stratified Cox regression with interaction. Factors that significantly influence the time of improvement of the mental condition of complementary therapy patients are age for the patients that ever had the interaction, gender and main complaint for the patients that have interaction history. The increasing tendency of patients to experience improvement in mental condition is indicated by interaction history variable. Lowest estimated hazard ratio when the patient has an interaction history, physical complaint and female. Male patient with physical complaint has the lowest estimated hazard ratio when he ever had an interaction history.

Acknowledgements

The research was funded by Directorate of Research, Technology and Community Service (Ministry of Education, Culture, Research and Technology) based on Master Contract for the Implementation of the State University Operational Assistance Program (Collaborative Research - Domestic) with Master Contract Number: 112/E5/PG.02.00. PL/2023 and Research Contract Number: 1967/PKS/ITS/2023.

.

References

1. Kementerian Kesehatan Republik Indonesia, Laporan Nasional Riskesdas 2018, Lembaga Penerbit Badan Penelitian dan Pengembangan Kesehatan (LPB), Jakarta, Indonesia (2019)
2. R. de Diego-Cordero, P. Suarez-Reina, B. Badanta, G. Lucchetti and J. Vega-Escano, Applied Nursing Research, **67**, (2022)
3. P. M. Barnes, B. Bloom, R. L. Nahin, National Health Statistics Reports, 12, (2008)
4. M. Mahjoob, J. Nejati, A. Hosseini, N. M. Bakhshani, J Relig Health, **55**, 38-42, (2014)
5. S. A. Alshammary, B. Duraisamy, F. al-Odeh, M. S. Bashir, M. R. Wedad Salah Alharbi, A. Altamemi, International Journal of Research Studies in Medical and Health Sciences (IJRSMHS), **3**, 1-5, (2018)
6. R. Sawatzky and B. Pesut, *Journal of holistic nursing: Official journal of the American Holistic Nurses' Association,* **23**, 19-33, (2005)
7. M. H. Aljawadi, A. T. Khoja, A. D. AlOtaibi, K. T. Alharbi, M. A. Alodayni, M. S. AlMetwazi, A. Arafah, S. A. Al-Shammari, T. A. Khoja, Evidence-Based Complementary and Alternative Medicine, **2020**, 1-14, (2020)
8. B. Jabbari, M. Mirghafourvand, F. Sehhatie and S. Mohammad-Alizadeh-Charandabi, J Relig Health, **59**, 544-554, (2020)
9. L. Breiman, Machine Learning, **45**, 5–32, (2001)
10. D. G. Kleinbaum and M. Klein, Survival Analysis A Self-Learning Text Third ed., (Springer, London, 2012)
11. D. Collett, Modelling Survival Data in Medical Research Third ed., (Chapman and Hall/CRC, London, 2015)

Learning Method Recommendation Based on VARK Model Using Certainty Factor Algorithm

Izzu Zantya Fawwas[1], *Casi* Setianingsih[1*], *Fussy* Mentari Dirgantara[1], *Ari* Cahya Saputra[1], *Ariana* Novanti[1], *Muhammad* Izzudin Islam[1], *Agustio*[1], and *Yusuf* Sulle[1]

[1]School of Electrical Engineering, Telkom University, Bandung, Indonesia

Abstract. In lecture activities, students are required to master several courses that have been determined based on their respective majors. In the learning process, students often have difficulty understanding lecture material. One factor is the mismatch between how students learn and the type of learning style of each student. It is important for each student to know their respective learning styles so that in the learning process can understand the material to the fullest. One way to find out the type of student learning style is with VARK modalities (Visual, Auditory, Read/Write, and Kinaesthetic). The VARK model classifies learning style types into four types. Everyone must have all four types of learning styles, but there must be one of the most dominant. By knowing the type of learning style, students can determine how to learn according to the type of learning style. This recommendation system is implemented using Certainty Factor algorithms involving the expertise of a psychologist in it, this system is built in the website platform. The system achieves an accuracy of 94.52%, so it is good enough to provide recommendations on how to learn properly for users.

Keywords: Learning Styles, VARK, Learning Methods, Certainty Factor, Educational Psychology

1 Introduction

In lecture activities, students are required to master several courses that have been determined based on their respective majors. In the learning process, students often have difficulty understanding the lecture material so that they get less than maximum results. Students' difficulty in learning lecture materials is caused by many factors, one of which is a mismatch between how to learn and the type of learning style of each student. It is important for each student to know their respective learning styles so that in the learning process can understand the material to the fullest.

The problem formulation in this Final Task is how to help students know how to learn according to their type of learning style and how certainty factor algorithms perform on a web-based learning recommendation system. The objective of this Final Task is to design and implement a system of recommendations on how to learn in a web platform using certainty factor algorithms, as well as testing the recommendation system of web-based learning methods on respondents to find out the accuracy of the system. The methods used in this final task are literature studies, consultation with guidance lecturers, consultation with experts. System design, system implementation, and system testing.

2 Related Work

Research on the identification of learning style types using Visual, Auditory, Read/Write, and Kinaesthetic (VARK) models has previously been conducted whose research results are in the form of decision tables that state the relationship between the type of learning style of the VARK model and its characteristics. But in the study, the system created has not been based online [1]. In this Final Task, designed and built a system of recommendations for how to learn based on the VARK model using the web-based Certainty Factor algorithm.

There is another study that talks about identifying learning style patterns. The expert system for identifying patterns of children's learning styles is an expert system designed as a tool for parents to identify patterns of children's learning styles. The VARK method (Visual, Auditory, Read/write, Kinaesthetic) is used to facilitate system performance in making conclusions. This expert system will display several questions as indicators of the characteristics of the child's learning style that are felt, then later arrive at the final question. The result of this study using the certainty factor method will be to show the characteristics of the children's learning styles. With the obtained from the visual learning style (0.144) with symptoms of MB (0.2) and MD (0.1) ; auditory learning style (0.28) with symptoms of watch videos and movies at MB (0.6) and MD (0,1); read / write learning style

* Corresponding author: casisetianingsih@telkomuniversity.ac.id

(0.31) with symptoms, material read study and summarize the material MB (0.2) and MD (0,1); and kinaesthetic learning style (0.38) with can move symptoms, the body (a) is positive (MB) (0.8) and MD (0.3) [2].

Utilization of Certainty Factor algorithm was done by a paper for Cirrhosis detecting purposes. A system for detecting a disease surely needs a core that utilizes the expertise of a doctor that has related knowledge about the disease. This paper describes the process for detecting liver disease based on a combination of symptom weights from the doctor and the user. The main system flow of this final project is inspired by the main system flow of this paper. Accurate and precise calculations are needed to diagnose symptoms so that they can infer the output using the Certainty Factor (CF). The result is this application be able to diagnose of Cirrhosis with 100% accuracy achieved [3].

Another implementation of Certainty Factor algorithm was written in a study about identification of dog disease. These expert systems can give information about diseases of dogs based on symptoms they have. Five options are given to answer the calculation question using each method: no, quite sure, sure enough, certain, and certainty sure. Accuracy Analysis of each method is tested by assessing the results of each analysis method based on user feedback. The purpose of this study is to implement the Certainty Factor method in the diagnosis system of canine diseases that can provide space in providing value confidence in knowledge. The result is each of the rules achieved more than 70% accuracy [4].

3 Research Methods

This section discusses the research methods used in this study.

3.1 Learning Style Type

Learning style is the way a person chooses to use his abilities (Santrock, 2010). Keefe states that a person's learning style influences the way he or she learns. A person must tend to use ways that learn that match his learning style in the hope that it will be easier to understand the material he learns. Everyone will feel a different easy learning style. According to Hamzah that whatever one's learning style, it is the fastest and best way for every individual to be able to absorb and process information from outside himself [5]. The type of learning style can be known as one of them with the VARK model. The VARK model is a simple instrument in determining one's preferences in receiving information. This model was put forward by Neil Fleming and Collen Mills in 2006 [6]. The authors of this model (Fleming and Mills) identified four main learning styles: Visuals have a preference for graphics, tables, charts as verbal representations rather than many words; Auditory is characterized by a preference for hearing information in the form of, audio recordings, conversations or exchanging opinions; Read/Write characterizes people who prefer information in written form (books, articles) and they use notes in various forms; Kinaesthetic prefers examples of material taught to look at relationships with real examples and have a tendency to experiment [7].

3.2 Certainty Factor Algorithm

Certainty Factor (CF) is a measure of belief in a fact / event based on existing evidence or from the consideration of an expert [8]. Certainty Factor was introduced by Shortliffe Buchanan [9]. Certainty Factor algorithms are useful in providing solutions by measuring how confident an analysis is in a case. This algorithm is often used in studies that use surveys about symptoms/indications that aim to diagnose something uncertain [10].

Here are the Certainty Factor equations for some conditions [3]:

3.2.1 CF for rules with a single premise:

$$CF_{combination}\,(H,E)=CF\,(user)*CF\,(expert) \quad (1)$$

$CF_{combination}$ (H, E) is a measure that quantifies the degree of certainty associated with a hypothesis based on the available evidence. It provides an assessment of how confident one can be in a particular hypothesis given the evidence at hand. CF (user) denoting the certainty factor or belief level assigned to a hypothesis based on the evidence provided by a user, offers insight into the user's perspective on the hypothesis's credibility. This factor encapsulates the user's judgment and interpretation of the evidence, shedding light on their level of conviction regarding the hypothesis. CF (expert) representing the certainty factor or belief level attributed to a hypothesis based on the evidence evaluated from an expert's viewpoint, offers a distinct perspective. This factor encapsulates the expert's assessment and interpretation of the evidence, presenting their degree of confidence or belief in the hypothesis based on their expertise and knowledge in the relevant field.

3.2.2 CF for rules that lead to the same conclusion: applies if CF [old] and CF [new] are positive.

$$CF_{final}\,[H,E]=CF[old]+CF[new]*(1-CF[old]) \quad (2)$$

CF_{final} (H, E) is a certainty factor based on multiple rules with a single premise. CF (old) is a certainty factor / confidence rule level in previous iterations. CF (new) is a certainty factor/ belief rule level in the current iteration.

3.3 Test Design

In this system, there are 16 questions used to detect the type of user's learning style. Table 1 below lists the questions to be used that come from previous research [1]. Code Q01 means question number 1 and so on. quations should be centred and should be numbered with the number on the right-hand side.

Table 1. Questionnaire

Code	Question
Q01	You want to help someone to go to the airport, to the city center, and to the train station, what are you going to do?
Q02	On the web there is a video showing how to create a good graphic, there are people talking, there are some to-do lists, and some diagrams. What are you supposed to do?
Q03	You're planning a vacation with friends, you want friends to respond to your plans, what are you going to do?
Q04	You want to cook something for someone, what are you going to do?
Q05	A group of tourists wants to learn about life in your area, what are you going to do?
Q06	You want to buy a digital camera or mobile phone at various prices, what decision will you take?
Q07	Do you remember how you learned something new? Avoid direct movements such as riding a bike, who will you learn from?
Q08	You have a problem with your feelings, what does the doctor want to help you with?
Q09	You want to learn new programs, skills or games on the computer. What are you going to do?
Q10	Do I like sites that have?
Q11	If you look at the price, what decision will influence you to buy a nonfiction book?
Q12	You use a book, CD, and site to learn to take photos with a new digital camera. What do you want to ask?
Q13	You choose to be a teacher or to be a host. You're going to see from whom?
Q14	You've finished taking a championship or test that wants the results. What result do you want?
Q15	You will choose food in a restaurant or café. What are you going to do?
Q16	You'll make an important speech at a conference or job interview. What are you going to do?

In Table 2 there will be as many as 64 features of learning styles categorized into VARK models equally. These features are designed to be the answer choice of questions mentioned in Table 1. The "Expert CF" column will contain the Certainty Factor (CF) value of a psychologist. This expert's CF value means that the expert (psychologist) believes as big as that value that the characteristics of the learning style mentioned indicate into the type in question. The value of the Expert CF is at a range ranging from -1 (not sure absolute) to 1 (absolute sure). If the value of Expert CF is getting closer to 1 then the higher the level of expert confidence and vice versa. Table 2 here presents a list of learning style traits [1]. Expert CF, whose source is a psychologist, assigns weight to each learning style attribute. Code V1 denotes a visual response for question number 1, A1 a verbal response for question number 1, R1 a read-and-write response for question number 1, K1 a kinesthetic response for question number 1, and so on.

Table 2. Learning Style Type Traits

Code	Characteristics of Learning Style	Type	Expert CF
V1	Draw or show or provide a map	Visual	1
A1	Explain direction orally	Auditory	0.95
R1	Write directions	Read/write	1
K1	Go with him	Kinesthetic	1
V2	Just look at the diagram	Visual	1
A2	Listen	Auditory	1
R2	Read sentences only	Read/write	1
K2	Watch it	Kinesthetic	1
V3	Use maps to see them beautiful places	Visual	0.9
A3	Call, text or email them	Auditory	0.89
R3	Give them a brochure about the place.	Read/write	1
K3	Explain some of the outlines they will experience	Kinesthetic	1
V4	Look on the internet and cookbooks	Visual	1
A4	Asking a friend for advice	Auditory	0.95
R4	Using recipe help	Read/write	1
K4	Cooking something you know without a recipe	Kinesthetic	1
V5	Showing maps and pictures on the internet	Visual	1
A5	Talk and compile information about it	Auditory	1
R5	Provide a guidebook on the life of the area	Read/write	1
K5	Get them to jump in there.	Kinesthetic	1
V6	See whether the design	Visual	1
A6	Listen to the explanation from the seller	Auditory	1
R6	Read in detail or check over the internet	Read/write	1
K6	Try or check first	Kinesthetic	1
V7	From a diagram, map, or chart to see the instructions	Visual	1
A7	Listen to someone's	Auditory	1

Code	Characteristics of Learning Style	Type	Expert CF
	explanation and ask		
R7	View from manuals or manuals	Read/write	1
K7	Just watch	Kinesthetic	0.8
V8	Shows which diagrams or parts are wrong	Visual	1
A8	Explain where the error lies	Auditory	1
R8	Give something to read and explain what's wrong.	Read/write	1
K8	Use a tool and show what's wrong	Kinesthetic	1
V9	Follow the instructions from the book	Visual	1
A9	Talk to people who know about the program.	Auditory	1
R9	Read the instructions	Read/write	1
K9	Using the internet	Kinesthetic	1
V10	With an attractive design and parts.	Visual	1
A10	From the web who can listen to music, from the radio or interviews	Auditory	1
R10	Very interesting explanation of the list and description	Read/write	0.9
K10	Something that is easy to open and try	Kinesthetic	1
V11	From his appearance	Visual	1
A11	From a friend who recommends	Auditory	1
R11	From the sections that are easy to read	Read/write	1
K11	From true stories, experiences, and examples	Kinesthetic	1
V12	Direct instructions from the camera and its parts	Visual	1
A12	Opportunity to ask questions and talk about the pictures on camera	Auditory	1
R12	You read and write from instructions about what you're going to do.	Read/write	1
K12	Examples of good photos	Kinesthetic	1
V13	From diagrams, charts, or graphs	Visual	0.9
A13	Or will do a Q&A, talk, or from a group discussion, or from a guest.	Auditory	1
R13	You will immediately read the book.	Read/write	1
K13	Learn directly	Kinesthetic	1
V14	View from the chart of the results you have received	Visual	1
A14	From someone who took the exam with you	Auditory	1
R14	Using an explanation of the results	Read/write	1
K14	Using examples of what you've done	Kinesthetic	1
V15	See others eat what or see menu pictures	Visual	1
A15	Listen to what your waiter and friend have to offer	Auditory	1
R15	Choose from the menu	Read/write	1
K15	Choose something you've eaten before	Kinesthetic	1
V16	Create diagrams and graphs that will help explain something	Visual	1
A16	Writing important points and memorizing	Auditory	0.9
R16	Rewrite and read it repeatedly	Read/write	1
K16	Collect examples and stories for easy presentation	Kinesthetic	1

The list of learning style characteristics will be mapped to the corresponding questions in Table 1. Each question will have 4 answer options, each answer choice indicates one of 4 types of learning styles (Visual, Auditory, Read/Write, and Kinesthetic). Users can choose more than one answer to each question. In determining the final CF value of each type, a certainty factor (CF) value is needed from the user which will be used to measure the CF combination of each learning style feature; therefore, the user will fill in the User's CF on each answer they choose. Table 3 shows the CF user selection.

Table 3. User CF Selection

Label	**Value**
Not	0
Less sure	0.2
Pretty sure	0.4
Believe	0.6
Very Confident	0.8

After the user answers the questions displayed, data will obtain a pair of data of the selected learning style traits along with the User CF value. The data will be grouped by VARK model type, so there are as many as 16 data entries per type. Based on the formula 1 each User CF is multiplied by the corresponding Expert CF resulting in a CF combination. Based on CF, the combination obtained uses formula 2 to calculate the magnitude of the final CF of each type of learning style. After getting the final CF value of each type of learning style (Visual, Auditory, Read / Write, and Kinesthetic), of the four types of learning styles will be selected the highest CF Final as the best type of learning style. Figure 1 is a diagram of how to determine the best type of learning style.

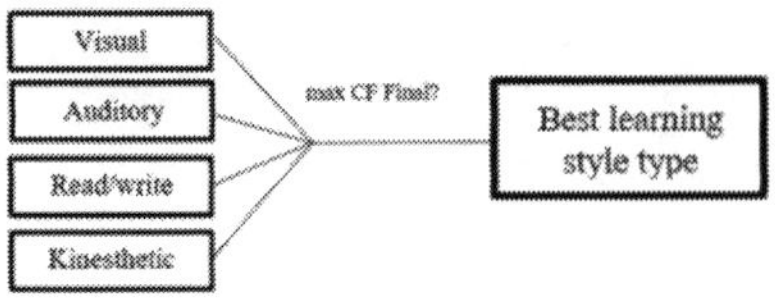

Fig. 1. Best Learning Type Determination

Based on the best type of learning style, users will be given recommendations on several appropriate ways of learning. Table 4 below is a list of ways to learn that will be recommended to users.

Table 4. Learning Method Recomendations

No.	Learning Style Type			
	V	A	R	K
1	Read diagrams , mind maps, charts	Debate	Reading a book	Learn by using real-life examples
2	Convert writing into diagrams , mind maps, charts	Discussion	Noted	Demonstration
3	Colored writing	Listen to podcasts	Summarize	Physical activity
4	Using different fonts	Learn while listening to the back sound	Paraphrases	Teaching others
5	Notes with a good layout	Seminar/ webinar	Describes graphics into writing	Trial and error
6	Notes with attractive designs	Audio book	Using headings and lists	Study outdoors
7	Animation	Story	Create a glossary	Ask someone else
8	Illustration images	Read aloud	Print digital learning materials	Designing something new

3.4 Web Planning

This system is created by having features, namely: registration, log in, fill out questionnaires, edit accounts, and delete accounts. The user interface built into this system is a landing page, an account creates page, an account sign-in page, a profile page, a profile edit page, a questionnaire content page, a results page.

In the world of software development, known various kinds of life cycles such as waterfall, agile, v-model etc. Each life cycle has its own advantages and disadvantages. The waterfall life cycle is suitable for software development whose specifications are known for sure. Agile life cycle is suitable for software development whose specifications often change over time and products must be completed in a short time. V-Models are suitable for software development whose specifications change frequently and involve testers in the early phases of development. In this Final Task, the system made has clear targets and specifications and most likely there are no frequent specification changes, therefore the author will use waterfall life cycle in the development of this system [11].

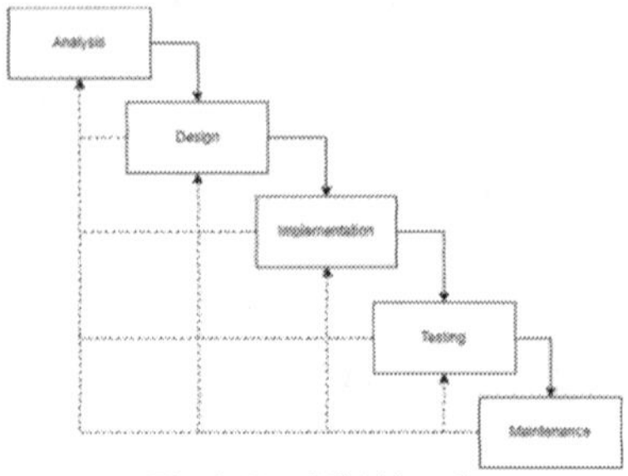

Fig. 2 Watefall Lifecycle

Waterfall model is a sequential software development process, which every process cannot do if the previous process has not been completed except for the first process, because each process requires inputs where the input is the output of the previous process. Figure above shows the stages of software development using waterfall life cycle. Waterfall consists of 5 stages, namely:

3.4.1 Analysis

The analysis stage is the stage of determining the specifications of the system to be built, both functional and non-functional specifications. Functional specifications are what features the user will use to interact with the system. Usually, functional specifications are defined on *a use case* diagram. Examples of functional specifications are features, *user interface* specifications, database specifications, etc. Non-functional specifications are specifications imposed on system design and operation. Examples of non-functional specifications are scalability, performance, quality standards, etc [12].

3.4.2 Design

The design stage is the stage of designing a system so that the system can solve the problem that is being faced based on the results of the analysis stage. The specifications that have been determined at the analysis stage are translated into system design diagrams. At this stage there is the process of designing algorithms, software architecture, *user interface design*, etc [12]. Figure 3 below is a flowchart diagram of the system.

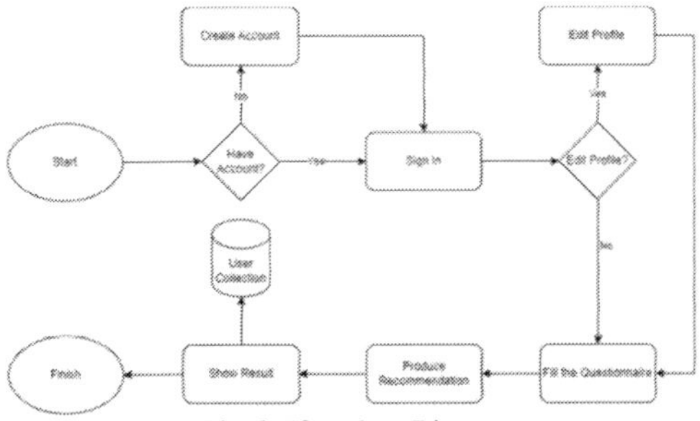

Fig. 3. Flowchart Diagram

3.4.3 Implementation

The implementation stage is the stage of realizing the design of the system that has been made at the design stage. The specifications that were in the system design, at this stage, were realized into features that can be used. The process in realizing system design is by coding and deployment process [12].

The programming language used to develop the interface is HTML (HyperText Markup Language) with the assistance of the React.js framework. Styling is done using CSS (Cascading Style Sheet) with the help of the Tailwind CSS framework. Meanwhile, on the backend side, the programming languages used are JavaScript and Nest.js. Data storage in this application uses a NoSQL type database, which allows storing data with different structures on the same collection so that it is more flexible. The backend and frontend sides of this application are stored on different servers. The backend side is stored on the Heroku.com platform, while the frontend side is stored on the Vercel.com platform. The two-exchange data using the API (Application Programming Interface) using the JSON format. This how-to learns recommendation application can be accessed at https://presisi.vercel.app

3.4.4 Testing

At this stage the system can be used but has not been tested whether it has met the goals set at the beginning and has met the specifications that have been predetermined. In the testing phase, the verification and validation process are carried out. Verification is the process of evaluating whether the system has fulfilled the purpose this system was created. Validation is the process of evaluating whether the system meets predefined specifications [12].

$$Accuracy = \frac{Amount\ of\ appropriate}{Amount\ of\ respondents}\ x\ 100\% \quad (3)$$

The respondents were university students, aged 18-22 years old, who had never done a test on how to study that suited them from a psychological perspective.

3.4.5 Maintenance

The treatment stage is an advanced process after the system has gone through the testing process. There are various processes at this stage such as improving system quality, adding new features, updating system components to newer versions, etc [12].

4 Result and Analysis

This chapter describes the results and discussion of the tests that have been carried out to determine the success of the system.

4.1 Web Implementation

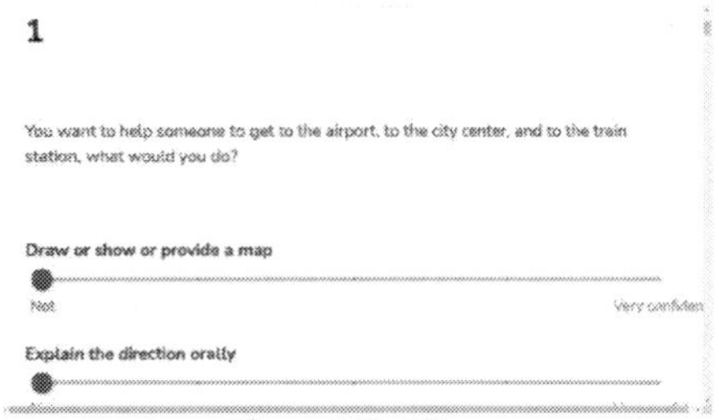

Figure 4. Test Page

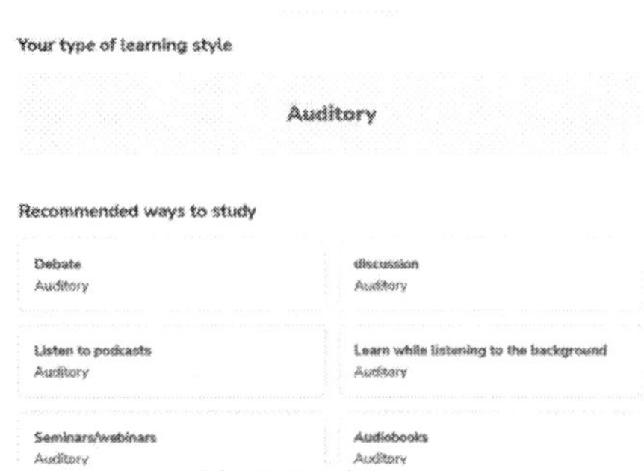

Fig. 5. Result page

The two figures above are the user interface of the main features of this application, namely the test page and the results page.

4.2 Accuracy of the Algorithms

Table 5. Result Validation

No.	Results (Application)	Results (Expert)	Conclusion
1	V	V	Appropriate
2	K	K	Appropriate
3	A	A	Appropriate
4	K	A	Not appropriate
5	A	A	Appropriate
6	K	K	Appropriate
7	K	K	Appropriate
8	A	A	Appropriate
9	A	A	Appropriate
10	K	K	Appropriate
11	A	A	Appropriate
12	K	K	Appropriate
13	K	K	Appropriate
14	V	V	Appropriate
15	K	K	Appropriate
16	A	A	Appropriate
17	A	A	Appropriate
18	K	K	Appropriate
19	K	K	Appropriate
20	V	V	Appropriate
21	K	A	Not appropriate
22	K	A	Not appropriate
23	K	ARK	Appropriate
24	A	A	Appropriate
25	R	R	Appropriate
26	K	K	Appropriate
27	K	K	Appropriate
28	K	K	Appropriate
29	K	K	Appropriate
30	R	R	Appropriate
31	K	K	Appropriate
32	K	K	Appropriate
33	A	A	Appropriate
34	A	A	Appropriate
35	K	K	Appropriate
36	V	V	Appropriate
37	K	K	Appropriate
38	R	R	Appropriate
39	A	A	Appropriate
40	K	K	Appropriate
41	A	A	Appropriate
42	V	V	Appropriate
43	K	K	Appropriate
44	V	VA	Appropriate
45	A	A	Appropriate
46	V	V	Appropriate
47	K	AK	Appropriate
48	V	VK	Appropriate
49	K	K	Appropriate
50	V	A	Not appropriate
51	V	V	Appropriate
52	R	AR	Appropriate
53	R	R	Appropriate
54	K	AK	Appropriate
55	K	K	Appropriate
56	K	K	Appropriate
57	A	A	Appropriate
58	K	AK	Appropriate
59	V	V	Appropriate
60	A	A	Appropriate
61	A	A	Appropriate
62	A	A	Appropriate
63	K	ARK	Appropriate
64	K	K	Appropriate
65	A	AR	Appropriate
66	K	K	Appropriate
67	A	A	Appropriate
68	K	K	Appropriate
69	K	K	Appropriate
70	R	R	Appropriate
71	K	K	Appropriate
72	A	A	Appropriate
73	K	AK	Appropriate

Data from the validation with psychologists is shown in Table 5. In total, there were 73 respondents involved in the study. The accuracy value, derived from formula 3, is 94.52%, and it is visually represented in Figure 6.

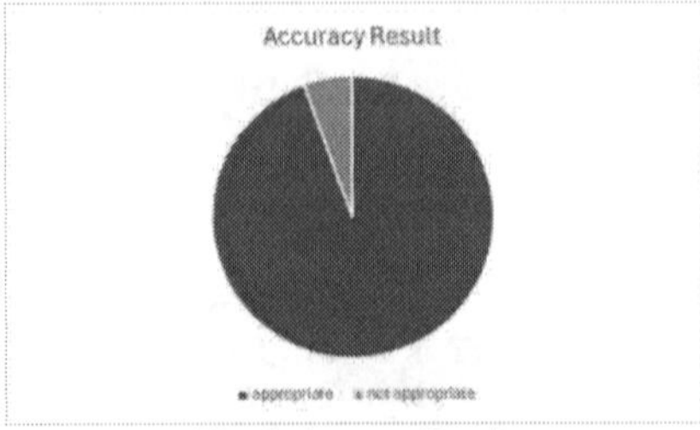

Fig. 6. Acuracy Result

$$Accuracy = \frac{69}{73}\, x\ 100\% = 94.52\%$$

5 Conclusion

A Certainty Factor algorithm implemented in this system achieves accuracy of 94.52% and has been validated by psychologists, so this system is quite good at providing recommendations for how to learn that is suitable for students based on the best learning style type.

References

1. E. D. S. Mulyani, Y. H. Agustin and I. Nur'aeni, *Aplikasi Pakar Untuk Mengidentifikasi Karakteristik Gaya Belajar Dengan Menerapkan Modalitas Vark*, Stmik Tasikmalaya, Tasikmalaya. (2018)

2. L. Marlinda, D. Saputra and W. Indrarti, *Expert System Identification Of Learning Patterns The VARK Method With Certainy Factor*, Journal Publications & Informatics Engineering Research, **3** (2019)
3. L. Safira, B. Irawan and C. Setianingsih, *Implementation of the Certainty Factor Method for Early Detection of Cirrhosis Based on Android*, in International Conference on Electronics Representation and Algorithm (2019)
4. L. Marlinda, Widiyawati, W. Indrarti and R. Widiastuti, *Dog Disease Expert System Using Certainty Factor Method*, Jurnal dan Penelitian Teknik Informatika, **4** (2020)
5. N. H. Jeanete Ophilia Papilaya, *IDENTIFIKASI BELAJAR MAHASISWA*, Jurnal Psikologi Undip, **15** (2016)
6. J. L. Espinoza-Poves, W. A. Miranda-Vílchez and R. Chafloque-Céspedes, *The Vark Learning Styles among University Students of Business Schools, Propósitos y Representaciones,* **7** (2019)
7. Faculty of Theology and Social Science, Cernica, *THE VARK MODEL INVESTIGATED AT THE STUDENTS FROM PPPE*, JOURNAL OF EDUCATION STUDIES, **1** (2019)
8. M. Raditya, Fauziah and E. T. Winarsih, *Expert System Testing To Recommend Diabetes Mellitus Using Web-Based Certainty Factor Method,* Jurnal Mantik, **3** (2020)
9. A. S. Sembiring, Sulindawaty, O. Manahan, M. Helentina, P. S. Hasugian, F. Riandari, R. Mahdalena, A. Simangunsong, Y. Utami dan H. Tamando, *Implementation of Certainty Factor Method for Expert System*, in The International Conference on Computer Science and Applied Mathematic, (2019)
10. I. Sumartono, D. Arisandi, A. P. U. Siahaan and M. Aan, *Expert System of Catfish Disease Determinant Using Certainty Factor Method,* International Journal of Recent Trends in Engineering & Research, **3** (2017)
11. S.Balaji and D. Murugaiyan, *Wateerfallvs V-Model Vs Agile: A Comparative Study On Sdlc,* International Journal of Information Technology and Business Management,**2** (2012)

N-Grams Modeling for Protein Secondary Structure Prediction: Exploring Local Features and Optimal CNN Parameters

Annisa Rizqiana[1] and *Afiahayati*[2*]

[1]Master Program of Computer Science, Department of Computer Science and Electronics, Universitas Gadjah Mada, Indonesia
[2]Department of Computer Science and Electronics, Universitas Gadjah Mada, Indonesia

Abstract. This study explores the potential of n-gram modeling in protein secondary structure prediction. Experiments are conducted on three datasets using bigrams, trigrams, and a combination of the best n-grams with PSSM profiles. Optimal parameters for Convolutional Neural Networks (CNNs) are investigated. Results indicate that bigrams outperform trigrams in Q8 accuracy. Adding another feature, that is, PSSM, can improve model performance. Deeper convolution layers and longer convolution sizes enhance accuracy. Both bigrams and trigrams demonstrate similar performance trends, with bigrams slightly more effective. The study offers insights into local feature extraction, which is n-grams for protein modeling. These findings contribute to protein structure analysis and bioinformatics advancements, facilitating improved protein function prediction.

Keywords: Protein secondary structure prediction, n-grams, sequence labeling, convolutional neural network, bioinformatics.

1 Introduction

Proteins are essential macromolecules that play crucial roles in the functioning of living organisms. Their diverse functions include catalyzing biochemical reactions, structural elements, aiding signal transmission, and regulating gene expressions. The specific role of a protein largely depends on its unique three-dimensional structure, which is determined by the linear sequence of amino acids in its polypeptide chain [1].

One of the critical aspects of protein structure is its secondary structure, which refers to the local spatial arrangement of amino acids in the protein chain. The primary types of secondary structures include alpha helices, beta sheets, and random coils or loops. Understanding protein secondary structure is vital because it provides insights into its stability, folding, and function [2]. Wherever the 3-state is extended to an 8-state, that could provide more detailed local structure information, this method is the most often used in PSSP.

Initially, experimental techniques like X-ray crystallography and nuclear magnetic resonance (NMR) spectroscopy were used to determine protein structures. While these methods provide highly accurate results, they are laborious, time-consuming, and often require purified protein samples, which can be challenging to obtain [3]. In response to these limitations, bioinformatics has played a vital role in developing computational methods for predicting secondary structures of proteins from their amino acid sequences.

Protein secondary structure prediction (PSSP) is a fundamental task in protein science and computational biology, and it can be used to understand protein 3-dimensional (3-D) structures and, further, to learn their biological functions. This 3-d protein structure can be obtained using cryo-EM methods [4]. In PSSP, each sequence element will be predicted and produced a secondary structure label for each amino acid position in the sequence, known as the sequence labeling task. In the past decade, many methods have been propose for PSSP [2]. Some computational techniques have been use to predict protein structures, like machine learning and neural networks. A standard method used for sequence labeling is Convolutional Neural Network (CNN). CNN is one of the neural networks approach that has emerged as a powerful tool in various image and sequence analysis tasks, including natural language processing. Their ability to automatically learn hierarchical patterns and feature representations from data makes them well-suited for protein secondary structure prediction. CNN is often used in various techniques for PSSP tasks, such as DeepPrime2Sec [1], Generative Stochastic Network [2], multi-input [3], and MUST-CNN [4]. MUST-CNN solves the problem of reducing the length of protein sequences by carrying out the shift and stitch technique, where predicting protein structure requires the same input and output length.

Other than that, protein features also contribute to increasing accuracy. Various protein features like PSSM profiles and physical properties [5] have been trying to improve the study's accuracy. Some embedding methods are also applied to represent amino acid sequences in different formats, ProtVec and ELMo Embedding [1].

* Corresponding author: afia@ugm.ac.id

This study will use the n-grams approach [10] to model amino acid sequences. N-grams are usually used in natural language processing tasks. In protein modeling, n-grams play a significant role in capturing short-range sequence patterns. By utilizing n-grams, which are contiguous amino acid sub-sequences, predictive models can identify recurring motifs that contribute to specific structural elements. In protein secondary structure prediction, n-grams refer to a sequence of *n* amino acids that are close together in a protein sequence, where n-grams can help identify patterns or motifs related to a particular secondary structure. However, choosing the correct value of n is critical to getting good results.

We propose a Convolutional Neural Network with n-grams protein modeling to predict secondary structure prediction for Q8 accuracy. We use CNN for PSSP because we saw this task as a sequence labeling task, and many previous studies also used CNN for their method for the PSSP task, as described above. N-grams modeling applies to CullPDB and CB513 datasets, whose *n* values are 2 and 3, commonly referred to as bigrams and trigrams. We investigate the effect of n-grams modeling on secondary structure prediction performance using metrics evaluation, that is, Q8 accuracy, precision, and recall.

This paper is organized as follows: Section 1 presents the introduction and motivation for this work, while Section 2 presents related work. Section 3 offers n-grams dataset modeling. Methods and materials are described in section 4. In section 5, we will discuss the result of the study. And section 6 concludes this paper with possible future work.

2 Related Work

On the other hand, deep learning has shown great promise in representation learning, allowing the discovery of practical features and their mappings directly from data, thereby overcoming the limitations of hand-designed features. In particular, convolutional neural networks (CNNs) have demonstrated remarkable success in image recognition and are being increasingly explored for various bioinformatics applications, including protein secondary structure prediction. By leveraging the hierarchical representations learned from data, CNNs offer the potential to better understand the local features and patterns in protein sequences, thus providing a promising avenue for enhancing the accuracy and performance of protein secondary structure prediction models [11].

Convolutional Neural Networks (CNNs) use in protein secondary structure prediction has been widely studied, with researchers employing different approaches. Among them, exploring local features using n-gram modeling has shown promise. [6] utilized CNN as a classifier for predicting protein secondary structure and compared its performance with PSSM protein profile features, along with a combination of PSSM and features extracted using a Generative Confrontation Network (GCN). Notably, GCN-based protein feature extraction significantly impacted the study.

[7] proposed a combination of CNN and SVM for protein secondary structure prediction using the CullPDB and CB513 datasets. They focused on the Shift and Stitch CNN architecture, which preserved the protein sequence length during training and output. The input dataset included orthogonal input profiles and PSSM profiles with a size of 42. [3] also employed the Shift and Stitch CNN approach, known as MUST-CNN (Multi-input Shift and Stitch CNN), for protein secondary structure prediction using one-hot encoding of protein sequences. The MUST-CNN model utilized 1-dimensional convolutions to preserve an output size equal to the input size through shift and stitch techniques, resulting in improved prediction accuracy.

Moreover, [1][3] investigated biophysical features of amino acids, including flexibility scores, instability, hydrophobicity, hydrophilicity, and surface accessibility, for protein secondary structure prediction. They also explored different amino acid embeddings, such as ProtVec and contextualized embeddings, and found that combining one-hot encoding and PSSM profiles yielded the best results.

In summary, while previous research has extensively explored CNN-based methods and various feature representations for protein secondary structure prediction, the potential of n-gram modeling in capturing local features remains an underexplored area. This study aims to fill this gap by focusing on the n-grams approach and its impact on accurately predicting protein secondary structures. By investigating n-grams as an effective method for extracting local features from protein sequences, this research seeks to contribute valuable insights to bioinformatics and protein structure analysis.

3 N-grams modeling

n-grams is the *n* character of a longer string. Regarding text processing, n-grams are used as a term for adjacent words. N-grams are used in various natural language processing applications, such as language models, sequence modeling, etc. N-grams can provide information about the occurrence and sequence patterns of units in a text, which can be used for prediction, analysis, and other text-related tasks.

N-grams refer to a sequential sequence of n units (such as words or characters) in text. Several *n*-grams include bi-grams ($n = 2$), tri-grams ($n = 3$), and others with a higher *n* value. For example, in the word "computer", the bigrams are 'co', 'om', 'mp', 'pu', 'ut', 'te', and 'er'. The trigrams are 'com', 'omp', 'mpu', 'put', 'ute', and 'ter' [8].

Because the protein sequence is the same as the text, which has a sequence pattern that can be learned for various tasks, it is a prediction task in this case. A protein's primary structure, consisting of several amino acids of various lengths, is then cut into overlapping pieces of *n*-grams. In *n*-grams where n is greater than or equal to 2, an empty padding will be added to each sequence's beginning and end so that no information is wasted. An illustration of how n-grams are applied in protein sequence shows in Table 1.

 https://doi.org/10.1051/bioconf/20237501008

Table 1. Example of *n*-grams modeling in protein sequence.

(a) Primary structure AENFDSRELE
(b) Bigrams ' A', 'AE', 'EN', 'NF', 'FD', 'DS', 'SR', 'RE', 'EL', 'LE', 'E '
(c) Trigrams ' AE', 'AEN', 'ENF', 'NFD', 'FDS', 'DSR', 'SRE', 'REL', 'ELE', 'LE '

The amount of padding is determined by Equation 1.

$$Padding = n - (sequence\_length \% n) \quad (1)$$

4 Data and Methodology

4.1 Dataset

The protein dataset used in these experiments is filtered CullPDB and CB513 dataset from [2] study. Those datasets contain no duplicates. The training and validation sets include 80% and 20% CullPDB, respectively. However, CB513 is used as a test set. Those datasets performed n-grams modeling before they trained to CNN.

The CullPDB dataset has 5365 protein sequences, where the shortest and the longest sequence have 12 and 696, respectively. The CB513 dataset is essential for checking the performance of the model. This dataset has 514 protein sequences, the shortest length is 20, and the longest is 700. The class distribution of both datasets is imbalanced, which will affect the model's accuracy.

4.2 Methodology

The dataset will be extracted to retrieve only the amino acid sequence features and PSSM profiles. The n-grams modeling was applied to a feature set of amino acid sequences with the size of 2 (bigrams) and 3 (trigrams). The two feature sets were tested and compared, and then the best n-grams were combined with the PSSM profile feature set. These steps are illustrated in Figure 1.

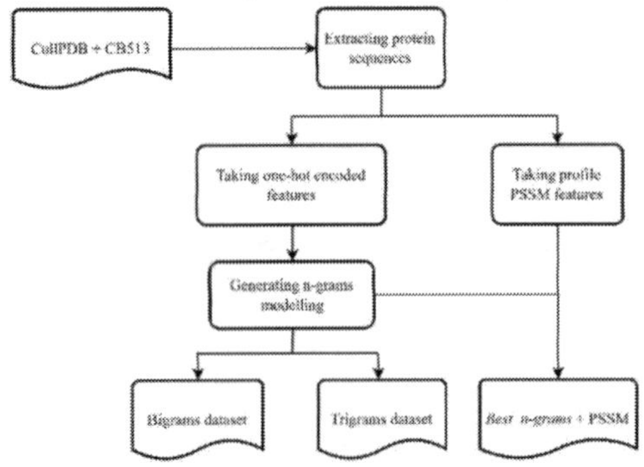

Fig. 1. Generating n-grams dataset

A detailed illustration of how to combine best n-grams with PSSP features can be seen in Figure 2. PSSM features are added at the end of n-grams features so that the total features can be calculated by $(n * i) + j$, where n is the value of n in n-grams, i is the number of feature sets produced by the best n-grams, and j is the number of PSSM feature sets.

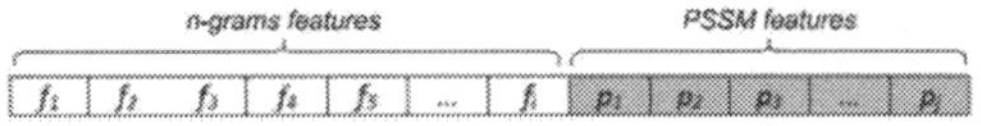

Fig. 2. Combination of n-grams and PSSM features

Shift and stitch CNN architecture, as described in the previous section, has good potential in the protein structure prediction study field because the input and output are expected to have the same length. Shift and stitch restore the length of the protein sequence by multiplying it and going through the processes of convolution and pooling. The architectural design is shown in Figure 3.

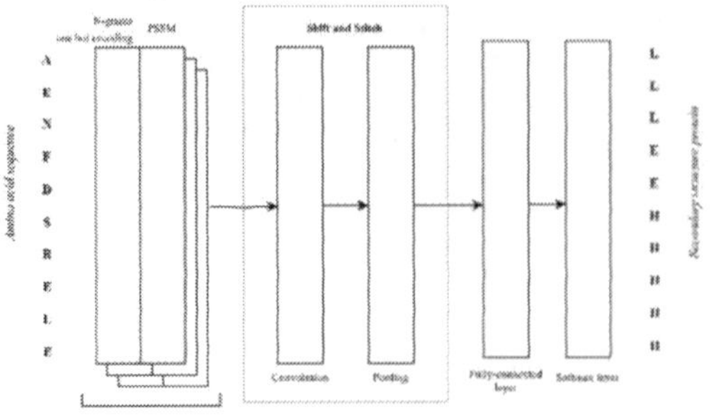

Fig. 3. Architecture for this study

The input layer will receive n-grams data and the best n-grams combination with the PSSM profile. The number of features input will differ depending on the size of n. The shape of the input layer is (700 x f), where f is the number of features. Then, in a shift layer, each amino acid sequence is duplicated and input to the convolution layer. In this layer, the two duplicate sequences will be padded so that the convolutional layer's output is not reduced. Next, the two amino acid sequences are pooled with size 2, where the two sequences will have half the length of the initial amino acid sequence, namely (350 x f), which are then combined through a shift layer to produce an output with the same length as the actual sequence length (700 x f). The result of this stage then becomes the input for the fully-connected and softmax layers, where the output layer will produce the probability prediction.

This study implemented the shift and stitch technique with TensorFlow and Keras libraries by tuning parameters to get the best result from the combination parameters. We then compare the two n-grams modeling to find the best n value for these problems.

5 Results

We conducted experiments on 3 data sets, respectively bigrams, trigrams, and a combination of the best n-grams with PSSM profile features, to find out the potential of n-gram modeling to understand local features of proteins in protein secondary structure prediction.

Table 2. Fixed parameters.

Parameter	Value
Pooling size	2
Fully connected layer	2
Dropout rate	0.2
Activation function	{ReLU, Softmax}

Table 3. Tuned parameters

Parameter	Value
Convolution layer	{2,3,4}
Convolution size	{3,5,7}
Feature maps	{32,64,128}

Table 4. Result in bigrams dataset

Conv Layer	Conv Size	Feature Map	Q8 Accuracy (%)
2	3	32	47.53730903
2	3	64	48.58373149
2	3	128	48.70288445
2	5	32	49.71273521
2	5	64	50.96089188
2	5	128	50.89482687
2	7	32	52.40252463
2	7	64	53.05137734
2	7	128	52.75172536
3	3	32	51.01633929
3	3	64	51.39857252
3	3	128	50.07373326
3	5	32	52.52875597
3	5	64	51.10599894
3	5	128	52.24444051
3	7	32	53.70258951
3	7	64	52.32348257
3	7	128	53.99398337
4	3	32	51.67462986
4	3	64	52.7823984
4	3	128	51.42452663
4	5	32	53.52209049
4	5	64	52.96643662
4	5	128	52.34117855
4	7	32	53.77809237
4	7	64	54.01993747
4	7	128	53.88780747

Before we analyze the potential of n-grams modeling, we first search for the optimal CNN parameters. We divide parameters into two types: fixed parameters, where the parameter values will not change during each training process, and tuned parameters. The set parameters include pool size, the number of fully-connected layers, dropout rate, and the activation function (see Table 2). While the tuned parameter will combine three parameters, including the depth of the convolution layer, the size of the convolution, and the number of feature maps (see Table 3). These three parameters will produce 27 models with 27 parameter combinations for each training dataset.

In this study, we calculate the Q8 accuracy for each experiment. We find that between bigrams and trigrams, bigrams produce the highest accuracy. So, we combine the bigrams dataset with PSSM profile features. The convolution layer parameter increases the Q8 accuracy in almost every depth layer, where the deeper the convolution layer, the higher the resulting accuracy. Details results are presented in Table 4, Table 5, and Table 6, respectively, the bigrams, trigrams, and combination of bigrams and PSSM profiles features.

Table 5. Results in trigrams dataset

Conv Layer	Conv Size	Feature Map	Q8 Accuracy (%)
2	3	32	50.24243497
2	3	64	50.2105822
2	3	128	50.31911756
2	5	32	51.40211172
2	5	64	51.60384593
2	5	128	51.67816906
2	7	32	52.61015749
2	7	64	53.06671386
2	7	128	51.79496254
3	3	32	52.18309444
3	3	64	49.38123046
3	3	128	52.1925323
3	5	32	52.58302365
3	5	64	52.47802749
3	5	128	52.4603315
3	7	32	53.37580369
3	7	64	53.21182092
3	7	128	51.72417861
4	3	32	53.14339645
4	3	64	52.98885153
4	3	128	53.20946145
4	5	32	53.55394325
4	5	64	53.22833717
4	5	128	53.36164691
4	7	32	53.51855129
4	7	64	53.69905032
4	7	128	53.61528933

From these results, it was observed that in the three datasets that were trained, the layer depth that produces the best Q8 accuracy is at depth 4. Likewise, with the convolution size parameter, where the length of the parameter size is directly proportional to the increase in Q8 accuracy. In this case, it can be said that the best kernel length is 7.

Unlike the two parameters mentioned, feature maps cannot provide consistent results for increasing accuracy. It can be observed from the three tables presented the number of feature maps determined by three values gives variations in the results of different Q8 accuracy. Therefore, the best value for the number of feature maps cannot be determined; it is necessary to try a combination of various feature maps to obtain the best Q8 accuracy.

Table 6. Results in bigrams combined with the PSSM dataset

Conv Layer	Conv Size	Feature Map	Q8 Accuracy (%)
2	3	32	62.8136613
2	3	64	63.22066891
2	3	128	63.00123872
2	5	32	62.30991565
2	5	64	63.07202265
2	5	128	63.49790598
2	7	32	64.20338583
2	7	64	64.74842211
2	7	128	64.8215655
3	3	32	64.10310859
3	3	64	63.71733616
3	3	128	63.62295759
3	5	32	61.24815667
3	5	64	65.02211998
3	5	128	64.57264201
3	7	32	65.11177963
3	7	64	63.58992509
3	7	128	65.40435321
4	3	32	64.56792308
4	3	64	65.21913526
4	3	128	63.95800153
4	5	32	58.10299062
4	5	64	63.23364596
4	5	128	64.26355217
4	7	32	65.68512948
4	7	64	65.39963428
4	7	128	64.07361529

As we explained earlier, the number of each data class in the dataset is imbalanced, especially in classes "I" and "B" with relatively small data. This will affect the precision and recall evaluation values of each class. Table 7 and 8 presents the precision and recall for each dataset, respectively.

Table 7. The precision of the best accuracy in each data

	Precision		
	bigrams	**trigrams**	**bigrams + PSSM**
L	0.46	0.45	0.53
B	0	0	0.75
E	0.54	0.58	0.67
G	0.29	0.28	0.39
I	0	0	0
H	0.63	0.60	0.81
S	0.39	0.74	0.47
T	0.38	0.39	0.50

The table shows that the class labels 'L', 'E', and 'H' produce higher precision and recall than other classes due to their appearance in large datasets. Whereas class labels 'B' and 'I' produce very little precision and recall, leading to zero because their appearance in the dataset is very small. This explains that the model failed to predict several minor classes but succeeded in predicting several major classes. This might happen because a minor class with a small number of classes does not have unique features that can distinguish it from other features.

Table 8. Recall of the best accuracy in each data

	Recall		
	bigrams	**trigrams**	**bigrams + PSSM**
L	0.47	0.50	0.58
B	0	0	0.003
E	0.66	0.59	0.84
G	0.03	0.03	0.17
I	0	0	0
H	0.82	0.85	0.91
S	0.03	0.01	0.12
T	0.39	0.36	0.48

We also provide the results of the n-grams modeling we used (bigrams and trigrams). For bigrams, the result of precisions ranges from 0.29 to 0.63 for different secondary structure classes, where the precision (0.63) is achieved for the "H" class, and the lowest precision (0) is obtained for the "B" and "I" classes. While in trigrams, the result of precisions ranges from 0.28 to 0.74 for different secondary structure classes, where the highest precision (0.74) is achieved for the "S" class, and the lowest precision (0) is obtained for the class "I" and "B" classes. Recall for bigrams ranges from 0.03 to 0.82, while trigrams range from 0.01 to 0.85. The highest recall for both n-grams is the "H" class and the lowest recall for are "B" and "I" classes.

6 Discussion

The results of our study show that the percentage of accuracy of the results obtained is consistently below 70%. We consider several things that affect the performance of the models we build.

In our analysis, we compare the use of bigrams and trigrams in the context of our model. The average training loss on bigrams shows slightly better results than on trigrams, with 0.4017 for bigrams and 0.4051 for trigrams. Along with that, we also consider the variability and complexity of the data. Variability in biological data refers to the natural variation in data whereby proteins with similar amino acid sequences can have different secondary structures. The complexity of the data used is multidimensional, thus providing a challenge in establishing an accurate model. We see that the use of trigrams may lead to more specific patterns but may cause the model to be less tolerant of variations in the elements in the data, particularly changes in amino acid sequence. This also affects the model in its less generalization ability. In this context, using bigrams results in a more adaptive model to a wider range of variations. This is shown from the evaluation Q8 accuracy results, which is 54.019% in bigrams and 53.699% in trigrams. Based on these considerations, we decided that better performance by the model resulted from using bigrams. Therefore, bigrams were combined with PSSM and increased the Q8 accuracy up to 65.69%.

Overall, both bigrams and trigrams modeling demonstrate similar performance trends, with bigrams performing slightly better than trigrams in precision and recall for most secondary structure classes. This indicates that the Bigrams method was more effective in correctly predicting the presence of specific secondary structure elements in protein sequences. Our approach may not fully address this problem, but these results provide new insights into the performance of n-grams in PSSP tasks with the constructed model. However, it is important to note that both methods face challenges in predicting secondary structure protein.

7 Conclusion

In conclusion, this study aimed to explore the potential of *n*-gram modeling in understanding local features of proteins for protein secondary structure prediction using a Convolutional Neural Network (CNN). Experiments were conducted on two datasets using bigrams, trigrams, and a combination of the best n-grams with PSSM features. The results showed that bigrams slightly outperformed trigrams regarding accuracy for the protein secondary structure prediction. So, bigrams were combined with PSSM profile features to further improve accuracy.

We conclude that from the analysis of n-grams, bigrams are more effective and have better accuracy than trigrams in protein secondary structure prediction tasks, especially in the model we build. The Q8 accuracy of both n-grams was less than 70%. However, it also underscores the need for further research and optimization, particularly for minor classes with limited instances, to achieve more balanced and accurate predictions. Addressing the class imbalance challenge highlighted within the study could involve employing techniques such as resampling methods, applying ensemble methods, and considering algorithms inherently robust to imbalanced classes.

References

1. P. D. Sun, C. E. Foster, and J. C. Boyington, "Overview of protein structural and functional folds." *Curr. Protoc. Protein Sci.*, vol. Chapter 17, pp. 1–189, 2004, doi: 10.1002/0471140864.ps1701s35.
2. Q. Jiang, X. Jin, S. J. Lee, and S. Yao, "Protein secondary structure prediction: A survey of the state of the art," *J. Mol. Graph. Model.*, vol. 76, pp. 379–402, 2017, doi: 10.1016/j.jmgm.2017.07.015.
3. Y. F. Huang and S. Y. Chen, "Extracting physicochemical features to predict protein secondary structure," *Sci. World J.*, vol. 2013, 2013, doi: 10.1155/2013/347106.
4. K. Murata and M. Wolf, "Cryo-electron microscopy for structural analysis of dynamic biological," *BBA - Gen. Subj.*, vol. 1862, no. 2, pp. 324–334, 2018, doi: 10.1016/j.bbagen.2017.07.020.
5. E. Asgari, N. Poerner, A. C. McHardy, and M. R. K. Mofrad, "DeepPrime2Sec: Deep Learning for Protein Secondary Structure Prediction from the Primary Sequences," *bioRxiv*, 2019, doi 10.1101/705426.
6. J. Zhou and O. G. Troyanskaya, "Deep supervised and convolutional generative stochastic network for protein secondary structure prediction," in *31st International Conference on Machine Learning, ICML 2014*, 2014, vol. 2, pp. 1121–1129.
7. S. I. Jalal, J. Zhong, and S. Kumar, "Protein Secondary Structure Prediction using Multi-input Convolutional Neural Network," in *Conference Proceedings - IEEE SOUTHEASTCON*, 2019, vol. 2019-April. doi: 10.1109/SoutheastCon42311.2019.9020333.
8. Z. Lin, J. Lanchantin, and Y. Qi, "MUST-CNN: A multilayer shift-And-stitch deep convolutional architecture for sequence-based protein structure prediction," in *30th AAAI Conference on Artificial Intelligence, AAAI 2016*, 2016, pp. 27–34.
9. B. Zhang, J. Li, and Q. Lü, "Prediction of 8-state protein secondary structures by a novel deep learning architecture," *BMC Bioinformatics*, vol. 19, no. 1, pp. 1–13, 2018, doi: 10.1186/s12859-018-2280-5.
10. P. Majumder, M. Mitra, and B. B. Chaudhuri, "N-gram : a language-independent approach to IR and NLP," in *ICUKL*, 2002, vol. 2.
11. S. Min, B. Lee, and S. Yoon, "Deep learning in bioinformatics," *Brief. Bioinform.*, vol. 18, no. 5, pp. 851–869, 2017, doi: 10.1093/bib/bbw068.
12. Y. Zhao, H. Zhang, and Y. Liu, "Protein secondary structure prediction based on generative confrontation and convolutional neural network," *IEEE Access*, vol. 8, pp. 199171–199178, 2020, doi: 10.1109/ACCESS.2020.3035208.
13. V. M. Sutanto, Z. I. Sukma, and A. Afiahayati, "Predicting Secondary Structure of Protein Using Hybrid of Convolutional Neural Network and Support Vector Machine," *Int. J. Intell. Eng. Syst.*, vol. 14, no. 1, pp. 232–243, 2020, doi: 10.22266/IJIES2021.0228.23.
14. W. Cavnar and J. Trenkle, "N-Gram-Based Text Categorization," 2001.

Upper Gastrointestinal Tract Bleeding as a Predictor of Mortality in COVID-19 Patients Admitted to RSUP Dr. Sardjito, Yogyakarta, Indonesia

Tabita Padmaya Setiawan[1], *Eko* Budiono[1], *Neneng* Ratnasari[1], *and Dhite* Bayu Nugroho[1,2,*]

[1]Department of Internal Medicine, Faculty of Medicine, Public Health, and Nursing, Universitas Gadjah Mada, Yogyakarta, Indonesia
[2]Centre of Epidemiology and Biostatistic Unit, Faculty of Medicine, Public Health, and Nursing, Universitas Gadjah Mada, Yogyakarta, Indonesia

Abstract. This retrospective cohort study explored the association between Upper Gastrointestinal Tract Bleeding (UGIB) and mortality in adult COVID-19 patients admitted to RSUP Dr. Sardjito Yogyakarta hospital from January 2021 to October 2022. Data, sourced from electronic medical records (EMRs) and analyzed using R Studio, aimed to discern if UGIB could predict mortality in COVID-19 patients, considering other relevant comorbidities. The univariate analysis identified several significant mortality-associated factors. Notably, UGIB presented an odds ratio (OR) of 2.14 (95% CI 1.48-3.11, $p < 0.001$) for increased mortality. Type 2 diabetes mellitus (OR 1.56, 95% CI 1.34-1.81), hypoalbuminemia (OR 2.05, 95% CI 1.70-2.48), hyperkalemia (OR 3.35, 95% CI 2.44-4.67), and renal impairment (OR 2.91, 95% CI 2.41-3.53) also exhibited significant associations. In contrast, being female reduced mortality risk (OR 0.78, 95% CI 0.69-0.90). The multivariate analysis, after adjusting for influential factors, indicated UGIB as an independent predictor with an OR of 1.68 (95% CI 1.02-2.79, $p = 0.042$). The results underscore UGIB's significance in predicting COVID-19 patient mortality, suggesting the need for proactive interventions to enhance patient management and outcomes.
Keywords: Upper gastrointestinal bleeding, COVID-19, mortality, claim-based registry, electronic medical records

1 Introduction

Cases of viral pneumonia associated with a severe acute respiratory syndrome were first reported in December 2019 in the city of Wuhan, China. Severe acute respiratory syndrome coronavirus 2 [SARS-CoV-2]), the causative agent of coronavirus disease 2019 (COVID-19), was identified in January 2020. The primary symptoms of COVID-19 include fever, dry cough, dyspnea, fatigue, myalgia, and headache. [1]

Similar to other coronaviruses, SARS-CoV-2 infects the gastrointestinal tract. Several case reports have described the occurrence of gastrointestinal bleeding in COVID-19 patients despite common gastrointestinal symptoms such as diarrhea, nausea, and vomiting. [2]

The use of mechanical ventilation, extracorporeal membrane oxygenation (ECMO), steroids, antiviral agents and anticoagulation in COVID-19 infection are also known to increase the risk of gastrointestinal bleeding significantly. [3]

Previous study said that individuals with COVID-19 were found to be at risk for gastrointestinal bleeding, especially upper gastrointestinal bleeding (UGIB). [2]

Another study said that UGIB was suspected in 62.5% of the COVID-19 patients and lower GI bleeding (LGIB) in 37.5% COVID-19 patients. There was also no statistically significant difference in ICU admission and mortality with the use of anticoagulation in COVID-19 patients. [4]

Meanwhile, previous systematic review and meta-analysis that aggregate data from 10 studies showed an overall gastrointestinal bleeding rate of 2%, of which 1% for UGIB and 1% for LGIB, respectively. [3]

A high prevalence of peptic ulcer disease complicated by bleeding was noticed in patients with moderate-to-severe acute respiratory distress syndrome caused by COVID-19. [2]

Previous study said that gastrointestinal bleeding was not the independent predictor of mortality in COVID-19 patients. Higher mortality in COVID-19 patients is likely secondary to respiratory failure and critical illness as seen in prior studies and may not be directly secondary to gastrointestinal bleeding. [2]

However, the real burden of gastrointestinal bleeding in COVID-19 patients still needs to be clarified. This study is conducted to know whether upper gastrointestinal tract bleeding can be a predictor of mortality in COVID-19 patients.

* Corresponding email: dhite@ugm.ac.id

2 Materials and Methods

2.1 Study Design and Setting

This study employed a retrospective cohort design. It utilized data extracted from the claim-based registry of electronic medical records (EMRs) at the RSUP Dr. Sardjito Yogyakarta, a tertiary referral hospital. The data utilized for the study spanned from January 2021 to October 2022.

2.2 Study Population and Sampling Method

The study cohort included all adult patients with a confirmed diagnosis of COVID-19 admitted to the hospital during the study period. The research utilized a total sampling method due to the retrospective nature of the study. This approach permitted the inclusion of all patients meeting the defined eligibility criteria within the specified timeframe.

2.3 Data Collection

Data was systematically extracted from the hospital's claim-based registry. The collected variables encompassed demographic factors (age, sex), comorbidities (as determined by ICD-10 codes, focusing on the top 10 comorbidities occurring in COVID-19 patients, as well as other relevant comorbidities deemed significant as potential predictors), the occurrence of UGIB, the level of care provided (intensive or non-intensive), and patient outcomes (including mortality status and length of stay). To ensure data integrity and efficient management, the Research Electronic Data Capture (REDCap) platform was employed. This secure and user-friendly application facilitated both effective data handling and subsequent analysis.

2.4 Statistical Analysis

Statistical analyses were performed using R Studio, specifically employing the 'tidyverse' and 'gtsummary' packages. The 'tidyverse' package was utilized for data cleaning and pre-processing, while 'gtsummary' was leveraged for generating analytical tables. Depending on data normality, assessed through the Shapiro-Wilk test, continuous data was presented as either a median with an interquartile range (IQR) or a mean with a standard deviation (SD). Independent t-tests, Wilcoxon tests, and Chi-square tests were used to evaluate differences in means, medians, and percentages, respectively. A multivariate logistic regression model was created for variable selection based on domain knowledge and a stepwise analysis method considering the p-value and the highest Akaike information criterion (AIC) index. The results of both univariate and multivariate analyses were displayed concurrently for comparative purposes.

2.5 Ethical Considerations

The research design adhered to the principles of the Declaration of Helsinki and Good Clinical Practice. As the study involved a retrospective analysis of existing data, the risk to patients was minimal. All patient information was anonymized before the analysis stage to ensure the protection of patient privacy. The study protocol was submitted for approval to the Institutional Review Board or Ethics Committee before initiating the research.

3 Results

Figure 1 presents the top ten comorbidities in a cohort of 3647 COVID-19 patients at the RSUP Dr. Sardjito Yogyakarta hospital. This large and diverse patient pool allowed for a comprehensive examination of the comorbidities associated with COVID-19.

The most prevalent comorbidity was type 2 diabetes mellitus (T2DM) without complications, followed by essential (primary) hypertension (HT). The third most common comorbidity was disorders of plasma-protein metabolism, such as hypoalbuminemia, and acute kidney failure, ranked fourth. The fifth most common comorbidity was hyponatremia. Nonspecific elevated levels of transaminase and lactic acid dehydrogenase took the sixth spot, with unspecified anemia ranking seventh. Hypokalemia was the eighth most common comorbidity, with urinary tract infection (UTI), site not specified, coming in ninth, and hyperkalemia rounding out the top ten.

In addition to these, upper GIT bleeding also emerged as a significant medical event within this cohort. There were 127 cases of UGIB in 3647 COVID-19 patients. The rest, 3520 cases did not experience UGIB.

These identified comorbidities, along with upper GIT bleeding, were subsequently employed as potential predictors in our study. The hypothesis was that the presence of these comorbidities might significantly influence patient outcomes, particularly mortality, within our COVID-19 patient cohort.

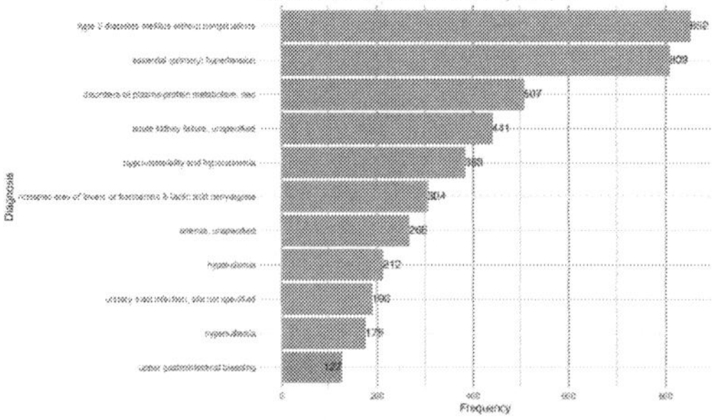

Fig. 1. Top ten comorbidities in a cohort of 3647 COVID-19 patients at the RSUP Dr. Sardjito Yogyakarta hospital, along with the incidence of upper GIT bleeding.

Table 1 provides a comprehensive overview of the demographic and clinical characteristics of the 3647 COVID-19 patient in this study, alongside the distribution of the primary predictors, which encompass various comorbidities and the occurrence of upper GIT bleeding.

In terms of demographic characteristics, the median age of patients was 55 years old, with an interquartile range (IQR) from 42 to 64 years old, signifying that the patient population was primarily middle-aged to elderly. The sex distribution was almost equal, with males representing 49.8% of the cohort.

The median length of stay (LOS) in the hospital was six days, ranging between three to ten days. The majority of patients (90.4%) were treated in non-intensive care units.

As for the presence of comorbidities, type 2 diabetes mellitus without complication was the most prevalent, affecting 23% of the patients, closely followed by hypertension. Hypoalbuminemia, renal impairment, elevated transaminase levels, hyponatremia and anemia affected 7.3-14% of the patients. Meanwhile hypokalemia, urinary tract infection, and hyperkalemia were observed in 4.8-5.8% of COVID-19 patients. Upper GIT bleeding was present in 3.5% of cases. The prevalence of cardiovascular conditions was also of interest. Stroke, coronary arterial disease (CAD), arrhythmia and heart failure (HF) were observed in 2.4-4.0% from the total sample.

These data provide an informative snapshot of the patient cohort, shedding light on the overall condition and comorbidities of patients that could potentially influence COVID-19 outcomes.

Table 1. Demographic and clinical characteristics data

Characteristic	Overall, N = 3647[1]
Age (years)	55 (42, 64)
Sex (% male)	1818 (49.8%)
Length of stay (days)	6.0 (3.0, 10.0)
Level of care (% intensive care)	349 (9.6%)
T2DM without complication (% yes)	852 (23.3%)
Hypertension (% yes)	809 (22%)
Hypoalbuminemia (% yes)	507 (14%)
Renal impairment (% yes)	441 (12%)
Hyponatremia (% yes)	383 (10.5%)
Elevated transaminase levels (% yes)	304 (8.3%)
Anemia (% yes)	266 (7.3%)
Hypokalemia (% yes)	212 (5.8%)
Urinary tract infection(% yes)	190 (5.2%)
Hyperkalemia (% yes)	176 (4.8%)
Upper gastrointestinal bleeding (% yes)	127 (3.5%)
Stroke (% yes)	145 (4.0%)
Coronary arterial disease (% yes)	130 (3.6%)
Heart failure (% yes)	88 (2.4%)
Arrhythmia (% yes)	110 (3.0%)
[1] n (%); Median (IQR)	

Table 2 compares the demographic and clinical characteristics of the patient cohort, segmented by outcome (alive versus deceased). A p-value of less than 0.05 signifies a statistically significant difference between the two groups for that particular characteristic.

The median age of patients who survived was 52 years old (IQR: 37-62), significantly younger than those who deceased, whose median age was 59 years old (IQR: 49-67) ($p < 0.001$). The sex distribution differed significantly between the two groups as well, with a lower proportion of males in the surviving group (47%) compared to the deceased group (53%) ($p < 0.001$).

There was a marked difference in the length of hospital stay, with survivors staying longer (median: 7.0 days, IQR: 5.0-11) than those who deceased (median: 4.0 days, IQR: 2.0-8.0) ($p < 0.001$). Intensive care was required in 0.4% of survivors versus 23% of deceased patients, a striking difference that was statistically significant ($p < 0.001$).

In terms of comorbidities, a higher proportion of deceased patients had type 2 diabetes mellitus without complication (25%) compared to survivors (22%) ($p < 0.001$). Hypoalbuminemia, renal impairment and hyponatremia were also significantly more prevalent in the deceased group ($p < 0.001$, $p < 0.001$, and $p = 0.020$, respectively). Elevated transaminase levels were more common in the deceased group as well (13% versus 5.2%, $p = 0.015$).

Notably, upper GIT bleeding was more than twice as prevalent in the deceased group (5.3%) compared to survivors (2.3%), a difference that was statistically significant ($p < 0.001$). Among cardiovascular conditions, heart failure and arrhythmia were significantly more prevalent in the deceased group ($p = 0.005$ and $p < 0.001$, respectively). However, the incidence of hypertension, anemia, hypokalemia, urinary tract infection, stroke, and coronary arterial disease showed no significant statistically differences between the two groups.

These findings highlight the variables that differ significantly between survivors and non-survivors and can be further analyzed as potential predictors of mortality in COVID-19 patients.

In our study examining predictors of mortality in COVID-19 patients, both univariate and multivariate analyses in Table 3 revealed multiple significant variables. Yet, among them, upper gastrointestinal tract bleeding emerged as a particularly crucial predictor, thereby meriting heightened attention.

Table 2. Comparison of demographic and clinical characteristics of the patient cohort, segmented by outcome (alive versus deceased)

Characteristic	**Alive, N = 2181**[1]	**Deceased N = 1466**[1]	**p-value**[2]
Age (years)	52 (37, 62)	59 (49, 67)	<0.001
Sex (% male)	1034 (47%)	784 (53%)	<0.001
Length of stay (days)	7.0 (5.0, 11.0)	4.0 (2.0, 8.0)	<0.001
Level of care (% intensive care)	9 (0.4%)	340 (23%)	<0.001
T2DM without complication (% yes)	478 (22%)	374 (25%)	<0.001
Hypertension (% yes)	636 (29%)	173 (12%)	0.13
Hypoalbuminemia (% yes)	225 (10%)	282 (19%)	<0.001
Renal impairment (% yes)	197 (9.0%)	244 (16%)	<0.001
Hyponatremia (% yes)	208 (9.5%)	175 (12%)	0.020
Elevated transaminase levels (% yes)	114 (5.2%)	190 (13%)	0.015
Anemia (% yes)	194 (8.9%)	72 (4.9%)	0.4
Hypokalemia (% yes)	127 (5.8%)	85 (5.8%)	>0.9
Urinary tract infection(% yes)	112 (5.1%)	78 (5.3%)	0.8
Hyperkalemia (% yes)	56 (2.6%)	120 (8.1%)	<0.001
Upper gastrointestinal bleeding (% yes)	50 (2.3%)	77 (5.3%)	<0.001
Stroke (% yes)	80 (3.7%)	65 (4.4%)	0.2
Coronary arterial disease (% yes)	76 (3.5%)	54 (3.7%)	0.8
Heart failure (% yes)	40 (1.8%)	48 (3.3%)	0.005
Arrhythmia (% yes)	49 (2.2%)	61 (4.2%)	<0.001
[1] n (%); Median (IQR)			
[2] Pearson's Chi-squared test; Fisher's exact test; Wilcoxon rank sum test			

The univariate analysis revealed that, several variables demonstrated a significant relationship with mortality. For instance, type 2 diabetes mellitus raised the odds of death by 56% (OR 1.56, 95% CI 1.34-1.81, $p < 0.001$), while hypoalbuminemia more than doubled this risk (OR 2.05, 95% CI 1.70-2.48, $p < 0.001$). Upper gastrointestinal tract bleeding also amplified the odds of death, showing an odds ratio of 2.14 (95% CI 1.48-3.11, $p < 0.001$). Furthermore, hyperkalemia and renal impairment posed particularly potent threats, tripling the likelihood of death (OR 3.35, 95% CI 2.44-4.67, $p < 0.001$ and OR 2.91, 95% CI 2.41-3.53, $p < 0.001$, respectively). Notably, elevated transaminase levels were a significant risk factor (OR 1.29, 95% CI 1.05-1.59, $p = 0.015$), whereas being female presented a protective effect, reducing the odds of death (OR 0.78, 95% CI 0.69-0.90, $p < 0.001$). Lastly, every additional year of age marginally increased the risk (OR 1.03, 95% CI 1.03-1.04, $p < 0.001$), while each day in hospital remarkably reduced this risk (OR 0.91, 95% CI 0.89-0.92, $p < 0.001$).

In the multivariate analysis, we factored in the intricate interplay between these variables. Even a midst this complex landscape, upper gastrointestinal tract bleeding steadfastly retained its significance as an independent predictor. When controlled for influential factors like type 2 diabetes mellitus, hypoalbuminemia, renal impairment, age, sex, and length of stay, the presence of upper gastrointestinal tract bleeding still augmented the odds of mortality by 68% (OR 1.68, 95% CI 1.02-2.79, $p = 0.042$).

Essentially, despite adjusting for several pivotal variables, upper gastrointestinal tract bleeding was consistently associated with a higher risk of death, thereby underscoring its critical role in the prognosis of COVID-19 patients. Our findings stress the need for enhanced vigilance and potentially aggressive interventions to manage patients with this condition and improve their outcomes. Moreover, these results exemplify the necessity for a comprehensive and multifaceted approach when assessing COVID-19 patient prognosis, reflecting the diverse and interconnected nature of this disease and its influences on patient outcomes.

Figure 2 provides a graphical representation of a 30-day survival analysis comparing patients with upper gastrointestinal tract bleeding to those without it, among the COVID-19 patients admitted to RSUP Dr. Sardjito.

The curve for patients with UGIB shows a slight decrease over time, indicating some mortality in this group. Similarly, the curve for patients without UGIB bleeding also decreases, reflecting mortality within this group. However, there does not appear to be a substantial difference between the two curves throughout the 30-day period.

This visual impression is confirmed by the provided p-value of 0.56, which is above the significance threshold of 0.05. This indicates that there is no statistically significant difference in the 30-day survival probabilities of COVID-19 patients with and without UGIB admitted to RSUP Dr. Sardjito.

Table 3. Univariate and multivariate analyses toward several factors as predictors of mortality in COVID-19 patients

Variables	Univariate				Multivariate		
	N	OR [1]	95% CI [1]	p-value	OR [1]	95% CI [1]	p-value
HT							
No	2838	—	—		—	—	
Yes	809	1.12	0.97, 1.29	0.13	0.86	0.71, 1.04	0.13
T2DM							
No	2795	—	—		—	—	
Yes	852	1.56	1.34, 1.81	<0.001	1.44	1.19, 1.75	<0.001
Stroke							
No	3502	—	—		—	—	
Yes	145	1.22	0.87, 1.70	0.2	0.72	0.47, 1.08	0.11
Hypoalbuminemia							
No	3140	—	—		—	—	
Yes	507	2.05	1.70, 2.48	<0.001	2.20	1.69, 2.88	<0.001
Hyperkalemia							
No	3471	—	—		—	—	
Yes	176	3.35	2.44, 4.67	<0.001	2.22	1.46, 3.39	<0.001
UTI							
No	3457	—	—		—	—	
Yes	190	1.04	0.77, 1.39	0.8	1.56	1.03, 2.37	0.036
CAD							
No	3517	—	—		—	—	
Yes	130	1.06	0.74, 1.51	0.8	0.59	0.38, 0.93	0.024
Elevated transaminase levels							
No	3343	—	—		—	—	
Yes	304	1.29	1.05, 1.59	0.015	1.47	1.13, 1.93	0.005
UGIB							
No	3520	—	—		—	—	
Yes	127	2.14	1.48, 3.11	<0.001	1.68	1.02, 2.79	0.042
Renal impairment							
No	3206	—	—		—	—	
Yes	441	2.91	2.41, 3.53	<0.001	2.13	1.66, 2.75	<0.001
Age (years)	3647	1.03	1.03, 1.04	<0.001	1.04	1.03, 1.04	<0.001
Sex							
Male	1818	—	—		—	—	
Female	1829	0.78	0.69, 0.90	<0.001	0.83	0.71, 0.98	0.030
Length of stay	3647	0.91	0.89, 0.92	<0.001	0.82	0.81, 0.84	<0.001
Level of care							
Intensive care	349	—	—		—	—	
Non intensive care	3298	0.01	0.01, 0.03	<0.001	0.00	0.00, 0.01	<0.001
Hyponatremia							
No	3264	—	—				
Yes	383	1.29	1.04, 1.59	0.021			
Hypokalemia							
No	3435	—	—				
Yes	212	1.00	0.75, 1.32	>0.9			
Anemia							
No	3381	—	—				
Yes	266	1.09	0.90, 1.32	0.4			
HF							
No	3559	—	—				
Yes	88	1.81	1.19, 2.78	0.006			

[1] OR = Odds Ratio, CI = Confidence Interval

This finding suggests that UGIB did not have a significant impact on the 30-day survival rate in this patient cohort.

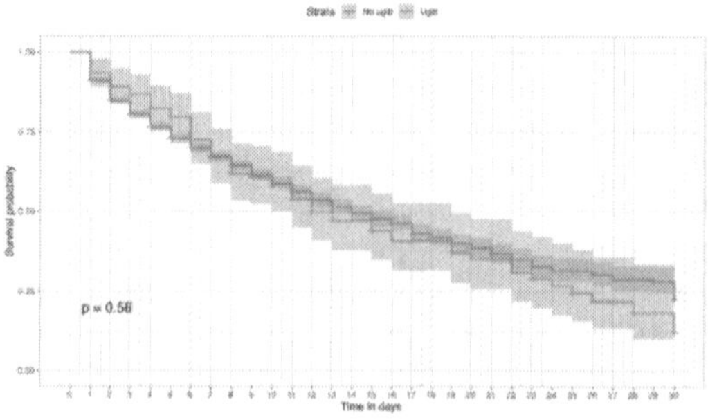

Fig. 2. Representation of a 30-day survival analysis comparing patients with upper gastrointestinal tract bleeding (UGIB) to those without UGIB.

4 Discussion

Our cohort study demonstrated that the median age of patients was 55 years old, with an inter quartile range (IQR) from 42 to 64 years old. The median age of patients who survived was 52 years old (IQR: 37-62), significantly younger than those who deceased, whose median age was 59 years (IQR: 49-67) ($p < 0.001$). Every additional year of age marginally increased the risk (OR 1.03, 95% CI 1.03-1.04, $p < 0.001$). This result can be explained as follows.

The cellular immune system of the aged is compromised on both an innate and adaptive level (immunosenescence), with a pro-inflammatory propensity (inflammaging), which appears to be increased during COVID-19, worsening the disease. [5,6]

An earlier study in 90 healthy people revealed that ciliary disarrangement and nasal mucociliary clearance time increased with age. The increasing incidence and severity of (any) lung infections may be partially explained by these age-related changes, which include diminished pulmonary reserve and airway clearance. [5,6,7]

Poorer patient outcomes including mortality and morbidity are also linked to higher frailty scores. [6]

Being female presented a protective effect in this study, reducing the odds of death (OR 0.78, 95% CI 0.69-0.90, $p < 0.001$). The sex distribution in this study differed significantly between alive and deceased group with a lower proportion of males in the surviving group (47%) compared to the deceased group (53%) ($p < 0.001$).

Males have 5% fewer heterozygous loci than females due to the fact that they only have one X chromosome. X chromosomal activation may cause variations in the methylation of the ACE2 gene. This finding may help to explain why men are more likely than women to get COVID-19 infection. [5,8,9]

Males with low testosterone levels may be more prone to thromboembolic events in COVID-19 because testosterone is essential for maintaining platelet and coagulation homeostasis. A lack of testosterone may also lead to an increase in the expression of the ACE2 receptor, which makes it easier for SARS-CoV-2 to enter host cells and cause more respiratory failure and lung damage. [8]

Increased estrogen levels have anti-inflammatory effects on endothelial cell function, boost T helper 2 and humoral immune responses, and decrease the pro-inflammatory innate immune response. Furthermore, female patients have higher protective SARS-CoV-2 immunoglobulin G (IgG) antibody levels than male patients do in patients with severe disease, which may help with better clinical outcomes. [8]

The median length of stay in the hospital in this study was six days, ranging between three to ten days. The majority of patients (90.4%) were treated in non-intensive care units, with only 9.6% requiring intensive care. There was a marked difference in the length of hospital stay in this study, with survivors staying longer (median: 7.0 days, IQR: 5.0-11.0) than those who deceased (median: 4.0 days, IQR: 2.0-8.0) ($p < 0.001$). Intensive care was required in 0.4% of survivors versus 23% of deceased patients, a striking difference that was statistically significant ($p < 0.001$). Each day in hospital remarkably reduced this risk (OR 0.91, 95% CI 0.89-0.92, $p < 0.001$).

An earlier study discovered that mild illnesses and longer LOS were related. A possible explanation is that, according to a previous report (SARS-CoV-2 Viral Load in Upper Respiratory Specimens of Infected Patients), viral loads in symptomatic and asymptomatic patients are equal. The "Diagnosis and Treatment Scheme for Novel Coronavirus Pneumonia (Trial), 6th Edition" in China states that patients with mild disease status were mostly separated from other people and treated with a few symptomatic medications. Nevertheless, the majority of these patients had a high viral load while they were in the hospital. [10]

This study found that a higher proportion of deceased patients had type 2 diabetes mellitus (T2DM) without complication (25%) compared to survivors (22%) ($p < 0.001$), statistically significant. The univariate analysis revealed that type 2 diabetes mellitus raised the odds of death by 56% (OR 1.56, 95% CI 1.34-1.81, $p < 0.001$).

Poor glycaemic management in T2DM patients has been linked in the literature to increased reactive oxygen species (ROS), pro-inflammatory cytokines, and alteration of several immune response components. People with T2DM had higher levels of intracellular furin and an increased expression of angiotensin-converting enzyme-2 (ACE2), a SARS-CoV-2 virus receptor, which made it simpler for the virus to enter cells and multiply, triggering an excessive inflammatory response and raising COVID-19 morbidity and mortality in those with type 2 DM. [11]

As a result, it has been proposed that T2DM may elevate the danger of infection, hospital admission, severe illness, and demise in COVID-19 patients. However, compared to the general population, COVID-19 patients with T2DM had more severe disease and a higher fatality rate. [11,12]

Hypoalbuminemia, renal impairment, and hyponatremia were also significantly more prevalent in the deceased group in this study ($p <0.001$, $p <0.001$, and $p =0.021$, respectively). The univariate analysis revealed that hypoalbuminemia raised the odds of death by more than double (OR 2.05, 95% CI 1.70-2.48, $p < 0.001$). Renal impairment posed particularly potent threats, tripling the likelihood of death (OR 2.91, 95% CI 2.41-3.53, $p < 0.001$). Hyponatremia raised the odds of death by 29% (OR 1.29, 95% CI 1.04-1.59, p 0.021). This result is consistent with the theory and earlier research.

An earlier retrospective analysis found that in COVID-19, a blood albumin level below 35 g/L at presentation independently increased the probability of death by at least six times. [13]

Hypoalbuminemia may therefore be used to assess the severity of epithelial-endothelial damage in COVID-19 patients. Neutrophil extracellular traps (NETs) plays a significant role in mediating tissue damage in inflammatory illnesses including COVID-19. Because serum albumin is known to prevent the development of NETs, this may help to explain why patients with hypoalbuminemia are more likely to experience severe respiratory failure and pass away. [3,14]

Previous study said that among patients with COVID-19 and AKI, high-inflammatory response and severe AKI were associated with significantly higher mortality. Pre-renal AKI is caused by the following factors: COVID-19-related hypovolemia; complement activation; cytokine storm; hypercoagulability; and microangiopathy; nephrotoxic drugs or contrast media; and comorbidities like type 2 diabetes mellitus and hypertension. [15]

According to a different study, pro-inflammatory cytokine levels are up in chronic kidney disease (CKD) patients, and this raises oxidative stress, which then triggers an inflammatory immunological response. [16]

Previous study said that hyponatremia was found to be significantly associated with increased odds for mortality (OR = 1.97 [95% CI, 1.50–2.59]), ICU admission (OR = 1.91 [95% CI, 1.56–2.35]), assisted ventilation need (OR = 2.04 [95% CI, 1.73–2.38]), and with increased LOS (SMD of 5.74 h [95% CI, 0.092–0.385]).

In a previous study, the most frequently reported causes of hyponatremia among SARS-COV-2 patients were SIADH, adrenal causes and hypovolemia. Although the precise mechanism by which SIADH-induced hyponatremia causes pneumonia is still unknown, one explanation for this is the compensatory hypoxic pulmonary vasoconstriction that results from a ventilation perfusion mismatch. [17]

Furthermore, hyperkalemia posed particularly potent threats, tripling the likelihood of death in this study (OR 3.35, 95% CI 2.44-4.67, $p < 0.001$).

SARS-CoV-2 can lead to both decreases and increases in serum potassium levels. Previous study said that compared to patients with COVID-19 who had a Ka+ level of 4.0 to 4.5 mmol/L, those with a Ka+ level 5.0 mmol/L had a significantly higher 30-day mortality. [18]

According to a systematic review and meta-analysis, patients with acute myocardial infarction had a greater risk of death for serum potassium levels that are both lower (3.5 mEq/L) and higher (4.5 mEq/L). Acid-base balance problems may have aberrant plasma potassium as one of its symptoms, which signals severe acute respiratory distress syndrome. [18]

Elevated transaminase levels occurred in 8.3% of the total patients in this study. Elevated transaminase levels were slightly more common in the deceased group in this study as well (13% versus 5.2%, p = 0.015). This phenomenon can be explained as follows.

The direct liver injury, related inflammatory responses, congestive hepatopathy, hepatic ischemia, drug-induced liver injury (DILI), and muscle breakdown are just a few possible contributory etiologies to increased liver enzymes in SARS-CoV-2 patients. [19]

Bile duct and liver epithelial cells also express ACE2, making it simple for SARS-CoV-2 to bind to ACE2-positive cholangiocytes and impair liver function. [20]

In this study, upper GIT bleeding was detected in 3.5% of cases. Upper gastrointestinal bleeding increased the risk of mortality in univariate analysis, with an odds ratio of 2.14 in this study.

The presence of upper gastrointestinal bleeding still augmented the odds of mortality by 68% when influential factors like type 2 diabetes mellitus, hypoalbuminemia, renal impairment, age, sex, and length of stay were controlled (OR 1.68, 95% CI 1.02-2.79, p = 0.042). Fundamentally, upper gastrointestinal bleeding was continuously linked to an increased probability of death, highlighting its crucial significance in COVID-19 patients' prognosis.

Some writers proposed that, in addition to the direct effects of the virus on the gastrointestinal mucosa, bleeding may also result through the development of an inflammation-induced coagulopathy and thromboinflammation. [2]

Due to the fact that the brush border of intestinal enterocytes expresses angiotensin-converting enzyme 2 at the highest level in the human body—the viral binding site—SARS CoV-2 is able to infect enteric cells. In the cytoplasm of gastric, duodenal, and rectal cells from COVID-19 patients who had SARS-COV-2 fecal shedding, SARS CoV-2 nucleocapsid proteins were found. Infection with SARS-CoV-2 resulted in inflammation of the gastrointestinal mucosa and a decrease in the functional mass of epithelial cells. Since bleeding mostly happened while patients were in hospitals, a complex explanation has been proposed. [2]

The majority of COVID-19 patients with symptoms used anticoagulants such as heparin, at least in preventive dosages, to prevent prothrombotic activity linked to COVID-19. Additionally, elevated levels of D-dimer and fibrinogen caused by the COVID-19-associated coagulopathy, which may raise the risk of thrombosis and explain the development of ischemic colitis, might cause gastrointestinal bleeding. Other possible causes of gastrointestinal bleeding in COVID-19 individuals include ulcers that develop during periods of extreme stress, including hospitalization, or disseminated intravascular coagulation, a hypercoagulable illness that also causes bleeding.[2]

Previous meta analysis suggested that the incidence of occult gastrointestinal bleeding was significantly higher in severe patients, probably as a result of stress-related mucosal disease (SRMD) in severe cases. SRMD can result from hypotension, hypovolaemia, elevated catecholamine levels, the release of pro-inflammatory cytokines, vasoconstriction, or hypotension. [3]

Coagulopathy has been identified as a risk factor for gastrointestinal bleeding. Prolonged PT may exacerbate gastrointestinal bleeding brought on by mucosal injury. Two crucial host defensive mechanisms are coagulation and a rise in the inflammatory response as the disease progresses, which could harm the host. In COVID-19 individuals, irregular coagulation is linked to a higher risk of death. [3]

In a recent study, autopsy results of COVID-19 patients revealed characteristic platelet-rich thrombus deposits in the tiny arteries of the lungs and other organs. It implies that the coagulopathy linked to COVID-19 is a fusion of localized pulmonary thrombotic microangiopathy and low-grade disseminated intravascular coagulation, which may have a major effect on organ functioning. [3]

In critically ill patients, cytokine storms characterized by high concentrations of proinflammatory cytokines and chemokines can be observed, and the release of tumor necrosis factor (TNF α) and interleukin can affect coagulation function. [3]

Nevertheless, this study found that there is no statistically significant difference in the 30-day survival probabilities of COVID-19 patients with and without upper GIT bleeding admitted to RSUP Dr. Sardjito (p-value 0.56). This result is possible because 30-day survival probabilities of COVID-19 patients are influenced by many factors, not just the incidence of upper GI tract bleeding. Age, sex, severity of COVID-19, accompanying comorbidities as explained above, greatly affect survival probabilities. These results exemplify the necessity for a comprehensive and multifaceted approach when assessing COVID-19 patient prognosis, reflecting the diverse and interconnected nature of this disease and its influences on patient outcomes.

Our study has some limitations including the retrospective design of the study which may have introduced bias in the study results. Secondly, the number of patients with upper GI tract bleeding is small in our study, and we suggest that larger studies need to be carried out in future. Thirdly, our study lacked data on drug use before and during COVID-19 treatment. Fourth, patients with COVID-19 were unable to perform comprehensive gastrointestinal examination to clarify gastrointestinal mucous damage and bleeding owing to the restriction of clinical conditions.

5 Conclusion

Upper gastrointestinal bleeding still augmented the odds of mortality by 68% (OR 1.68, 95% CI 1.02-2.79, p = 0.042) when influential factors like type 2 diabetes mellitus, hypoalbuminemia, renal impairment, age, sex, and length of stay were controlled. Nevertheless, there is no statistically significant difference in the 30-day survival probabilities of COVID-19 patients with and without upper GIT bleeding admitted to RSUP Dr. Sardjito (p-value 0.56)

References

1. M. Junior, S. Augusto,Y.Elias, C. Costa, P. Neder, Arq Bras Cir Dig. **34**,3 (2021)
2. G. Marasco, M. maida, G. Morreale, M. Licata, M. Renzulli, C. Cremon, V. Stanghellini, G.Barbara, Can J Gastroenterol Hepatol. **2021**, 2534975 (2021)
3. X. Zhao, M. Tao, C. Chen, Y. Zhang, Y. Fu, Infect Drug Resist. **14**,4217-4226 (2021)
4. U.Iqbal, P.Patel, C.Pluskota, A.Berger, H.Khara, B.Confer, Gastroenterology Res. **15**,1 (2022)
5. F.Perrotta, G.Corbi, G.Mazzeo, M.Boccia, L.Aronne, V.Agnano, K.Komici, G.Mazzarella, R.Parrella, A.Bianco, Aging clin. exp. res.. **32**,1599–1608 (2020)
6. A. Smorenberg, E.Peters, P.Daele, E.Nossent, M.Muller, Eur J Intern Med. **83**,1-5 (2021)
7. L.Wong, S.Perlman, Nat. Rev. Immunol. **22**, 47–56 (2022)
8. S.Wray, S.Arrowsmith, Front Physiol. **12**, 627260 (2021)
9. R.Fan, S.Mao, T.Gu, F.Zhong, M.Gong, L.Hao, F.Yin, C.Dong, L.Zhang, Mol Med Rep. **15**,3905-3911 (2017)
10. A.Guo, H.Tan, Z.Kuang, Y.Luo, T.Yang, J.Zu, J.Yu, C.Wen, A.Shen, Sci. Rep. **11**,7310 (2021)
11. M.Kusumawati, R.Koesoemadinata, Z.Fatma, E.Susandi, H.Permana, N.Soetedjo, A.Soeroto, B.Bestari, B.Andriyoko, B.Alisjahbana, Y.Hartantri, PLoS One.**18**,6 (2023)
12. P.Sharma, T.Behl, N.Sharma, S.Singh, A.Grewal, A.Albarrati, M.Albratty, A.Merraya, S.Bungau, Biomed Pharmacother. **151**, 113089 (2022)
13. J.Huang, A.Cheng, R.Kumar, Y.Fang, G.Chen, Y.Zhu, S.Lin, J Med Virol. **10**, 2152-2158 (2020)
14. V.Zerbato, G.Sanson, M.Luca, S.Bella, A.Masi, P.Caironi, B.Marini, R.Ippodrino, R.Luzzati, Infect Dis Rep.**14**, 278-286 (2022)
15. T.Sabaghian, A.Kharazmi, A.Ansari, F.Omidi, S.Kazemi, B.Hajikhani, R.Harami, A.Tajbakhsh, S.Omidi, S.Haddadi, A.Bonjar, M.Nasiri, M.Mirsaedi, Front Med (Lausanne). **9**, 705908 (2022)
16. R.Cai, J.Zhang, Y.Zhu, L.Liu, Y.Liu, Q.He, Int Urol Nephrol. **53**, 1623–1629 (2021)
17. R.Khidir, B.Ibrahim, M.Adam, R.Hassan, A.Fedail, R.Abdulhamid, S.Mohamed, Int J Health Sci (Qassim). **16**, 69-84 (2022)
18. M.Noori, S.Nejadghaderi , M.Sullman, K.Chahhoud, M.Ardalan, A.Kolahi, S.Safiri, Mol Biol Rep. **48**, 6655-6661 (2021)
19. A.Moon, A.Barritt, Dig Dis Sci. **66**, 1767-1769 (2021)
20. R.Clark, B.Waters, A.Stanfill, Nurse Pract.**46**, 21-26 (2021)

Biochemical and Molecular Characterization of Eel Fish Trypsin (*Anguilla bicolor* McClelland) as Potential Candidates Protease Enzyme

Yuni Kulsum[1], and Husna Nugrahapraja[2*]

[1] Graduate Student in Departement of Biology, School of Life Sciences and Technology, Institut Teknologi Bandung, Indonesia
[2] School of Life Science and Technology, Institut Teknologi Bandung, Indonesia

Abstract. Trypsin is one alkaline protease type widely used in various industry fields. One type of potential fish trypsin source is *Anguilla bicolor*. This study aims to characterize biochemical and molecular characterization of eel fish trypsin (*Anguilla bicolor* McClelland) as a possible candidate protease enzyme. The method used in this research is experimental research consisting of biochemical and molecular characterization. Fish Trypsin Extract was isolated from the digestive organs and then crushed using an electric homogenizer. During the pulverization process, 50 mM Tris-HCl buffer was added at a ratio of 1: 8 (w/v). The supernatant was then collected and can be stored at -80°C to measure enzyme activity. The treatment was given to juveniles and adults with stadia of *Anguilla bicolor*. While the molecular method was carried out using In Silico analysis in the analysis of the diversity of trypsin sequences in various fish species, preparation of specific primers, and analysis of Whole Genome Sequencing diversity of different species of Anguilla Spp. After that, extraction of *Anguilla bicolor* DNA, optimization of primer annealing temperature, DNA amplification, fish trypsin DNA fragments using the Sanger and Nanopore methods, and analysis of sequencing and phylogenetic results. The result of the protein content of the trypsin extract in the juvenile stage of *Anguilla bicolor* had an average of 0.488 ± 0.004 g/dL, and the adult stage of *Anguilla bicolor* had an average of 1.778 ± 0.080 g/dL. The highest trypsin activity was obtained in the juvenile stadia, 0.529 ± 0.016 (U/mL), and in the adult stadia, 0.399 ± 0.009 (U/mL). Trypsin activity increases with increasing temperature used and reaches a maximum of 40°C. The molecular character of the fish enzyme *Anguilla bicolor* shows that the sequence analyzed tend to be close to the Trypsinogen and Trypsin-like genes from Anguilla japonica, Anguilla anguilla, and Megalops cyprinoides.
Keywords: *Anguilla bicolor*; biochemical; Characterization; Enzyme; Molecular; Trypsin

1 Introduction

The demand for environmentally friendly products in the modern world requires that enzymatic methods replace industrial products produced by chemical processes. One of the most important groups of enzymes used for various industrial developments is alkaline proteases. Alkaline proteases are one of the most important groups of enzymes which industrially and scientifically account for about 65% of the annual enzymes market. Alkaline proteases have a history of application in the food and detergent industries, holding the largest share of the enzyme market worldwide [1], [2]. One type of alkaline protease that is generally used in various fields is trypsin.

Trypsin is a serine protease group digestive enzyme that hydrolyzes proteins on the carboxyl side of the amino acids lysine or arginine. Trypsin is responsible for protein hydrolysis in the digestive system into smaller peptides or even amino acids [3]. Alkaline proteases such as trypsin can be used as ingredients in detergents, the leather industry, medical diagnostics, vaccines, textiles, mining, the food and feed industry, and many more. Due to its widespread application, many industries have started their production at a commercial level. Enzymes are needed to meet the increasing global demand in industrial markets [4], [5]. Meeting the needs of the enzyme can be explored from the existing biological resources in Indonesia.

Indonesia's biological resources are abundant, one of which is fish. According to data from the Ministry of Maritime Affairs and Fisheries in 2015, Indonesia has around 400,000 species of animals and fish. It is estimated that 8500 species of fish or strata, with a total of 45% of the number of species scattered in various parts of the world, live in Indonesian waters. The data shows that around 1,300 species from the data presented occupy areas in Indonesian freshwaters.

On the other hand, the high level of aquatic biodiversity, especially fish resources, also creates environmental problems such as increased water waste, one of which is fish offal waste [6], [7]. In contrast, fish offal as a by-product of the fishery industry has been recognized as a potential source of different enzymes, especially protease enzymes. One is trypsin, also known

* Corresponding author: nugrahapraja@sith.itb.ac.id

as fish trypsin [8], [9]. Only a little is known about fish trypsin, so its use in various fields is minimal.

Fish trypsin of various species and its potential applications in multiple industries have been developed. Fish trypsin applications include as an ingredient in detergents, extracting carotene proteins from shrimp waste, producing protein hydrolysates as food ingredients, and various applications in the food industry [10]. The number of potential fish species in Indonesia creates an opportunity to develop fish enzymes. According to [11], this predatory fish has excellent potential to produce high-quality alkaline proteases. One example of a superior predatory fish in Indonesia is the eel, or one of its scientific names, *Anguilla bicolor*.

The eel (*Anguilla bicolor*) is a type of fish with a catadromous nature; its life cycle occurs in two environmental conditions, namely being born in the sea and then growing in freshwater areas and when the maturation phase for reproduction returns to the sea [12]. Information on the enzyme activity of eel (*Anguilla bicolor* McClelland) using specific substrates is still unavailable. According to [13], when enzymes are used in industrial processes and analytical procedures, evaluation of enzyme activity becomes very important because it relates to the amount of enzyme needed to carry out the process correctly, the duration of the reaction, the amount of substrate to be converted, the number of conditions where the reaction occurs, and the overall cost of the process. So it is necessary to research the molecular characterization and enzyme activity of fish trypsin in eel (*Anguilla bicolor* McClelland) to develop its use in various applications.

2 Method

2.1 Experimental Design

The method used in this research is experimental research consisting of biochemical and molecular characterization. The biochemical method uses a completely randomized design (CRD) with six types of treatment in the form of temperature differences and four repetitions. Determination of the number of repetitions is based on the repetition calculation formula according to [14] as follows:

$$t\,(r-1) \geq 15 \qquad (1)$$

Information:
t = treatment (number of treatments)
r = replication (number of repetitions)
15 = general degrees of freedom

2.2 Object of Research

The sample of research is *Anguilla bicolor* obtained from PT. Laju Banyu Semesta (LABAS Sidat) Bogor City, West Java. The selected fish have different body sizes and weights based on juvenile and adult life stages. The number of each type of fish used in this study was 48 individuals representing juveniles (24 fish) and adults (24 fish). Each fish is placed in a different container according to the type of stadia and quarantined for one week by giving a commercial feed.

2.3 Sample Preparation and Morphometric Measurements

Body morphometric measurements of *Anguilla bicolor* were carried out on each eel sample in two stadia. Measurements were made using a ruler and calipers with an accuracy of 0.01 mm, while the weight of the fish was weighed with a digital scale with an accuracy of 0.001 gram. The morphometric characters measured included 11 characters. Terms and abbreviations of the characters measured include standard length (SL), total length (TL), head length (HL), head height (HD), head width (HW), muzzle length (SNL), the distance between eyes (IW), eye diameter (ED), body height (BD), body weight (TW) and body width (BW). *Anguilla bicolor* morphometric scheme can be seen in Figure 1.

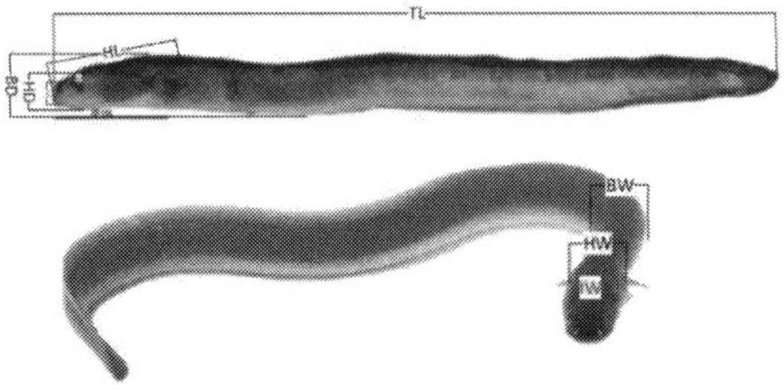

Fig. 1. Schematic of measuring the morphometric characters of *Anguilla bicolor*

Eel samples were isolated from the digestive organs by dissecting the ventral part of the eel's body. It begins with anesthetizing the fish by immersing it in ice water. Then the digestive organs are separated from the body, followed by the separation of the intestine and pyloric caeca and put into a particular container. The intestine that has been taken is then cleaned, and measurements are taken, including the intestine's width, the intestine's weight, and the intestine's length. Measurements were made using a ruler and calipers with an accuracy of 0.01 mm, while the weight of the intestine was weighed with a digital scale with an accuracy of 0.001 gram. Next, the intestine is cut into smaller parts of about 2 to 4 cm. After that, put it in a film bottle to be stored in the refrigerator at a temperature of -20°C to minimize protease autolysis.

2.4 Preparation of Fish Trypsin Extract

Samples of intestinal organs were isolated and then crushed using an electric homogenizer. During the pulverization process, 50 mM Tris-HCl buffer was added at a ratio of 1: 8 (w/v). The resulting homogenate was collected and transferred to a 1.5 mL tube, then centrifuged for 15 minutes at a speed of 12,000 rpm at a temperature of 4°C. The supernatant was then collected and transferred to another tube. The supernatant can be stored at -80°C and measured to measure trypsin enzyme activity [15].

2.5 Determination of Protein Content

The working reagent (R1), albumin calibrators solution, and samples to be tested are stored at room temperature before use. Enter the albumin calibrators solution and working reagent solution into each well on the 96-well plate microplate to make a standard curve with the proportions according to Table 1.

Table 1. Preparation of standard albumin curve

Well No.	Working Reagent (uL)	Albumin Calibrators Solution (uL)	Final Concentration (g/L)
1	150	0	0
2	150	0,25	2,16
3	150	0,5	4,32
4	150	1	8,64
5	150	2	17,28

A total of 0.5 µL of the sample was put into the well, and 150 µL of the Working Reagent Solution was added to each sample. Then the solution was incubated at 15-25°C for 5 minutes. The absorbance of the solution was measured at a wavelength of 630 nm using the Multimode reader TECAN Infinite M200.

2.6 Fish Trypsin Activity Test

Trypsin enzyme activity test can be determined using a specific type of trypsin substrate, namely BAPNA. This activity test was applied to all supernatants from three types of stadia. Following the procedure from [16], making a BAPNA solution with the composition in 1 mL of DMSO dissolved 0.0435 g of BAPNA powder. Then it was dissolved in 100 mL with 0.05 M Tris-HCl containing 0.02 M CaCl2.2H2O. Then 0.05 mL of the sample was added with 2.5 mL of BAPNA solution. Then the sample tubes were incubated at 10°C, 20°C, 30°C, 40°C, 50°C and 60°C for 10 minutes. Then after completion of the incubation period, the reaction was stopped by adding 1 mL of 30% acetate solution. The incubation process was continued for 10 minutes at 37°C. The same procedure was carried out on the blank tube except that adding the enzyme extract was carried out after administration of 1 mL of 30% acetate solution. The reaction mixture was mixed until homogeneous using a vortex. The absorbance of the solution was checked using a spectrophotometer at a wavelength of 410 nm. Enzyme activity is calculated based on the equation:

$$\textbf{Enzym Activity} = \frac{(\text{Asampel} - \text{Ablanko}) \times \text{total volume after reaction} \times 1000}{8800 \times \text{time of inkubation} \times \text{volume of the enzyme reacted}}$$

2.7 In Silico Primary Design for Trypsin Gene Amplification in Eel (Anguilla bicolor)

2.7.1 Nucleotide Sequence Retrieval and Gene Library Construction

The gene encoding the trypsin in *Anguilla bicolor* is currently unavailable in the GenBank database. Therefore, the primer design was carried out by referring to the trypsinogen gene sequences in other species. The number of sequences used for the preliminary design came from 2 species with ten sequence variations (Table 2). For this study, [17] suggested a gene nucleotide search be performed on the NCBI database (https://ncbi.nlm.nih.gov). The whole trypsinogen gene sequences obtained, both partial and full-length sequences, were downloaded in FASTA format. All sequences obtained were selected from authenticated species known to possess the trypsinogen gene. A Multiple Sequence Alignment process was performed using Mega-X Software to determine the conservation area as the primary candidate selection area. Alignment is made using two algorithms, CLUSTALW and MUSCLE, to know which algorithm is optimal. The sustainable area is determined, and a primary degenerate is designed to amplify the area. The primers used to strengthen the sustainable regions are shown in Table 2. In addition to MSA, all sequences were constructed to form a phylogeny tree to analyze the trypsinogen gene's affinity level.

Table 2. Sequence identity for primary design

No.	Taxa Name	Locality	GenBank Accession No.
1	*Anguilla japonica*	China	KR827547.1
2	*Anguilla anguilla*	Republik Ceko	XM_035422887
3	*Anguilla anguilla*	Republik Ceko	XM_035422739
4	*Anguilla japonica*	Japan	AB519643.1
5	*Anguilla japonica*	Japan	AB070720.1
6	*Anguilla anguilla*	Republik Ceko	XM 035429595.1
7	*Anguilla anguilla*	Republik Ceko	XM 035428317.1
8	*Anguilla anguilla*	Republik Ceko	XM 035429594.1
9	*Anguilla anguilla*	Republik Ceko	XM 035429592.1
10	*Anguilla anguilla*	Republik Ceko	XM 035429593.1

2.7.2 In-Silico Evaluation of Primary Properties with Primer3Plus

Forward and reverse primary candidates that have been determined based on the results of Multiple Sequence Alignment were analyzed using Primer3Plus at the address https://www.bioinformatics.nl/cgi-bin/primer3plus/primer3plus.cgi. The results are the three best primer candidates from forward and reverse primers (Table 3) with different nucleotide base arrangements. This device displays primary properties such as % GC, hairpin value, base length, self-dimer, melting temperature (Tm), and cross dimer. To confirm the results of the primary property evaluation, check again at https://www.ncbi.nlm.nih.gov/tools/primer-blast/.

Table 3. Primer design of fish trypsin *Anguilla bicolor*

Primer Name	Sequence 5'→3'	Length (bp)	Tm (°C)	GC%	SC
AF	GGCTCTGGATGATGATAAGATTG	23	59,9	43,5	0,0
AR	AGCTCACCGTTACACACCAC	20	58,6	55,0	0,0
BF	ATGAGGTCTCTGGTTTTTATTCTGC	25	61,2	40,0	2,0
BR	CGTATCCCCAGGACACAACA	20	61,8	55,0	0,0
CF	CTGTGGCTCTGGATGATGATAA	22	60,1	45,5	3,0
CR	CGTATCCCCAGGACACAACA	20	61,8	55,0	0,0

2.7.3 Isolation of DNA, Primer Optimization using Gradient PCR, and Electrophoresis Process

The Isolation of DNA procedure complies with TIANGEN's TiaNamp DNA Extraction KIT protocol. 1 mg of the sample was crushed using a mortar and put into a 1.5 mL Eppendorf tube. First, 500 ul of GMO1 buffer was added, then 20 µL of Proteinase K was added slowly and homogenized within 1 minute. Incubate the solution for 1 hour at 56°C and vortex every 15 minutes. 200 µL of GMO2 Buffer was added, then incubated at room temperature for 10 minutes. The sample was centrifuged for 5 minutes at 12,000 rpm and then transferred the supernatant to a new tube. Then add 0.7 volume of isopropanol and centrifuge again for 3 minutes at 12,000 rpm. Then discard the supernatant. For pellets, continue adding 700 µL of 70% alcohol, centrifugation for 1 minute at 12,000 rpm. Repeat twice in the previous step. After completion, the pellets were incubated at room temperature for 10 minutes. Dry the shells at room temperature, then add 50 µL of TE Buffer. Store DNA isolates at -20°C to -80°C.

The PCR amplification technique used a DNA template isolated from *Anguilla bicolor*. Optimization of the PCR profile at the primary annealing temperature is necessary to obtain the best conditions for amplifying the *Anguilla bicolor* trypsinogen gene. The composition of the PCR reaction (Table 4) is as follows:

Table 4. Composition of PCR reaction components using 2x PCR Master Mix Solution (i-MAXTM II)

No	Components	Concentration
1	Mastermix i-Max Intron (1X)	7,50 µl
2	Forward Primer (10 µM)	0,60 µl
3	Reverse Primer (10 µM)	0,60 µl
4	DNA Template	1,50 µl
5	Nuclease Free Water (NFW)	Up to 15 µl

The amplification process begins with an initial denaturation step for 5 minutes at 95°C, then 30 cycles of reaction are carried out consisting of 30 seconds of denaturation at 95°C, 30 seconds of annealing at 56-61°C (temperature gradient for temperature optimization), and 1 minute. DNA elongation at 72°C. Then the final elongation was carried out for 60 seconds at 72°C (Figure 2).

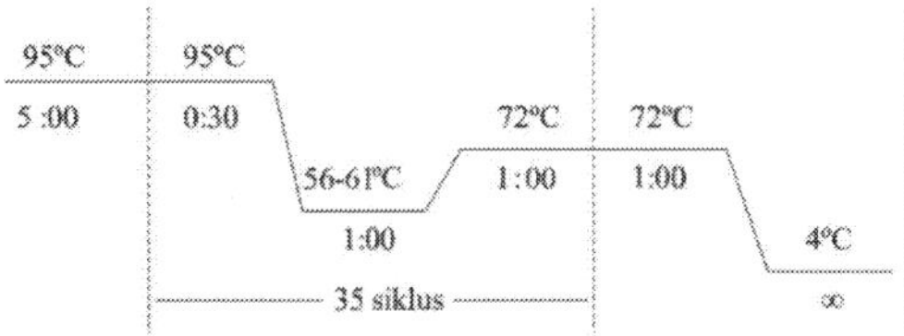

Fig. 2. PCR profile using 2x PCR Master Mix Solution (i-MAXTM II)

Electrophoresis begins with the preparation of 2% agarose gel by dissolving 2 g of agarose in 100 ml of 1x TAE buffer. 100 µL of SYBR safe was added, homogenized, and poured into the Scie Plas HU10 Mini-Plus electrophoresis tray (10 x 11.5 cm). 5 µl of sample and 1 µl of loading dye were put into the electrophoresis chamber using a micropipette. The electrophoresis process was carried out using a power of 80 volts for 40 minutes. Then the electrophoresis results were visualized using a UV transilluminator.

2.7.4 Sanger Sequencing

The Sanger sequencing method used an Automatic DNA Sequencer with the dye terminator labeling method. The stages of DNA sequencing carried out in this study included preparing the DNA, followed by the amplification process via PCR using primers, DNA purification, electrophoresis, and electrophoretic reading of the sequenced results.

The data obtained shows the nucleoid sequence marked by a different color and is stored as an electropherogram in ABI file format. Green color indicates adenine base type, black color indicates guanine base type, red color indicates thymine base type and blue color indicates cytosine base type. The amplification of *Anguilla bicolor* trypsinogen DNA, which PCR had successfully confirmed, was then sequenced to determine the base sequence. Direct sequencing uses degenerate primers with two-way reading, namely forward and reverse. The entire base sequencing process was done using the Applied Biosystems 3730xl DNA Analyzer by Macrogen Inc. (Singapore).

2.8 Data Analysis

2.8.1 Analysis of Biochemical Data

The biochemical data obtained were analyzed using ANOVA at a 95% confidence level using SPSS version 21.0 to determine the effect of differences in incubation temperature on the activity of fish trypsin in the digestive tract of eels. A follow-up test followed the significantly different ANOVA results in the form of a DMRT (Duncan's Multiple Range Test).

2.8.2 Analysis of PCR Result Data

PCR amplification results were analyzed using GelAnalyzer 19.1 to determine the bp length of the bands obtained. Obtaining a number indicating the length of bp will help analyze the differences between fish trypsin in eel (*Anguilla bicolor*) and other types of trypsin.

2.8.3 Data Analysis of Sanger Sequencing Results

Sequence alignment was performed using the Clustal W Program. The base sequencing results were analyzed using the BioEdit program to obtain the base sequence of the base-sorted fragments. Sanger sequencing results were then analyzed using the BLAST program available on the NCBI website (www.ncbi.nlm.nih.gov), which was used to look for the similarity of a nucleotide or protein sequence (query sequence) to the database sequence (subject sequence).

3 Results And Discussion

3.1 Body and Intestinal Morphometrics Anguilla bicolor

Fish morphometrics describes characters related to fish body parts or sizes, for example, standard length, total length, and other length measurements [17]. The eels used in the study were almost uniform in length and body weight. Morphometric measurements of each sample of eel from three stages were measured using calipers with an accuracy of 0.01 mm, and the weight of the fish was weighed with a digital scale with an accuracy of 0.001 gram which included 11 characters. The results of *Anguilla bicolor* morphometric measurements are as follows:

Table 5. Results of measurements of the body morphometric characters (mm) of *Anguilla bicolor*

Characters	Stadia	
	Juvenile	Adult
TL	310,3±16,792	464,08±27,673
SL	308,4±12,244	450,09±16,557
HL	21,2±5,241	40,9±2,511
HW	13,3±1,733	31,4±2,246
HD	8,3±1,542	26,4±3,275
SNL	4,1±0,016	6,8±0,031
ED	1,02 ± 0.094	1,81 ± 0.756
IW	2,01 ± 0.022	3,81 ± 0.633
BD	10,8±0,057	21,58±0,617
BW	16,8±0,254	27,58±0,179
TW (g)	66,08±12,497	186,79±35,413

The results of *Anguilla bicolor* body morphometric measurements (Table 5) show that the juvenile stage has a body weight (of 66.08 ± 12.497 g), and the adult stage has a body weight (of 186.79 ± 35.413 g). Currently, available information regarding the morphometrics of eels in Indonesia was provided by [11], who conducted research on local eels in the Segara Anakan area, Cilacap, by classifying the catches of collectors into eels with a size of 41.25 ± 0.898 g included in the elver stage, eels with a size of 319.8 ± 4.666 g were included in the yellow eel stage and eels with a size of 569.5 ± 9.150 g were included in the pre-silver stage.

The weight of the eel from PT. Laju Banyu Semesta (LABAS Sidat) is less valuable than local eels caught directly from the ocean. It is because the type of treatment for eels in captivity is different from eels that live in the ocean. In addition, the age difference between the samples taken from PT. Laju Banyu Semesta (LABAS Sidat) is different from the age of eel fish that live in the ocean. [18] states that the proportion of fish body parts has different values, influenced by several factors, including fish size, water conditions or habitat, fish species, fish physiological condition, and the handling process from catching to preparation. In contrast, differences in fish growth are influenced by several factors such as age, sex, food availability, and heredity.

Table 6. Results of measuring the morphometric characters (mm) of the intestines of *Anguilla bicolor*

Measurement Parameters	Stadium	
	Juvenile (n=24)	Adult (n=24)
Total Length (mm)	310,30±16,79	464,08±27,67
Body Weight (g)	66,080±12,500	186,790±35,413
Intestine Length (mm)	128,20±11,32	277,71±13,47
Intestine Weight (g)	0,014±0,373	0,038±0,914
The Ratio Intestinal Length/ Total Body Length (RGL)	0,413±0,674	0,598±0,0,486
The Ratio of Intestinal Length/Body Weight (ISI)	1,941±0,924	1,487±0,0813

The results of measuring the morphometric characters (mm) of the intestinal tract of *A. bicolor* (Table 6) show that the eel has a Gut Length Ratio (RGL) of 0.27 ± 0.99 at the larval stage, 0.58 ± 0.92 at the juvenile stage, and 0.62 ± 0.81 in the adult stage. The results of measuring the morphometric characters (mm) of the intestinal tract of *A. bicolor* showed that the eel had an Intestinal Somatic Index (ISI) of 2.106 ± 0.992 at the larval stage, 1.941 ± 0.924 at the juvenile stage, and 1.487 ± 0.0813 at the adult stage.

The older the stadium, the greater the Gut Length Ratio (RGL) value, but the older the stadium, the smaller the Intestinal Somatic Index (ISI). According to [19] Gut Length Ratio (RGL) is calculated as the ratio of intestinal length to total length. This ratio is used to determine the eating habits of fish, namely herbivores (RGL > 3), carnivores (RGL < 1), or omnivores (RGL = 1-3). Meanwhile, ISI is used to understand morphometric studies and research on fish physiology. This ratio can provide information about the adaptation and function of the fish's digestive system, mainly how the fish's intestine develops according to the body's metabolic needs as well as the health and nutritional conditions of the fish. Fish that have a high ISI have a more designed and functioning gut that has an excellent ability to digest and absorb nutrients. Conversely, a decrease in ISI can indicate a health problem or digestive disorder in the fish.

According to [20] based on how they eat, eel is a carnivore, so it has a shorter gut size when compared to other fish, which are herbivorous and omnivorous. [21] explained that one type of fish can be distinguished based on the morphology of the digestive system, which compares the ratio of digestive tract length to body length for fish in different food categories. In addition, the quantification of these morphological measurements is assessed based on the trophic position of a species.

3.2 Protein Content of Anguilla bicolor Trypsin Extract

The protein content of the enzyme extract (Figure 3) in the larval, juvenile, and adult stages of *Anguilla bicolor* with three repetitions had an average of 0.0893 ± 0.106 g/dL, 0.488 ± 0.114 g/dL, and 1.778 ± 0.480g/dL. The results of the DMRT test showed that there was a significant difference in the average between the types of treatment (stadia) on protein content ($P<0.05$). *A. bicolor* belongs to a kind of fish with a high protein content. These results follow research from [22], which revealed that the protein content of Anguilla spp. as a food ingredient for fulfilling nutrition has a high protein content ranging from 15-20%.

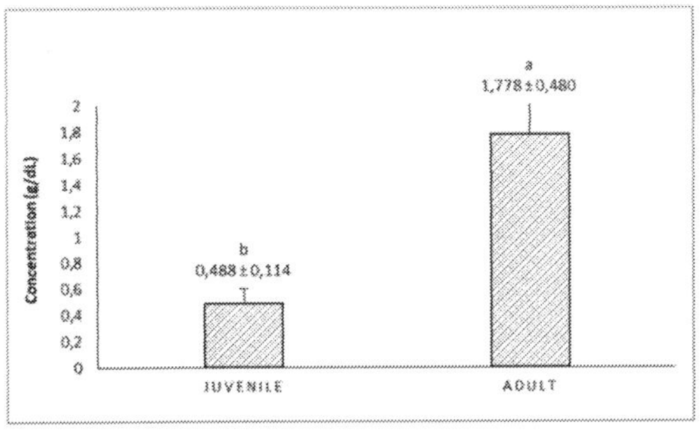

Fig. 3. Mean (±SD) intestinal protein levels of *Anguilla bicolor*

Compared with other types of eel, as stated by [22] the optimum protein content in the juvenile phase is 44% for the *Anguilla japonica* type. In contrast, for the Anguilla marmorata eel size 2.29 g, the optimum feed content is 50%, and a height of 21.97 g is 45%. According to[11], the protein requirement for carnivorous fish is more significant than for herbivorous and omnivorous fish due to the supporting factors described by [23], which stated that the protein content of fish is strongly influenced by age, stadia, type, size, feed protein quality, feed digestibility, and environmental conditions.

The body's need for protein is closely related to the work of enzymes because protein will be able to be absorbed by the body if enzymes have broken it down into simpler forms. Enzyme work control can take the form of regulation in producing active enzymes from inactive enzymes (inactive precursors). The precursors for proteases are called preprotein or proenzymes. Proenzymes can be activated into active enzymes by other already active enzymes [24].

3.3 Enzyme Activity of Anguilla bicolor Fish Trypsin

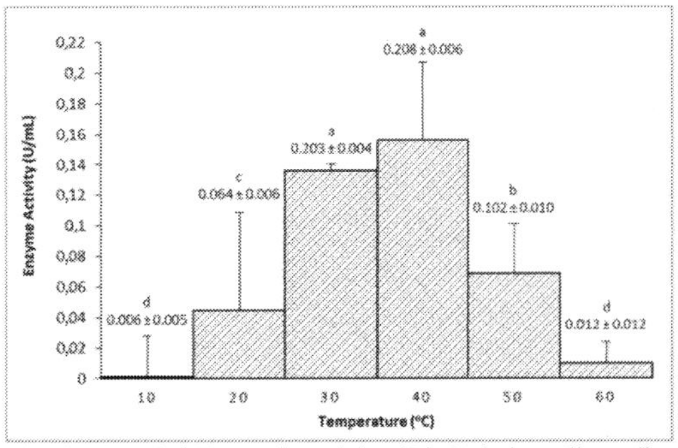

Fig. 4. Mean (±SD) trypsin activity (U/mL) in juvenile stadia at different incubation temperatures (°C)

The results of the measurement of enzyme activity (Figure 4) showed that the juvenile stadia *Anguilla bicolor* had an average enzyme activity at temperatures of 10°C (0.006 ± 0.027 U/mL), 20°C (0.064 ± 0.064 U/mL), 30°C (0.203 ± 0.004 U/mL), 40°C (0.208±0.051 U/mL), 50°C (0.102±0.032 U/mL), and 60°C (0.012±0.012 U/mL). The results of the DMRT test showed that there was a significant difference ($P<0.05$) between the treatments (incubation temperature (°C)) in the juvenile stage.

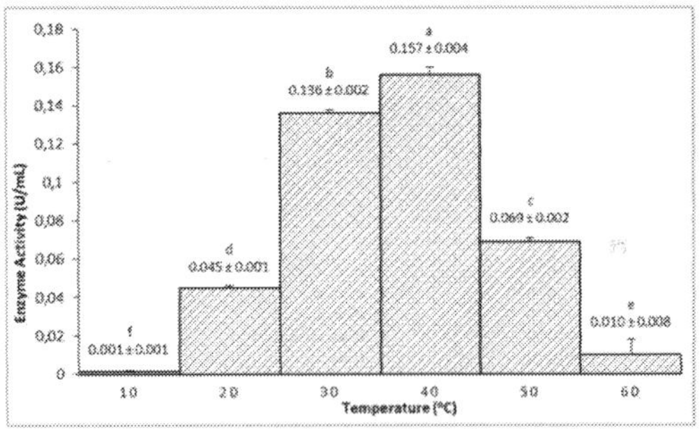

Fig. 5. Average (±SD) trypsin activity (U/mL) in adult stadia at different incubation temperatures (°C)

The results of measurements of enzyme activity (Figure 5) showed that the adult stage of *Anguilla bicolor* had an average enzyme activity at temperatures of 10°C (0.001 ± 0.001 U/mL), 20°C (0.045 ± 0.001 U/mL), 30°C (0.136 ± 0.002 U/mL), 40°C (0.157±0.004 U/mL), 50°C (0.069±0.002 U/mL), and 60°C (0.010±0.008 U/mL). The results of the DMRT test showed that there was a significant difference ($P<0.05$) between the treatments (incubation temperature (°C)) in the adult stage.

The enzyme activity measurements showed differences in trypsin digestion capacity, which was described as trypsin activity between eels with different stages. The younger the type of stadia or the smaller the size of the eel, the higher the value of enzyme activity and total activity. In line with research from [11], which informed that eels with a size of 41.25 ± 0.898 g belong to the elver stage, which has higher trypsin-like activity compared to the size of 319.8 ± 4.666 g of the stadia yellow eel and 569.5 ± 9.150 g pre-silver stage.

This high enzyme activity is related to the role of digestive organs and glands in secreting enzymes [26]. If the function of digestive organs and glands is optimal in secreting enzymes, then the digestive process will be optimal. The optimal digestive process will help to support growth [25]. Another factor that affects the activity of fish trypsin is feed, especially protein.

Protein is a source of energy for fish, especially if the availability of carbohydrates and fats in the feed cannot meet energy needs. Protein is needed continuously because the body needs amino acids to form new proteins during growth [26]. The development will run if energy and protein intake from the feed is sufficient or excessive. Optimum fulfillment of protein needs can affect fish growth [27]. In elver stadia eel, it is suspected that it requires protein to support its growth and encourages the digestive organs and glands to secrete large amounts of enzymes to compensate for the high protein requirements. As a result, protease activity, especially alkaline protease, was higher in the elver stage eel compared to the yellow eel and pre-silver stages [28].

3.4 Optimum Temperature of Fish Enzyme Activity Anguilla bicolor

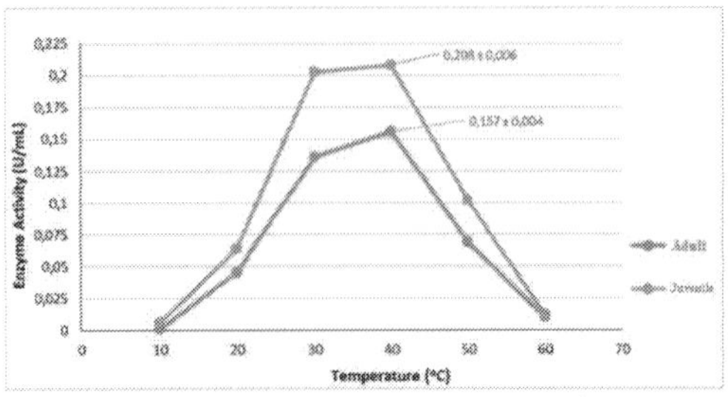

Fig. 6. Optimum temperature (°C) of fish enzyme activity in larval (blue line), juvenile (grey line), and adult (orange line) stages of *Anguilla bicolor*

The optimum temperature for fish enzyme activity (Figure 6) in juvenile stadia *Anguilla bicolor* was 40°C with an enzyme activity value of 0.208 ± 0.006 U/mL. The optimum temperature for fish enzyme activity in juvenile stadia *Anguilla bicolor* was 40°C with an enzyme activity value of 0.157 ± 0.004 U/mL. According to [13], the temperature generally affects reaction rates, interfering with the solubility of reagents, enzyme stability, and kinetic constants. Two opposing mechanisms of activation (rate constant increases with increasing temperature) and denaturation (quaternary thermal opening and tertiary structure of the enzyme) coincide as the reaction temperature increases. Generally, at 50-60°C, the reaction rate (activation) increase exceeds the thermal denaturation limit. Except for thermophilic enzymes, denaturation predominates at temperatures above 60°C, and the reaction rate slows and stops around 80-90°C. The optimum temperature of an enzyme is the temperature where the amount of substrate is the most in a unit change of time.

The effect of temperature on enzyme activity is determined at temperatures ranging from 10 to 70°C. The optimum temperature for all protease sources was 40°C, with the highest activity of 90.61 U ml-1 obtained from the viscera of yellow tuna. The protease activity decreased to 54 U ml-1 at 50°C and sharply reduced at 60-70°C (13 U ml-1). However, proteases are more stable at lower temperatures (37-40°C) [28] which is supported by [29] who stated that proteases, especially trypsin, are active at 35-45°C, whereas protease activity is unstable at lower temperatures and extreme pH.

Enzyme activity can change depending on several factors, including pH, enzyme concentration, substrate concentration, and temperature. Enzyme and substrate concentrations affect the rate of enzymatic reactions. Changes in pH will involve changes in charge on substrates or enzymes, which can cause changes in the structure of enzymes and substrates [28], [30].

Enzymes with protein structures will be significantly affected by temperature. If there is an increase of 10°C above the minimum temperature, the enzyme activity will undoubtedly increase twofold until it reaches optimum conditions. An increase in temperature will generally increase the speed of enzymatic chemical reactions, but an increase in temperature that exceeds the optimum limit can cause enzyme denaturation [31]. Raising the temperature to the optimum point will cause an increase in the speed of the enzyme reaction due to the addition of kinetic energy of the molecules in the substrate and the enzyme so that contact between the enzyme and the substrate can occur. Too high a temperature causes the enzyme to lose its three-dimensional structure and catalytic ability [30]. [32] and [16] reported that fish belonging to the predatory species, for example, *Anoplarchus purpurescens*, had a faster increase in trypsin activity when compared to other types. When enzymes are used in industrial processes and analytical procedures, the evaluation of their activity becomes very important. According to [13], focusing on the industrial scale, the decision to use or not an enzyme in a process must consider several things, such as the amount of enzyme needed to run the process correctly, the duration of the reaction, the amount of substrate to be converted, the number of conditions where the reaction occurs, and the overall cost of the process. The success of the enzymatic process depends on the optimization of three factors, namely the amount of enzyme needed, operating conditions (pH, temperature, and agitation), and the reaction results.

3.5 Molecular Characteristics of Enzyme Fish Trypsin Anguilla bicolor

3.5.1 In-Silico Analysis of Trypsin of Various Species

Searching for the diversity of trypsin gene sequences for various species was carried out through the NCBI gene bank page (https://www.ncbi.nlm.nih.gov/) by writing the names of the targeted genes with the keywords trypsin and trypsinogen. Data were collected, including accessions, nucleotide lengths, and nucleotide sequences in FASTA format. All DNA sequences collected from NCBI were aligned using CLUSTALW in the MEGA X software. It aims to determine the level of homology and identify sequences that have the potential to be used as barcodes. Sequences that have the potential to be barcodes are different and distinctive compared to the others.

Search results for the trypsin gene of various species were carried out through the NCBI gene bank page, obtained 4-four full-length DNA sequences from four species, 16 partial-length DNA sequences from two species, 51 protein sequences from 40 species, 40 partial-length mRNA sequences from 37 species, and 32 full-length trypsin mRNA sequences from 32 different species.

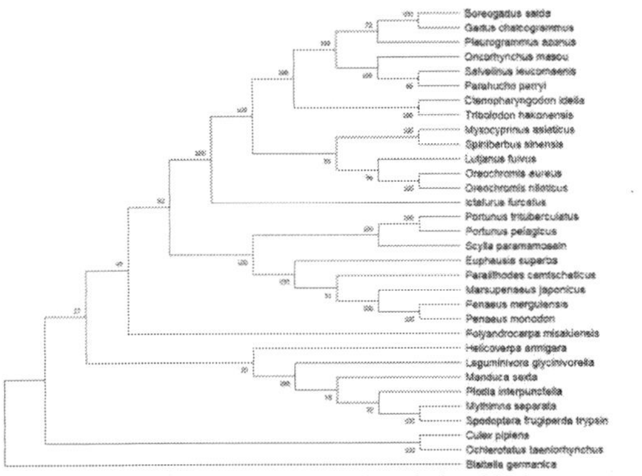

Fig. 7. Phylogenetic tree showing the relationship between trypsin mRNA sequences of various species obtained from NCBI constructed using the Maximum Likelihood method with the GTR+G+I model.

After analyzing the alignment process, these sequences contain many gap areas and only have a few conservation areas. The phylogenetic tree showing the relationship between the trypsin mRNA sequences of various species (Figure 7) illustrates that the trypsin genes of multiple species have a very high sequence diversity. Only a few references provide information about the character sequences of the trypsin gene, so the information is still minimal.

According to [33] and [34], the gene encoding the serine protease enzyme group is the PRSS1 gene. This gene encodes trypsinogen, a member of the serine protease trypsin family. This enzyme is secreted by the pancreas and converted to its active form in the small intestine. Several other trypsinogen genes are localized at the T-cell receptor beta locus on chromosome 7. These genes can provide instructions for making an enzyme called cationic trypsinogen. The PRSS1 gene is TRP1, TRY1, TRY4, and TRYP1.

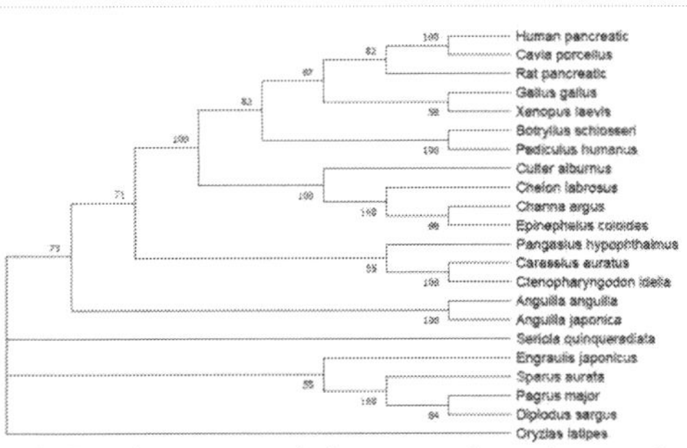

Fig. 8. Trypsinogen gene phylogeny tree for various species constructed using the Maximum Likelihood method with the GTR+G+I model.

Searching for the trypsin gene through the NCBI gene bank page has not shown any species close to the *Anguilla bicolor* type. So that the search was expanded by looking for trypsinogen genes of various species to obtain trypsinogen sequences that are closely related to *Anguilla bicolor*. It can be done to get specific degenerate primers by looking for the closest reference sequence. According to [35], designing a primer for a new gene or organism can be done by aligning the most relative species in one genus and determining the primary candidate in a consensus conservation area. Fragmented reads showing low query cover (≤ 30%) of the entire length of each sequence can be retrieved to minimize poor alignment quality. Species-specific genes are determined based on the query cover level (%) and then used to design primary sets through various existing software.

The results of searching the trypsinogen genes of various species were carried out through the NCBI gene bank page, obtained 4-four full-length DNA sequences from four species and 22 trypsin full-length mRNA sequences from 22 different species. The phylogenetic tree showing the relationship between the trypsinogen mRNA sequences of various species (Figure 8) illustrates that the trypsinogen genes of multiple species have a very high sequence diversity. After analyzing the alignment process, these sequences contain a lot of gap areas and only have a few conservation areas

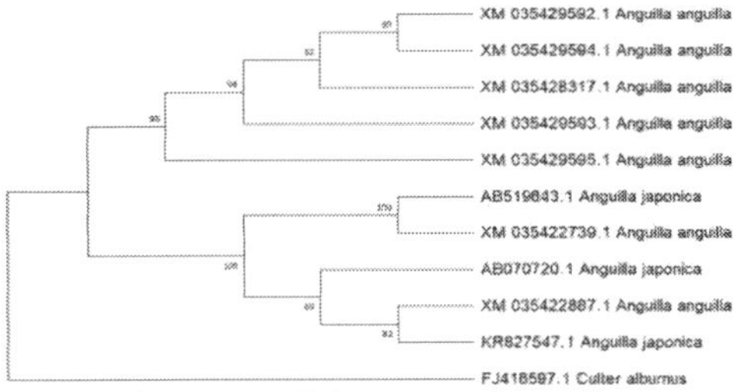

Fig. 9. Trypsinogen gene phylogeny tree of Anguilla spp. was constructed using the Maximum Likelihood method with the GTR+G+I model.

Searching for the trypsinogen gene through the NCBI gene bank page shows that some species are close to the *Anguilla bicolor* type, namely *Anguilla japonica* and *Anguilla anguilla*, which are in the same genus. Therefore, the mRNA sequence of the trypsinogen gene of the Anguilla genus was chosen (Figure 9) to be aligned in forming the *Anguilla bicolor* trypsinogen primary candidate. There are ten full-length trypsinogen mRNA sequences from two Anguilla species which show few gap areas and many conservation areas with nucleotide lengths between 800-900 bp. According to [36], one type of enzyme that plays an essential role in the growth process is the trypsin, where the appearance of trypsin and trypsinogen has been identified in the early stages of larval development of fish species.

3.5.2 Fish Trypsin Primer and Phylogenetic Analysis

The conventional PCR process requires a pair of specific primers to amplify certain parts of the genome. Primers are oligomeric components designed to limit the template and complete the target PCR amplicon's final sequence and the initial site for DNA chain synthesis [37], [38]. According to [39], primers generally have 50-60% guanine-cytosine content and a base length of 15-25 nucleotides. The primers used in PCR are oligonucleotides identical to one of the template DNA chains (5'-phosphate) and oligonucleotides identical to the other template (3'-OH). Each of the PCR primers can complement a single strand that is different from the double-stranded target.

Data processing results with trimming, contig, and alignment stages using the BioEdit program[32]. Data processing based on these steps resulted in the DNA sequence of the *Anguilla bicolor* trypsinogen gene. Alignment was performed on the DNA of the *Anguilla bicolor* trypsinogen gene with the sequences in the GenBank database using the BLAST program. The assembly results gave a sequence length of 1030, 1422, and 1459 bp for the same sample using primer options A, B, and C.

The resulting contig sequences were identified taxonomically using BLAST with the "nucleotide collection" database (https://blast.ncbi.nlm.nih.gov/Blast.cgi). Sample sequences A, B, and C with the highest percentage of similarity and total score are referred to as the taxon for the input sequences. Next, the top 14 taxa with the highest similarity values were selected to be used in phylogenetic construction. The sequence identities used for phylogenetic construction are as shown in Table 7.

Table 7. Sequence identity for phylogenetic construction

No.	Taxa Name	Locality	GenBank Accesion No.
1	*Anguilla japonica*	China	KR827547.1
2	*Anguilla anguilla*	Republik Ceko	XM_035422887
3	*Anguilla anguilla*	Republik Ceko	XM_035422739
4	*Anguilla japonica*	Japan	AB519643.1
5	*Anguilla japonica*	Japan	AB070720.1
6	*Megalops cyprinoides*	Singapura	XM 036515540.1
7	*Megalops cyprinoides*	Singapura	XM 036538626.1
8	*Megalops cyprinoides*	Singapura	XM 036537958.1
9	*Megalops cyprinoides*	Singapura	XM 036537957.1
10	*Anguilla anguilla*	Republik Ceko	XM 035429595.1
11	*Anguilla anguilla*	Republik Ceko	XM 035428317.1
12	*Anguilla anguilla*	Republik Ceko	XM 035429594.1
13	*Anguilla anguilla*	Republik Ceko	XM 035429592.1
14	*Anguilla anguilla*	Republik Ceko	XM 035429593.1

Analysis of the maximum likelihood method of the GTR+G+I model with Bootstrap 1000 has been carried out and confirmed using Molecular Evolutionary Genetics Analysis (MEGA 11) software to obtain cross-species reconstruction based on branch line length. Different line lengths indicate each species' evolution level [40]. Based on the phylogram, it looks like a tree showing the distance in evolutionary time. The longer the line, the farther the evolutionary distance, while the shorter the line, the closer the evolution of the species.

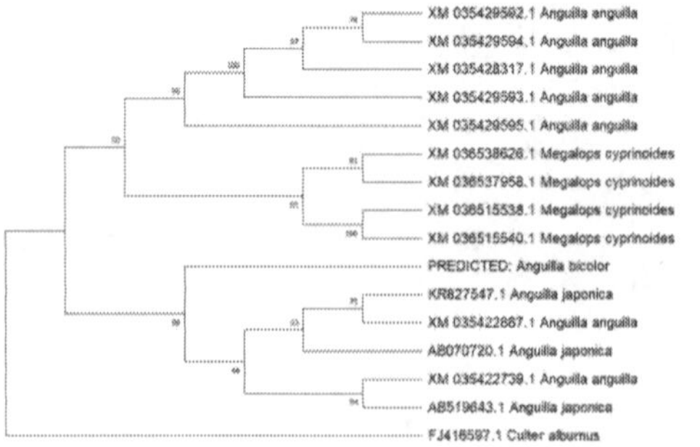

Fig. 10. The tree topology was constructed using the Maximum Likelihood method with the GTR+G+I model.

The *Anguilla bicolor* trypsinogen gene sequences analyzed were in the same group. They tended to be close to the trypsinogen and trypsin-like genes from *Anguilla japonica* and *Anguilla anguilla* with a bootstrap value at the node of 92 (Figure 10). This shows that the *Anguilla bicolor* kinship based on the trypsin gene sequence has the closest kinship with other Anguilla species in the same genus. The phylogenetic tree shows that the in-group groups are grouped within each clade based on the similarity of the trypsin gene sequence.

Conclusion

1. The protein content of the enzyme extract in the juvenile stage of *Anguilla bicolor* had an average of 0.488 ± 0.004 g/dL, and the adult stage of *Anguilla bicolor* had an average of 1.778 ± 0.080 g/dL. The highest enzyme activity was obtained in the juvenile stage, 0.529 ± 0.016 (U/mL), and in the adult stage, 0.399 ± 0.009 (U/mL). Enzyme activity increases with increasing temperature used and reaches a maximum of 40°C.
2. Trypsin sequences of *Anguilla bicolor* analyzed tend to be close to the Trypsinogen and Trypsin-like genes from *Anguilla japonica*, *Anguilla anguilla*, and *Megalops cyprinoides*.

References

1. R. S. Prakasham, C. Subba Rao, R. Sreenivas Rao, and P. N. Sarma, "Alkaline protease production by an isolated Bacillus circulans under solid-state fermentation using agroindustrial waste: Process parameters optimization," *Biotechnol. Prog.*, vol.

21, no. 5, pp. 1380–1388, 2005, doi: 10.1021/bp050095e.

2. D. S. Ningthoujam, P. Kshetri, S. Sanasam, and S. Nimaichand, "Screening , Identification of Best Producers and Optimization of Extracellular Proteases from Moderately Halophilic Alkalithermotolerant Indigenous Actinomycetes," *World Appl. Sci. J.*, vol. 7, no. 7, pp. 907–916, 2009.
3. B. S. Kaphalia, "Biomarkers of acute and chronic pancreatitis," *Biomarkers Toxicol.*, pp. 279–289, 2014, doi: 10.1016/B978-0-12-404630-6.00016-6.
4. H. Sundus, H. Mukhtar, and A. Nawaz, "Industrial Applications and Production Sources of Serine Alkaline Proteases: A Review," *J. Bacteriol. Mycol. Open Access*, vol. 3, no. 1, pp. 191–194, 2016, doi: 10.15406/jbmoa.2016.03.00051.
5. Y. Zhang, Q. Liang, C. Zhang, J. Zhang, G. Du, and Z. Kang. "Improving production of Streptomyces griseus trypsin for enzymatic processing of insulin precursor," *Microb. Cell Fact.*, vol. 19, no. 1, pp. 1–11, 2020, doi: 10.1186/s12934-020-01338-9.
6. T.- Nurhayati, E.- Salamah, - - Cholifah, and R.- Nugraha, "Optimasi Proses Pembuatan Hidrolisat Jeroan Ikan Kakap Putih," *J. Pengolah. Has. Perikan. Indones.*, vol. 17, no. 1, pp. 42–52, 2014, doi: 10.17844/jphpi.v17i1.8136.
7. R. G. La Apu, "PEMANFAATAN LIMBAH JEROAN IKAN CAKALANG (Katsuwonus pelamis) SEBAGAI BAHAN SUBTITUSI TEPUNG IKAN PADA PERTUMBUHAN IKAN NILA (Oreochromis niloticus)," *J. Sains dan Teknol. Perikan.*, vol. 1, no. 2, pp. 13–24, 2021.
8. V. Venugopal, "Enzymes from Seafood Processing Waste and Their Applications in Seafood Processing," *Adv. Food Nutr. Res.*, vol. 78, pp. 47–69, 2016, doi: 10.1016/bs.afnr.2016.06.004.
9. S. P. Kumari and R. Reshma, "Effect of alkaline protease produced from fish waste as substrate by bacilluclausii on destaining of blood stained fabric," *J. Trop. Life Sci.*, vol. 11, no. 1, pp. 59–66, 2021, doi: 10.11594/jtls.11.01.08.
10. K. Jesús-de la Cruz, C. A. Álvarez-González, E. Peña, J. A. Morales-Contreras, and Á. Ávila-Fernández, "Fish trypsins: potential applications in biomedicine and prospects for production," *3 Biotech*, vol. 8, no. 4, 2018, doi: 10.1007/s13205-018-1208-0.
11. F. A. Larassagita, Hana, and U. Susilo, "Aktivitas Tripsin-Like dan Kimotripsin-Like pada Ikan Sidat Tropik *Anguilla bicolor* McClelland," vol. 5, no. 1, pp. 55–60, 2018, doi: https://doi.org/10.20884/1.SB.2018.5.1.789.
12. H. Y. Sugeha and M. U. Genisa, "External and internal morphological characteristics of glass eels *Anguilla bicolor* bicolor from the Cibaliung River Estuary , Banten , Indonesia," *Oseanologi dan Limnol. di Indones.*, vol. 41, no. 1, pp. 37–48, 2015.
13. K. A. Gaidhani, M. Harwalkar, and P. S. Nirgude, "World Journal of Pharmaceutical ReseaRch SEED EXTRACTS," *World J. Pharm. Res.*, vol. 3, no. 3, pp. 5041–5048, 2014, doi: 10.20959/wjpr20202-16660.
14. N. Salkind, "American Statistical Association," *Encycl. Res. Des.*, vol. 51, no. 276, pp. 667–669, 2012, doi: 10.4135/9781412961288.n9.
15. K. Rungruangsak-Torrissen, R. Moss, L. H. Andresen, A. Berg, and R. Waagbø, "Different expressions of trypsin and chymotrypsin in relation to growth in Atlantic salmon (Salmo salar L.)," *Fish Physiol. Biochem.*, vol. 32, no. 1, pp. 7–23, 2006, doi: 10.1007/s10695-005-0630-5.
16. T. Nurhayati, R. Nugraha, and D. Lihuana, "Characterization of Ammonium Sulphate Fraction Tripsin Isolated from Intestine of Little Tuna," *Jphpi*, vol. 23, pp. 372–382, 2020.
17. C. Turan, "A note on the examination of morphometric differentiation among fish populations: The Truss System," *Turkish J. Zool.*, vol. 23, no. 3, pp. 259–263, 1999.
18. A. B. Kusuma, "Komposisi Nutrisi Ikan Sidat *Anguilla bicolor* bicolor dan Anguilla marmorata," *Jphpi*, vol. 21, no. 3, pp. 504–512, 2018.
19. F. S. Kertikasari, N. Cokrowati, and A. W. Puspitasari, "Gut content analysis of tilapia (," vol. 040007, 2019.
20. Y. Sugianti, M. R. A. Putri, and S. . Purnamaningtyas, "Eel fish species (Anguilla spp.) and its migratory habitat characteristics in Cikaso River, Sukabumi, West Java.," *Limnotek Perair. darat Trop. di Indones.*, vol. 27, no. 1, pp. 39–54, 2020.
21. D. L. Kramer and M. J. Bryant, "Intestine length in the fishes of a tropical stream: 2. Relationships to diet - the long and short of a convoluted issue," *Environ. Biol. Fishes*, vol. 42, no. 2, pp. 129–141, 1995, doi: 10.1007/BF00001991.
22. B. C. Sungchul Bai, K. Katya, and D.-J. Kim, "Japanese eel aquaculture in Korea," vol. 2011, no. November 2012, pp. 1–6, 2012, [Online]. Available: https://www.aquaculturealliance.org/advocate/japanese-eel-aquaculture-in-korea/?headlessPrint=AAAAAPIA9c8r7gs82oWZBA
23. D. Sanjayasari and Kasprijo. "Estimasi Nisbah Protein-Energi Pakan Ikan Senggaringan (*Mystus Nigriceps*) Dasar Nutrisi Untuk Keberhasilan Domestikasi," *J. Perikan. dan Kelaut.*, vol. 15, pp. 89–97, 2010.
24. L. D. Nelson and M. M. Cox, "Principle of Biochemistry," *Cell Biol. Physarum Didymium*, pp. 393–435, 1982, doi: 10.1016/b978-0-12-049601-3.50017-1.
25. D. Y. . Airin and C. Lumenta, "Pakan diameter berbeda bagi pertumbuhan benih sidat (Anguilla sp)," *e-Journal Budid. Perair.*, vol. 3, no. 3, pp. 30–41, 2015, doi: 10.35800/bdp.3.3.2015.10409.

26. C. D. Webster and W. Oxon, "Nutrient requirements and feeding of finfish for aquaculture C.D.," vol. 214, pp. 419–420, 2002.
27. Z. Yandes, R. Affandi, and I. Mokiginta, "Pengaruh pemberian selulosa dalam pakan terhadap kondisi biologis benih ikan gurami (Osphronemus gourami Lac)," *J. Iktiologi Indones.*, vol. 3, no. 1, pp. 27–33, 2003.
28. E. Liviawaty and E. Afrianto, "Pakan Ikan," *Yogyakarta: Kanisius*, no. September, pp. 69–82, 2012.
29. A. Bougatef, "Trypsins from fish processing waste: Characteristics and biotechnological applications - Comprehensive review," *J. Clean. Prod.*, vol. 57, pp. 257–265, 2013, doi: 10.1016/j.jclepro.2013.06.005.
30. K. Arief, F. Nisa, dan U. Murdiyatmo, J. Teknologi Hasil Pertanian, F. Teknologi Pertanian, and U. Brawijaya, "Partial Characterization of Crude Protease Extracted from Bacillus amyloliquefaciens NRRL B-14396," *J. Teknol. Pertan.*, vol. 7, no. 2, pp. 96–105, 2006.
31. L. M. Shuler and F. Kargi, "Edition, Bioprocess Engineering Basic Concepts Second," *Prentice Hall books Up. Saddle River, NJ 07458*, vol. 22, no. 3, p. 293, 2022, doi: 10.1016/0168-3659(92)90106-2.
32. M. A. K. Bahrin, M. F. Othman, N. H. N. Azli, and M. F. Talib, "Industry 4.0: A review on industrial automation and robotic," *J. Teknol.*, vol. 78, no. 6–13, pp. 137–143, 2016, doi: 10.11113/jt.v78.9285.
33. C. Férec, O. Raguénès, R. Salomon, C. Roche, J. P. Bernard, M. Guillot, I. Quéré, C. Faure, B. Mercier, M. P. Audrézet, P. J. Guillausseau, C. Dupont, A. Munnich, J. D. Bignon, L. Le. Bodic "Mutations in the cationic trypsinogen gene and evidence for genetic heterogeneity in hereditary pancreatitis," *J. Med. Genet.*, vol. 36, no. 3, pp. 228–232, 1999.
34. Hu C, Wen L, Deng L, Zhang C, Lugea A, Su HY, Waldron RT, Pandol SJ, Xia Q, "The Differential Role of Human Cationic Trypsinogen (PRSS1) p.R122H Mutation in Hereditary and Nonhereditary Chronic Pancreatitis: A Systematic Review and Meta-Analysis," *Gastroenterol. Res. Pract.*, vol. 2017, 2017, doi: 10.1155/2017/9505460.
35. I. You and E. B. Kim, "Genome-based species-specific primers for rapid identification of six species of Lactobacillus acidophilus group using multiplex PCR," *PLoS One*, vol. 15, no. 3, pp. 1–9, 2020, doi: 10.1371/journal.pone.0230550.
36. A. Kvåle, A. Mangor-Jensen, M. Moren, M. Espe, and K. Hamre, "Development and characterisation of some intestinal enzymes in Atlantic cod (Gadus morhua L.) and Atlantic halibut (Hippoglossus hippoglossus L.) larvae," *Aquaculture*, vol. 264, no. 1–4, pp. 457–468, 2007, doi: 10.1016/j.aquaculture.2006.12.024.
37. D. A. Hewajuli and N. Dharmayanti, "The Advance of Technology of Reverse Transcriptase-Polymerase Chain Reaction in Identifying the Genome of Avian Influenza and Newcastle Diseases," *Indones. Bull. Anim. Vet. Sci.*, vol. 24, no. 1, pp. 16–29, 2014, doi: 10.14334/wartazoa.v24i1.1022.
38. M. Ehtisham, F. Wani, I. Wani, P. Kaur, and S. Nissar, "Polymerase Chain Reaction (PCR): Back to Basics," *Indian J. Contemp. Dent.*, vol. 4, no. 2, p. 30, 2016, doi: 10.5958/2320-5962.2016.00030.9.
39. R. R. Garafutdinov, A. A. Galimova, and A. R. Sakhabutdinova, "The influence of quality of primers on the formation of primer dimers in PCR," *Nucleosides, Nucleotides and Nucleic Acids*, vol. 39, no. 9, pp. 1251–1269, 2020, doi: 10.1080/15257770.2020.1803354.
40. C. Lambré J. M. Barat, C. Bolognesi, and P. S. Cocconcelli, "Safety evaluation of food enzyme trypsin from porcine pancreas," *EFSA J.*, vol. 19, no. 6, 2021, doi: 10.2903/j.efsa.2021.6637.

Classification of finger pulse oximeter based on their response time using quantitative analysis

Septia Khairunnisa[1,4*], *Indah* Soesanti[2], and *Dyah* Listyarifah[3]

[1]Biomedical Engineering, The Graduate School, Universitas Gadjah Mada
[2]Electrical Engineering and Information Technology Department, Faculty of Engineering, Universitas Gadjah Mada
[3]Department of Dental Biomedical Sciences, Faculty of Dentistry, Universitas Gadjah Mada
[4]Loka Pengamanan Fasilitas Kesehatan Banjarbaru

Abstract. The measurement of response time of pulse oximeter lacks standardized method and proper thresholds and alternative measurement methods are needed to minimize error and improve efficiency in calibration method of pulse oximeter due to the increasing number of pulse oximeter variations distributed due to the covid 19 pandemic. This study aims to measure the Response Time (RT) of a finger pulse oximeter with 6 different types of conditioning to determine the RT mean and standard deviation in order to classify the pulse oximeter based on their types. We evaluated the response time of 50 finger pulse oximeters (20 patient monitor type, 8 handheld type, and 22 fingertip type) using 6 saturation and desaturation conditioning methods with a SpO2 Simulator. Quantitative analysis used to determine the initial threshold value. From 50 pulse oximeters found the fastest response pulse oximeter are fingertip type with mean time are 9.71 seconds and the most stable of each conditioning pulse oximeter are patient monitor type with the average RT 3.47 seconds, and Handheld type are put in the middle. With the conclusion that Patient monitor type are classified in monitoring class, Handheld type are classified both in monitoring and diagnostic class and Fingertip both classified in diagnostic and preventive class.

Keywords: *Response time; Calibration; Quantitative Analysis; Patient Monitor; Handheld; Fingertip*

1 Introduction

Blood oxygen saturation measured by a pulse oximeter is the interpretation of oxygen bound by hemoglobin in arterial blood vessels. Reduced vascular oxygen saturation occurs due to cardiovascular and respiratory disruptions, leading to insufficient oxygen supply to organs and potential malfunction. Brain malfunction poses a high risk since it regulates overall body functions, potentially resulting in organ failure and death [1].

Pulse oximeters play a vital role in assessing patients' physical condition and have seen increased use due to the COVID-19 pandemic [2]. Originally used in critical care settings, such as operating rooms and intensive care units, they are now commonly used in the community for diagnosis and preventive measures [2]. Despite their widespread use, medical personnel often lack knowledge about the limitations and principles behind their function, In addition, providers do not know the basic principles behind the mechanism of its function, that it has limitations, which can lead to errors in oxygen saturation readings. Understanding these limitations and ensuring proper testing and calibration is crucial [4].

Testing is the process of physically examining, measuring, and assessing the function of a medical device by comparing it to a standard device to determine any measurement errors. On the other hand, calibration is performed to verify the accuracy of a device's indicator value. This procedure is carried out by the Health Facility Inspection Centre (Balai Pengamanan Fasilitas Kesehatan / BPFK) and private calibration agencies in accordance with the work methods issued by the Directorate General of the Ministry of Health of the Republic of Indonesia [6]. In the case of pulse oximeter devices, testing involves conducting physical, functional, and electrical safety checks. Calibration is performed by measuring heart rate and oxygen saturation parameters at specific measuring points [7]. One of the ways to check the device's function is by examining the response time.

The response time of a pulse oximeter refers to the delay between the appearance of arterial hypoxemia (low oxygen levels in the blood) and its detection by the pulse oximeter. Several factors influence the response time, including blood perfusion, skin and nail color, hemoglobin levels, body movement, sensor placement, sensor type, and the quality of device components and connections [8] [4]. This delay in response can lead to errors in oxygen saturation readings, particularly in situations where rapid changes in oxygen saturation are

* Corresponding author: septiakhairunnisa1996@mail.ugm.ac.id

expected, such as during cardiopulmonary resuscitation, when handling patients with compromised airways, in pediatric patients, delivery mothers with congenital heart or lung disease, or for the early detection of happy hypoxia in COVID-19 patients [8] [9] [10]. Therefore, it is necessary to have a shorter response time for detecting hypoxemia with pulse oximeters [8].

The response time of different pulse oximeters can vary. Some devices have an average response time of 8-15 seconds in default mode, 130 seconds under normal conditions, and 215 seconds when the patient has mild hypothermia [10]. The response time is closely related to the circulation time of blood. Pulse oximeters placed in large blood vessels near the heart, such as the head, earlobes, and trachea, have a faster response time compared to peripheral parts like fingers and toes [11], [12]. To summarize, response time in pulse oximeters refers to the delay in detecting arterial hypoxemia. Factors such as blood perfusion, skin and nail color, hemoglobin levels, body movement, sensor placement, sensor type, and device quality can influence the response time. A shorter response time is necessary for accurate detection of hypoxemia in various clinical situations. Different pulse oximeters have varying response times, with faster detection in devices placed closer to the heart.

Measurement of pulse oximeter response time in the pre-calibration function check process is currently carried out qualitatively with subjective considerations from officers in determining whether the device has a fast or slow response, based on comparisons between pulse oximeters. This is due to the absence of guidance on the method of determining the pulse oximeter response time threshold, from the Director General of Healthcare in 2018. In recent years, the use of pulse oximeters in the community has increased, so that new brands appear on the market that are not well controlled for quality, including the quality of the device's response time. This can reduce the validity of the saturation number indicated by the device, which in turn can lead to errors in determining saturation results and taking necessary medical actions. Therefore, in addition to testing and calibration, proper threshold of response time needs to be done. This is to minimize the possibility of oxygen saturation reading errors by calibration officers and medical personnel, due to pulse oximeters that have a slow response time, and improve the efficiency of the testing and calibration process.

Previous research related to pulse oximeter response time, using a pulse oximeter with a brand and a limited number of devices [4], as well as in certain conditions of hypoxemia in human patients that cause a decrease in the accurate response of the pulse oximeter [13], indicate the need for an alternative method to measure response time of pulse oximeters, as the accuracy of oxygen saturation values can be compromised by uncontrolled variables. One proposed solution is to use a simulator or artificial finger instead of human patients, reducing the influence of uncontrolled variables and allowing for multiple measurements without risking patient harm The classification model derived from this research divides data into three classes: preventive, diagnostic, and monitoring, each tailored to the specific use of the pulse oximeter.

2 Method

2.1 Pulse Oximeter Calibration

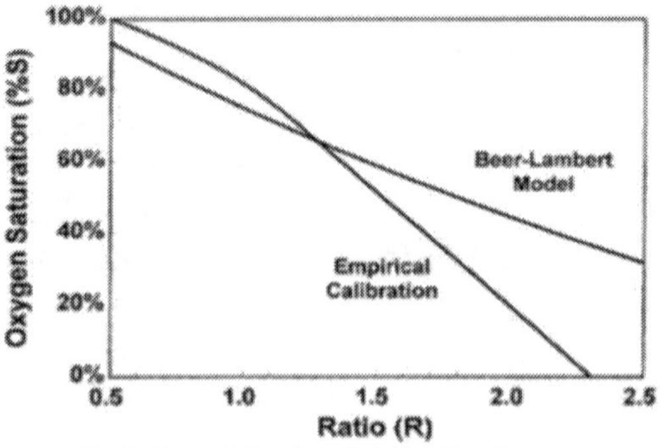

Fig. 1. Two relationships between R-ratio and oxygen saturation of patients [12]

$$SaO_2 = \frac{C_O}{C_O + C_O} = \frac{\alpha_{r2}R - \alpha_{r1}}{(\alpha_{r2} - \alpha_{o2})R - (\alpha_{r1} - \alpha_{o1})} \quad (1)$$

Pulse oximeters originally employed equation (1) during their early 1980s manufacturing phase to compute arterial SaO2. However, the use of Beer-Lambert's law as the calculation basis inadequately considered the scattering of light by red blood cells. Despite utilizing an alternate technique, oximetry only partially compensates for scattering due to wavelength variations. Equation (1) is an oversimplification. Figure 1 illustrates two relationships: one based on Beer-Lambert's law and another from empirical data, connecting the R ratio and patient oxygen saturation. Devices following Beer-Lambert's law often inaccurately estimate oxygen saturation, particularly below 85% SaO2 values. Over time, methods have emerged to incorporate scattering into the theory. Today, many pulse oximeters rely on lookup tables derived from calibration studies involving healthy volunteers with invasively measured oxygen saturation [12].

2.2 SpO2 Simulator

SpO2 Simulator is a device used as a reference for pulse oximeters, with adjustable parameters (oxygen saturation percentage, heart rate, and blood perfusion percentage) that includes an artificial finger part for testing. Annex FF of ISO 80601-2-61 standard clarifies the distinction between "calibrator" and "simulator." A calibrator is a primary standard with higher accuracy than the Unit Under Test (UUT), while a simulator is a transfer standard that serves as a validated reference. As described in figure 3, Pulse oximeters use two light wavelengths and the ratio of pulsatile and non-pulsatile signals to determine oxygen saturation. The monitor firmware calculates the R value and displays the oxygen saturation percentage and pulse rate with the comparisons between these values as describe in figure 2. Objective performance verification of pulse oximeters has been challenging [13].

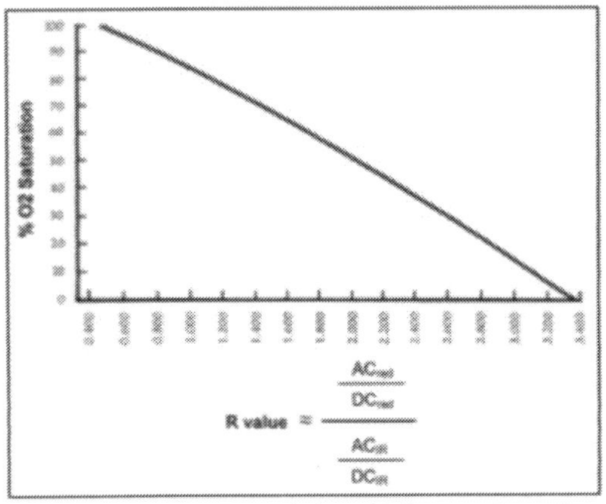

Fig. 2. R-curve, correlation with O_2 saturation with R value [13]

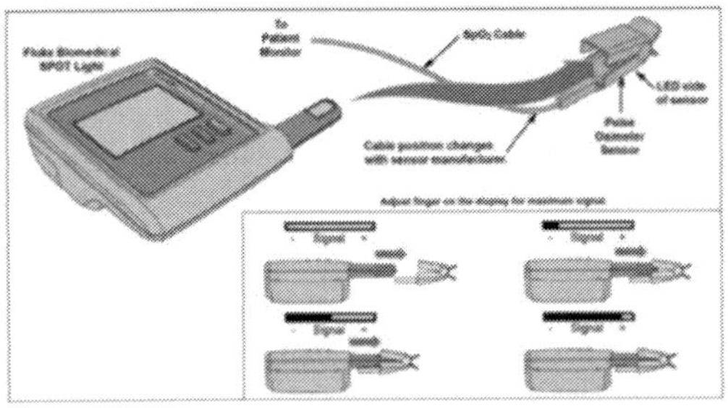

Fig. 3. SpO2 Simulator and how to use it in a pulse oximeter sensor [14]

2.3 Materials

The study used 50 finger pulse oximeter with 3 different types of devices consist of 20 patient monitor type (PM), 8 handheld type (HH) and 22 fingertip type (FT) (figure 2). The purpose of selecting these three types of equipment as measuring objects is because they are commonly used in various settings, including hospitals, health centers, and by the general public. Each type of equipment has specific specifications that cater to different consumer needs. For example, patient monitors are frequently found in hospitals, particularly in intensive care units and operating rooms, as they can be used continuously. Fingertip pulse oximeters, on the other hand, have a small and lightweight design and are relatively inexpensive, making them suitable for use by the general public and in health centers. Handheld pulse oximeters have higher specifications than fingertip oximeters and are easier to carry compared to patient monitors, making them ideal for intense usage in emergency departments and intensive care units for measurements at any time.

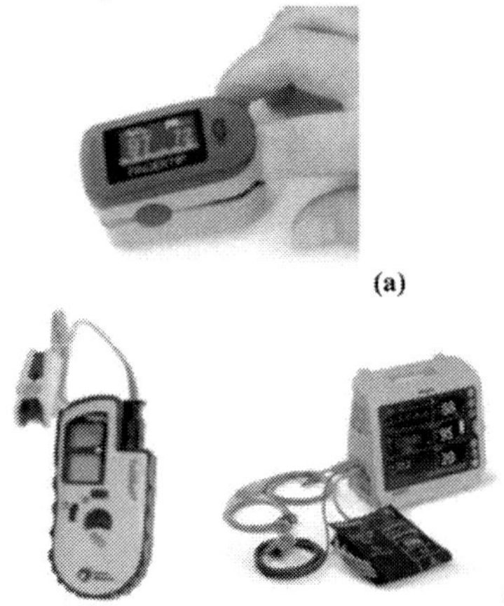

(a)

(b) (c)

Fig. 4. (a) Pulse oximeter fingertip[14]. **(b)** Pulse oximeter Handheld[15]. **(c)** Pulse oximeter patient monitor[16]

To ensure the suitability of the pulse oximeter as a measuring object, established reliability standards for the device. The tested device must meet accuracy and precision standards based on the calibration method. This includes the difference between the device's readability and a 1% difference in oxygen saturation within the range of 100-85% oxygen saturation. It is even better if the device possesses a certificate of fitness for use issued by an authorized testing and calibration institution. The tools using in this research are described in the table 1.

Table 1. The Tools

Tools	Brand / Type	Parameter	Units
SpO_2 Simulator	Fluke / Spotlite	Oxygen Saturation	%
		Heart Rate	Beat Per Minute
		Perfusion Index	%
Stopwatch	Extech	Time	second
Thermo hygrometer	Greisinger or Extech	Temperature	°C
		Relative Humidity	%RH

2.4 Method

The method of data acquisition involves taking samples by measuring the response time of the 50 finger pulse oximeter in two ways: placing the device on the index finger or thumb, as shown in Figure 5a, and measuring time using a stopwatch; and placing the device on the SpO2 Simulator and measuring time using a stopwatch, as shown in Figure 5b. Sampling data from the pulse oximeter response time is performed to measure the accuracy and precision of the measurements at different oxygen saturation conditions. The following conditions are considered:

1) The pulse oximeter is turned on but not yet placed on the finger until a stable result is obtained.
2) The pulse oximeter is turned on but not yet installed on the SpO2 Simulator until stable results are achieved with specific value settings on the simulator such as: 100 %, 99 %, 98 %, 97 %, 95 %, 90 %, 85 %, 80 %, 75%, and 75 %.
3) The pulse oximeter is turned on and attached to an SpO2 Simulator with high to low oxygen saturation values set (desaturation), measured between points such as : 100-99 %, 99-98 %, 98-97 %, 97-95 %, 95-90 %, and 90-85 %.
4) The pulse oximeter is turned on and attached to an SpO2 Simulator with low to high oxygen saturation value settings (re-saturation), measured between points such as : 85-90 %, 90-95 %, 95-97 %, 97-98 %, 98-99 %, and 99-100 %.
5) The pulse oximeter is turned on and attached to an SpO2 Simulator with oxygen saturation values set from normal to hypoxemia, with three different oxygen saturation conditions. Such as : 100-95 %, 100-90 %, 100-85 %, 99-95 %, 99-90 %, 99-85 %, 98-95 %, 98-90 %, and 98-85 %.
6) The pulse oximeter is turned on and attached to an SpO2 Simulator with hypoxemia oxygen saturation values set to normal, with three different oxygen saturation conditions. Such as 95-100 %, 9 %0-100 %, 85-100 %, 95-99 %, 90-99 %, 85-99, 95-98 %, 90-98 %, and 90-85 %.

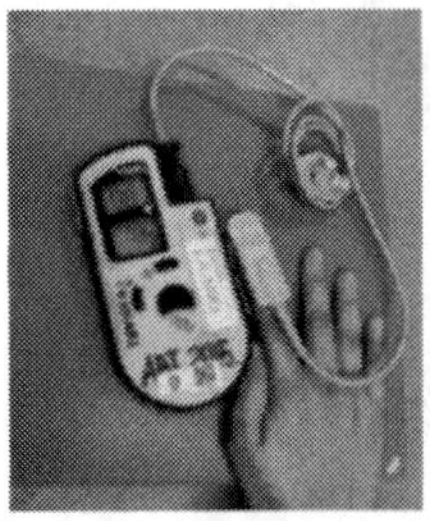
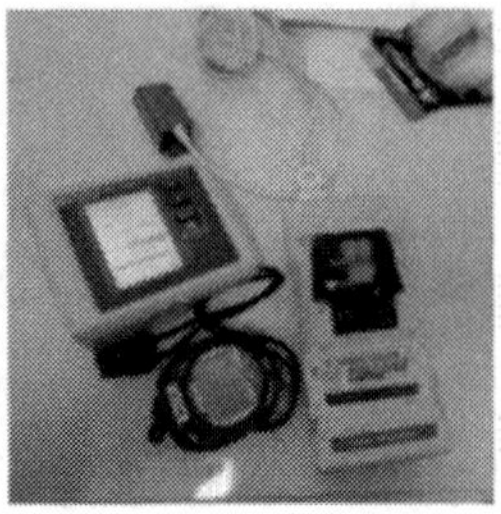

Fig. 5. (a) Oxygen saturation measurement on index finger using Pulse Oximeter
(b) Example of Finger Pulse Oximeter Installation on Simulator, response time measured using Stopwatch

The data is presented in form of a graph of oxygen saturation values towards response time, with certain saturation setting conditions according to the tables 1-6. And the research location is at calibration laboratory of LPFK Banjarbaru, as well as at hospitals and health facilities that are willing to collect sampling data in kalimantan.

This research begins with the process of acquiring Response Time (RT) data from 50 pulse oximeters. This RT value is obtained by measuring the pause time in the oxygen saturation measurement process on the pulse oximeter attached to the researcher's finger and artificial finger on the simulator, with 6 different conditions according to the explanation in section 3.1. After the RT value of each condition is measured, the measurement data will be analyzed using quantitative analysis in accordance with the explanation in the theoretical basis section 2.4 to obtain the type of distribution of the data set. The results of this analysis will then be compared with references from other medical research journals related to pulse oximeter RT, to see if there is any relevance of the measurement results of the RT measured using the previous research method with the method used by the researcher.

The results of this analysis were also compared with medical journal references related to the time lag threshold of pause hypoxemia detection, especially in medical conditions related to anaesthesia and respiratory abnormalities. This is to assess whether the response time of the current pulse oximeter meets the time lag threshold of hypoxemia detection or not.

The output of this analysis process will result in the grouping of RT values into 3 groups, namely monitoring, diagnosis, and preventive groups. This division is made based on the use of the pulse oximeter in health services, with the fastest response in the monitoring group, and the longest response in the preventive group. In brief, the division of these groups is divided with the following considerations:

1. Monitoring: intended for pulse oximeters that are used continuously and are expected to detect changes in oxygen saturation values in patients quickly and precisely to speed up the treatment process if there are indications of hypoxemia. For example, use in the ICU room, in the operation process in the OK room, and so on.
2. Diagnosis: intended to detect the oxygen saturation value at the beginning of the measurement. As a tool to help doctors diagnose a disease related to oxygen saturation values. It is expected that the pulse oximeter used in this process brings up an accurate value at the beginning of the measurement, thus minimizing the risk of taking pulse oximeter measurements that do not match the actual oxygen saturation value. For example, use in Emergency Units, inpatient units, and so on.
3. Preventive: intended for the screening process or homecare in patients with indications that may experience a decrease in oxygen saturation at any time. It is hoped that the use of this pulse oximeter can help detect early if there is

hypoxemia in the patient. For example, it is used in patients who are undergoing self-isolation due to Covid-19 infection, screening before the Covid-19 antigen or PCR examination process, and in the medical check-up process.

2.5 Data Analysis

Pulse oximeter data were analyzed quantitavively with descriptive and inferential analysis using Microsoft Excel applications.

3 RESULTS

Most of the measured pulse oximeter devices are from Idaman Banjarbaru district hospital from various treatment rooms. Meanwhile, the other data was obtained from devices sent to the laboratory from various healthcare facilities in Kalimantan. The difference in the quantity of these three types of devices is due to the difference in the number of devices available in hospitals and other healthcare facilities. The most commonly used equipment is patient monitors and fingertip oximeters, while the availability of Handhelds tends to be limited in hospitals. This is because fingertip oximeters are easy to use due to their small size, and patient monitors have comprehensive features for examining patients, such as Electrocardiograph (ECG), temperature, Non-invasive blood pressure (NIBP), Respiration rate (RR), oxygen saturation, and other features. On the other hand, Handhelds type only have one feature and are not as compact as fingertip oximeters. This is why Handhelds are not widely used for measuring oxygen saturation in healthcare facilities.

An overview of the distribution of average values and standard deviations obtained from 50 devices, which were used to measure the direct oxygen saturation on the finger. The range of average values spans from 6 to 18 seconds, while the range of standard deviations ranges from 0 to 4 seconds. Device 6, FT exhibited the highest average value and standard deviation. On the other hand, device 36, HH displayed the lowest average value, whereas device 4, FT had the lowest standard deviation. Notably, devices number 5 and 6, both manufactured by the brand Onecare, demonstrated average values and standard deviations that fell outside the overall range of the devices. The overall average of the average values across all devices amounted to 8.64 seconds, with a total standard deviation of 0.95 seconds. The results of analysis of first conditioning data acquisition shows in terms of average, the smallest one is FT at 8.30 seconds and the highest is HH at 12.05 seconds. In terms of standard deviation, the smallest to the highest are PM (1.89 seconds), HH (3.38 seconds), and FT (4.02 seconds). From the clustering graph in figure 6a, based on the average and standard deviation, there are 2 instruments that have results outside the group, namely instrument 6 with an average of 17.92 and a standard deviation of 4.26 seconds, and instrument 13 with an average of 13.32 and a standard deviation of 4.14 seconds.

An overview of the distribution of average values and standard deviations obtained from 50 devices from second conditioning was found that each device displayed fluctuating averages and standard deviations. The device with the highest average RT was number 17, which had a value of 15.48 seconds and belonged to the PM. On the other hand, device number 36 had the highest standard deviation, measuring 9.46 seconds, and belonged to the Handheld category. Device number 3, also HH, exhibited the lowest averages and standard deviations. The overall average measurement of RT was 8.69 seconds, with a corresponding average standard deviation of 2.24 seconds.

The additional graph 5b indicates that the highest average is possessed by PM, while the other two types have averages below the overall average. The lowest average is found in the FT. Meanwhile, the standard deviation of PM shows the most stable value compared to the other two. In the standard deviation graph of the HH, there is an increase in RT starting from an oxygen saturation of 98% and it experiences a decline when the oxygen saturation approaches 97% and then there is a gradually increase until it reaches its peak at the 85%-point mark within 10 seconds. Afterward, it experiences a decline when the oxygen saturation approaches 70%. Additional figure 2b revealed that three devices, namely 16, 36, and 44, had values that deviated from the average and standard deviation graph. Devices 36 and 44 were FT, while device 16 belonged to the FT type.

From the results of measurement of 3rd conditioning there are quite a few devices that exhibit fluctuating values compared to the overall average of 11.67 seconds. Therefore, in this measurement, the range of results has a wide range, with the highest average value of 21.44 seconds obtained from device 27 and the lowest average value of 4.90 seconds obtained from device 11. The highest standard deviation is observed in device 16 with a value of 12.70, while the lowest is obtained by device 1 with a value of 0.50. Devices 1 and 27 belong to the PM type, while devices 11 and 16 belong to the FT type.

In graph in additional figure 1c, the average values for each point for the Handheld and patient monitor types exhibit similar characteristics. There is a gradual increase in RT as the oxygen saturation value decreases. However, there is a decrease in RT values at a saturation range of 90-85%. On average, FT has lower values compared to the overall average. Meanwhile, the FS and PM types have average values above the overall average. The highest average value is attributed to the FS type with a value of 13.36 seconds, while the lowest average belongs to the FT type with a value of 9.52 seconds. The lowest standard deviation is observed in the HH type with a value of 3.45 seconds, whereas the patient monitor type has the highest standard deviation with a value of 5.12 seconds. In terms of standard deviation, the patient monitor type has the highest average value compared to other types, but the graph for each point shows that the patient monitor type has a relatively constant standard deviation. On the other hand, other types exhibit fluctuations and have the highest values at the saturation range of 97-95%.

According to the clustering results shown in Figure 2c, which compare the averages and standard deviations, there are 4 tools that do not fall into any data group: tools 3, 12, 13, and 16. Tool 3 belongs to the FS type, while tools 12, 13, and 16 belong to the FT type. This measurement reveals the existence of 2 distinct groups: one with an average range of 5-15 seconds and a standard deviation below 6 seconds, and another with an average of 20 seconds and a standard deviation below 2 seconds. The group with an average of 20 seconds shares a common characteristic, which is being classified as the PM type and having the same brand, Philips.

The results of measurements of 4th conditioning indicates that the average range of these measurements is 6 to 16 seconds. As for the standard deviation, it falls within the range of 0 to 6 seconds, with some instruments exceeding this range of values. The overall average of the 50 values is 11.07 seconds, with a standard deviation average of 3.12 seconds. The instrument with the highest average is instrument 26, with a value of 17.57 seconds, while the lowest is instrument 6, with a value of 5.53 seconds. The highest standard deviation is found in instrument 12, with a value of 11.19 seconds, and the lowest is in instrument 43, with a value of 0.81 seconds.

In graph 5d, it can be observed that the fourth conditioning measurement exhibits an inverted curve compared to the third conditioning. Thus, it can be generally concluded that higher saturation values result in lower response time. The average characteristic of fingertip is lower than the overall average, while the average for PM and HH is higher than the overall average. However, there is a difference in the curve for PM, where there is no increase in response time values at the saturation of 90-95%, unlike FT and FS. The standard deviation generated by FT and HH experiences significant fluctuations at points 90-95% and 97-98%. On the other hand, the curve generated by PM tends to be constant across the point changes and below the average standard deviation value. The highest average response time is for PM type with a value of 13.45 seconds, and the lowest is for FT type with a value of 9.04 seconds. Meanwhile, the highest standard deviation is for FT type with a value of 4.53 seconds, and the lowest is for PM type with a value of 3.42 seconds.

After performing clustering by comparing two values, namely the average and standard deviation shown in figure 6d, the range of results falling within the linear line is from an average range of 5 to 20 seconds and a standard deviation below 6 seconds. There are 4 instruments that have measurements outside of the cluster, namely instruments 12, 16, 36, and 50. Instrument 16 is of type FS, while the rest are of type FT.

From the measurement results of 50 instruments with 5th conditioning, the range of average values obtained is 6 to 20 seconds. There is an instrument, namely instrument 5 (FT-type), that has a relatively high average and standard deviation outside the range of other instruments, with an average value of 27.81 seconds and a standard deviation of 22 seconds. The smallest average value is found in instrument 4 (FT type) with a value of 6.56 seconds, and the smallest standard deviation is in instrument 25 with a value of 0.71 seconds (PM type). The total average value is 14.23 seconds, and the average value of the standard deviation is 2.92 seconds.

In additional figure 1e, the measurement results of the average value for FT-type instruments show lower values compared to other instrument types and the overall average. On the other hand, PM and FS types have values above the overall average. However, for each data point, PM exhibits a tendency of constant values for each given change, while other types show significant variations, especially for FT-type instruments. Similar to the previous graph, the standard deviation for each data point is more stable for PM compared to other instrument types, with different characteristics for each type of instrument. The highest average value is found in FS with a value of 15.70 seconds, while the lowest is in FT instruments with a value of 12.27 seconds. As for the standard deviation, the highest value is observed in FT-type instruments with a value of 7.27 seconds, and the lowest value is in PM-type instruments with a value of 4.29 seconds.

According to the graph shown in figure 6e, there is 1 instrument that has a significantly different value compared to its group, which is instrument 5 of FT type. Additionally, there are 2 instruments whose standard deviation values are outside the range of their respective groups, namely instrument 6 of FT type and instrument 3 of FS type. There are instruments that have an average range above 20 seconds, with a very small standard deviation (approaching 0). These 5 instruments share the same characteristics, as they have the same brand and type (Philips PM type). Instruments 5 and 6 belong to the onecare brand.

The results of measurement of 6th conditioning indicates the range of average values obtained is 6 – 20 seconds. There are 3 devices with values above this range, namely devices 13, 16, and 50 with device type FT. On the other hand, the standard deviation has a range between 0 – 10 seconds, with the two highest values being device 6 with a standard deviation of 10.11 seconds and device 50 with a standard deviation of 8.35 seconds.

According to the information shown in additional figure 1f, the average results for PM exhibit a stable curve and higher values compared to the other 2 types of instruments, surpassing the overall average value. The curve tendency for all types tends to decrease as the normal saturation value setting decreases. The highest average is observed in PM with a value of 14.81 seconds, while the lowest is in FT with a value of 11.57 seconds. The highest standard deviation is found in FT with a value of 5.91 seconds, and the lowest is in FS with a value of 1.22 seconds. Figure 6e shows that the range of average values is from 6 to 20 seconds.

The clustering results shown in the plotting diagram in figure 6f, where the average values are compared with the standard deviation values, indicate measurement results within the range of 5-20 seconds for the average values and 0-6 seconds for the standard deviation. There are 4 devices that have values outside this range, namely devices 6, 13, 16, and 50 with fingertip device type.

4 DISCUSSION

PM exhibits the highest average value the first conditioning and the longest in the fifth and sixth conditioning, The graph shows a higher RT average value compared to other devices, with a more linear shape in comparison. The fastest standard deviation occurs in the first conditioning, whereas the longest is observed in the third data. However, the RT of the PM is more stable compared to other types. Therefore, PM is suitable for and classified as monitoring pulse oximeter for conditions that require consistent RT for each change in oxygen saturation. Despite not responding as quickly as other devices, PM provides better stability compared to other types, along with the capability to measure other required parameters in the monitoring process, especially in condition necessitating continuous monitoring of vital patients such as surgical rooms and intensive care units.

For FT type, the fastest average value and standard deviation were found in the second data conditioning and the longest in the fifth data conditioning. These showed that for the initial response to the value of each oxygen saturation, the FT type has fastest response to read the oxygen saturation value, but slowest response in desaturation oxygen changes from normal to hypoxemia. Subsequently, FT type is very suitable to be classified and used for preventive and diagnostic pulse oximeter, considering that the speed of response is superior to other pulse oximeter type. In addition, FT physical form is small and compact, making it suitable and handly for high mobility usage such as in emergency rooms, outpatient rooms, screening patients for medical check-up, and etc. this oximeter type also appropriate for home use for the wider community which only takes measurements at a certain time for monitoring their oxygen saturation.

For HH type, the fastest mean value is found in the second data, and the longest mean value is found in data 3 and 5. The most unstable standard deviation is found in data 2, while the most stable standard deviation is found in data 4 and 6. It can be concluded that just like the FT type, Handheld is appropriate for use in the initial examination because it has an initial response to different saturation values that are fast. As for saturation conditions from high to low values, for measurements between points and normal oxygen saturation values to hypoxemia handhelds tend to detect longer when compared to those values measured from low to high oxygen saturation values, but in general this type tends to be stable for measurements with oxygen saturation values that rise and fall between points and from normal to hypoxemia values. This makes it suitable for initial screening as well as for continuous screening. From its use in the hospital. This tool is placed in the NICU, ICU and partly in the emergency room. With the aim of being able to take measurements during non-real-time periodic monitoring carried out by medical personnel to record data on critical patients and emergency departments every hour. Due to the large battery capacity and display of the tool, but the size of this type of tool is not too large when compared to the PM type, allowing the use of this tool with high mobility but with the position of the tool remaining standby around the room where it is used. So this tool is possible if it is included in the monitoring and diagnostic class when viewed from the response and characteristics of this type of tool.

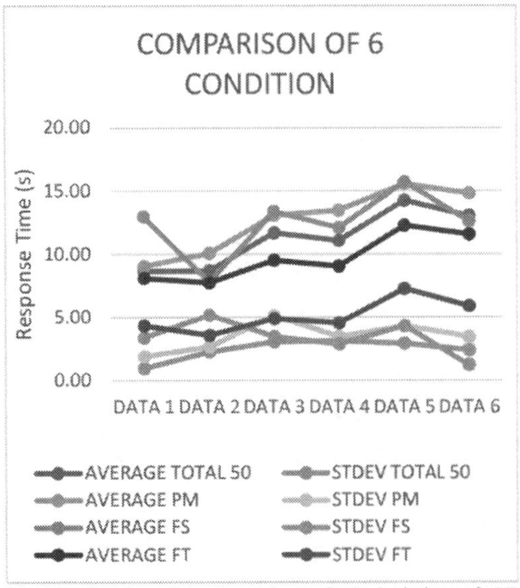

Fig. 6. Graphs of mean and standard deviation of RT measurement results for each data

For the total device, the fastest average value was found in data 1, while the longest average value was found in data 5 and 6. The most unstable standard deviation was found in conditions 3 and 4. This is in line with the principle that the higher the distance from the decrease or increase in oxygen saturation, the longer the device response will be. It was also found that the response on the finger was much faster when compared to the pulse oximeter response on the simulator. This is because the simulator uses an artificial finger which produces a smaller and more stable signal when compared to a human finger which produces a larger and fluctuating signal.

5 CONCLUSION

Response time (RT) of the three types of sensors, namely patient monitor, fingertip, and handheld have different characteristics. RT for the type of pulse oximeter on the patient monitor has a longer average response when compared to the other two types, but RT remains stable in all types of measurements both at high and low saturation. In the fingertip type there is a fast RT with an average of 3 seconds, but the lower the measured oxygen saturation value, the RT also slows down. And there are several tools that have RTs that are quite high from the average RT of other tools and RTs that are inconsistent in several measurement parameters. In the practice of data collection, the author found that some of these inconsistent devices were generally pulse oximeters without brands or were brands that had just spread in the communities.

For the overall RT range as shown in table 2, the mean range of the RT 50 pulse oximeter measurements is 4 - 22 seconds, while the standard deviation is 0 - 23 seconds. By continuing to include data that is outside the range of each conditioning, for the average range with the highest value of 22 seconds is still a fairly fast value when compared to the highest RT value of around 27 seconds. While for standard deviation, 23 seconds is a range far enough that it is necessary to add pulse oximeter data to ensure this range is still included in the average or not.

6 ACKNOWLEDGMENTS

Alhamdulillahi rabbil alamin, I thank Allah جل جلاله for all the favors that the author has received and the ease of working on this report, as well as both parents who always encourage and finance my studies. Then I would like to thank the supervisor and Mrs. Rini and Mrs. Arni who have always helped me from the beginning until now.

7 REFERENCES

1. World Health Organization, Pulse Oximetry Training Manual. 2011. Accessed: Apr. 16, 2023. [Online]. Available: https://cdn.who.int/media/docs/default-source/patient-safety/pulse-oximetry/who-ps-pulse-oxymetry-training-manual-en.pdf?sfvrsn=322cb7ae_6
2. Das, D. M., Gupta, A., Srivastava, A., Vidwans, A., Ahmad, M., Shelke, A., Kale, S., Ananthapadmanabhan, J., Sharma, D. K., & Baghini, M. S. (2018). A pulse oximeter system, OxiSense , with embedded signal processing using an ultra-low power ASIC designed for testability. Microelectronics Journal, 72, 1–10. https://doi.org/10.1016/j.mejo.2017.12.001
3. A. Jubran, "Pulse oximetry," Crit Care, vol. 19, no. 1, Jul. 2015, doi: 10.1186/s13054-015-0984-8.
4. Kementerian Kesehatan Republik Indonesia, Pengujian dan Kalibrasi Alat Kesehatan. Indonesia, 2015.
5. A. Saguni, "METODE KERJA PENGUJIAN DAN ATAU KALIBRASI ALAT KESEHATAN," Jakarta, 2018.
6. Choi, S. J., Ahn, H. J., Yang, M. K., Kim, C. S., Sim, W. S., Kim, J. A., Kang, J. G., Kim, J. K., & Kang, J. Y. (2010). Comparison of desaturation and resaturation response times between transmission and reflectance pulse oximeters. Acta Anaesthesiologica Scandinavica, 54(2), 212–217. https://doi.org/10.1111/j.1399-6576.2009.02101.x
7. Widysanto, A., Wahyuni, T. D., Simanjuntak, L. H., Sunarso, S., Siahaan, S. S., Haryanto, H., Pandrya, C. O., Aritonang, R. C. A., Sudirman, T., Christina, N. M., Adhiwidjaja, B., Gunawan, C., & Angela, A. (2020). Happy hypoxia in critical COVID-19 patient: A case report in Tangerang,

Indonesia. Physiological Reports, 8(20). https://doi.org/10.14814/phy2.14619

8. McLeod, D. B., Cortinez, L. I., Keifer, J. C., Cameron, D., Wright, D. R., White, W. D., Moretti, E. W., Radulescu, R., & Somma, J. (2005). The Desaturation Response Time of Finger Pulse Oximeters During Mild Hypothermia. Survey of Anesthesiology, 49(5), 284. https://doi.org/10.1097/01.sa.0000177174.00499.44
9. D. Young, C. Jewkes, M. Spittal, C. Blogg, J. Weissman, and D. Gradwell, "Response Time of Pulse Oximeters Assessed Using Acute Decompression," Crit Care, vol. 74, pp. 189–195, 1992.
10. L. Wang, W. Wei, M. Gong, and L. Mu, "A comparison of response time to desaturation between tracheal oximetry and peripheral oximetry," J Clin Monit Comput, vol. 24, no. 2, pp. 149–153, Apr. 2010, doi: 10.1007/s10877-010-9227-3.
11. R. M. A. Rizzo, "Accuracy and Response Time of a Portable Pulse Oximeter The Pulsox-7 with a Finger Probe," 1991.
12. D. J. McMahon, "There's no such thing as a SpO2 simulator," 2013. Accessed: Dec. 09, 2022. [Online]. Available: https://www.flukebiomedical.com/sites/default/files/resources/Prosimspot_whitepaper_nosuchthing_A_W.PDF
13. I. Medical Equipment Research, "Medical Equipment Resource." Accessed: Apr. 09, 2023. [Online]. Available: https://medicalequipmentresource.com/medsource-labs-fingertip-pulse-oximeter/
14. Soma Technology, "GE Datex Ohmeda Tuffsat." Accessed: Apr. 09, 2023. [Online]. Available: https://www.somatechnology.com/Pulse-Oximeters/GE-Datex-Ohmeda-Tuffstat.aspx
15. Coast Biomedical Equipment, "Philips SureSigns VM6 Patient Monitor With ECG, NIBP, SPO2, & Temp – Refurbished." Accessed: Apr. 09, 2023. [Online]. Available: https://coastbiomed.com/product/philips-vm6-vital-signs-monitor-with-ecg-nibp-spo2-temp-refurbished/

Gingival inflammation induction in pregnant Sprague-Dawley rats: A pilot study

Friska Ani Rahman[1,2], *Siti* Sunarintyas[3], *Ronny* Martien[4], *Ahmad* Syaify[5*]

[1]Doctoral Program, Faculty of Dentistry, Universitas Gadjah Mada, Indonesia.
[2]Dental Hygiene Program, Faculty of Dentistry, Universitas Gadjah Mada, Indonesia.
[3]Department of Dental of Biomaterials, Faculty of Dentistry, Universitas Gadjah Mada, Indonesia.
[4]Department of Pharmaceutical Technology, Faculty of Pharmacy, Universitas Gadjah Mada, Indonesia
[5]Department of Periodontics, Faculty of Dentistry, Universitas Gadjah Mada, Indonesia

Abstract. Pregnancy gingivitis is an inflammation of the gingiva caused by dental plaque and exacerbated by hormonal changes. Gingival inflammation is often induced in laboratory animal models, such as Sprague-Dawley rats used for experimental models. This investigation was to establish inducing gingival inflammation in pregnant rats. The purpose of this study was to establish a reproducible method for inducing gingival inflammation in both incisive mandibles of pregnant Sprague-Dawley rats using ligation and dietary manipulation. Pregnant Sprague-Dawley rats were used as experimental subjects. As ligation, 4-0 non-resorbable silk thread was utilized and affixed using a "8"-shaped knot technique. The ligatures were inserted between the mandible's central incisors. The operation was carried out under anaesthesia. Once the rat's ligature was removed after 5th,7th,10th and 14th days. The clinical appearance and radiographic examination were evaluated. Gingival inflammation induction by ligation and dietary intake modification caused inflammation of gingival tissue, was seen at 5th day. Clinical examination showed getting worst at 14th day. In our study, gingival inflammation on pregnant rat was achieved five days after ligation and dietary intake modification.
Keywords: gingival inflammation, induction, ligation, dietary intake, animal model

1 Introduction

Animal models are of utmost importance in advancing information in the field of medical sciences, including periodontology. The experimental models provide notable advantages owing to their capacity to accurately reproduce the cellular features and reactions that take place in humans in vivo. The utilization of animal models in the study of periodontal disease is of utmost importance as it provides a necessary foundation for the establishment of a scientific framework aimed at understanding the underlying pathogenic mechanisms [1]. Commonly used animal models in periodontal research are rats and other rodents. The gingival anatomy of rats exhibits similarities to that of humans, characterized by a shallow gingival sulcus and the adherence of junctional epithelium to tooth surfaces. Furthermore, the junctional epithelium present in both gingivae functions as a pathway for the transportation of exogenous chemicals, bacterial toxins, and inflammatory cell exudates [2].

In the Classification of Periodontal and Peri-Implant Diseases and Conditions 2017, gingivitis is inflammation of the gingival tissue at one or more sites and is characterised by bleeding on probing. Bleeding on probing is the main parameter in diagnosing gingivitis [3]. Gingivitis is a mild periodontal disease that causes redness and swelling (inflammation) of the gingiva. Gingivitis is a disorder that primarily affects the gingival area and does not entail damage to the periodontal tissues. However, it is important to note that gingivitis has the potential to progress into periodontitis, a more serious condition compared to gingivitis [4]. Pregnancy gingivitis is an inflammation of the gingiva caused by dental plaque and exacerbated by hormonal changes, particularly during the second and third trimesters [5]. Estrogen and progesterone are hormones that undergo significant fluctuations during pregnancy. These hormonal changes will influence the clinical appearance of periodontal tissue [6].

Periodontal diseases can be created in rats through the inoculation of bacteria, administration of a diet rich in carbohydrates, and the placement of ligatures around the teeth [7]. Male rodents had their second maxillary molars ligated and were fed moist food and 10% sucrose water. Two weeks later, the gingival inflammation model was effectively established [8]. Not much has been research regarding the induction of periodontal diseases in pregnant rats. An experimental study [9] was conducted periodontitis on pregnant rats by ligation around second upper molars until the last day of breast feeding. Another experiment using female rats were received an oral inoculation containing 1×10^9 CFU of *Porphyromonas gingivalis* for 4 consecutive days/week.

*Corresponding author: ahmad.syaify@ugm.ac.id

https://doi.org/10.1051/bioconf/20237502003

After 24 inoculations over a 12-week, female rats mated with male, and pregnancy was determined by vaginal swab [10].

Due to the lack of information about conducting experiments on pregnant rats, it is proposed to modify the existing model by ligating the central incisor and manipulating the food intake. The present study ligated the cervical region of the lower incisors using silk thread. The incisor is utilized more frequently due to its simple accessibility and straightforward operatory technique [1, 7, 11, 12].

The objective of this study was to establish a reliable experimental method for producing inflammation in the gingival tissue surrounding the incisive mandible teeth of pregnant Sprague-Dawley rats. This was achieved by employing a ligation approach in combination with a modified nutritional regimen.

2 Material and methods

2.1 Animals modeling

All procedures were conducted after approval from the ethics committee of the Faculty of Dentistry-Prof Soedomo Dental Hospital, Universitas Gadjah Mada with the number 2/UN1/KEP/FKG-RSGM/EC/2023. The experimental protocol was carried out in accordance with current animal welfare and experimentation ethics laws. Four female adults of Sprague-Dawley rats obtained from Laboratory animal facility at LPPT unit IV Universitas Gadjah Mada, aged three months with a weight of (200 to 250) g, were included in this study. A controlled environment was established to house a group of five rats in individual wire cages. The room was maintained at specific temperature and humidity levels, and a 12-hour light/dark cycle was implemented. The rats were provided with normal rat pellets and had unrestricted access to water. After a week of acclimation, female rats in estrus stage were placed overnight with a male for mating. Vaginal examinations determined the cycle of estrus or pregnancy. The following morning, if spermatozoa were detected in the vaginal smear as seen in Figure 1, pregnancy was determined [10, 13]. Spermatozoa were observed using a microscope.

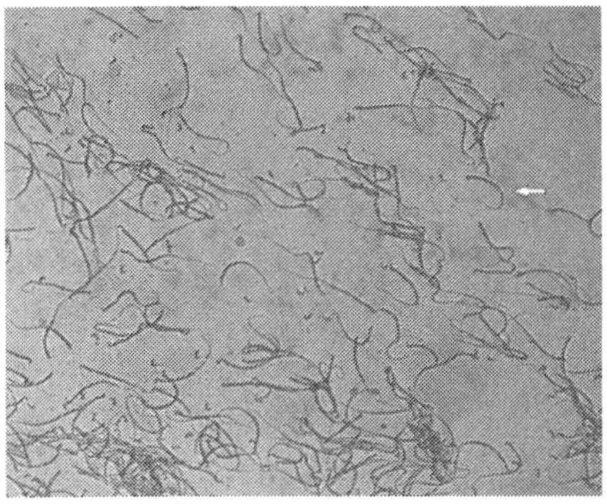

Fig. 1. Spermatozoa on vaginal smear as shown by the arrow

Following the confirmation of pregnancy, general anesthesia was induced through intramuscular injection using a solution consisting of Ketamine 10% and Xylazine 2% in a ratio of 2:1, at a dosage of 0.12 ml per 100 g of body weight. The animals were positioned on an operating table to enable unobstructed access to the rats' mouths, hence facilitating maintenance of the teeth. Both mandibular incisive were ligated with 4-0 silk thread non resorbable. The silk thread was knotted around three circles with "8"-shaped and pressed into the subgingival as much as possible (Figure 2). After placing ligatures, the animals were kept in cage. The animals were feed with moist feed and drank 10% sucrose water. Control of ligatures and animal models was performed daily.

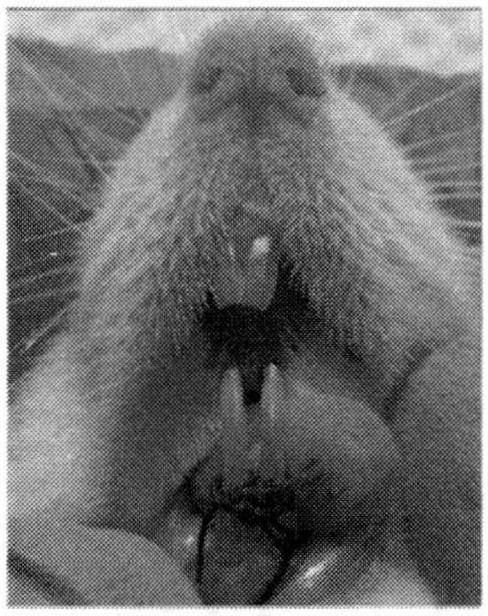

Fig. 2. The macroscopic appearance after the placement of the ligature.

2.2 Clinical and radiographic examination

The rats were divided into five groups. Before the extraction of the silk thread, a standard procedure of general anesthesia was administered through intramuscular injection using a solution consisting of 10% Ketamine and 2% Xylazine in a ratio of 2:1. The dosage administered was 0.12 ml per 100 grams of weight. On days 5th, 7th, 10th, and 14th, ligatures were removed in each rat. Clinically, the examination was made on day 5th until day 14th. The parameters gingival index, bleeding on probing, and probing depth were evaluated. Gingival Index (GI) as described by Lőe and Silness was used, with scores ranging from 0 to 3, score 0: healthy gingiva; score 1: mildly inflamed, slight change in color, slight edema, no bleeding on probing; score 2: moderately inflamed, redness, edema, glazing, bleeding on probing; score 3: severely inflamed, marked redness and edema, ulceration, spontaneous bleeding [14]. The probing depth (PD) and bleeding on probing were evaluated using a periodontal probe (UNC-15 probe, Osung, Korea). Lower incisors were examined in the gingival sulcus and their depth was measured. The bleeding on probing (BOP) was determined to be positive if bleeding occurred within thirty seconds of probe placement [15]. All parameters were documented in the chart for each subject.

After clinical examination, radiographic examination was taken using a digital periapical radiograph machine for animals. Conus was positioned upright to the mandible of rat.

3 Results

3.1 Clinical

The pregnant gingivitis model in Sprague-Dawley rats was established experimentally with the use of silk thread ligatures and dietary modifications. Observable clinical changes become apparent during a span of five days subsequent to the initiation of induction. The gingival tissue exhibited erythema and edema, with a probing depth of 1 millimeter. The gingival tissue exhibits bleeding upon probing. Indicating that the gingivitis model was effectively established. At two weeks, the gingival tissues were swollen, bleeding was observed during BOP examination, PD was 3 mm and gingival recession occurred on the mandibular left incisor (Figure 3 and Table 1).

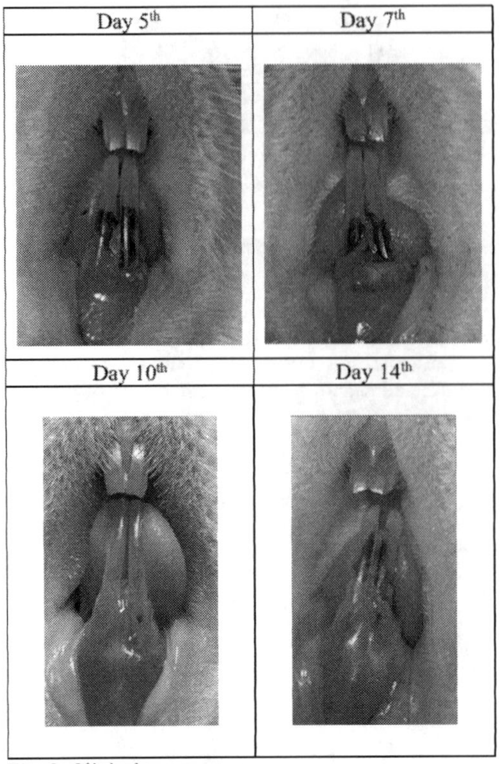

Fig. 3. Clinical appearances

Table 1. Clinical finding on pregnant rat

Indicators	Days			
	5th	7th	10th	14th
GI*	2	2	2	2
BOP	+	+	+	+
PD (mm)	1	1	2	3

*GI by Lõe & Silness; + bleeding

Alterations were seen on the fifth day following the application of the ligature, during which the gingival tissue exhibited signs of deviation from its typical appearance and structure. The coloration of the gingiva transitioned from pink to intense red. The presence of plaque accumulation was seen in nearby areas of the ligated silk thread. The detected modifications exhibited a noticeable increase from the fifth day to the fourteenth day, at which point the subjects were euthanized.

3.2 Radiographic

Rats' mandibles were evaluated using digital periapical radiographs for radiographic examination. The decrease in radiopacity and widening of periodontal ligament was seen at fourteenth day, as illustrated in Figure 4.

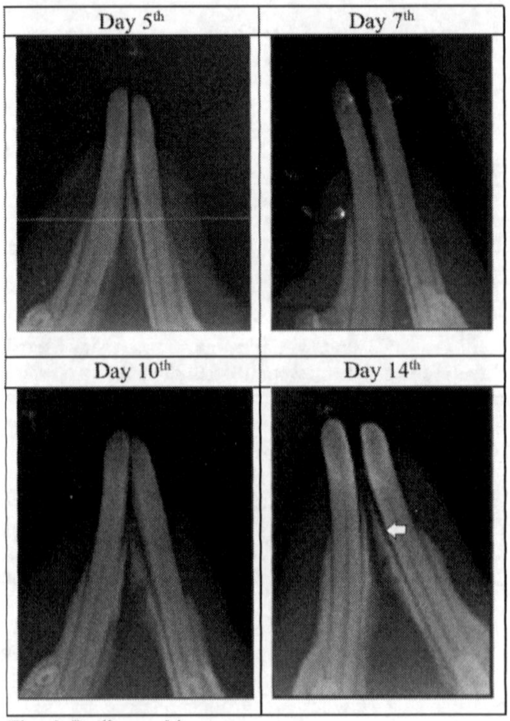

Fig. 4. Radiographic appearances

4 Discussion

The experimental model of gingivitis in pregnant rats was generated with the implementation of ligation, a diet consisting of moist food, and the provision of sugary beverages. The ligation line serves as a structural framework facilitating the aggregation and adherence of bacteria [1, 16]. In addition, the process of ligation leads to gingival irritation. Moist feed exhibits a notable level of viscosity, which facilitates its adherence to the tooth surface and thus enhances the adhesion of dental plaque [16].

The gingivitis model was established by daily administration of moist feed and sugary drinks. On the fifth day following the establishment of the gingivitis animal model, the rat mandibular incisor exhibited red and swollen gums, as well as gingival bleeding during exploration. Bleeding on probing is primary indicator to set the threshold for gingivitis [3]. The research demonstrates that the gingival inflammation models were success. Besides that, rat pregnancy gingivitis may be influenced indirectly by progesterone and its receptor [11].

In this present study, clinical changes of gingival inflammation can be seen on fifth day after induction started. Consistent with the findings of previous research, the obtained data were in accordance with those of earlier studies [12]. although was performed ligation on male rats with additional injection of bacteria. Studies conducted by Shi *et al.* [11] showed that gingival inflammation on pregnant rats were established after ligation for two weeks.

Visual manifestations of gingival inflammation are highly responsive indicators of early-stage gingivitis. Consequently, gingival index that focus on bleeding have been given significant emphasis [17]. The current investigation demonstrated a positive correlation between the gingival index and the presence of blood on probing. The clinical assessment revealed that there was an increase in probing depth values when the duration of ligation was extended.

The present study has limitations by a low sample size and reliance solely on clinical and radiographic assessments. Consequently, further investigation is warranted to assess the histopathological examination, weight, and hormone levels in pregnant rats. In the context of this study, gingival inflammation was achieved five days after ligature placement and dietary intake modification.

5 Conclusion

The present study elucidates the clinical and radiographic assessment of gingival inflammation provoked in pregnant rats with the application of ligation in conjunction with food modification. The short-term induction demonstrated gingival inflammation that occurred on the fifth day.

The authors express appreciation for the financial assistance provided by The Indonesian Endowment Funds for Education (LPDP) Number 20200921304644.

References

1. D.T. Graves, D. Fine, Y.T.A. Teng, T.E. Van Dyke, and G. Hajishengallis, "The use of rodent models to investigate host-bacteria interactions related to periodontal diseases," J Clin Periodontol **35**(2), 89–105 (2008).
2. X. Struillou, H. Boutigny, A. Soueidan, and P. Layrolle, "Experimental Animal Models in Periodontology: A Review," Open Dent. J. **4**(1), 37–47 (2010).
3. J.G. Caton, G. Armitage, T. Berglundh, I.L.C. Chapple, S. Jepsen, K.S. Kornman, B.L. Mealey, P.N. Papapanou, M. Sanz, and M.S. Tonetti, "A new classification scheme for periodontal and peri-implant diseases and conditions - Introduction and key changes from the 1999 classification," J. Clin. Periodontol. **45**(Suppl 20), S1–S8 (2018).
4. S. Komine-Aizawa, S. Aizawa, and S. Hayakawa, "Periodontal diseases and adverse pregnancy outcomes," J. Obstet. Gynaecol. Res. **45**(1), 5–12 (2019).
5. J.S. Gehrig, and D.E. Willmann, *Foundations of Periodontics for the Dental Hygienist*, 4th ed. (Wolter Kluwer, 2016).
6. M. Wu, S.W. Chen, and S.Y. Jiang, "Relationship between gingival inflammation and pregnancy," Mediators of Inflamm. **2015**, 1–11 (2015).
7. A. Ionel, O. Lucaciu, M. Moga, D. Buhatel, A. Ilea, F. Tabaran, C. Catoi, C. Berce, S. Toader, and R.S. Campian, "Periodontal disease induced in Wistar rats-experimental study," International Journal of the Bioflux Society **7**(2), 90–95 (2015).
8. Y. Wang, Q. Fan, Y. Xu, F. Zeng, X. Liu, D. Zhao, L. Zhang, and G. Bai, "Effect of Eucommia water extract on gingivitis and periodontitis in experimental rats," BMC Oral Health **22**(1), (2022).
9. M.P. Galvão, C.K. Rösing, and M.B. Ferreira, "Effects of ligature-induced periodontitis in pregnant Wistar rats.," Pesqui Odontol Bras **17**(1), 51–55 (2003).
10. M.S. Walkenhorst, L. Reyes, G. Perez, A. Progulske-Fox, M.B. Brown, and P.L. Phillips, "A Uniquely Altered Oral Microbiome Composition Was Observed in Pregnant Rats with Porphyromonas gingivalis Induced Periodontal Disease," Front. Cell. Infect. Microbiolol. **10**(92), 1–10 (2020).
11. Y.Z. Shi, S. Jin, H. Qin, H.B. Jiang, G.H. Song, and S.C. Qin, "Hydrogen-rich water ameliorates rat placental stress induced by water restriction," Med Gas Res **8**(3), 79–84 (2018).
12. S. Suryono, F. Resha Wulandari, H. Andini, J. Widjaja, and T. Dwisetyo Nugraheni, "Methodology in Wistar rats periodontitis induction: A modified ligation technique with injection of bacteria," Int. J. Oral Health Sci. **10**, 36–40 (2020).
13. S.G. Çekiç, E.G., Çekiç, D.E. Baydar, M. Tuncer, and G. Soydan, "Vascular responses of aortic, renal, and uterine arteries in suramin-induced preeclampsia-like syndrome in rats," Turk J Med Sci **48**(6), 1328–1339 (2018).
14. P. Boyle, A. Koechlin, P. Autier, J. Wiley, and R. Allé, "Mouthwash Use and the Prevention of Plaque, Gingivitis and Caries," Oral Dis **20**(Suppl. 1), 1–68 (2014).
15. L. Checchi, M. Montevecchi, V. Checchi, and F. Zappulla, "The Relationship Between Bleeding on Probing and Subgingival Deposits. An Endoscopical Evaluation," Open Dent J **3**, 154–160 (2009).

16. Y. Wang, F. Zeng, X. Liu, Y. Tian, and G. Bai, "Effect of a new mouthwash based on Tea polyphenols and chlorogenic acid on dental caries and gingivitis," Pharmacological Research - Modern Chinese Medicine **3**, (2022).
17. B. Ashwath, M. Shanmugam, E. Agila, V. Anitha, D. Aishwarya, and A. Zafrin, "Comparison of gingival inflammatory parameters and pain perception on periodontal probing -a cross sectional study, part II," Journal of Oral Research **11**(5), (2022).

In Silico Destruction of *Porphyromonas gingivalis* Fimbriae by *Streptomyces* sp. Strain GMY02

Fanni Kusuma Djati[1,2], *Dewi* Agustina[3], Mustofa[4], *Hera* Nirwati[5], *Jaka* Widada[6], *Ema* Damayanti[7], *Heni* Susilowati[8*]

[1]Doctoral Program, Faculty of Dentistry, Universitas Gadjah Mada, Yogyakarta, Indonesia
[2]Department of Dental Medicine, Faculty of Medicine, Jenderal Soedirman University, Purwokerto, Indonesia
[3]Department of Oral Medicine, Faculty of Dentistry, Universitas Gadjah Mada, Yogyakarta, Indonesia
[4]Department of Pharmacology and Therapy, Faculty of Medicine, Public Health and Nursing, Universitas Gadjah Mada, Yogyakarta, Indonesia
[5]Department of Microbiology, Faculty of Medicine, Public Health and Nursing, Universitas Gadjah Mada, Yogyakarta, Indonesia
[6]Department of Microbiology, Faculty of Agriculture, Universitas Gadjah Mada, Yogyakarta, Indonesia
[7]Research Center for Food Technology and Processing, Research Center for Food Technology and Processing, National Research and Innovation Agency, Gunungkidul, Indonesia
[8]Department of Oral Biology, Faculty of Dentistry, Universitas Gadjah Mada, Yogyakarta, Indonesia

Abstract. *Porphyromonas gingivalis*, a keystone pathogen in chronic periodontitis, has fimbria as one of the most important virulence factors because it plays a vital role in the pathogenesis of *P. gingivalis* infection. This bacterium has 2 fimbriae: the major (FimA) and minor (Mfa1) fimbriae. Fimbriae are attractive targets for anti-infective therapy for periodontal disease. The aim of this study was to analyze the interactions of several compounds thought to be present in *Streptomyces sp.* GMY02 strain against FimA and Mfa1 proteins in *P. gingivalis* fimbriae in silico. A total of 8 ligands were docked to FimA and Mfa1 fimbriae using AutoDock Vina in University of California, San Francisco (UCSF) Chimera 1.16. All of the selected ligands had higher free energy values than metronidazole as well. In conclusion, the compounds suspected to be present in *Streptomyces sp.* strain GMY02 has the potential to destruct *P. gingivalis* fimbriae.
Keywords: *Streptomyces*, *Porphyromonas gingivalis*, fimbriae, molecular docking, biofilm

1 Introduction

Porphyromonas gingivalis, a keystone pathogen in chronic periodontitis, has virulence factors including capsule, fimbriae, lipopolysaccharide (LPS), protease (gingipain), and outer membrane protein [1-3]. Although *P. gingivalis* expresses several potential virulence factors, the fimbriae are of particular importance because they play a vital role in host cell attachment and invasion, colonization of *P. gingivalis* with other bacteria and host tissues, biofilm formation, bacterial motility, and protein and DNA transport across the cell membrane [4]. *Porphyromonas gingivalis* has 2 fimbriae, namely the major (FimA) and minor (Mfa1) fimbriae, which consist of the protein polymers FimA and Mfa1, and are encoded by the fimA and mfa1 genes, respectively [5]. According to its DNA sequence, the fimA genotype of *P. gingivalis* can be divided into six categories (I, Ib, II, III, IV, V). Type I fimA exhibiting a solid correlation with plaque formation. The major role of type I fimbriae (FimA) of *P. gingivalis* is to mediate biofilm formation, adherence to saliva-coated surfaces, and adherence to gingival epithelial cells, it can also trigger an inflammatory response. FimA genotype I has also been observed at high frequency in patients with severe periodontitis. Given the important role of the fimbria in the pathogenesis of *P. gingivalis*, the fimbria is an attractive target for anti-infective therapy to prevent or treat periodontal disease [6,7].

Several new approaches to treat bacterial infections are by inhibiting biofilms without inducing microbial dysbiosis from the oral cavity. These new approaches include the use of nanomaterials, quaternary ammonium salts, small molecules, arginine, and natural materials [8]. One of the natural materials currently being developed as an anti-biofilm is Streptomyces sp. strain GMY02, is a potential bacterium isolated from marine sediment samples from Krakal Beach (8°8′44″S, 110°35′59″E), Yogyakarta, Indonesia. Identification, annotation and analysis of gene clusters involved in secondary metabolite biosynthesis in *Streptomyces sp.* the GMY02 strain was carried out with antiSMASH 6.0 [9].

Molecular docking is an approach used extensively in modern drug designing and development. This method is mainly used in drug design to explores the conformations of ligands within the macromolecular target binding site, providing an estimation of receptor-ligand binding free energy for all different conformations. Molecular docking is an established in silico structure-based method widely used in drug discovery. Nowadays, in silico

* Corresponding author: henisusilowati@ugm.ac.id

methodologies have become a crucial part of the drug discovery process. In silico is a research method that utilizes computing and database technology to develop further research. The use of the in silico method in drug development can save costs and time because it can predict drug structure through mathematical equations, visualization in three dimensions, and can evaluate interactions between compounds and targets before synthesizing these structures into drugs [10-12]. Based on this background, this study aims to analyze the interactions of several compounds thought to be present in *Streptomyces sp.* GMY02 strain against FimA and Mfa1 proteins in *P. gingivalis* fimbriae in silico.

2 Materials and Methods

2.1. Ligand Preparation

The ligand chosen in this study was metronidazole as a positive control and compounds suspected to be present in *Streptomyces* sp. strain GMY02. The candidate compounds produced by *Streptomyces* sp. GMY02 were predicted by using genome mining tool, AntiSMASH version 7 (available online in https://antismash.secondarymetabolites.org) (antiSMASH bacterial version) with NCBI accession number of whole genome sequence CP077658. PubChem as SMILES files converted to PDB format using NovoPro (https://www.novoprolabs.com /tools/smiles2pdb), then converted to mol2 format with UCSF Chimera software (version 1.16). The ligands tested and their structures can be seen in Table 1.

Table 1. List of compounds to be docked with FimA and Mfa1 fimbriae and their structures

COMPOUND	MOLECULAR FORMULA	SMILES	STRUCTURE
sangivamycin	$C_{12}H_{15}N_5O_5$	NC(=O)C1=CN([C@@H]2O[C@H](CO)[C@@H](O)[C@H]2O)C2=C1C(N)=NC=N2	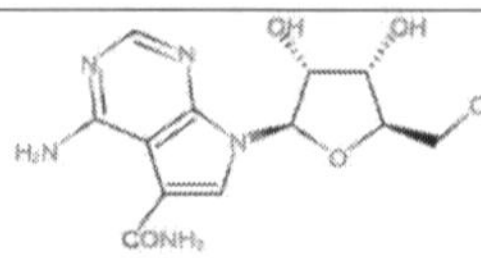
toyocamycin	$C_{12}H_{13}N_5O_4$	NC1=NC=NC2=C1C(=CN2[C@@H]1O[C@H](CO)[C@@H](O)[C@H]1O)C#N	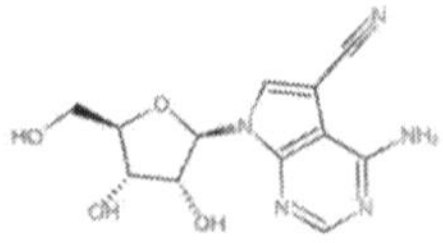
reductasporine	$C_{22}H_{20}N_3^+$	c1ccc2c(c1)c1c([nH]2)c2c(c3c1C[N+](C3)(C)C)c1c([nH]2)cccc1	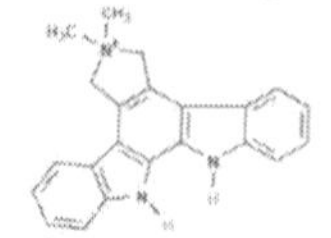
hopene	$C_{32}H_{54}O_1$	CC(CCCO)[C@H]1CC[C@@]2(C)[C@H]1CC[C@]1(C)[C@@H]2CC[C@@H]2[C@@]3(C)CCCC(C)(C)[C@@H]3C=C[C@@]12C	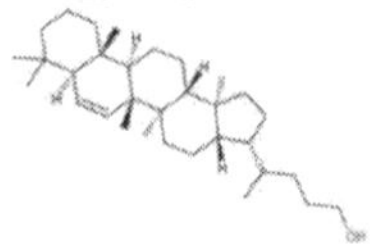
2-methylisoborneol	$C_{11}H_{20}O$	CC1(C2CCC1(C(C2)(C)O)C)C	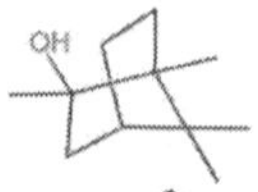
ectoine	$C_6H_{10}N_2O_2$	CC1=NCC[C@H](N1)C(=O)O	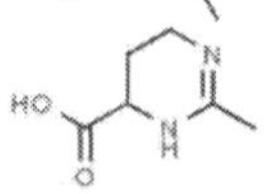
alkylresorcinol methylresorcinol	$C_7H_8O_2$	CC1=C(C=CC=C1O)O	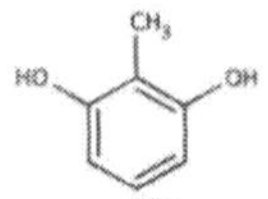
ethyresorcinol	$C_8H_{10}O_2$	CCC1=C(C=CC=C1O)O	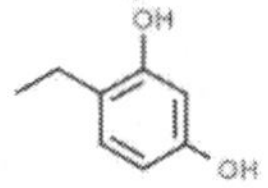
propylresorcinol	$C_9H_{12}O_2$	CCCC1=C(C=C(C=C1)O)O	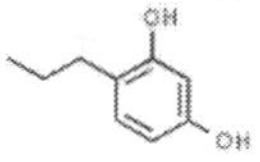

2.2 Protein Preparation

Crystal structure of FimA protein in *P. gingivalis* fimbriae (PDB ID: 6JZK) and Mfa1 (PDB ID: 5NF2), taken from the Protein Data Bank (PDB) (http://www.rcsb.org/pdb) as PDB files. The crystal structure of the FimA protein has a resolution of 2.10 Å, while that of the Mfa1 protein has a resolution of 1.73 Å. The crystal structure of the FimA and Mfa1 proteins can be seen in Figure 1. Protein preparation was carried out by remòving all unique ligands, all atoms, water molecules, and ions to obtain native receptor proteins, then polar hydrogen and charges were added, The protein preparation procedure was carried out using Chimera (version 1.16).

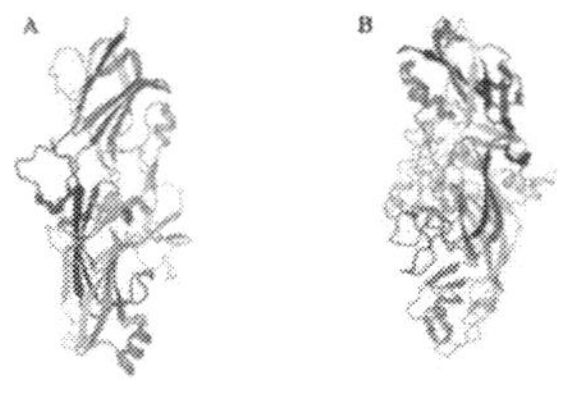

Fig. 1. The 3D structure, A. FimA protein complex from *P. gingivalis* with SRT (PDB ID: 6JZK); B. Mfa1 protein complex from *P. gingivalis* with ACT (PDB ID: 5NF2).

2.3 Molecular Docking

Protein and ligand preparation, energy minimization, and molecular docking were performed using UCSF Chimera (version 1.16) and AutoDock (version 1.5.7) software. The molecular docking simulation method was validated using the root-square deviation (RMSD) calculations by redocking the native ligand. The best conformation of the docked native ligand was retrieved and superimposed with the native protein prior to docking, and RMSD was calculated. An acceptable RMSD value should be less than 2.0 Å [13]. Docking was performed with a grid box that includes all receptor structures with dimensions of 50×50×50 Å and centered at 1.287, -3.755, and 92.921 respectively the X, Y, and Z coordinates for the FimA protein, and dimensions of 50x50x70 Å and centered on 29.787, 3.436, and 84.082 respectively the X, Y, and Z coordinates for the Mfa1 protein, and the Genetic Algorithm (GA) Run number of 50.

2.4 Data Visualization

Visualization of the interaction results of docking of the best ligand conformation which has the highest binding affinity of the compounds suspected to be present in *Streptomyces* sp. strain GMY02 with protein, using LigPlot plus. In addition, with LigPlot Plus it can also be seen the amino acid residues around the interactions that occur, the hydrogen bonds, and the hydrophobicity.

3 Results and Discussion

3.1 Docking Analysis

The results of the docking of compounds suspected to be present in *Streptomyces sp.* strain GMY02 with FimA protein in *P. gingivalis* fimbriae is shown in Table 2.

Table 2. List of docking results for compounds suspected to be present in *Streptomyces sp.* strain GMY02 with FimA fimbria

Compound or Ligand	Binding Energy	Inhibition constant (nM)
SRT	-2,66	11,23.10^{6}
Metronidazole	-3,66	2,09.10^{6}
Sangivamycin	-5,46	99,73.10^{3}
reductasporine	-10,04	44
2-methylisoborneol	-5,04	201,42.10^{3}
hopene	-6,27	25,18.10^{3}
ectoine	-4,07	1,04.10^{6}
alkylresorcinol	[redacted]	
methylresorcinol	-4,62	412,65.10^{3}
Ethylresorcinol	-4,08	1,02.10^{6}
Propylresorcinol	-3,96	1,26.10^{6}

Table 2 shows all the compounds tested have a higher binding energy compared to SRT whose value was -2.66 kcal/mol and metronidazole (positive control) which has a binding energy of -3.66 kcal/mol. The highest binding energy of the compounds suspected to be found in *Streptomyces sp.* the GMY02 strains with FimA protein are reductasporine (-10.04 kcal/mol), hopene (-6.27 kcal/mol), and sangivamycin (-5.46 kcal/mol) and RMSD is 1,99. The results of the docking of compounds suspected to be present in *Streptomyces sp.* strain GMY02 with Mfa1 protein in the fimbriae of *P.gingivalis* is shown in Table 3.

Table 3. List of docking results for compounds suspected to be present in *Streptomyces sp.* strain GMY02 with Mfa1 fimbriae

Compound or Ligand	Binding Energy	Inhibition constant (nM)
ACT	-2,84	8,35.10^{6}
Metronidazole	-3,68	2,02.10^{6}
Sangivamycin	-1,47	83,82.10^{6}
reductasporine	-9,14	199,21
2-methylisoborneol	-4,08	1,03.10^{6}
hopene	-6,76	11,13.10^{3}
ectoine	-3,20	4,55.10^{6}
alkylresorcinol	[redacted]	
methylresorcinol	-5,39	111,5.10^{3}
Ethylresorcinol	-5,66	71,53.10^{3}
Propylresorcinol	-5,86	50,69.10^{3}

Table 3 shows all the compounds tested, except sangivamycin and ectoine, which have a higher binding energy compared to ACT which is -2.84 kcal/mol and metronidazole (positive control) which has a binding energy of -3.68 kcal/mol. The highest binding energy of the compounds suspected to be found in *Streptomyces sp.* the GMY02 strain with Mfa1 protein are reductasporine (-9.14 kcal/mol), hopene (-6.76 kcal/mol), and alkylresorcinol: propylresorcinol (-5.86 kcal/mol). RMSD is 0,15.

Two compounds that have the highest binding energy with the FimA and Mfa1 proteins (reductasporine and hopene) were further analyzed to determine their interactions with the active sites of the FimA and Mfa1 target proteins. Figure 2 illustrates the interaction display of the best ligand conformation binding results of reductasporine and hopene with FimA and Mfa1 proteins using LigPlot plus software.

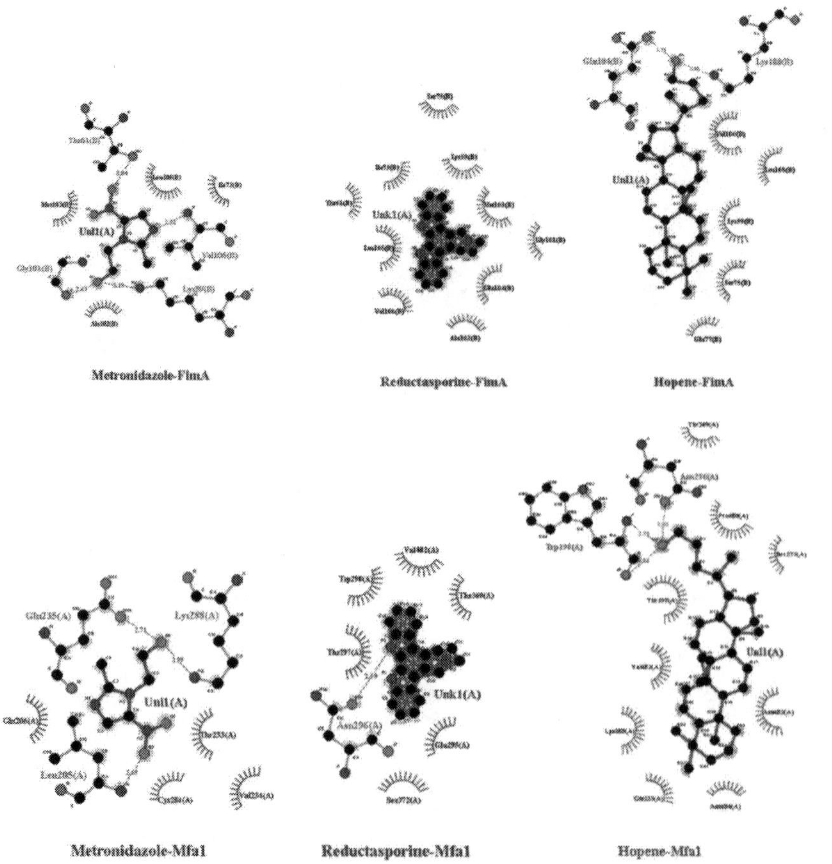

Fig. 2. Interaction of metronidazole and the best ligand conformation binding results from reductasporine and hopene with FimA and Mfa1 proteins using LigPlot plus software

Reductasporine contains a novel tryptophan dimer (TD) core structure. Reductasporine contains an indolocarbazole pyrrolinium core structure which may be a key new bioactivity profile. This new core structure gives reductasporin a bioactivity profile that is different from other TD core structures. TD has biological activity as an antibacterial and antifungal. Several tryptophan derived from marine alkaloids show strong and promising antimicrobial activity, whether against bacteria, fungi, or viruses [14,15].

Hopene is an important precursor for synthesizing bioactive hopanoids. Hopanoids are a group of pentacyclic triterpenoids consisting of a hopene skeleton and side chain sequences. Hopanoids which are natural products play an important role in stabilizing the structure of bacterial membranes such as the effect of sterols on eukaryotes. Hopanoids interact with glycolipids on the bacterial outer membrane to form a highly ordered bilayer in a manner similar to the interaction of sterols with sphingolipids on eukaryotic plasma membranes [16,17].

Reductasporine and hopene are 2 compounds that are thought to be present in *Streptomyces sp*. GMY02 strain had the highest binding energy with *P. gingivalis* fimbriae, even higher than metronidazole. The binding energy generated by the ligand when it binds to the target protein site can trigger a specific biological response, the more negative the binding score, the higher the effect on the activity of the target protein, it can be indicate that the compounds suspected to be present in *Streptomyces sp.* strain GMY02 have the potential to destruct *P. gingivalis* fimbriae.

This work suggests the application of the mentioned ligands or compounds to control *P. gingivalis* biofilm found in the oral cavity. By approving their potency in silico, it is necessary to carry out in vitro experiment to confirm the computational predictions that have been made.

4. Conclusion

Based on the data obtained from this study, it can be concluded that *Streptomyces sp.* strain GMY02 have the potential to destruct the *P. gingivalis* fimbriae.

References

1. G. Hajishengallis, R.P. Darveau, M.A. Curtis, Nat Rev Microbiol., **10**, 717-725 (2012). http://dx.doi.org/10.1038/nrmicro2873.
2. K.Y. How, K. Song, K. Chan, Front Microbiol., **7**, 53 (2016). https://doi.org/10.3389/fmicb.2016.00053
3. P.M. Preshaw, *Periodontal Disease Pathogenesis* in Newman and Carranza's Clinical Periodontology 13 Ed., Elsevier, Philadelpia (2019)
4. M. Enersen, K. Nakano, A. Amano, J Oral Microbiol., **5**, 20265 (2013). https://doi.org/10.3402/jom.v5i0.20265
5. K. Nagano, Y. Hasegawa, Y. Abiko, Y. Yoshida, Y. Murakami, F. Yoshimura, PLoS One., **7**(9), e43722 (2012). https://doi.org/10.1371/journal.pone.00437222
6. K. Nagano, Y. Abiko, Y. Yoshida, F. Yoshimura, Mol. Oral Microbiol., 28 (5), 392-403 (2013). https://doi.org/10.1111/omi.12032
7. S.R. Alaei, J.H. Park, S.G. Walker, D.G. Thanassi, Infect Immun. **87**(3), e00750-18 (2019). https://doi.org/10.1128/IAI.00750-18
8. X. Kuang, V. Chen, X. Xu, Biomed Res. Int. **2018**, 6498932 (2018). https://doi.org/10.1155/2018/6498932
9. J. Widada, E. Damayanti, Mustofa, Microbiol Resour Announc., **10**(40), e0068121 (2021). https://doi.org/10.1128/MRA.00681-21
10. L. Pinzi L., G. Rastelli, Int. J. Mol. Sci, **20**(18), 4331 (2019). https://doi.org/10.3390/ijms20184331
11. S.S. Butt, Y. Badshah, M. Shabbir, M. Rafiq, JMIR Bioinform Biotech, **1**(1), e14232 (2020). https://doi.org/10.2196/14232
12. Y. Chang, B.A. Hawkins, J.J. Du, P.W. Groundwater, D.E. Hibbs, F. Lai, Pharmaceutics, **15**(1), 49 (2022). https://doi.org/10.3390/pharmaceutics15010049
13. E.W. Bell, Y. Zhang, J Cheminform, **11**, 40 (2019). https://doi.org/10.1186/s13321-019-0362-7
14. F.Y. Chang, M.A. Ternei, P.Y. Calle, S.F. Brady, J Am Chem Soc. **137**(18),6044-52 (2015). https://doi.org/10.1021/jacs.5b01968
15. M.C. Almeida, D.I.S.P. Resende, P.M. da Costa, M.M.M. Pinto, E. Sousa, Eur J Med Chem., **209**:112945 (2021). https://doi.org/10.1016/j.ejmech.2020.112945
16. J.P. Sáenz, D., Grosser, A.S. Bradley, T.J. Lagny, O. Lavrynenko, M. Broda, K. Simons, *Hopanoids as functional analogues of cholesterol in bacterial membranes*, in Proceedings of the National Academy of Sciences of the United States of America, 112(38), 11971–11976 (2015). https://doi.org/10.1073/pnas.1515607112
17. Y. Wen, G. Zhang, A. Bahadur, Y. Xu, Y. Liu, M. Tian, W. Ding, T. Chen, W. Zhang, G. Liu, Microorganisms., **10**(12), 2408 (2022). https://doi.org/10.3390/microorganisms10122408

Sequence Conservation Analysis and Gene Relationships of Nucleocapsid (N) Gene in Orthocoronavirinae Subfamily

Husna Nugrahapraja[1*,2], *Adi* Nugraha[1], and *Alidza* Fauzi[1]

[1]School of Life Sciences and Technology, Institut Teknologi Bandung, Bandung, 40132, West Java, Indonesia
[2]University Center of Excellence for Nutraceuticals, Bioscience and Biotechnology Research Center, Institut Teknologi Bandung, Bandung, 40132, West Java, Indonesia

Abstract. Coronavirus (CoV) is a virus that causes respiratory and gastrointestinal diseases in animals and humans. It belongs to the Orthocoronavirina. The nucleocapsid protein (N) plays multiple roles in virus assembly, RNA transcription, and interaction with host cells. This study aimed to analyse the N protein by identifying conserved residues and exploring the gene and protein relationships within the Orthocoronavirinae. Therefore the results of this study are expected to help identify conserved regions of N protein in SARS-CoV-2 which can be used as probes for the virus identification process and can be used as target areas in vaccine development. We used 159 N gene and protein sequences, including 64 from Alpha, 51 from Beta-, 11 from Delta-, and 20 from Gammacoronavirus genera of the Orthocoronavirinae. Three sequences from Tobaniviridae were used as outgroups. Multiple sequence alignment (MSA) and phylogenetic tree analysis were performed using the neighbour-joining and Maximum Likelihood. The MSA results revealed several conserved residues, ranging from 18 to 41, were located in the N-terminal and C-terminal domains, the linker region, Nuclear Localization Signal (NLS), Nuclear Export Signal (NES) motifs, and Packing Signal (PS) binding sites. The phylogenetic tree analysis indicated that Gammacoronavirus and Deltacoronavirus were closely related to Betacoronavirus, while Alfacoronavirus showed the most distant relationship. Furthermore, the study identified 23 conserved residues involved in RNA binding, including amino acids such as Ser89, Val111, Pro112, Gly124, Tyr125, Phe150, Tyr151, Gly154, Thr155, Gly156, Trp180, Val181, Gly409, Arg411, Asn419, Gly421, and Pro443. These residues interacted with phosphate groups, nitrogenous bases, and pentose sugars and exhibited non-specific interactions with RNA. In summary, this study investigated the N protein in the Orthocoronavirinae subfamily, providing insights into its function, structure, and evolutionary relationships.
Keywords: Coronavirus, Motives, Phylogenetics, Mutations, Conserved Residues

1 Introduction

Coronaviruses are a group of viruses with RNA and membrane genetic material. Coronaviruses have a wide distribution, and are distributed among mammals and birds. Coronavirus is a member of the Orthocoronavirinae subfamily in the Coronaviridae family. The characteristics of the coronavirus are its large positive sense single-stranded genome which is around 26.4-31.7 kilobases, polyadenylated and has a stamp at the 5' end, has a viral membrane, and a spike protein shaped like a beater [1].

Coronavirus infection in animals was discovered since the early 20th century. One of the earliest known coronaviruses was the transmissible gastroenteritis virus (TGEV) which infected pigs in the early 20th century. TGEV is causing epidemics and pig deaths in the United States, especially because the mortality rate is up to 100%. This causes economic losses to the pig industry [2, 3]. After being discovered in pigs, in 1930 a coronavirus was discovered in birds, namely the infectious bronchitis virus (IBV) which causes infectious bronchitis in birds [4]. Until now, IBV is spread throughout the world and affects the ability of poultry to produce meat and eggs [5]. This causes economic losses, especially in countries that have intensive poultry industries [6].

TGEV and IBV are two examples of the many coronaviruses that cause disease in mammals and birds. In mammals, the coronavirus also infects mice, cats, dogs, cattle, beluga sharks, porcupines, camels, horses and civets [7 – 15]. In poultry, the coronavirus infects poultry, turkeys, and some birds, namely the nightingale, thrush, bondol, white-eye, finches, magpie robin, night heron, wigeon, and common morhen [16, 17] Several coronaviruses in mammals and poultry have caused massive economic losses, threatening zoonoses, especially in several viruses that are spread throughout the world (IBV, MHV, FIPV, CCoV, BCoV, FCoV, MERS-CoV, PHEV and PDCoV), which have caused epidemics (TGEV, PHEV, PEDV, PDCoV, SeACoV, TCoV, and CCoV), and those with hosts with close human interactions (MHV, LRNV, FIPV, FCoV, CCoV, CRCoV, MERS-CoV, FRSCV, FRECV, and ECoV). [2,

* Corresponding author: nugrahapraja@sith.itb.ac.id

5, 9, 10, 13, 15, 18-27]. MERS-CoV is a zoonotic product of camels which causes epidemics with a case fatality rate of 28-35%. In addition, studies by Lednicky and Vlasova revealed successive infections of PDCoV and CRCoV in humans [28, 29] although, it is necessary to re-examine whether these viruses cause disease in infected patients, and can transmitted to other humans or not. Therefore, coronaviruses in animals can have a negative impact on the economy, health, and even human survival.

The earliest coronaviruses found to cause infection in humans were HCoV-229E and HCoV-OC43 in 1966 and 1967 respectively. Both of them together with HCoV-NL63 which was discovered in 2004 caused mild respiratory infections. HCoV-229E, HCoV-OC43, and HCoV NL63 had case fatality rates of 25%, 9.1%, and 12.5%, respectively [30, 31]. HCoV-HKU1 was found in a patient in Hong Kong in 2004 with a low case fatality rate, usually death occurs because the patient has other serious illnesses or his immune system is very weak or not functioning properly [32].

This coronavirus is associated with several respiratory diseases, from the mildest is common fever to the most severe are pneumonia and bronchitis. SARS-CoV and MERS-CoV caused epidemics, while SARS-CoV-2 caused pandemics. All three cause acute respiratory system syndrome (SARS) and respectively have case fatality rates of 11%, 28-35%, and 7.3% [33-35]. Of the seven coronaviruses that cause disease in humans, HCoV-229E and HCoV-NL63 belong to the Alfacoronavirus genus, while the others belong to the Betacoronavirus genus. In addition, HCoV-OC43 and HCoV-HKU1 have natural hosts in rodents, while the other five coronaviruses have natural hosts in bats. In humans, the SARS-CoV-2 pandemic with a small case fatality rate can affect economic stability [33].

Allegations regarding the re-infection of the coronavirus in the future both to humans and other animals became clearer when there were cases handled by Lednicky and Vlasova regarding PDCoV and CRCoV infections to humans [28-29] and epidemics SeACoV in pigs in 2017 [25].

Interspecies transmission or cross-species transmission is the transmission of an infectious pathogen from one species to another. When a pathogen has been contracted from another species, the pathogen can cause disease in that species. Even these pathogens can be transmitted to their own species and cause epidemics or pandemics. Two transitional stages are necessary for the emergence of interspecies transmission, namely human contact with infectious agents, and interspecies transmission of these agents. In addition, there are two transitional stages that are important for a pathogen to cause a pandemic but do not occur in many pathogens that have occurred zoonoses, namely human-to-human transmission that supports, and genetic adaptation to the host [36]

Interspecies transmission of coronaviruses is supported by their long-term presence in nature, their rapid mutagenesis, high diversity, the evolution of coronaviruses within the host [37], as well as human interactions with several coronavirus hosts [38]. Therefore, as a long-term solution to coronavirus infection, the characterization of the genetic and biological components of the coronavirus becomes very important.

The coronavirus genome is single-stranded RNA, positive sense, and is about 26.4-31.7 kilobases (International Committee on Taxonomy of Viruses, 2012). The coronavirus genome encodes four structural proteins, namely spike protein (S), envelope protein (E), matrix protein (M), and nucleocapsid protein (N). In addition, the coronavirus genome also encodes 16 nonstructural proteins (nsp1-16) that form the replicase-transcriptase complex (RTC) and accessory proteins [39].

The nucleocapsid (N) protein is a structural protein measuring 43–46 kDa. The N protein plays a role in the packaging of the RNA genome because it forms a ribonucleoprotein, the efficiency of transcription and processing of the viral RNA genome through its interaction with nsp3, the assembly of viruses through its interaction with the M protein, and influences host cells and host cell cellular mechanisms by blocking the G1/S phase transition [33, 40, 41]. Important characteristics of N protein that can be used as vaccine candidates, a good inhibitor target is highly immunogenic, expressed in large quantities during infection, and can induce protective immunity against SARS-CoV and SARS-CoV-2 [42].

Comprehensive characterization of the protein N coronavirus in terms of sequence, phylogenetics, and implications for its structure and function can provide insight into potential treatment targets, epitopes for vaccines, inhibitor targets, etc. that can be used long-term for coronavirus infection in humans and animals. in the future. Therefore, this study aims to identify which residues are the most conserved from the N protein sequences in the Orthocoronavirinae subfamily, analyze the effect of these residues on the function and structure of the domain, motif or sequence region in the N protein, and analyze the kinship. Subfamily Orthocoronavirinae using N protein and gene sequences.

2 Materials and Methods

2.1 Classification and Sequence Alignment of Current Orthocoronavirinae

Data retrieval from the National Center for Biotechnology Information (NCBI) website is based on only the N gene sequence manually. A total of 156 cDNA sequences of the N gene were taken from members of the four genera in the Orthocoronaviriae subfamily. Sampling was carried out for each complete sequence, annotated on the N gene, and differed in terms of host and country of origin of the sample. Based on data collection of samples taken, 64 sequences out of 156 sequences came from the genus Alfacoronavirus, 61 sequences came from Betacoronavirus, 11 sequences came from the genus Deltacoronavirus, and 20 sequences came from the genus Gammacoronavirus.

Next translation of the N gene sequence was carried out using EMBOSS Transeq on the EMBL-EBI website. After the translation is done, check again whether the reading frame is correct with Jalview 2.11.0.

Furthermore addition of outgroups is carried out only for compilation of sequences that will be used for the construction of phylogenetic trees, while for compilation of sequences used for the identification of sustainable amino acid residues it is not added. As many as 3 outgroup individuals, namely bovine torovirus from the Tobaniviridae family. The Tobaniviridae family is in the same order as the Orthocoronavirinae subfamily, namely Nidovirales.

Additionally multiple Sequence Alignment of gene and protein N sequences was carried out both in sequence compilation to identify sustainable amino acid residues and in constructing phylogenetic trees using ClustalOmega on the EMBL-EBI website.

2.2 Construction and Visualization of Phylogenetic Tree

The construction of the phylogenetic tree was carried out by compiling cDNA and protein sequences with the addition of outgroups. The construction of phylogenetic trees was carried out using MrBayes and MEGA X. The method used in MrBayes is Bayesian, while the methods used in MEGA X are Neighbor Joining (NJ) and Maximum Likelihood (ML). The writer choose MEGA X because MEGA X is a versatile tool that covers a broader spectrum of molecular biology tasks, including sequence alignment and basic phylogenetic analysis. Moreover it is known for its user-friendly interface. On the other hand, Mr. Bayes is a specialized software specifically designed for Bayesian phylogenetic inference, offering a high level of flexibility and accuracy [43].

The parameters used in MEGA X for building NJ trees are gamma rates, including transitions and transversions, and 1000 times bootstrap with default configuration. The difference between the model for cDNA and protein sequences is that the model for cDNA sequences uses Maximum Composite Likelihood (MCL), while the model for protein sequences is the Jones-Taylor-Thornton (JTT) model. In building ML trees, the parameters used are gamma distributed rates with invariant sites (G+I), use of all sites, the ML heuristic method is Nearest Neighbor-Interchange (NNI), and bootstrap 1000 times. The ML tree of cDNA sequences was constructed using the GTR model and estimation of the pairwise distance matrix using MCL, while for protein sequences using the JTT model and estimation [43].

Label color modification based on genus, appearance of bootstraps, and leaf sorting on phylogenetic trees were used using iTOL

2.3 Identification of Conserved Amino Acid Residues

Identification of conserved amino acid residues was carried out using Jalview 2.11.0 on the MSA-produced protein sequences. The sequence identity thresholds used were 80%, 90%, 95%, 97.5%, and 100%. After that, using WebLogo 2.8.2, a visualization of the sustainability of each amino acid residue in the protein sequence from MSA was made.

2.4 Analysis and Comparison between Phylogenetic Tress with Reference

Lastly from Figure 1 analysis of the conserved residue data was carried out by identifying the domains, motifs, and regions present in the N protein sequence that have conserved residues. After that, estimation of the location of the domains, motifs, and regions contained in the N protein sequence was carried out using the results of MSA and the characterization of the N protein in several coronaviruses. Some of the coronaviruses whose characterization results are used as references are MERS-CoV, HCoV-229E, IBV, MHV, NL63, OC43, PEDV, and HKU1. Then, a literature study was carried out regarding the effect of these conserved amino acid residues on the structure and/or function of a domain, motif, and region contained in the N protein sequence. Comparison of phylogenetic trees was carried out between phylogenetic trees that had been made and with reference trees that had been made by Tabibzadeh [43].

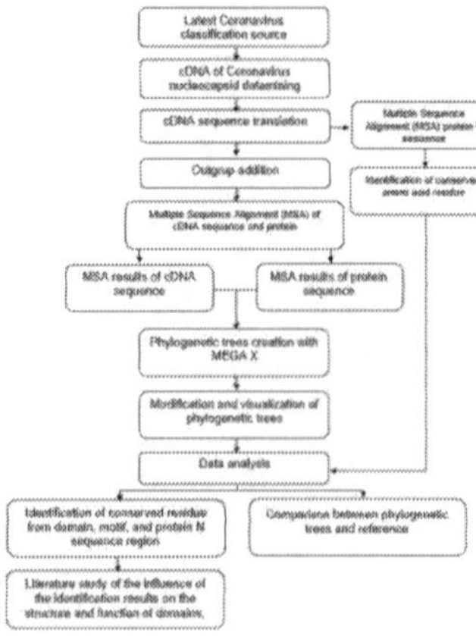

Fig. 1. Research Workflow

3 Results and Discussion

3.1 Predicted Conserved Amino Acid Residues

Preservation analysis is one of the most widely used methods in the prediction of functionally important residues in protein sequences [44]. From Figure 2 of the N protein sequence length of around 400 residues in the analyzed coronavirus, residues that exceed the 80% sequence identity threshold have 41 residues or around 10%. In addition, 18 residues with 100% sequence identity were found in the sample sequences tested. By knowing that the N protein has a low sequence identity but has the same modular organizational structure, this can prove that there are conserved residues used by N protein to maintain its function.

When compared with the MSA performed by Laude & Masters, it can be seen that the residues that were sustainable in Laude & Masters [45] but not preserved in the results of the MSA performed were two arginine residues, leucine, histidine, two alanine residues, aspartic acid, and valine.

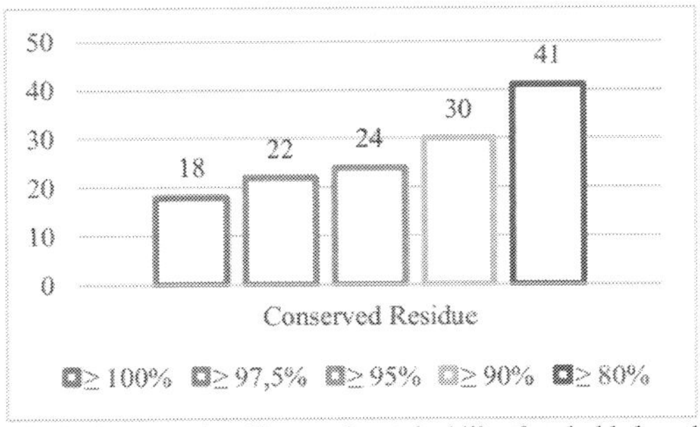

Fig. 2. Conserved residues and sustainability thresholds based on MSA results

3.2 Implication of Conserved Residues on Function and Structure

As we can see in Figure 3. The visualization of the MSA in the N-terminal domain, serine residues in the 89th site and arginine in the 200th site in the visualization of the MSA results are Ser64 and Arg164 residues in HCoV-OC43 which play a role in direct interaction with the 2'-hydroxyl group in pentose sugar RNA. The Tyr residue at the 151st site is also the Tyr126 residue in HCoV-OC43 which plays a role in its interaction with nitrogenous bases [46].

In addition, this residue along with the arginine residue at the 133th site is a Tyr94 and Arg76 residue in IBV which play a role in binding RNA. The aromatic properties of the tyrosine residues and the alkaline properties of the arginine residues play a role in binding RNA by creating a large surface that is in contact with the viral gRNA [47].

Although not confirmed, we can see that in the Table 1. Tyr residues at sites 125, 151, 152, and tryptophan at site 180 which correspond to residues Tyr87, Tyr112, Tyr113, and Trp133 in SARS-CoV are on the same β-sheet surface and play a role in packaging RNA and is a residue that plays a role in forming hydrophobic pockets in HCoV-OC43 which can orient nitrogenous bases on the protein surface, rather than selecting a protein-RNA sequence [46, 48]. The β-hairpin structures are similar in structure but vary in electrostatic surface and topology, which may indicate a specific adaptive function. As we can see from the Table 2. This structure motif plays a functionally important role in the N-NTD for binding RNA [46] and neutralizing the phosphate group [47], in this structure there is an arginine residue at the 133th site and glycine at the 133th site. 139th site. The β-sheet core consists of the secondary structure β1β2β5β6β7. In this structure there are conserved residues that occupy β5 and β6. The role of the β-sheet core structure is to "hold" the RNA by neutralizing the phosphate groups of the RNA, and the aromatic amino acid residues in this section interact with the base portion of the RNA [47, 49] In the border area between NTD and LKR, namely at sites 213 to 245 on MSA results, Schuster (2020) said that this area is the result of recombination that also occurs in several coronaviruses, namely Pangolin-CoV MP789, Bat-CoV RaTG13, and bat-SL-CoVZXC21.

Table 1. Conserved residues in the domain in the N protein based on their location on the MSA result site

Domain/Motif/Region	Residue
NTD (N-terminal domain)	SER89, VAL111, PRO112, GLY124, TYR125, TRP126, ARG133, GLY139, TRP148, PHE150, TYR151, TYR152, GLY154, THR155, GLY156, PRO157, GLY177, TRP180, VAL181, GLY185, ALA186, GLY197, ARG199, PHE213, PRO220, SER237
LKR (linker region)	*SER241, ARG242, LEU330*
CTD (C-terminal domain)	*LYS393, ARG394, PHE408, GLN409, ARG411, ASN419, PHE420, GLY421, GLY429, ALA439, PRO443, ALA447, PRO489, ALA504*

Table 2. Conserved residues in the motifs and regions in protein N based on their location on the MSA result site

Motif/Region	Residue
NLS1 (Nuclear Localization Signal 1)	LYS393, ARG394
NES1 (Nuclear Export Signal 1)	PHE213
NES2	LEU330
Putative PS-BS (Packing Signal-Binding Site)	LYS393, ARG394, PHE408, GLN409, ARG411

3.3 Gene and Protein N Phylogenetic Tree Analysis

As we can see in the Figure 4 the phylogenetic tree obtained from the cDNA data, that the two trees have high similarity. This can happen because the resulting NJ tree can produce the correct topology [50] and the use of the Maximum Composite Likelihood (MCL) model on the Neighbor Joining tree is robust in consistency with the Maximum Likelihood Estimation (MLE), also in terms of efficiency and overall the computation. When compared with the Tabibzadeh tree [43], which uses cDNA data using the Neighbor Joining method, you can see the similarities to the two trees from the cDNA data generated. The similarity is that the Betacoronavirus kinship is close to Gammacoronavirus, and Alfacoronavirus is the genus that is most distantly related. Meanwhile, the difference is in the tree topology which can be caused by the number of different samples, outgroups and Deltacoronavirus samples which are not present in the Tabibzadeh tree [43] but are present in the created tree

https://doi.org/10.1051/bioconf/20237502005

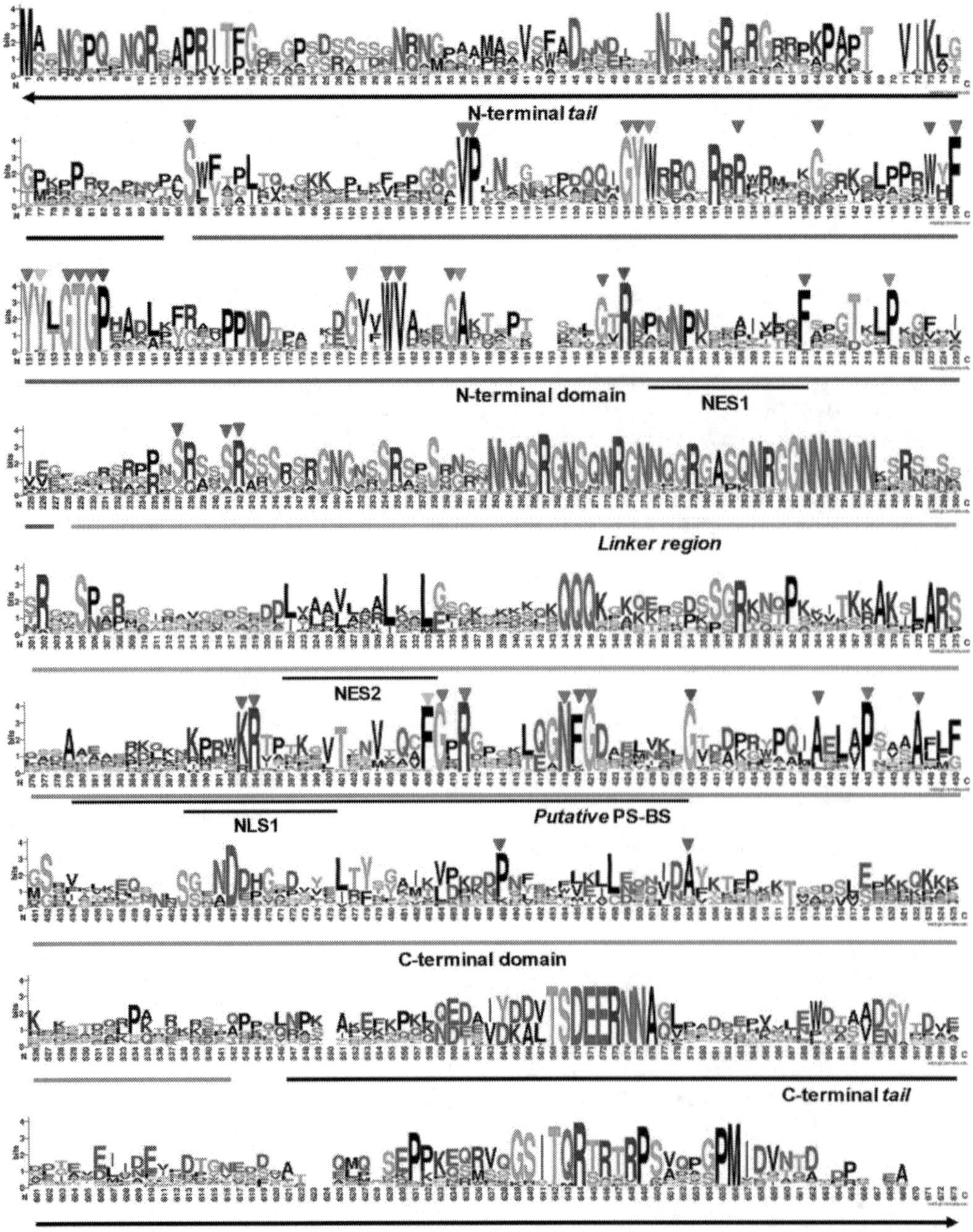

Fig. 3. Visualization of conserved residues in MSA results with WebLogo. Conserved residues are indicated by the triangle above the residue illustration. The color of the triangle above the residue indicates that the residue has exceeded the residue identity threshold a) blue: 80%, b) green: 90%, c) orange: 95%, d) purple: 97.5%, and e) red: 100% Motifs and domains that have conserved residues are mentioned in the visualization. The site numbering on the MSA results is below the residuals.

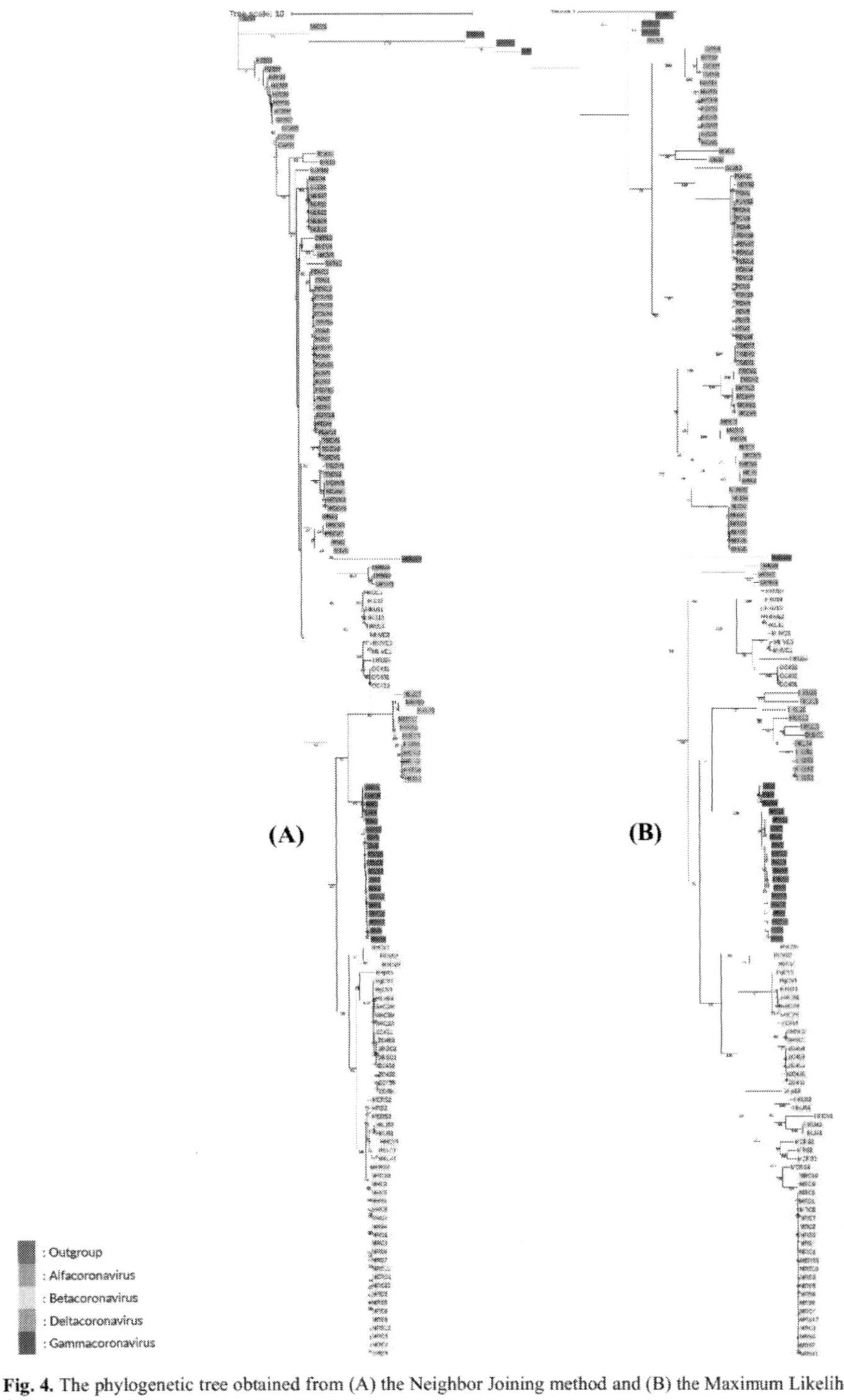

Fig. 4. The phylogenetic tree obtained from (A) the Neighbor Joining method and (B) the Maximum Likelihood method from cDNA data.

https://doi.org/10.1051/bioconf/20237502005

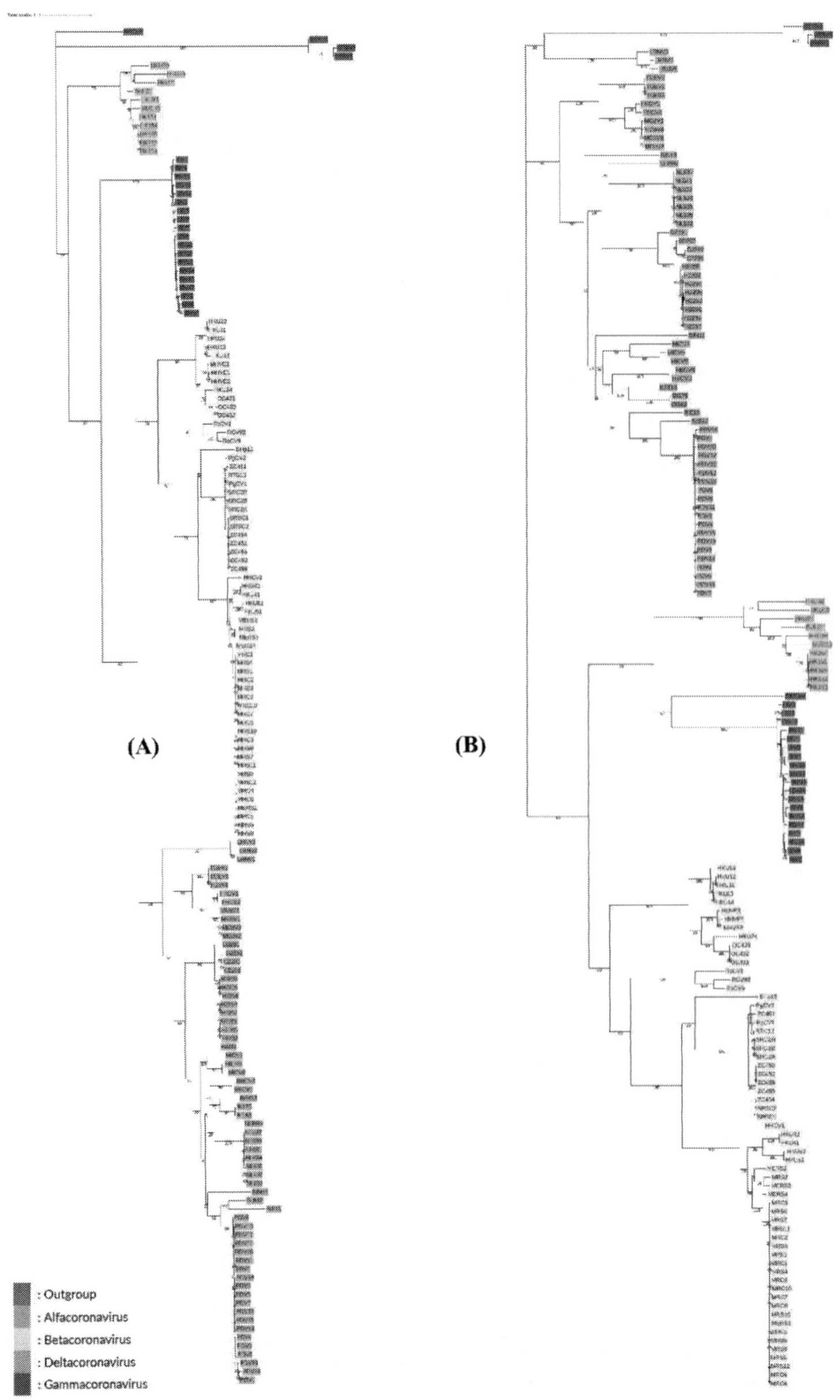

Fig. 5. The phylogenetic tree obtained from (A) the Neighbor Joining method and (B) the Maximum Likelihood method from protein data.

In the phylogenetic tree obtained from the protein data, it can be seen from Figure 5 that the two trees tend to be dissimilar. The similarities between the two trees are only found in their kinship structure, where Gammacoronavirus is closely related to Deltacoronavirus, then both are close to Betacoronavirus, and the most distantly related is Alfacoronavirus. However, the difference with other trees is that the genera closest to the outgroup, the cDNA tree and the ML tree from the protein data show that the closest to the outgroup is Alfacoronavirus, not Deltacoronavirus as in the NJ tree obtained from protein data. Errors in the Neighbor Joining method are caused significantly by zero-length branches in the tree [51]. In the generated NJ tree, a large number of zero-length branches can be seen, especially in sequences with the same virus species.

The ML tree obtained from the protein data is similar to both the Tabibzadeh tree [43] and the tree constructed with cDNA data in terms of the genus arrangement in the tree. This is because the resulting tree from the ML method has the advantage of having low variance compared to other methods, being robust against violations of assumptions in the evolutionary model, being able to outperform the performance of the parsimony or distance methods even though the sequences used are very short, evaluating different tree topologies, using all the information on the sequence. Compared to the distance method, and is better for calculating branch length [52].

4 Conclusions

It is concluded that the most conserved residues in the nucleocapsid protein of the coronavirus are Ser89, Val111, Pro112, Gly124, Tyr125, Phe150, Tyr151, Gly154, Thr155, Gly156, Trp180, Val181, Gly409, Arg411, Asn419, Gly421, and Pro443. Second, 2. Ser89 and Arg200 play a role in their interaction with the 2'-hydroxyl group pentose sugar. Tyr151 plays a role in its interaction with nitrogenous bases. Tyr125, Tyr151, and Trp180 play a role in orienting bases on the protein surface and RNA packaging. Phe150, Tyr151, and Trp180 and Val181 can "grasp" RNA by neutralizing the phosphate group of RNA. Arg411 plays a role in binding negatively charged oligonucleotides based on non-specific interactions. Gly409, Arg411, Asn419, Gly421 play a role in binding packing signal (PS). Pro443 plays a role in hydrophobic interactions. Finally, the genera Deltacoronavirus and Gammacoronavirus are the most closely related, followed by the genus Betacoronavirus, while the genus Alfacoronavirus is the most distant. The results of this study are expected to help identify conserved regions of N protein in SARS-CoV-2 which can be used as probes for the virus identification process and can be used as target areas in vaccine development.

Acknowledgements

The authors thank the School of Life Sciences and Technology for partially funding this research under the Biology Study Program.

Conflicts of interest statement

The authors declared that they have no conflicts of interest in this study.

References

1. International Committee on Taxonomy of Viruses, 2012. *Virus Taxonomy: Ninth Report of the International Committee on Taxonomy of Viruses,* Oxford: Elsevier.
2. Spickler, A. R., Roth, J. A., Galyon, J. & Brown, G., 2016. *Transboundary and Emerging Diseases of Animals.* 1st ed. Ames: Iowa State University.
3. Xia Xia, L., Yang, Y., Wang, J., Jing, Y., & Yang, Q. (2018). Impact of TGEV infection on the pig small intestine. *Virol. J., 15*(1), 1-7.
4. Jackwood, M. W., 2012. Review of Infectious Bronchitis Virus Around the World. *Avian Disease,* Volume 56, 634-641.
5. The Pirbright Institute, n.d. *Infectious bronchitis virus.* [Online] Available at: https://www.pirbright.ac.uk/viruses/infectious-bronchitis-virus [Accessed 28 August 2021].
6. Cavanagh, D., 2007. Coronavirus avian infectious bronchitis virus. *Veterinary Research,* **38**, 281-297.
7. Wang, W., Lin, X. D., Guo, W. P., Zhou, R. H., Wang, M. R., Wang, C. Q., ... & Zhang, Y. Z. 2015. Discovery, diversity and evolution of novel coronaviruses sampled from rodents in China. *Virology*, **474**, 19-27.
8. Bailey, O. T., Pappenheimer, A. M., Cheever, F. S., & Daniels, J. B. 1949. A murine virus (JHM) causing disseminated encephalomyelitis with extensive destruction of myelin: II. Pathology. *J.Exp.Med., 90*(3), 195-212.
9. Pratelli, A., 2006. Genetic evolution of canine coronavirus and recent advances in prophylaxis. *Veterinary Research,* **37**(2), 191-200.
10. Centre for Agriculture and Bioscience International, 2019. *bovine coronavirus infection.* [Online] Available at: https://www.cabi.org/isc/datasheet/9170 [Accessed 28 August 2021].
11. Mihindukulasuriya, K. A., Wu, G., St. Leger, J., Nordhausen, R. W., & Wang, D. 2008. Identification of a novel coronavirus from a beluga whale by using a panviral microarray. *J. Virol.*, **82**(10), 5084-5088.
12. Corman, V. M., Kallies, R., Philipps, H., Göpner, G., Müller, M. A., Eckerle, I., & Drexler, J. F. , 2014. Characterization of a Novel Betacoronavirus Related to Middle East Respiratory Syndrome Coronavirus in European Hedgehogs. *Virol.J.,* **88**(1), 717-724.
13. Müller, M. A., Corman, V. M., Jores, J., Meyer, B., Younan, M., Liljander, A., & Drosten, C., 2014. MERS coronavirus neutralizing antibodies in camels, Eastern Africa, 1983–1997. *Emerg. Infect. Dis.,* **20**(12), 2093–2095.
14. Pusterla, N., Vin, R., Leutenegger, C. M., Mittel, L. D., & Divers, T. J. , 2018. Enteric coronavirus infection in adult horses. *Vet. J.,* **231**, 13-18.

15. Murray, J., Kiupel, M. & Maes, R. K., 2010. Ferret Coronavirus-Associated Diseases. *Vet Clin North Am Exot Anim Pract.,* **13**(3), 543-560.
16. Woo, P. C., Lau, S. K., Lam, C. S., Lai, K. K., Huang, Y. I., Lee, P., & Yuen, K. Y., 2009. Comparative analysis of complete genome sequences of three avian coronaviruses reveals a novel group 3c coronavirus. *Virol.J.,* **83**(2), 908-917
17. Woo, P. C., Lau, S. K., Lam, C. S., Lau, C. C., Tsang, A. K., Lau, J. H., & Yuen, K. Y. , 2012. Discovery of Seven Novel Mammalian and Avian Coronaviruses in the Genus Deltacoronavirus Supports Bat Coronaviruses as the Gene Source of Alphacoronavirus and Betacoronavirus and Avian Coronaviruses as the Gene Source of Gammacoronavirus and Deltacoronavi. *Virol.J.,* **86**(7), 3995-4008.
18. Huang, D. S., Emancipator, S. N., Fletcher, D. R., Lamm, M. E., & Mazanec, M. B., 1996. Hepatic pathology resulting from mouse hepatitis virus S infection in severe combined immunodeficiency mice. *Lab. Anim. Sci.,* **46**(2), 167-173.
19. Holzworth, J., 1963. Some important disorders of cats. *The Cornell Veterinarian,* Volume 53, 157-160.
20. Taharaguchi, S., Soma, T. & Hara, M., 2012. Prevalence of Feline Coronavirus Antibodies in Japanese Domestic Cats during the Past Decade. *J. Vet. Med. Sci.,* **74**(10), 1355-1358.
21. Mora-Díaz, J. C., Piñeyro, P. E., Houston, E., Zimmerman, J., & Giménez-Lirola, L. G., 2019. Porcine Hemagglutinating Encephalomyelitis Virus: A Review. *Front. vet. sci,* **6**(53)
22. Boley, P. A., Alhamo, M. A., Lossie, G., Yadav, K. K., Vasquez-Lee, M., Saif, L. J., & Kenney, S. P., 2020. Porcine Deltacoronavirus Infection and Transmission in Poultry, United States. *Emerg. Infect. Dis.,* **26**(2), 255-265.
23. Snelson, H., 2014. *Comprehensive Discussion of PEDV, Slide 3, American Association of Swine Veterinarians presentation to the 2014 Pork Management Conference.* [Online] Available at: http://www.slideshare.net/trufflemedia/dr-harry-snelson-pedv-lessons-learned [Accessed 28 August 2021].
24. Pérez-Rivera, C., Ramírez-Mendoza, H., Mendoza-Elvira, S., Segura-Velázquez, R., & Sánchez-Betancourt, J. I., 2019. First report and phylogenetic analysis of porcine deltacoronavirus in Mexico. *Transbound. Emerg Dis,* **66**(4), 1436-1441.
25. Yang, Y.-L., Yu, J.-Q. & Huang, Y.-W., 2020. Swine enteric alphacoronavirus (swine acute diarrhea syndrome coronavirus): An update three years after its discovery. *Virus Res,* 285.
26. Peterson, E. & Hymas, T., 1951. Antibiotics in the Treatment of an Unfamiliar Turkey Disease. *Poult. Sci.,* **30**(3), 466-468.
27. Priestnall, S. L., Brownlie, J., Dubovi, E. J. & Erles, K., 2006. Serological prevalence of canine respiratory coronavirus. *Vet. Microbiol,* **115**(1), 43-53.
28. Lednicky, J. A., Tagliamonte, M. S., White, S. K., Elbadry, M. A., Alam, M. M., Stephenson, C. J., & Morris Jr, J. G., 2021. Emergence of porcine delta-coronavirus pathogenic infections among children in Haiti through independent zoonoses and convergent evolution. *medRXiv. Medicine,* **90**(3), 195-212.
29. Vlasova, A. N., Diaz, A., Damtie, D., Xiu, L., Toh, T. H., Lee, J. S. Y., ... & Gray, G. C. , 2021. Novel Canine Coronavirus Isolated from a Hospitalized Patient With Pneumonia in East Malaysia. *Clin. Infect. Dis.,* Issue ciab456.
30. Krvavac, A., Patel, T. P., Karle, E. M., Epstein, N. B., Reznikov, E. A., Gates, L. G., & Holliday, Z. M., 2020. Increased Incidence, Morbidity, and Mortality in Human Coronavirus NL63 Associated with ACE Inhibitor Therapy and Implication in SARS-CoV-2 (COVID-19). *Missouri Medicine,* 117(4), 346-354.
31. Choi, W. I., Kim, I. B., Park, S. J., Ha, E. H., & Lee, C. W., 2021. Comparison of the clinical characteristics and mortality of adults infected with human coronaviruses 229E and OC43. *Scientific Reports,* **11**(4499).
32. Kanwar, A., Selvaraju, S. & Esper, F., 2017. Human Coronavirus-HKU1 Infection Among Adults in Cleveland, Ohio. *Open Forum Infect. Dis.,* **4**(2).
33. Artika, I. M., Dewantari, A. K. & Wiyatno, A., 2020. Molecular biology of coronaviruses: current knowledge. *Heliyon,* **6**(8), e04743.
34. Chang, C. K., Hou, M. H., Chang, C. F., Hsiao, C. D., & Huang, T. H. , 2014. The SARS coronavirus nucleocapsid protein – Forms and functions. *Antiviral Research,* **103**, 39-50.
35. Ritchie, H., 2021. *Mortality Risk of COVID-19.* [Online] Available at: https://ourworldindata.org/mortality-risk-covid [Accessed 5 Agustus 2021].
36. Childs, J. E., Richt, J. A. & Mackenzie, J. S., 2007. Introduction: Conceptualizing and Partitioning the Emergence Process of Zoonotic Viruses from Wildlife to Humans. In: S. Rallison & A. Clauss, eds. *Wildlife and Emerging Zoonotic Diseases: The Biology, Circumstances and Consequences of Cross-Species Transmission.* Heidelberg: Springer, 1-31.
37. Zhou, Z., Qiu, Y. & Ge, X., 2021. The taxonomy, host range and pathogenicity of coronaviruses and other viruses in the Nidovirales order. *Animal Diseases,* **1**(5).
38. Parrish, C. R., Holmes, E. C., Morens, D. M., Park, E. C., Burke, D. S., Calisher, C. H., & Daszak, P., 2008. Cross-Species Virus Transmission and the Emergence of New Epidemic Diseases. *Microbiol. Rev.,* **72**(3), 457-470.
39. Hartenian, E., Nandakumar, D., Lari, A., Ly, M., Tucker, J. M., & Glaunsinger, B. A., 2020. The molecular virology of coronaviruses. *J. Biol. Chem.,* **295**(37), 12910–12934.
40. Lu, S., Ye, Q., Singh, D., Cao, Y., Diedrich, J. K., Yates III, J. R., & Corbett, K. D. , 2021. The SARS-CoV-2 nucleocapsid phosphoprotein forms mutually exclusive condensates with RNA and the membrane-associated M protein. *Nat. Commun.,* **12**(502).
41. Gao, T., Gao, Y., Liu, X., Nie, Z., Sun, H., Lin, K., & Wang, S., 2021. Identification and functional analysis of the SARS-COV-2 nucleocapsid protein. *BMC Microbiology,* **12**(58).

42. Surjit, M. & Lal, S. K., 2008. The SARS-CoV nucleocapsid protein: a protein with multifarious activities. *Infect. Genet. Evol.*, **8**, 397-405.
43. Tabibzadeh, A., Esghaei, M., Soltani, S., Yousefi, P., Taherizadeh, M., Safarnezhad Tameshkel, F., ... & Karbalaie Niya, M. H., 2020. Evolutionary study of COVID-19, severe acute respiratorysyndrome coronavirus 2 (SARS-CoV-2) as an emerging coronavirus: Phylogenetic analysis and literature review. *Vet. Med. Sci.*, **7**(2), 1-13.
44. Capra, J. A. & Singh, M., 2007. Predicting functionally important residues from sequence conservation. *Bioinformatics,* **23**(15), 1875-1882.
45. Laude, H. & Masters, P. S., 1995. The Coronavirus Nucleocapsid Protein. In: S. G. Siddell, ed. *The Coronaviridae.* New York: Plenum Press, 141-163.
46. Kang, S., Yang, M., Hong, Z., Zhang, L., Huang, Z., Chen, X., & Chen, S., 2020. Crystal structure of SARS-CoV-2 nucleocapsid protein RNA binding domain reveals potential unique drug targeting sites. *Acta Pharm. Sin. B.*, **10**(7), 1226-1238.
47. McBride, R., Zyl, M. v. & Fielding, B. C., 2014. The Coronavirus Nucleocapsid Is a Multifunctional Protein. *Viruses,* **6**(8), 2991-3018.
48. Chan-Yeung, M. & Xu, R.-H., 2003. SARS: epidemiology. *Respirology,* 8(Supplementary 1), S9-S14.
49. Schuster, N., 2020. Using the nucleocapsid protein to investigate the relationship between SARS-CoV-2 and closely related bat and pangolin coronaviruses. *bioRxiv,* 1-27.
50. Saitou, N. & Nei, M., 1987. The Neighbor-joining Method: A New Method for Reconstructing Phylogenetic Trees. *Molecular Biology and Evolution,* **4**(4), 406-425.
51. Kumar, S. & Gadagkar, S. R., 2000. Efficiency of the Neighbor-Joining Method in Reconstructing Deep and Shallow Evolutionary Relationships in Large Phylogenies. *J. Mol. Evol.*, **51**, 544-533.
52. Will, K., 2012. *Principles of Phylogenetics - University of California, Berkeley.* [Online] Available at: http://ib.berkeley.edu/courses/ib200a/lect/ib200a_lect11_Will_likelihood.pdf [Accessed Agustus 2021].

A Machine Learning-Based Virtual Screening for Natural Compounds Potential on Inhibiting Acetylcholinesterase in the Treatment of Alzheimer's Disease

Ulfah Nur Azizah[1], *Eri* Dwi Suyanti[1], *Muhammad* Rezki Rasyak[2], *Yekti* Asih Purwestri[1], and *Lisna* Hidayati[1]
[1]Faculty of Biology, Universitas Gadjah Mada, Jl. Teknika Selatan, Sekip Utara, Yogyakarta, 55281
[2]Eijkman Center for Molecular Biology, National Research and Innovation Agency, Jakarta, Indonesia

Abstract. Alzheimer's disease (AD) is a progressive neurodegenerative disease caused by neural cell death, characterized by the overexpression of acetylcholinesterase (AChE) and extracellular deposition of amyloid plaques. Currently, most of the FDA-approved AChE-targeting drugs can only relieve AD symptoms. There is no proven treatment capable to stop AD progression. Many natural products are isolated from several sources and analyzed through preclinical and clinical trials for their neuroprotective effects in preventing and treating AD. Therefore, this study aims to explore and determine potential candidates from natural bioactive compounds and their derivatives for AD treatment targeting AChE. In this study, feature extraction was carried out on 1730 compounds from six plants resulting from literature studies with limitations on international journals with a minimum publication year of 2018 and database searches, then classified using machine learning algorithms: Random Forest (RF), Logistic Regression (LR), and Support Vector Machine (SVM). Hit compounds predicted to be active and inactive in the selected model were then processed through ensemble modelling. From 1730 compounds, there are 986 predicted active compounds and 370 predicted inactive compounds in the LR and RF ensemble modelling. Quercetin, Kaempferol, Luteolin, Limonene, γ-Terpinene, Nerolidol, and Linalool predicted active found overlapping in two to three plants in both LR and RF models.
Keywords: AChE inhibitor, Alzheimer's disease, machine learning, natural bioactive compounds

1 Introduction

Alzheimer's Disease (AD) is a neurodegenerative disorder associated with increasing age and is included in progressive dementia. AD is clinically one of the main causes of dementia which can trigger memory, cognitive, executive dysfunction, and behavioural changes that lead to mental disorders in patients [1]. According to WHO (2022), AD is a global disease contributing to the number of dementia cases by 60-70% compared to other neurodegenerative diseases. The incidence of AD until now continues to increase every year, where every 3 seconds there is 1 person in the world experiencing AD. The prevalence of dementia caused by AD globally reaches 55 million cases and it is estimated that there are 10 million new cases each year. In Indonesia, cases of death due to AD in 2020 reached 27,054 people [2]. The high number of AD cases is influenced by the accumulation and aggregation of β-amyloid in the brain [3].

β-Amyloid is a peptide that accumulates abnormally in brain tissue and forms extracellular plaques that can induce neurodegeneration [4]. AD is formed from the accumulation of Aβ40 and Aβ42 peptides which are the result of an abnormal process of amyloid protein precursors between β-secretase and **γ**-secretase and an imbalance in production and synthesis pathways [5]. Several enzymes are known to be involved in increasing neurodegenerative disorders including cholinesterase (Acetylcholinesterase (AChE) and Butyrylcholinesterase (BuChE)), Prolyl endopeptidase (PEP) or oligopeptidase (POP), and the cleavage enzyme APP β (BACE1) [6]. AChE is found mainly in blood and nerve synapses [7]. Therefore, this enzyme is a suitable target for the treatment of AD. AChE is strongly suspected of interacting directly with Aβ plaque formation. This confirms that AChE inhibitors play an essential role in curing AD rather than as a palliative measure [6].

Commercial drugs currently used as AChE inhibitors are tacrine, rivastigmine, and donepezil. However, these drugs have several side effects causing nausea, vomiting, diarrhea, bleeding, and shrinking of brain tissue [8]. Seeing these problems, it is necessary to find alternative natural ingredients that are safer with minimal side effects. As a tropical country, Indonesia has many potential natural ingredients that have the potential to become candidates for anti-Alzheimer's drugs.

Currently, the discovery and development of new drugs for AD treatment required a long time and are quite expensive. It takes between 10 and 15 years of research and testing. Therefore, the approved drug therapies for AD that temporarily relieve the symptoms and slow down the disease progression could only be countered by hand [9]. It is indicated that drug discovery efforts for Alzheimer's treatment still need enhancement [10]. In the traditional discovery of natural chemical compounds, the compounds were isolated randomly, then their biological activity was identified by a simple test. In addition, in the wet lab experimental tests, not all isolated compounds were tested for their therapeutic activity. With the development of information technology, the drug discovery process can be simulated in silico more quickly and accurately through virtual screening. Machine learning-based virtual screening can be an alternative way to select natural product compounds more effectively than compounds that contain the desired activity [11].

Previous research by Periwal *et al.* (2022) used a machine-learning approach in the form of a trained classification model to explore natural compounds and their derivatives more quickly on a much larger scale [10]. In this study, the similarity was predicted between the approved drug and its natural compounds.

Therefore, this study aims to explore and predict Indonesian natural product compounds that can potentially become acetylcholinesterase inhibitors as AD treatment.

2 Material and Methods

2.1 Data set and preparation

Indonesian herbal plants that have the potential to have anti-Alzheimer's activity were searched through literature studies with limitations on international journals with a minimum publication year of 2018 (Table 1). The compounds contained in each plant are then searched through the ChEMBL Database (https://www.ebi.ac.uk/chembl/) and KNaPSAcK (http://www.knapsackfamily.com/). Each SMILES ID is taken and saved with Notepad++ in .smi format.
Via DUD-E Docking (https://dude.docking.org/targets) the active and decoy compounds of Acetylcholinesterase (AChE) (Code: ACES) were downloaded. At this stage, the SMILES ID of the active and decoy compounds was stored in Notepad++ in .smi format.

2.2 Environment used

This study needs Java programming language to obtain optimal results through the software that is used. The environment variables are set by adding Java SE Development Kit (JDK) and Java SE Runtime Environment (JRE).

2.3 Feature extraction

Feature extraction is performed on each active and decoy compound with PaDEL Descriptor software using PubChem Fingerprint. At this stage, remove salt, aromaticity detector, and standardize nitro groups are selected.

PubChem has 881 binary structural keys that indicate the presence or absence of a certain group of chemical features in a compound. Compared to other fingerprints that use a floating point number and require 32 bits for one feature, PubChem fingerprint only requires one bit of storage for each feature in the compound. The small bit of fingerprint can speed up the machine learning process. PubChem Fingerprint uses the 2D structure of the compound which is used as a measure of the similarity of the compound to the compounds that have been found on the website http://pubchem.ncbi.nlm.nih.gov.

The fingerprint results of the active and decoy compounds are then combined into one big data in .csv format. Each natural product compound is given a class label. The active compound is labeled 1, while the decoy compound is labeled 0.

2.4 Machine Learning

Machine learning (ML) is a subfield of Artificial Intelligence (AI) that focuses on developing algorithms and statistical models that enable computers to learn and improve their performance on specific problems without being explicitly programmed [11]. Machine learning has become an increasingly valuable tool in research due to its ability to analyze large and complex datasets, identify patterns, and make predictions.

This article utilized a machine-learning approach to extract compounds from the data obtained from ChEMBL and KNaPSaCK databases. This approach has developed a set of algorithms that have been optimized for performance and accuracy using advanced computational techniques. These algorithms can process large volumes of data efficiently and extract valuable information from it.

In this study, a machine learning approach was used with supervised learning. The selected algorithms include Random Forest (RF), Logistic Regression (LR), and Support Vector Machine (SVM). All three have different ways of classifying objects (compounds) into their classes. RF is a classifier that consists of a collection of tree-structured classifiers. RF uses multiple trees to average (regression) or calculates the most votes (classification) in the terminal leaf nodes to make a prediction. In decision trees, each node is separated by the best separation among all variables. While in RF, each node is divided by the best among the predictor subsets chosen randomly at that node. On decision trees, at each decision node, the features are split into two branches and are repeated until the leaf nodes are reached to make the final prediction. compared to RF, decision trees can't generalize well the unseen data and are notorious for overfitting. RF is able to overcome overfitting by using a decision tress ensemble where the values are random and independent. RF is suitable for medium to large datasets. In RF, not all predictor variables are used at once so that when the number of independent variables is greater than the number of observations, this algorithm can run, unlike the LR algorithm which will not run because the predicted parameters exceed the number of observations [12].

Logistic Regression (LR) is one of the most commonly utilized linear statistical models in which the response variable is quantitative. The response variable in LR is in the form of a log of the possibilities that are classified in group I of binary responses or multi-class responses. LR makes several assumptions such as independence, the responses (logits) at each subpopulation level of the variables are normally distributed, and the constant variance between the responses and all explanatory value variables. a transformation to the variable is applied over the output classes between 0 and 1, called "logistic" or "sigmoid". LR has a vulnerability to underfitting and has low accuracy [13]. In addition, the LR algorithm also has problems with class imbalance in datasets with high dimensions [14].

Support Vector Machine (SVM) is a technique for finding hyperplanes that can separate two data sets from two different classes. SVM has the basic principle of a linear classifier. Even so, SVM can also work on non-linear problems through the kernel concept in high-dimensional space. In a high-dimensional space, we will look for a hyperplane that can maximize the margin between data classes. In addition, SVM uses structural risk minimization (SRM), which is the inductive principle of nature having a learning model from a limited set of training data. The advantage of this model is that it is able to determine the distance to the support vector so that the computational process runs faster.

The machine-learning approach was conducted on Lenovo Ideapad Slim 3i, Intel Core i3-14IGL05, RAM 5 GB, SSD 256 GB. Analysis of the machine learning approach was carried out using the Orange Data Mining application with 75% fixed proportion data and 5- cross-validation. The results obtained are then evaluated based on ROC Analysis and Confusion Matrix to select the algorithm to be used for the next process (Figure 1).

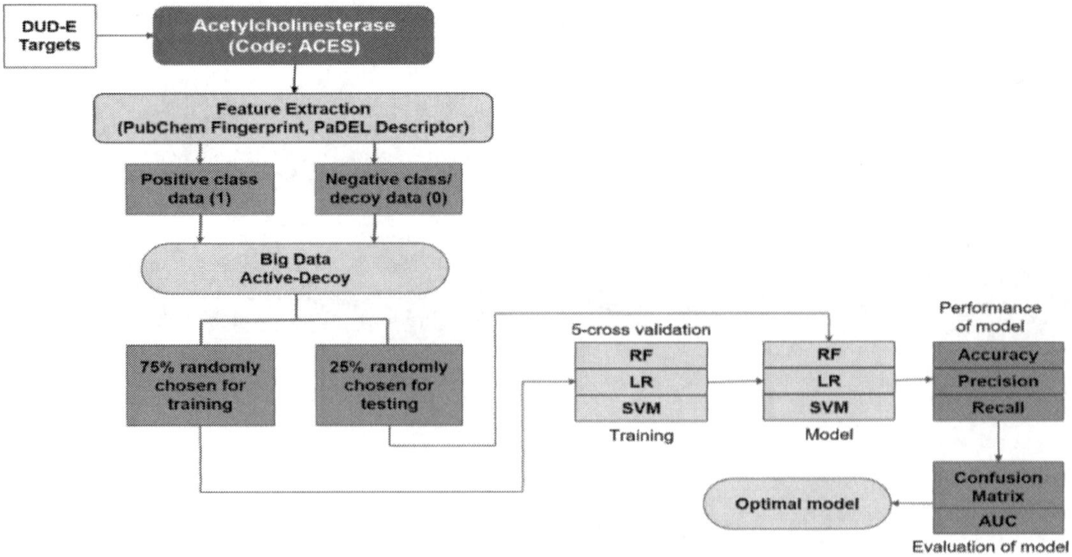

Fig. 1. A schematic representation of the machine learning approach to choose the optimal and validated model for predicting natural compounds

2.5 Predicting Indonesian Herbal Compounds

Prediction of Indonesian medicinal compounds is investigated with selected models. The natural ingredient compounds from the 6 selected plants do not have labels. Therefore, each plant compound is predicted to be active and decoy. From the classification results obtained, an ensemble was performed on compounds that were predicted to be active and inactive in both models (Figure 2).

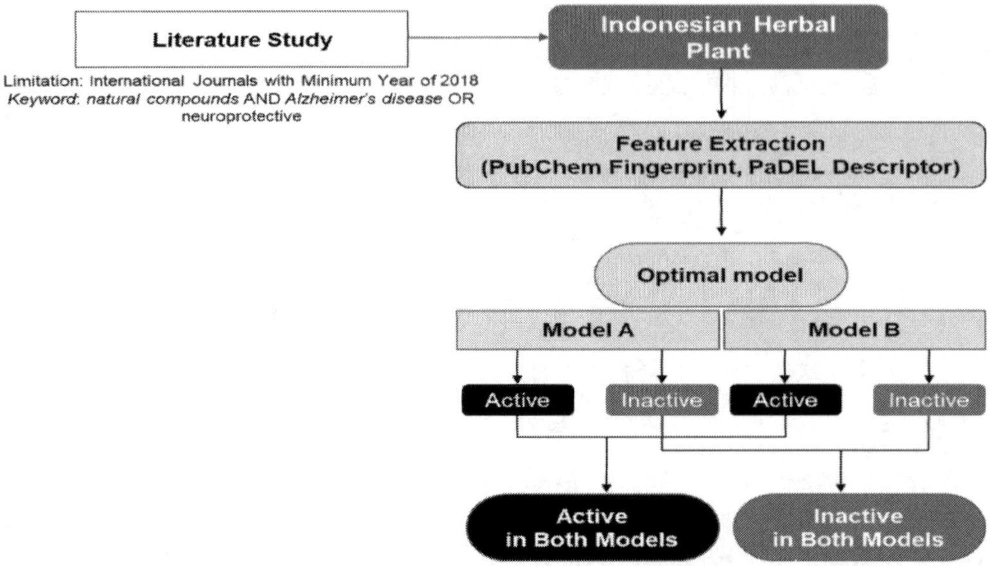

Fig. 2. The scheme of predicting natural compounds using the optimal and validated model

2.6 Pharmacokinetic Analysis

The pharmacokinetic tests of the compounds included the HIA and toxicity (BBB) test through PreADMET (https://preadmet.webservice.bmdrc.org/adme/) and the Lipinski Rule of Five (Ro5) test through SwissADME (http://www.swissadme.ch/).
In addition, Predictions of Activity Spectra for Substances (PASS) were carried out via the http://www.pharmaexpert.ru/passonline/predict.php page. This test was carried out to determine the activity potential of a compound based on the relationship between the structure of the compound and its biological activity (Structure Activity Relationship/SAR).

3 Results and Discussion

3.1 Data source

Based on the results of a literature review search, there are 6 candidate plants that had the potential as anti-Alzheimer's (Tabel 1). Based on ChEMBL and KNaPSAcK, a total of 1730 compounds were obtained from 6 selected plants. There are 14 compounds in *Moringa oleifera*, 31 compounds in *Zingiber officinale*, 43 compounds in *Allium sativum*, 1292 compounds in *Annona crassiflora*, 142 compounds in *Citrus aurantium*, and 208 compounds in *Annona muricata*.

Table 1. List of Indonesian herbal compound related to anti-Alzheimer.

Plants	Bioactivity	Reference
Moringa oleifera	Inhibitor of butyrylcholinesterase (BChE); inhibitor of acetylcholinesterase (AChE); lower the glycemic index, total cholesterol, triglycerides, and low-density lipoprotein cholesterol (LDL-C) level; increase high-density lipoprotein cholesterol (HDL-C) in plasma	[15]
Zingiber officinale	Antioxidant, anti-inflammatory, increase expression of nerve growth factor (NGF)	[16]
Allium sativum	Antioxidant, neuroprotective agent, inhibitor of AChE and BChE enzymes, anti-neuroinflammatory	[17]
Annona crassiflora	Antibacterial, antimutagenic, anti-inflammatory, antinociceptive, hepatoprotective, and antitumoral, inhibitor of AChE	[18]
Citrus aurantium	Anti-oxidative, antihypertensive, anti-hyperlipidemia, anti-diabetic, anti-inflammatory, and hepato-protective potentials, neuroprotective	[19]
Annona muricata	Anticancer, antioxidant, antiviral, anti-haemolytic, sedative and neuroprotective	[20]

3.2 Machine learning

Through the machine learning approach, we conducted each model with 75% fixed proportion and 5-cross validation to obtain the optimal prediction model. The performance prediction model calculated using the validation dataset is shown in Figure 3.

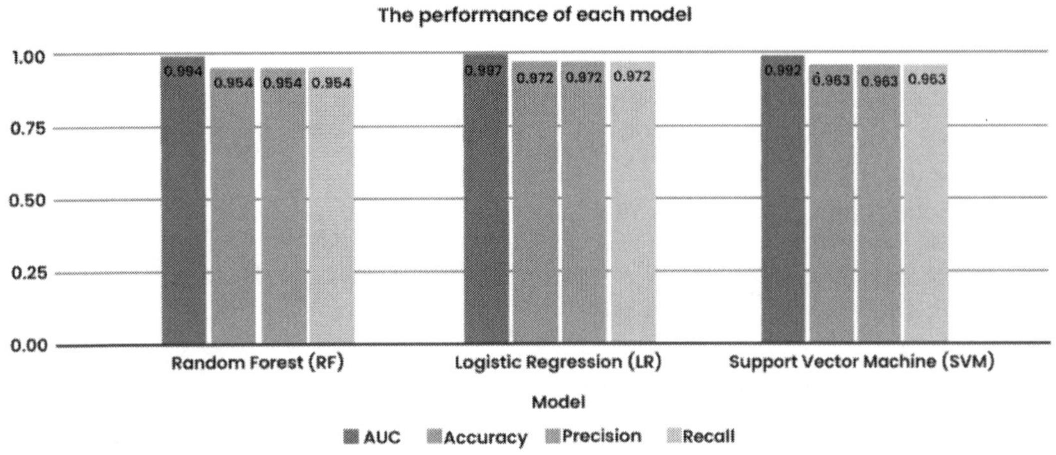

Fig. 3. The performance prediction of each model

The performance prediction model contains AUC, accuracy, precision, and recall. Area Under Curve (AUC) is the area under the ROC curve. If the value is close to 1, it means that the model obtained is more accurate. Accuracy is true positive and true negative divided by a total number of positive and negative. Precision is the number of true positive divided by the total number of positive predictions. Recall is the number of true positive divided by the total number of true positive and false negative [21]. Both precision and recall are focusing on positive examples and prediction.

Based on Figure 3, the LR model has the highest accuracy value (0.972), followed by SVM (0.963) and RF (0.954), likewise with the value of precision and recall. The LR model also has the highest AUC value. The higher the AUC value, the better the model performance in distinguishing positive and negative classes.

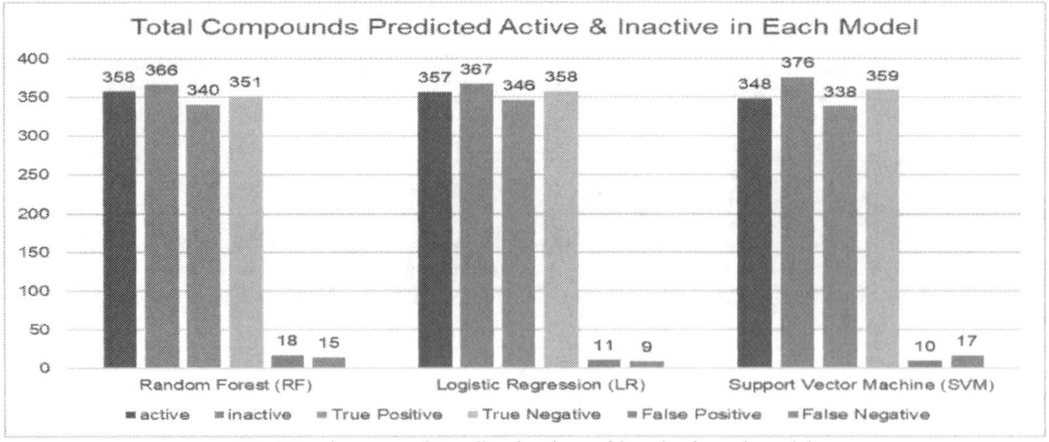

Fig 4. Total compounds predicted active and inactive in each model

The confusion matrix has four categories. They are True Positive (TP), True Negative (TN), False Positive (FP), and False Negative (FN). True Positive (TP) means that the actual data is predicted correctly as positive. True Negative (TN) means that the actual data is predicted correctly as negative. Sometimes, the system can make a mistake by predicting the actual negative value as a positive value (False Positive/FP) or the actual positive value is predicted as a negative value (False Negative) [21].

As mentioned in Figure 4, the LR model has the highest TP value (346) and the SVM model has the lowest TP value (338).

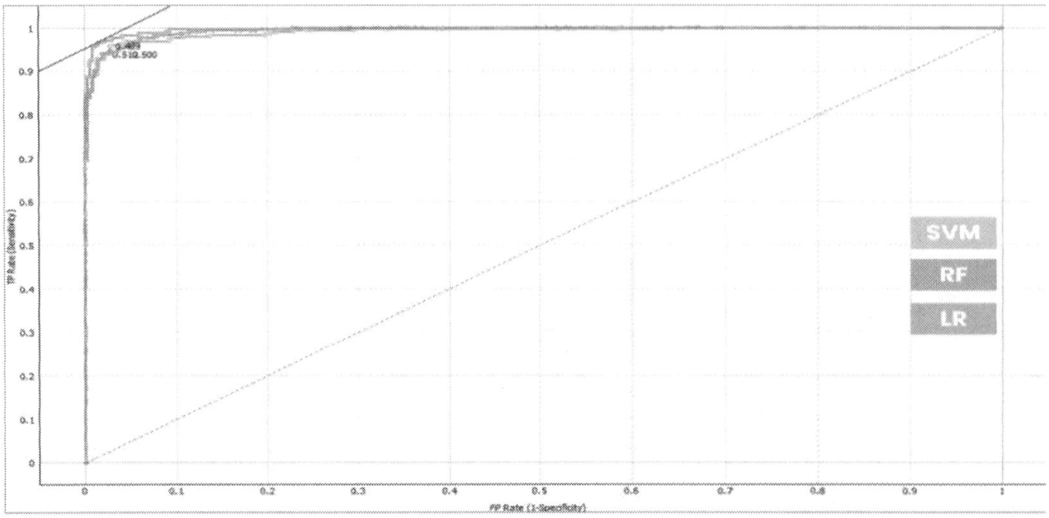

Fig. 5. ROC curve of each model

Provos *et al.* (1998) have argued that the results can be misleading if we look only at the accuracy results. When evaluating binary decision problems, it is recommended to use Receiver Operating Characteristic (ROC) [22]. ROC is the cross-validated performance measurement for the classification model [23]. ROC curve is fixed. An ROC curve begins at the (0,0) coordinate which is the decision threshold at which all test results are negative, and forms diagonal line ending at the (1,1) coordinate which is the decision threshold at which all test results are positive. This diagonal line called the "chance diagonal" and represents the ROC curve of a prediction test with no ability to distinguish positive and negative results [22].

ROC curve has False Positive Rate (FPR/specificity/recall) on the x-axis and True Positive Rate (TPR/sensitivity/precision) on the y-axis. The FPR measures wrongly the actual negative value as a positive value. The TPR measures the actual positive value correctly as a positive value. A significant change in the number of false positives just makes a small change in the FPR in ROC analysis [22].

Figure 5 shows that at a specificity of 0–0.05, the LR model has the highest sensitivity, followed by SVM, then RF. At a specificity of 0.05–0.1, the LR model has the highest sensitivity, followed by RF, then SVM. The higher the sensitivity, the better the positive class is classified correctly. The discrimination has a better particular model if the curve is more convex and approaches the upper left corner [24]. While looking at Figure 5, the LR model appears to be fairly close to optimal.

Compared to the confusion matrix, ROC has several advantages. ROC can be extended to multi-class through one vs. one (OVO) or one vs. all (OVA) methodologies. ROC depends on TPR and FPR which are calculated against true positive and true negative independently, so the change of class distribution will not impact the ROC curve. Obuchowski (2004) states that for classifying the results, the basic measures of accuracy require a decision rule or positivity threshold. ROC curves do not depend on the decision threshold although constructed from sensitivity and specificity [25].

To obtain the optimal results, here ensemble model is used. The ensemble model refers to combining multiple models into one. Based on the ROC curve, the LR and RF models are chosen here to predict the active and inactive compounds of the six plants above. This is intended to minimize the bias that can occur when only referring to one model [26].

3.3 Predicting Indonesian Herbal Compounds

Compounds of each plant are predicted using LR and RF models by its fingerprint. The results can be shown in Figure 6.

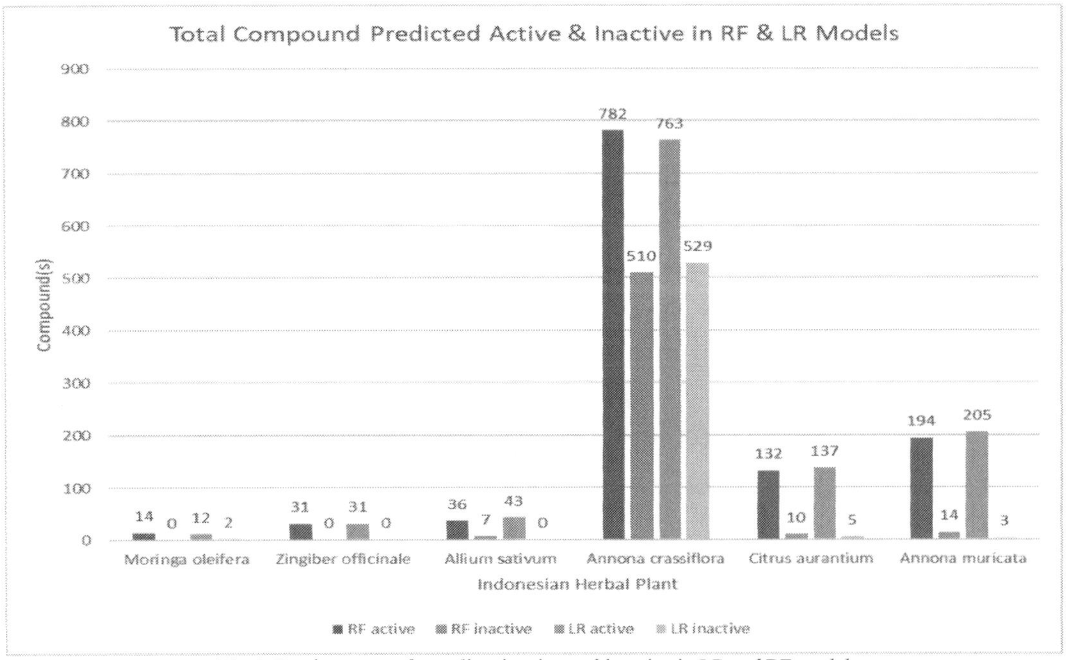

Fig 6. Total compounds predicted active and inactive in LR and RF models

Based on Figure 6, all compounds of *Moringa oleifera* are predicted as active compounds in the RF model, but there are only 12 compounds predicted as active compounds in the LR model. All compounds of *Zingiber officinale* are predicted as active compounds in both RF and LR models. All compounds of *Allium sativum* are predicted as active compounds in LR model, but there are only 36 compounds predicted as active compounds in the RF model. In *Annona crassiflora,* there are 782 predicted active compounds in the RF model and 763 compounds predicted active compounds in the LR model. In *Citrus aurantium*, there are 132 active compounds in the RF model and 137 active compounds in the LR model. In *Annona muricata,* there are 194 active compounds in the RF model and 205 active compounds in the LR model.

Compounds that are predicted active in both LR and RF models then are collected. From 1730 compounds, there are 986 predicted active compounds and 370 predicted inactive compounds in the LR and RF ensemble modeling. From these compounds, there are lots of predicted active compounds found overlapping in two to three plants in both LR and RF models (Table 2).

Table 2. List of Indonesian herbal compound related to anti-Alzheimer.

Compounds	Molecular Formula	SMILES ID	Plants
Quercetin 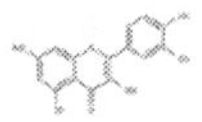	C15H10O7	c1(cc(c2c(c1)oc(c(c2=O)O)c1ccc(c(c1)O)O)O)O	*Allium sativum, Annona muricata,* and *Citrus aurantium.*
Kaempferol 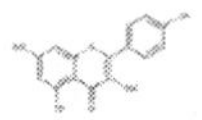	C15H10O6	c1(cc(c2c(c1)oc(c(c2=O)O)c1ccc(cc1)O)O)O	*Allium sativum* and *Annona muricata*
Luteolin 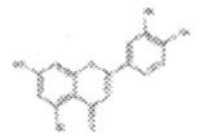	C15H10O6	c1(cc(c2c(c1)oc(cc2=O)c1cc(c(cc1)O)O)O)O	*Annona muricata* and *Citrus aurantium*
Limonene 	C10H16	C1=C(CC[C@H](C1)C(=C)C)C	*Annona muricata* and *Citrus aurantium*
γ-Terpinene 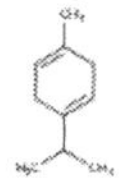	C10H16	C1C=C(CC=C1C)C(C)C	*Allium sativum* and *Citrus aurantium*
Nerolidol 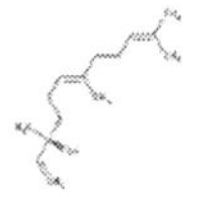	C15H26O	CC(=CCC/C(=C/CC[C@](C=C)(C)O)/C)C	*Allium sativum* and *Citrus aurantium*
Linalool 	C10H18O	CC(=CCC[C@](C=C)(O)C)C	*Annona muricata* and *Citrus aurantium*

Table 3. Toxicity and HIA Analysis

Compounds	HIA (%)	BBB
Quercetin	77.21	-
Kaempferol	79.44	0.286076
Luteolin	79.43	0.367582
Limonene	100.00	8.27823
γ-Terpinene	100.00	8.03745
Nerolidol	100.00	13.9838
Linalool	100.00	6.12506
Donepezil	97.95	0.187923

3.4 Pharmacokinetics Analysis

The seven overlapping compounds were then further analyzed to determine the HIA value and their toxicity through PreADMET (Table 3).

Based on the predicted results of Human Intestinal Absorption (HIA) in Table 3, it shows that the Quercetin, Kaempferol, and Luteolin compounds have a value below 70%. While Limonene, γ-Terpinene, Nerolidol, and Linalool have HIA values of 100%. HIA in the range of 80% -100% indicates that the compound has good absorption in the intestinal wall [27].

The Blood Brain Barrier (BBB) values for the compounds shown in Table 3 obtained the highest to lowest BBB values respectively Nerolidol (13.9838), Limonene (8.27823), γ-Terpinene (8.27823), Linalool (6.12506), Luteolin (0.367582), and Kaempferol (0.286076). All compounds have a BBB value higher than Donepezil as a positive control. The BBB value indicates the absorption of the compound into the blood-brain barrier [27].

Compounds in Table 2 are then analyzed further with the Lipinski Rule of Five (Ro5) test to determine the similarity of drugs or chemical compounds with their pharmacological properties. Several parameters considered in the Lipinski test include molecular weight (< 500 Da), lipophilicity (iLOGP, XLOGP3, WLOGP, MLOGP, SILICOS-IT) (< 5), number of donor hydrogen atoms (< 5, number of OH and NH), and the number of acceptor hydrogen atoms (< 10, the number of N & O) [28].

Table 4. Lipinski Rule of Five of Selected Compounds.

Compounds	Molecular Weight (g/mol)	H-bond acceptors	H-bond donors	LogP
Quercetin	302.24	7	5	1.63
Kaempferol	286.24	6	4	1.70
Luteolin	286.24	6	4	1.86
Limonene	136.23	0	0	2.72
γ-Terpinene	136.23	0	0	2.73
Nerolidol	222.37	1	1	3.64
Linalool	154.25	1	1	2.71
Donepezil	379.49	4	0	4.00

Based on Lipinski test in Tabel 4, it shows that all compounds have no violation and have fulfilled the Lipinski Rule of Five. The molecular weight of all compounds is lighter than donepezil as a positive control. Compounds that have a molecular weight of less than 500 Da, indicate that the compounds are orally active. If the molecular weight of a compound is high, the permeability of the compound in the intestine and central nervous system is lower [29].

The hydrogen-bond acceptor of all compounds has compiled the Lipinski Rule of Five (H-bond acceptor ≤ 10), which means that all compounds bind well with solvents such as water. The hydrogen-bond donor of all compounds has compiled the Lipinski Rule of Five (H-bond donor ≤ 5), which means that all compounds have the ability to penetrate the bilayer membrane. The higher the hydrogen bonding capacity of a compound/molecule, the higher the energy required for the absorption process [30].

The lipophilicity values of all compounds showed a value of no more than 5 and complied with the Lipinski rule. All compounds have lower lipophilicity than donepezil. The lipophilicity value indicates the degree of absorption of the compound and is the algorithm of the ratio of the drug partitioning to the organic phase which is in the aqueous phase. The positive lipophilicity values indicate that the compounds easily penetrate the lipid bilayer membrane [29].

3.5 PASS (Prediction of Activity Spectra for Substances) Analysis

The PASS test is performed to predict pharmacological effects, types of biological activity, mechanism of action, and specific toxicity of different chemical compounds. The predicted activity spectrum in PASS is presented by a list of activities with the probability of active (Pa) and being active inactive (Pi). Pa value > 0.7 indicates that the compound is very likely to show activity in the experiment. A Pa value of 0.5 < Pa < 0.7 indicates that the compound is likely to show activity in experiments, but is less likely and unlike any known pharmaceutical agent. Then, if the Pa value <0.5 then the compound may not show activity in the experiment [31].

The PASS web server predicts various biological activities of the compounds, but the focus of the research here is on the prediction of the Acetylcholine neuromuscular blocking agent.

Table 5. Biological Activity of Each Compound Related to Alzheimer's Disease.

Compounds	Acetylcholine neuromuscular blocking agent	CYP3A4 substrate	CYP2D6 substrate	CYP3A4 inhibitor
Quercetin	0.512	0.617	0.41	0.294
Kaempferol	0.545	0.623	0.425	0.275
Luteolin	0.57	0.567	0.429	0.223
Limonene	0.743	0.309	0.254	-
γ-Terpinene	0.556	0.383	0.258	-
Nerolidol	0.261	-	0.353	-
Linalool	0.34	0.198	0.314	-
Donepezil	0.561	0.231	0.188	-

Several cholinesterase inhibitors currently used in the treatment of Alzheimer's disease are metabolized via CYP-related enzymes [32]. This drug can interact with many other drugs that are substrates, inhibitors or inducers of the CYP system. Some cholinesterase inhibitors (tacrine, donepezil, galantamine) are metabolized via CYP-related enzymes, especially CYP2D6, CYP3A4, and CYP1A2 [33].

Based on Table 5, Quercetin, Kaempferol, Luteolin, and γ-Terpinene are likely to show activity in experiments, but are less likely and unlike any known pharmaceutical agent ($0.5 < Pa < 0.7$) of Acetylcholine neuromuscular blocking agent, same as Donepezil. Only Limonene is very likely to show activity in the experiment to this biological activity. When looking at the CYP system, Quercetin, Kaempferol, and Luteolin are also likely to show activity in experiments, but are less likely and unlike any known pharmaceutical agent ($0.5 < Pa < 0.7$) for CYP3A4 substrate. For CYP2D6 substrate, all compounds do not show activity in the experiment.

From a drug metabolism standpoint, when a compound acts as a substrate the bioavailability will be reduced due to these compounds will be metabolized into relatively more polar compounds to make it easier excreted. However, if compounds play a role as an inhibitor that inhibits the action of cytochromes P450, then the compound causes bioavailability of other compounds will increase that can cause toxicity [34].

4 Conclusions

In this study, a machine learning approach was used to look for candidate compounds that have the potential to become acetylcholinesterase inhibitors for the treatment of Alzheimer's disease. Analysis through machine learning and evaluation through the ROC curve shows that the RF and LR models are optimal models in differentiating active and inactive classes. From 1730 compounds, there are 986 predicted active compounds and 370 predicted inactive compounds in the LR and RF ensemble modeling. Quercetin, Kaempferol, Luteolin, Limonene, γ-Terpinene, Nerolidol, and Linalool predicted active found overlapping in two to three plants in both LR and RF models. Based on the results of pharmacokinetic analysis, Limonene, γ-Terpinene, Nerolidol, and Linalool showed optimal results. Nevertheless, further research both in silico and in vitro study still needs to be developed.

Acknowledgement

The author thanks to MBKM research grant from Faculty of Biology Universitas Gadjah Mada.

References

1. L. Xue., F. Xiaojin, S. Xiaodong, H. Ningning, H. Fang, L. Yongping. Global, regional, and national burden of Alzheimer's disease and other dementias, 1990-2019, Frontiers in Aging Neuroscience (2022)
2. World Health Organzation. Dementia. Accessed 2 June 2023. URL: www.who.int/news-room/fact-sheets/detail/dementia (2022)
3. S. Xiaojuan, C. Wei-Dong, W. Yan-Dong. β-Amyloid: the key peptide in the pathogenesis of Alzheimer's disease, Frontiers in Pharmacology **6,** 221 (2015)
4. T. Elena., G. Michela., V. Vasciaveo, T. Massimo. Oxidative stress and beta amyloid in Alzhemimer's disease, which comes first: the chicken or the egg?, Antioxidants **10**, 9 (2021)
5. D. Ture, A. Michael, D. Dennis. The neuropathological diagnosis of Alzheimer's disease, Molecular Neurodegeneration **14,** 32 (2015)

6. S. Tomas, A. Marketa, O. Lubomir, C. Lucie., J. Daniel, H. Martina, K. Jiri, C. Jakub. Cholinesterase and prolyl oligopeptidase inhibitory activities of alkaloids from Argemone platyceras (Papaveraceae), Molecules **22**, 1181 (2017)
7. B., Buket, U. Duygu, N. Nurlu, K. Gulen, U. Nehir. Chemical profile, acetylcholinesterase, butyrylcholinesterase, and prolyl oligopeptidase inhibitory activity of Gaucium corniculatum subsp. Refractum, Brazilian Journal of Pharmaceutical Sciences **58,** (2022)
8. G. Marucci, M. Buccioni, D. Ben, C. Lambertucci, R. Volpini, F. Amenta. Efficacy of acetylcholinesterase inhibitors in Alzheimer's disease, Nuropharmacology, **190,** (2021)
9. P. Long, P. Quan. Virtual screening strategies in drug discovery: A brief overview, Vietnam Journal of Science and Technology **59**, 4 (2021)
10. V. Periwal, S. Bassler, S. Andrejev, N. Gabrielli, K. Patil, A. Typas, K. Patil. Bioactivity assessment of natural compounds using machine learning models trained on target similarity between drugs, PLOS Computational Biology **18**, 4 (2022)
11. I. Fernandez, J. Peters. Machine learning and deep learning in medicine and neuroimaging, Annals of the Child Neurology Society **1,** 2 (2023)
12. A. Liaw, M. Wiener. Classification and regression by random forest, R News **2**, 3 (2002)
13. Harrington, P. Machine Learning in Action. Manning Publications Co. (2012)
14. C. Lin, R. Weng, S. Keerthi. Trust Region Newton Method for Large-Scale Logistic Regression, Journal of Machine Learning Research. (2008)
15. A. Ademosun, G. Oboh, O. Ajeigbe. Influence of Moringa (Moringa oleifera) enriched ice creams on rats' brain: Exploring the redox and cholinergic systems, Current Research in Food Science **5**, (2002)
16. R. Arcusa, D. Villano, J. Marhuenda, M. Cano, B. Cerda, P. Zafrilla. Potential Role of Ginger (Zingiber officinale Roscoe) in the Prevention of Neurodegenerative Diseases, Front. Nutr. **9**, (2022)
17. P. Tedeschi, M. Nigro, A. Travagli, M. Catani, A. Cavazzini, S. Merighi, S. Gessi. Therapeutic Potential of Allicin and Aged Garlic Extract in Alzheimer's Disease, Int. J. Mol. Sci. **23**, 6950 (2022)
18. M. Barbosa, A. Justino, M. Martins, K. Belaz, F. Ferreira, R. de Oliveira, A. Danuello, F. Espindola, M. Pivatto. Cholinesterase inhibitors assessment of aporphine alkaloids from Annona crassiflora and molecular docking studies, Bioorganic Chemistry **120**, (2022)
19. P. Shayan, A. Amir. Evaluation of antioxidant and inhibitory properties of Citrus aurantium L. on the acetylcholinesterase activity and the production of amyloid nano–bio fibrils, International Journal of Biological Macromolecules **182,** (2021)
20. W. Kim, Y. Kim, E. Cho, E. Byun, W. Park, H. Song, K. Kim, S. Park, E. Byun. Neuroprotective effect of Annona muricata-derived polysaccharides in neuronal HT22 cell damage induced by hydrogen peroxide, Bioscience, Biotechnology, and Biochemistry **84**, (2020)
21. D. Chicco, G, Jurman. The advantages of the Matthews correlation coefficient (MCC) over F1 score and accuracy in binary classification evaluation, BMC Genomics. (2020)
22. J. Davis, M. Goadrich. The relationship between precision-recall and ROC curves. (2006)
23. P. Ananat, P. Gupta. Application of machine learning in understanding bioactivity of beta-lactamase AmpC, Journal of Physics: Conference Series **2273**, (2022)
24. K. Gajowniczek, T. Zabkowski. Estimating the ROC curve and its significance for classification models' assessment, Quantitative Methods in Economics **15**, 2 (2014)
25. N. Obuchowski. Fundamental of Clinical Research for Radiologists, American Journal of Roentgenology **184**, (2005)
26. T. Mahesh, D. Kumar, V. Kumar, J. Asghar, B. Bazezew, R. Natarajan, V. Vivek. Blended ensemble learning prediction model for strengthening diagnosis and treatment of chronic diabetes disease, Computational Intelligence and Neuroscience (2022)
27. T. Hou. J. Wang, Y. LiADME evaluation in drug discovery: the prediction of human intestinal absorption by a support vector machine, J. Chem. Inf. Model **47**, 6 (2007)
28. D. Sen, K. Nandi, D. Saha. Rule of five: The five men army to cross the blood brain barrier for therapeutically potent, World Journal of Advance Healthcare Research **5**, 3 (2021)
29. M. PollastriOverview on the rule of five, Current Protocols in Pharmacology **49,** (2010)
30. T. Altamash, A. Amhamed, S. Aparicio, M. Atilhan. Effect of hydrogen bond donors and acceptors on CO2 absorption by deep eutectic solvents, Processes **8,** (2020)
31. M. Basanagouda, J. Jadhav, M. Kulkarni, R. Rao. Computer Aided Prediction of Biological Activity Spectra: Study of Correlation between Predicted and Observed Activities for Coumarin-4-Acetic Acids, Indian J Pharm Sci. **73**, 1 (2011)
32. R. Cacabelos, R. Llovo, C. Frail, L. Fernández-Novoa L. Pharmacogenetic aspects of therapy with cholinesterase inhibitors: the role of CYP2D6 in Alzheimer's disease pharmacogenetics, Curr Alzheimer Res. **4,** 4 (2007)
33. S. Ruangritchankul, P. Chantharit, S. Srisuma, L. Gray. Adverse Drug Reactions of Acetylcholinesterase Inhibitors in Older People Living with Dementia: A Comprehensive Literature Review, Ther Clin Risk Manag **17,** (2021)

34. J. Hakkola, J. Hukkanen, M. Turpeinen, O. Pelkonen. Inhibition and induction of CYP enzymes in humans: an update, Springer Science and Business Media Deutschland GmbH **94,** (2020)

Metabolite Profile and Antibacterial Potential of Leaf And Stem Extract *Castanopsis tungurrut* (Blume) A.DC. against *Escherichia coli* and *Staphylococcus aureus*

Irfan Ilyas[1], *Muhammad Imam* Surya[2*], *Intani Quarta* Lailaty[2*], *Frisca* Damayanti[2*], and *Tri Rini* Nuringtyas[1*]

[1]Faculty of Biology, Gadjah Mada University, 55281 Yogyakarta, Indonesia
[2]Research Center for Plant Conservation, Botanic Gardens and Forestry, BRIN, 16911 West Java, Indonesia

Abstract. Giving antibiotics or antibacterial drugs is one of the therapeutic strategies used to overcome the problem of infectious diseases caused by microorganisms. The emergence of bacteria that are resistant to various types of antibacterials can delay the recovery period from infectious disorders and cause treatment with antibiotics to become ineffective to use. For this reason, research to find new alternative antibacterial drugs from natural materials needs to be carried out. This study aims to determine the potential of *Castanopsis tungurrut* leaf and stem extracts as natural antibacterials against *Escherichia coli* and *Staphylococcus aureus* bacteria. The research method was extraction of samples by maceration with ethanol, ethyl acetate and aquadest. Antibacterial test with Kirby-Bauer method or disc diffusion. Identification of secondary metabolite compounds by TLC method and continued with metabolite profiling by spectrophotometric method. The results showed the highest antibacterial activity was found in 70% ethanol extract of stem at a concentration of 400 mg/mL with an inhibition zone diameter of 9.06 mm. The 70% ethanol leaf extract was detected to contain phenolics and flavonoids and the ethyl acetate leaf extract was detected to contain triterpenoid and phenolic compounds evaluated by TLC. The metabolite profile showed that ethyl acetate extract had maximum absorbance at 400-600 nm. The 70% ethanol extract has maximum absorbance at 270-350 nm. The distilled water extract had maximum absorbance at 205-250 nm.

Keywords: *Castanopsis tungurrut,* Antibacterial, Secondary Metabolite, Spectrophotometry, Thin Layer Chromatography, Chemometric

1 Introduction

Gram-positive bacteria like *Staphylococcus aureus* and gram-negative bacteria like *Escherichia coli* are the most frequent bacteria that infect and cause health issues in the population. In order to reduce the incidence of serious diseases caused by microorganisms, the administration of antibiotics or antibacterial drugs is one of the therapy strategies used to combat the problem of infectious diseases. Antibiotics in the forms of amoxicillin, penicillin, tetracycline, and other antibiotics are extensively produced and widely used[1].

The onset of numerous side effects, such as diarrhea, vomiting, allergic responses, and other digestive diseases, is a common concern among[2]. The emergence of bacteria that are resistant to various types of antibacterials can delay the recovery period from infectious disorders and cause treatment with antibiotics to be no longer effective for use [3]. According to Jaja et al. (2020)[4] *E. coli* and *S. aureus* bacteria have high antibiotic resistance in clindamycin, ampicillin, rifampicin, streptomycin, and amoxicillin antibiotics.

Antibiotic resistant microorganisms are a key issue in the development of antibiotic medications today, necessitating the use of alternative therapies. One of these is the use of antibacterial agents derived from plant secondary metabolites. The use of secondary metabolites from plants has been extensively used in traditional medicine. Traditional medicine is recognized as having less negative effects than synthetic drugs and uses materials that are widely available [5]. Secondary metabolites found in plants include tannins, terpenoids, alkaloids, and flavonoids which have been shown to provide antibacterial activity [6].

Castanopsis genus is one recognized plant with antibacterial properties. According to the findings of Khan et al. (2001)[7], the leaves and bark of *Castanopsis acuminatissima* contain many secondary metabolites with antibacterial activity, including flavonoids, tannins, triterpenoids, and saponin. Some Indonesian *Castanopsis* species, such as *Castanopsis tungurrut* and *Castanopsis argentea*, have not been examined extensively about their antibacterial properties.

Based on the background described, this study aims to determine the potential of *C. tungurrut* extract as a

*Corresponding author: muha108@brin.go.id inta009@brin.go.id fris001@brin.go.id tririni@ugm.ac.id

natural antibacterial in *E. coli* and *S. aureus*. Followed by characterization of the extracts used in this study using Thin Layer Chromatography (TLC) and metabolic fingerprinting using UV/Vis spectrophotometric scanning.

2 Methodology

2.1 Materials

The materials used in this study were leaves and stems of *C. tungurrut* obtained from Kebon Raya Cibodas, West Java. The solvents used for extraction of samples were ethyl acetate, ethanol, and aquadest. All solvents used were analytical grades. The bacteria including *E. coli* (FNCC 0091) and *S. aureus* (FNCC 0047) were obtained from PAU UGM.

2.2 Methods

2.2.1 Extraction of leaves and stems of Castanopsis tungurrut

Powder samples were weighed as much as 250 g. Samples were macerated with several polarity solvents, including ethyl acetate, 70% ethanol, and water with a sample to solvent ratio of 1:10 (w/v). Maceration with ethanol and ethyl acetate solvents was carried out for three days and 4 nights with occasional shaking. Extraction with distilled water was carried out by mixing the dry ingredients in distilled water and heating it at 50 °C for 50 minutes in a water bath. The mixture was incubated in the refrigerator for 24 hours. All the mixture is then filtered using a funnel and filter paper. The filtrate was evaporated using a rotating vacuum evaporator at 40 °C for ethyl acetate and ethanol solvents and 50 °C for water solvents to form a thick filtrate.

2.2.2 Antibacterial assay

Antibacterial assay was performed by Kirby-Bauer method (Filter Paper Disk Agar Diffusion Method) using Agar disk diffusion. The sample is diluted to 100 - 400 mg/mL with 10% DMSO as the sample to be tested. 1 ml of 10% DMSO without sample was used as a negative control. Amoxicillin was dissolved in 10% DMSO to 1 mg/mL as a positive control. Pure isolates were subculture on MHB medium and turbidity levels were measured using a 600 nm OD spectrophotometer in the absorbance range of 0.1 – 0.5 A. Bacteria *E. coli* and *S. aureus* were inoculated into MHA media in petri dishes using the swab method. Each of the 6 paper discs was dripped with 20 µL of 1 mg/mL amoxicillin; DMSO 10%; sample solution 100 mg/mL; sample solution 200 mg/mL; and 400 mg/mL sample solutions were tested in bacterial culture. The bacterial culture was incubated for 24 hours at 37 °C.

2.2.3 Thin layer chromatography (TLC)

The sample selected for the detection of secondary metabolites is a leaf sample with 70% ethanol and ethyl acetate solvent. Samples were diluted to 0.5 mg/mL in each solvent. The content of active compounds was identified using TLC with a stationary phase of GF254 silica gel plates (Merck). The mobile phase with n-hexane and ethyl acetate eluents = 7:3 for the ethyl acetate and toluene, ethyl acetate and glacial acetic acid extract samples = 7:2:0.5. A silica plate with a size of 2 cm 10 cm was added with an upper mark of 0.5 cm and a lower mark of 1 cm. The plate was then activated using an oven with a temperature of approximately 80°C - 100°C for 15 minutes. Eluent saturation was carried out by inserting a rectangular filter paper measuring 18 cm x 2 cm into the TLC vessel. The sample is placed on the right of the plate and the reference compound is placed on the left of the plate. TLC plates were run until the separated sample reached the upper limit mark. The TLC plate was baked for 5 minutes at 80°C - 100°C. The TLC plates were sprayed with reagent ($FeCl_3$, vanillin sulfate, citroborate, dragendorff) to detect phenolic compounds, terpenoids, flavonoids and alkoloids. Observed under UV light with a wavelength of 254 nm, 366 nm and visible light.

2.2.4 Compound profile of secondary metabolites

Identification of compound content in ethyl acetate extract, ethanol, and aqueous leaves and stems *of C. tungurut* was carried out with a UV-Vis spectrophotometer (Genesys 150). The wavelength is set at 200 nm-800 nm. The sample solution is diluted to 0.5 mg/mL as much as 10 mL inserted into the test tube and vortex for approximately 5 minutes. Solution blanks are used with customized individual sample solvents. A total of approximately 3 mL of sample is inserted into the quartz cuvette and the absorbance is read with a UV-Vis spectrophotometer until a chromatogram appears.

2.2.5 Data analysis

Antibacterial activity data were analyzed using ANOVA (Analysis of Variance) method and post hoc analysis of DMRT (Duncan's Multiple Range Test) Test at a signification level of 0.05 to determine the difference in significance between data. Profile analysis of secondary metabolite compounds was carried out using the MetaboAnalyst website (www.metaboanalyst.ca)

3 Result and discussion

3.1 Extraction of leaves and stems of *Castanopsis tungurrut*

The solvent used in this study was ethyl acetate, 70% ethanol and aquadest because all three have different polarity properties. Soluble compounds will follow the principle of "Like Dissolved Like", polar compounds will dissolve in polar solvents and vice versa nonpolar compounds will dissolve in nonpolar solvents. Ethyl acetate polarity level is 0.228; ethanol 0.654; and aquadest 1,000. The higher the level of polarity of the solvent, the more polar the solution will be. It can be seen that aquadest have a very polar level of polarity, ethanol is semipolar, and ethyl acetate is nonpolar[8].

Table 1. Results of Sample Extracts of *Castanopsis tungurrut* Leaves and Stems

Sample	Solvent	Simplisia Weight (g)	Extract Weight (g)	Yield (%)
Leaves	**Ethyl Acetate**	25,00	2,36	9,44
	Ethanol 70%	25,00	5,06	20,24
	Aquadest	25,00	3,24	12,96
Stems	**Ethyl Acetate**	25,00	2,23	8,92
	Ethanol 70%	25,00	3,53	14,12
	Aquadest	25,00	3,24	12,96

The extraction results obtained using 70% ethanol solvent have the highest yield of around 20.24% and 14.12% of the other three solvents, and ethyl acetate has the lowest yield of 9.44% and 8.92% as seen in the (Table 1). Alcohol has a high enough level of polarity to dissolve polar and nonpolar compounds. Aquades also have high polarity properties and can dissolve more polar compounds. Ethyl acetate is more likely to be nonpolar so that it only extracts secondary metabolite compounds that are nonpolar [9].

Leaf and stem samples extracted with ethyl acetate solvent have wet properties and tend to be green in color rather than extracts with ethanol and aqueous solvents. Ethyl acetate tends to be nonpolar so it can extract more nonpolar compounds such as essential oils. Based on research conducted by Lohani *et al.* (2015)[10] essential oil is found in ethyl acetate extract of Canola (*Brassica napus*). Extraction samples with 70% ethanol solvent have the same color as aquadest but the ethanol texture is slightly wet compared to aqueous with a dry texture. According to research by Yulianti *et al.* (2020)[11] ethanol solvents include semipolar solvents that can extract more phenolic compounds in cherry fruit (*Mutingia calabura*) leaf samples. Polar aquadest that can extract more polar compounds such as tannins and glycosides [12]

3.2 Antimicrobial activity

Antibacterial assay in this study used the Kirby-Bauer method to order disk diffusion. The samples tested were made at 3 different concentrations, namely concentrations of 100 mg/mL, 200 mg/mL and 400 mg/mL. Antibacterial activity is characterized by the formation of a clear zone around the disc. The wider the clear zone, the higher the antibacterial activity [13].

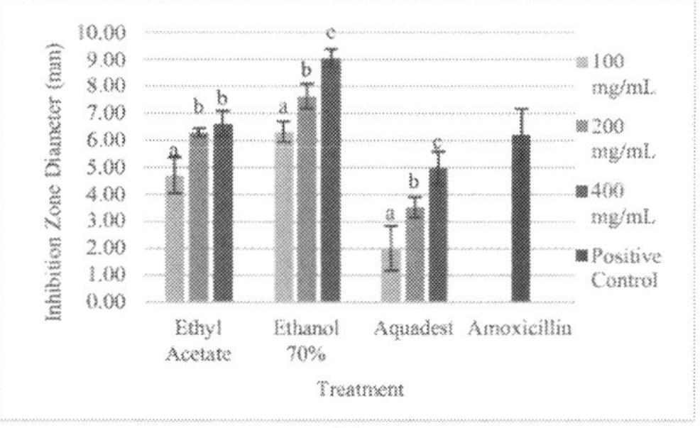

(a)

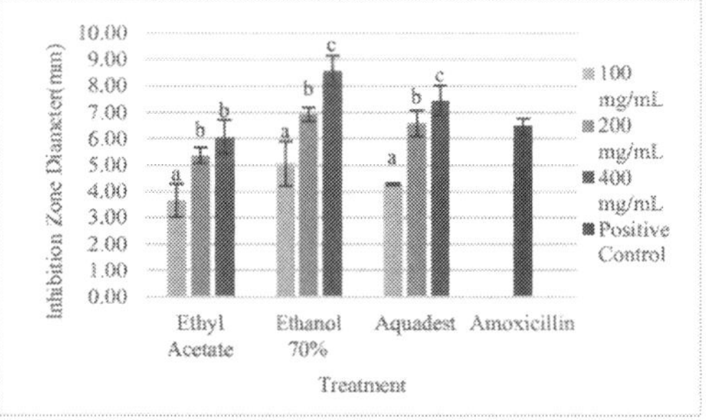

(b)

Fig. 1. Antibacterial Activity against *E. coli* Bacteria in *C. tungurrut* Leaf (a) and *C. tungurrut* Stems samples (b)

The results of antibacterial assay samples against *E. coli* bacteria are seen in (Figure 1). The difference in letters shows a real difference based on Duncan's post hoc ANOVA analysis with a significance value of $p < 0.05$. The solvent with the highest activity was recorded in 70% ethanol solvent, the concentration of 400 mg/mL leaf samples had an inhibitory zone diameter of 9.06±0.33 mm and the stem sample had an inhibitory zone diameter of 8.58±0.57 mm. Extracts of compounds with alcohol solvents showed the presence of secondary metabolite compounds that more actively inhibited bacterial growth compared to other solvents.

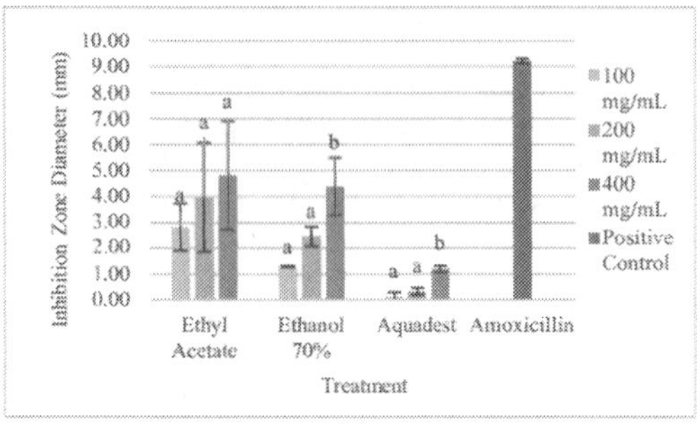

(a)

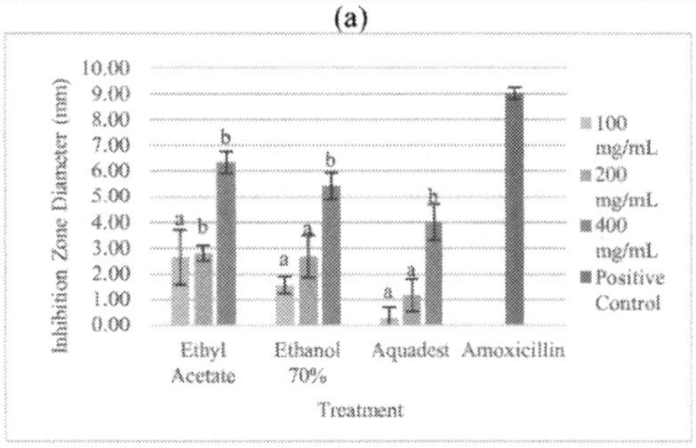

(b)

Fig. 2. Antibacterial Activity against *S. aureus* Bacteria in *C. tungurrut* Leaf (a) and *C. tungurrut* Stems samples (b)

The results of antibacterial assay samples against *S. aureus* bacteria are seen in (Figure 2). The results showed a significant difference in each concentration in 70% ethanol solvent and aquadest in stem organs. The difference in letters shows a real difference based on Duncan's post hoc ANOVA analysis with a significance value of $p < 0.05$. The solvent with the highest activity was recorded in 70% ethanol solvent, 400 mg/mL concentration, leaf samples had an inhibitory zone diameter of 4.39±1.11 mm, and stem samples had an inhibitory zone diameter of 5.43±0.52 mm. In ethyl acetate solvent concentration of 400 mg/mL, leaf samples have an inhibitory zone diameter of 4.82±2.09 mm and stem samples have an inhibitory zone diameter of 6.35±0.43 mm. But the increase in concentration in ethyl solvent does not show any significant difference. Extracts of compounds with alcohol and ethyl acetate solvents showed the presence of secondary metabolite compounds that more actively inhibited the growth of *S. aureus* bacteria compared to other solvents.

The difference in antibacterial activity results between *E. coli* and *S. aureus bacteria* can be influenced by microbial defense mechanisms against antibacterial compounds. Based on a scientific article by Ebbensgaard *et. al.* (2018) [14] gram-negative bacteria have an outer membrane (OM) equipped with a lipopolysaccharide structure (LPS). This structure allows gram-negative bacteria such as E. *coli* to block the entry of compounds toxic to cells. Thick peptidoglycan in *S. aureus* bacteria can decrease the permeability of toxic compounds to enter and damage cells [15].

3.3 Thin layer chromatography

Thin-layer chromatography is performed to detect the presence of compounds that have antibacterial activity such as alkaloids, phenolics, flavonoids, and triterpenoids [16]. The sample used for the thin layer chromatography test was selected which had the highest antibacterial activity found in the sample using 70% ethanol and ethyl acetate solvent. TLC Rf value can be seen in (Table 2).

Based on the results of TLC testing in (Figure 3 and Figure 4), 70% ethanol leaf extract samples showed positive results on flavonoid and phenolic compounds. Flavonoid compounds were found to be positive for yellow fluorescence at Rf 0.10 and 0.27 after spraying with citroborate reagents. Phenolic compounds were also found at rf 0.18 after spraying marked with a blue-black color after spraying with $FeCl_3$ reagent. TLC test results on ethyl acetate leaf extract samples showed positive results on triterpenoid and phenolic compounds in (Figure 5 and Figure 6). The triterpenoid compound was found positive in Rf 0.97 purple spots in visible light after spraying with vanillin sulfate reagent. Phenolic compounds are found at rf 0.83 and 0.91 in visible light marked with a blue-black color.

Table 2. Rf TLC Value

Sample	Target Compunds	Rf Value
Ethanol 70% Leaf Extract	Alkaloid	-
	Phenolic	0.18
	Flavonoid	0.10
		0.27
	Triterpenoid	-
Ethyl Acetate Leaf Extract	Alkaloid	-
	Phenolic	0.83
		0.91
	Flavonoid	-
	Triterpenoid	0.97

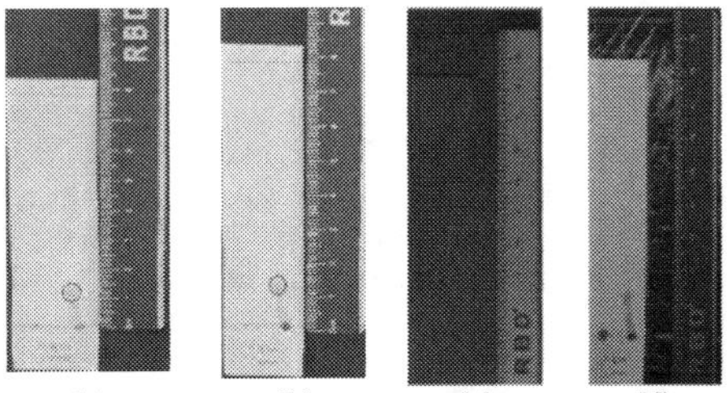

(a) (b) (c) (d)

Fig. 3. TLC Visualization of Phenolic Compounds of 70% Ethanol Leaf Samples in Visible Light Before Spraying (a), After Spraying (b); UV light 366 nm After Spraying (c); and UV Light 254 nm After Spraying (d)

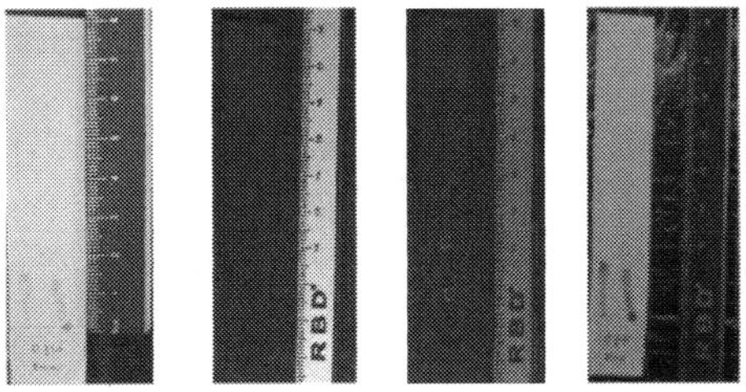

(a) (b) (c) (d)

Fig. 4. TLC Visulization of Flavonoid Compounds Ethanol Leaf Samples 70% on Visible Light (a), UV Light 366 nm Before Spraying (b); UV light 366 nm After Spraying (c); and UV Light 254 nm After Spraying (d)

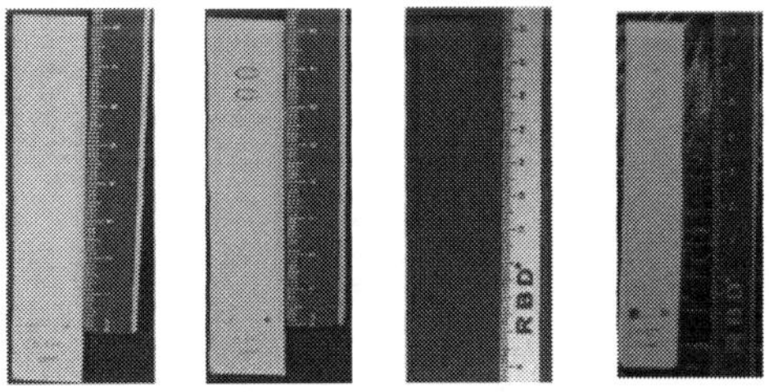

(a) (b) (c) (d)

Fig. 5. Visulization of TLC Phenolic Compounds of Ethyl Acetate Leaf Samples in Visible Light Before Spraying (a), After Spraying (b); UV light 366 nm After Spraying (c); and UV Light 254 nm After Spraying (d)

(a) (b) (c) (d)

Fig. 6. TLC Visulization of Triterpenoid Compounds of Ethyl Acetate Leaf Samples in Visible Light Before Spraying (a), After Spraying (b); UV light 366 nm After Spraying (c); and UV Light 254 nm After Spraying (d)

3.4 Profile of Secondary Metabolites of *Castanopsis tungurrut*

Profiling of secondary metabolite compounds of *C. tungurrut* leaf and stem extracts with ethyl acetate, ethanol, and aqueous solvents was carried out using UV-Vis Spectrophotometry and chromatograms were obtained from each sample. The chromatogram is analyzed by the chemometric method. A commonly used chemometric method is Principal Component Analysis (PCA) [17].

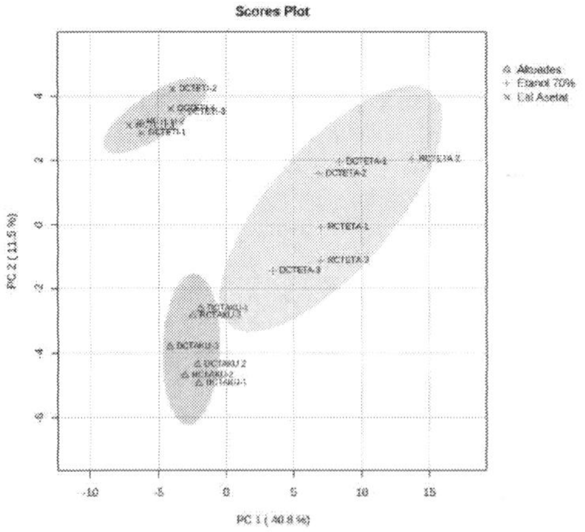

Fig. 7. Visualization *Scores Plot* between Solvent Ethyl Acetate (Blue), Ethanol 70% (Green), Aquades (Red)

The results of the analysis using the PCA method showed a grouping of data on the content of secondary metabolite compounds found in leaf samples and *C. tungurrut* stem is based on solvents in the form of ethyl acetate, ethanol and aquadest. Based on Figure 7, the results of Visualization Scores Plot of *C. tungurrut* leaf and Stem extracts based on the solvent were obtained. The results showed that PC 11 had a value of 40.8% and PC 2 had a value of 11.5%. Each sample is separate and grouped based on its own solvent, showing that each solvent can extract different metabolite compounds depending on their respective polarities.

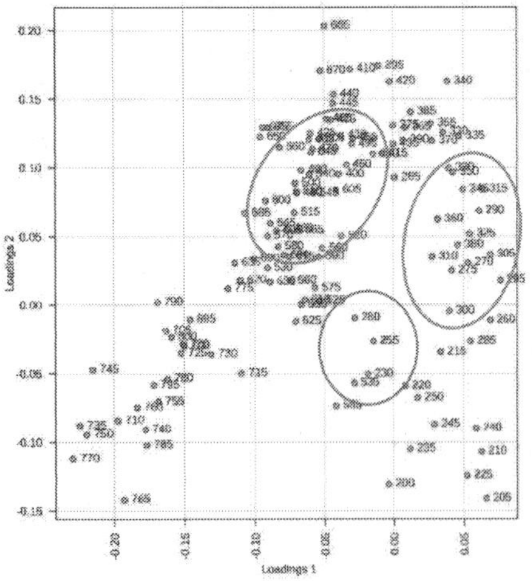

Fig 8. Visualization of Loading Plot between Ethyl Acetate Solvent (Blue), 70% Ethanol (Green), and Aquadest (Red)

Figure 8 shows the results of loading *C. tungurrut* leaves and Stems on each solvent. Each solvent groups and separates based on its spectrum. The first group of ethyl acetate solvents marked with a blue circle showed secondary metabolite compounds capable of absorbing UV-Vis light at wavelengths of 400-600 nm. Based on the absorbance of compounds at wavelengths of 400-600 nm, sample extracts with ethyl acetate solvent contain alkaloids and terpenoids. It is known that alkaloids have absorbance values at wavelengths of 400 – 650 [18]. Terpenoids have wave absorbance at lengths of 520-530 nm [19].

The group marked with a green line shows a grouping of secondary metabolite compounds with 70% ethanol solvent. Grouping based on absorbance wavelength 270 – 350 nm. Based on the absorbance at the wavelength, the sample extract with 70% ethanol solvent contains metabolite compounds in the form of tannins, phenolics and flavonoids. Known Flavonoids and Tannins have absorbance at wavelengths of 210 – 280 nm. Phenolic compounds have absorbance at wavelengths of 200 nm – 400 nm.

Groups marked in red indicate the grouping of secondary metabolite compounds with aqueous solvents. Grouping occurs at wavelength absorbances of 205-230 and 535 nm. Based on absorbance at these wavelengths, sample extracts with aqueous solvents contain secondary metabolite compounds in the form of phenolics, tannins, alkaloids, and triterpenoids [18].

4 Conclusion

The results of this study showed the highest antibacterial activity observed in leaf extracts with 70% ethanol solvent in *E. coli* bacteria. The highest activity was in stem extract with ethyl acetate solvent in *S. aureus* bacteria. Identification of secondary metabolite compounds from ethyl acetate leaf extract found triterpenoid compounds and 70% ethanol leaves found flavonoid compounds. The results of profiling of secondary metabolite compounds of *C. tungurrut* leaf and Stem extracts found alkodaloid and terpenoid compounds with maximum absorption at wavelengths of 400-600 nm. Leaf and Stem extracts from ethanol and aqueous solvents found compounds in the form of flavonoids, phenolics and tannins with maximum absorption at successive wavelengths of 270-350 nm and 205-250 nm.

5 Acknowledgement

The author thanks all parties who have helped and participated in the process of data collection and processing. The author also thanks the Cibodas Botanical Garden for its support in providing research samples.

References

1. Ventola, C.L. 'The antibiotic resistance crisis: part 1: causes and threats', *Pharmacy and Therapeutics.* **40**(4), 277-283 (2015)
2. Mohsen, S., Dickinson, J.A., Somayaji, R. 'Update on the adverse effects of antimicrobial therapies in community practice', *Canadian Family Physician,* **66**, 651- 659 (2020).
3. Tzaneva, V., Mldenova, I., Todorova, G., and Petkov, D. 2016. 'Antibiotic treatment and resistance in chronic wounds of vascular origin', *Clujul Medical.* **89**(3), 365-370 (2016)
4. Jaja, I.F., Jaja, C.I., Chigor, N.V., Anyanwu, M.U., Maduabuchi, E.K., Oguttu, J.W. and Gree, E. 'Antimicrobial resistance phenotype of staphylococcus aureus and escherichia coli isolates obtained from meat in the formal and informal sectors in south africa', *BioMed Research Internasional.* **1**, 1-3 (2020)
5. Zhang, J., Onakpoya, I.J., Posadzki, P., and Eddouks, M. 'The safety of herbal medicine: from prejudice to evidence', *Evid Based Complement Alternat Med.* **2015**,316706. (2015)
6. Cowan, M.M. 'Plant products as antimicrobial agents'. *Clinical Microbiology Reviews.* **12**(4), 564-582 (1999)

7. Khan M.R., Kihara, M. and Omoloso, A.D. Antimicrobial activity of *Castanopsis acuminatissima*. *Fitoterapia*, **72**, 174-176 (2001)
8. Abubakar, A.R and Haqe, M. 'Preparation of medicinal plants: basic extraction and fractionation procedures for experimental purposes', *Journal of Pharmacy and Bioallied Sciences*. **15**(1), 1-10 (2020)
9. Yunita, E., Destasary, E.M., and Wicaksana, F.H. 'The effect of different solvent extraction on chemical content and quercetin levels of ketapang (terminalia cattapa l.)', *Proceedings International Conference on Healthcare*. **2**(1), 1-14 (2021)
10. Lohani, U.C., Fallahi P., and Muthukumarappan, K. 'Comparison of ethyl acetate with hexane for oil extraction from various oilseeds', *Journal of the American Oil chemists' Societi*. **92**, 743-754 (2015)
11. Yulianti, W., Ayuningtiyas, G., Martini, R., and Resmeiliana. 'Pengaruh metode ekstraksi dan polaritas pelarut terhadap kadar fenolik total daun kersen (muntingia calabura l)', *Jurnal Sains Terapan*. **10**(2); 41-49 (2020)
12. Obenu, N.M. 'Ekstraksi dan identifikasi kandungan metabolit fraksi diklorometana dan aquades ekstrak metanol daun sirsak (annona muricata linn)', *Jurnal Saintek Lahan Kering*. **2**(1), 17-19 (2019)
13. Coorevits L., Boelens, J., and Claeys, G. 'Direct susceptibility testing by disk diffusion on clinical samples: a rapid and accurate tool for antibiotic stewardship', *Eur J Clin Microbiol Infect Dis*. **34**, 1207–1212 (2015)
14. Ebbensgaard, A., Mordhorst, H., Aarestrup, F.M. and Hanses, E.B. 'The role of outer membrane proteins and lipopolysaccharides for the sensitivity of escherichia coli to antimicrobial peptides', *Frontiers in Microbiology*. **9**, 1-13 (2018)
15. Guo, Y., Guanghui, L., Sun, M., Wang, J., and Wang, Y. 'Prevalence and therapies of antibiotic-resistance in staphylococcus aureus', *Frontiers in Cellular and Infection Microbiology*. **10**,107 (2020)
16. Compean, K.L. and Ynalvez, R.A. 'Antimicrobial activity of plant secondary metabolites', *Research Journal of Medicinal Plant*. **8**(5), 204-213 (2014)
17. Orche, A.E., Adade, C.A., Mefetah, H., Cheikh, A., Karrouchi, K., Karbane, M.E., and Bouatia, M. 'Chemometric analysis of uv-visible spectral fingerprints for the discrimination and quantification of clinical anthracycline drug preparation used in oncology', *BioMed Research Internasional*. **2021**:8 (2021)
18. Raji P., Samrot, A.V., Rohan, D.B., Kumar, M.D., Geetika, R., Sharma, V.K. and Keerthana, D. 'Extraction, characterization and invitro bioactivity evaluation of alkaloids, flavonoids, saponins and tannins of cassia alata, thespesia populnea, euphorbia hirta and wrightia tinctoria', *Rasayan Journal of Chemistry*. **12**(1), 123-137 (2019)
19. Talamond, P., Verdeil, J., and Conéjéro, G. 2015. 'Secondary metabolite localization by autofluorescence in living plant cells'. *Molecules*. **20**, 5024-5037 (2015)

Double-Coated Nanoparticle of Ribosome Inactivating Protein (RIP) from *Mirabilis jalapa* L. prepared from Chitosan-Sodium Tripolyphosphate and Alginate-Calcium Chloride: The New Strategy for Protein Drug in Oral Delivery.

Amalia Miranda[1], *Hilda* Ismail[1*], *Ronny* Martien[1], *Ummi* Hadiba Ciptasari[1], *Ariyani* Kusniasari[1], *Dewa* Ayu Arimurni[1], *Made* Dwi Pradipta Wahyudi S.[1], and *Sismindari*[1]

[1]Faculty of Pharmacy, Universitas Gadjah Mada, Yogyakarta

Abstract. Oral delivery of protein drugs is challenging due to the instability of the compound and structural barrier exists in the gastrointestinal (GI) tract. Nanoparticle technology is known as a promising drug delivery strategy to ensure drug bioavailability. This study aims to formulate an oral delivery system of a potential anticancer agent named *Ribosome Inactivating Protein from Mirabilis jalapa* L.-C (RIP MJ-C) through double-coated nanoparticles prepared from Chitosan-Sodium Tripolyphosphate (TPP) and Alginate-Calcium Chloride ($CaCl_2$). Nanoparticles were prepared through the ionic gelation method, with the core nanoparticle (RMJCN-1) formulated in the pH of 3.5-5.5 using 0.3-0.5 % w/v of chitosan and 0.03 % w/v TPP. The RMJCN-1 optimum formula was selected to be subsequently coated with the second layer of alginate and $CaCl_2$, called RMJCN-2, with a concentration of 0.3% w/v and 0.1-0.3 %, respectively. The sample was characterized by the entrapment efficiency (EE), physical appearance, particle size, polydispersity index (PI), and potential zetta. The result showed the optimum RMJCN-1 formula with of EE value of 57.10 ± 0.04 % was obtained by formulating 0.5 % w/v chitosan and 0.3 % w/v STPP in pH 5.5. The optimum RMJCN-2 was obtained by the combination of alginate 0.3 % w/v and $CaCl_2$ 0.1% w/v in the outer layer. This final formula produces nanoparticles with a zeta potential of -14.4 mV, 739.8 nm in size, with good stability during 7 days at room temperature. This study has successfully developed a formulation of double-coated nanoparticles from Chitosan-TPP and Alginat-$CaCl_2$ for RIP MJ-C, leads to a safe nanocarrier system for oral delivery of RIP MJ-C that ensures its bioavailability.

Keywords: Nanoparticle, RIP MJ-C, chitosan, alginate, double coated.

1 INTRODUCTION

Protein based drug therapy is considered to be a promising strategy for cancer treatment due to its ability to selectively destroy the cancer cell without significantly damaging healthy cells [1]. Ribosome Inactivating Proteins (RIPs) is a plant enzyme having high potential as anticancer agent through ribosomes inactivation mechanism. RIPs cleavages the specific adenine *N*-glycosidic chain that causes inhibition of the prolongation factor in ribosome, leading to the termination of protein synthesis [2]. Previous research by Sudjadi et al. (2007) reported that the unbound protein fraction from *Mirabilis* jalapa L. leaves, has shown a cytotoxic effect on HeLa, myeloma, and T47D cancer cells. [3]. This unbounded protein is further known as the acidic fraction or the negative protein fraction from *Mirabilis* jalapa L., named *RIP MJ-C*.

Among all drug administration routes, oral administration is perceived as the most convenient one. However, maintaining drug bioavailability passing through the GI tract still become the main obstacles of oral drug delivery, especially for the protein-based drug. GI tract's environment may cause degradation and denaturation of proteins. In consequences, most of the peptide drugs is currently administered by parenteral routes. Developing a protein-based drug formulation that can safely permeate intestinal and cellular barriers is highly required. Here nanoparticle formulation appear as an alternative for its ability to protect compounds from premature degradation, enhancing absorbtion, regulate drug's pharmacokinetics and distribution profile, and to improve the GI tract and intracellular penetration [4] [5].

Chitosan is a cationic biopolymer that is widely used in nanoparticle preparations, and has been explored to improve the bioavailability of oral protein delivery [6]. However, the use of chitosan for oral administration is limited by its hidrophylicity and high solubility in gastric pH that might lead to drug degradation. Hence, another layer is required to protect a chitosan based nanoparticle. An alternative polymer with similar capability is sodium alginate. Sodium alginate was reported as having ability

*Corresponding email: hilda_fa@ugm.ac.id

to deliver bioactive drugs in a sustained and controlled release manner, without risk of mucosal damage [7], [8]. Study from Bagre et al., 2013 explored a double-coated nanoparticle prepared by chitosan-TPP and alginate-$CaCl_2$ could improve the release profile and enhance the oral bioavailability of enoxaparin [9]

In this study, RIP MJ-C will be formulated into double coated nanoparticle prepared by chitosan-TPP and alginate-$CaCl_2$. This double coated design is expected to protect the protein drug from enzymatic and degradation in the GI tract. Hence, the aim of this study is to obtain a double layer nanoparticle formulation for RIP MJ-C that can maintain its bioavailability at the site of action.

2 MATERIALS AND METHODS

2.1. Materials

Mirabilis jalapa leaves crude extract was obtained through the extraction of fresh leaves in phosphate buffer pH 6.5. The unbound protein fraction of *Mirabilis jalapa* L. (RIP MJ-C) was separated from the crude extract by chromatography column using CM-Sepharose CL-6B, and confirmed by BCA Protein Assay.

Bicinchoninic Acid (BCA) Protein Assay Kit, Low Molecular Weight Chitosan (LMW-Chitosan), alginate, calcium chloride, sodium tripolypohsphate (STPP), and CM-Sepharose were purchased from Sigma-Aldrich Co.

2.2. Preparation of RMJCN-1 (the RIP MJ-C-Chitosan Nanoparticle, layer 1)

The RMJCNP-1 was prepared by ionic gelation methods. A certain amount of LMW-chitosan was stirred to dissolve in 1.5% v/v acetic acid at 50°C. Subsequently, solution of acetate buffer with pH of 3.5; 4.5; and 5.5 respectively was added to the chitosan solution to make series of LMW-chitosan concentration of 0.3 – 0.5 % w/v for each pH. In another erlenmeyer, an amount of RIP MJ-C was mixed with chitosan and TPP with the ratio of 2:1:1 (method of Sekarningtyas, 2015) [10]. Following, the RIP MJ-C solution (0.04 % w/v) was gradually added to each chitosan solution, then homogenized by vortex for 30 seconds. Finally aqueous solution of SSTPP (0.03 % w/v) was added dropwise and mixed under vortex (50 second), leads to the formation of RIP MJ-C nanoparticle (RMJCN-1) in water. Selection for the optimum formulation was done based on the entrapment efficiency (EE) value and stability of nanoparticle formed.

2.3. Preparation of RMJCN-2 (RMJCN-1-Alginate Nanoparticle, layer 2)

The RMJCN-1 obtained form the optimum formula was separated by ultracentrifugation with RCF of 3270, for 2 hours at 4°C. The obtained sedimen was then redispersed with water, then filtered through 0.45 μm microfilter to remove any fallout fragments and non-nanoparticle substances. In other erlenmeyer, alginate was stirred to dissolve in phosphate buffer pH 5.5 (0.2 % w/v) for 2 hours at room temperature. Series of calcium chloride solution in water were prepared with the concentration of 0,1; 0,2; and 0,3 % w/v.

The double-layered nano particle (RMJCN-2) was made from a mixture of RMJCN-1, alginate solution, and calcium chloride solution in the ratio of 1.4:9:1 respectively. The RMJCN-1 solution in water (0.5 % w/v) was added dropwise into alginate solution then mixed under stirrer for 10 minutes. Following is the addition of each concentration of calcium chloride solution dropwise, and mixing toroughly by vortex for 30 seconds.

2.4. Entrapment Efficiency Analysis (EE)

The solution of RMJCN-1 was ultracentrifugated at 15,000 rpm at 4°C for 50 minutes. The supernatant was collected, then the existing free protein in the supernatant was measured using BCA Protein Assay Kit. A mixture of chitosan-STPP was used as the blank

Result of the BCA analysis that shows the amount of free protein, then used to calculate the %EE according to the following equation:

$$\%EE = \frac{total\ amount\ of\ protein - free\ protein}{total\ amount\ of\ protein}\ x\ 100\%$$

2.5. Visual Observation

Stability of nanoparticle was measured by observing the physical appearance of solution against a black background on the 1st, 3rd, and 7th day of storage in room temperature. Nanoparticle formation was indicated by the *opaque* solution with the absence of neither sediment nor aggregate.

2.6. Percent Transmittance (%T)

RMJCN-1 samples were analysed under UV-VIS Spectrophotometer at 610 nm. Percent transmittance determine the amount of light transmitted as it passes the solution, prescribing the presence or absense of particle agregates in the solution. Water was used as the blank sample.

2.7. Particle Size, Size Distribution and Zetta Potential Characterization

The measurement of particle size, size distribution, and zetta potential were performed with Particle Size Analyzer (HORIBA). The analysis condition was set at 24,8°C with scattering angle of 90.

https://doi.org/10.1051/bioconf/20237504001

3 RESULTS AND DISCUSSIONS

3.1. Results

3.1.1. Entrapment Efficiency of RMJCPN-1

RMJCN-1 were formulated through ionic gelation technique with variations in chitosan concentration and pH environment. The calculation of %EE from formula (F1) to Formula 9 (F9) is shown in **Table 1** and **Figure 1**.

Table 1. Entrapment efficiency (EE) of RMJCN-1

Code	Medium pH	Chitosan concentration (% w/v)	EE (%)
F1		0.3	46.57 ± 0.05
F2	3.5	0.4	49.85 ± 0.04
F3		0.5	47.83 ± 0.03
F4		0.3	49.85 ± 0.02
F5	4.5	0.4	49.57 ± 0.01
F6		0.5	49.14 ± 0.04
F7		0.3	55.52 ± 0.02
F8	5.5	0.4	55.03 ± 0.01
F9		0.5	57.10 ± 0.04

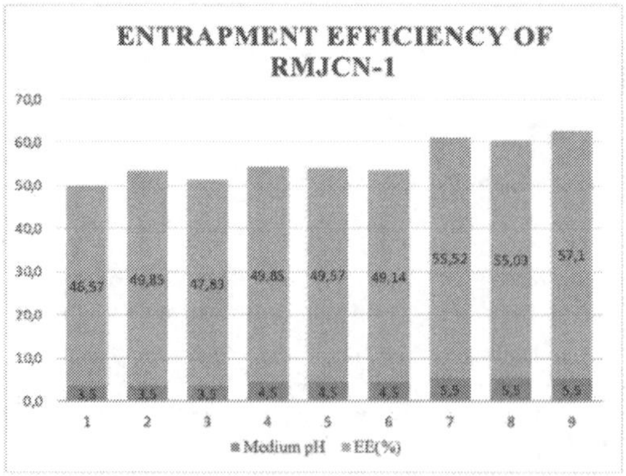

Fig. 1. Entrapment Efficiency of F1 – F9 for RMJCN-1

The EE value represents the encapsulation capacity of nanoparticle system. The high EE indicates that large amount of protein has been successfully carried in the nanocarrier system, while value low EE indicates the major loss of protein during the formulation process. Result shown in Table 1 indicate that F1 (medium pH of 3.5) has the lowest EE value (46.57 ±0.05%.), while F9 (medium pH of 5.5) was the highest (57.10 ± 0.04%).

3.1.2. Visual Observation of RMJCN-1 and RMJCN-2 Stability

The results for observations on RMJCN-1 physical stability is provided in **Table 2** below.

Table 2. Physical Stability of RMJCN-1

Code	Medium pH	Parameter	Observation Result		
			Day 1	Day 3	Day 7
F1	3.5	Agregates*	n.o.	n.o	n.o.
		%T	93.33 ± 0.31	95.06 ± 0.50	93.43 ± 0.15
F2		Agregates*	n.o.	n.o	n.o.
		%T	92.80 ± 0.80	95.36 ± 0.55	93.83 ± 0.25
F3		Agregates*	n.o.	n.o	n.o.
		%T	93.03 ± 0.60	95.33 ± 0.15	93.80 ± 0.10
F4	4.5	Agregates*	n.o.	n.o	n.o.
		%T	93.20 ± 0.44	94.50 ± 2.08	94.10 ± 0.20
F5		Agregates*	n.o.	n.o	n.o.
		%T	93.00 ± 0.36	95.63 ± 0.15	93.63 ± 0.12
F6		Agregates*	n.o.	n.o	n.o.
		%T	93.13 ± 0.31	95.60 ± 0.30	94.03 ± 0.23
F7	5.5	Agregates*	n.o.	n.o	n.o.
		%T	92.26 ± 0.32	95.16 ± 0.21	93.53 ± 0.40
F8		Agregates*	n.o.	n.o	n.o.
		%T	92.60 ± 0.40	95.10 ± 0.10	93.40 ± 0.26
F9		Agregates*	n.o.	n.o	n.o.
		%T	92.00 ± 0.20	94.50 ± 0.10	93.06 ± 0.38

*Done visual observation; n.o. = not observed; + = observed

Table 3. Physical Stability of RMJCN-2

Formula	$CaCl_2$ Concentration (% w/v)	Stability Parameter	Observation Results		
			Day 1	Day 3	Day 7
EG1	0.1	Agregates*	n.o.	n.o	n.o.
		%T	93.20 ± 0.46	93.07 ± 0.25	90.90 ± 1.31
EG2	0.2	Agregates*	n.o.	n.o	n.o.
		%T	93.73 ± 0.15	92.73 ± 1.01	90.40 ± 2,82
EG3	0.3	Agregates*	n.o.	+	+
		%T	96.03 ± 0.46	91.70 ± 1.32	93.9.0 ± 0.36

*Done by visual observation; n.o. = not observed; + = observed

The nanoparticle formation was indicated by observing the visual appearance and measuring the % Transmitance to see the turbidity of the solution. It is shown that all of the formulation depicted an *opaque* yellowish-brown liquid. The *opaque* appearance of the solution indicates the nanosize particles have been formed.

It appears that no formula formed any visible aggregates, proofing that a stable dispersion system was successfully formed. This result confirmed by the Transmittance

value from all of the samples that were above 90%. All of the NPMJ-1 formula with concentration of 0.3%-0.5% chitosan in pH of 3.5-5.5 were able to produce stable nanoparticle system in the room temperature.

From all formula for RMJCN-1, F9 (pH of 5.5 and concentration of chitosan of 0,5 %w/v) was found to be the optimum formula, giving the highest EE value and physical stability of the solution observed. This optimum formula was used for the second encapsulation step with alginate and calcium chlorida to prepare RMJCN-2.

3.1.3 Stability of RMJCN-2

The evaluation on stability of the double coated nanoparticle (RMJCN-2) was carried out as the evaluation of RMJCN-1. The results were shown in **Table 3**.

All of the RMJCN-2 show high %T value of >90%. Those indicated a low level of turbidity that would be visually observed as opaque solution. The results reveal that the difference of 0.1% w/v of calcium chloride gave no significant effect on the transmittance. Nevertheless, formula of EG3 ($CaCl_2$ of 0.3%) has shown a big fluctuation ini %T value from day 1 to day 7 of observation

3.1.4. Preparation and initial characterization of RMJCN-2

Visual observation of RMJCN-2 is shown in **Figure 3.** All formulas produce colorless and opaque solution. Based on the observation of nanoparticle physical stability, EG1 and EG2 showed stable physical appearance since during 7 days of observation. However, EG3 indicated an unstable nanoparticle system through the appearance of white aggregates which occupied ~ ½ part of solution

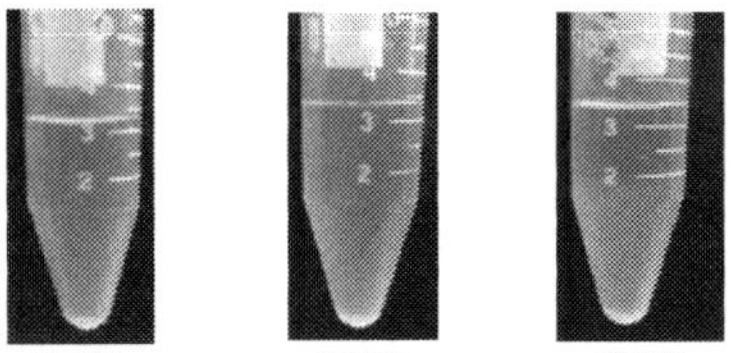

EG1 EG2 EG3

Fig. 3. Visual observation of NPMJ-2 samples on the 7th day of observation.

3.1.5. Characterization of RPMJCN-1 and RPMJCN-2

3.1.5.1. Particle Size and Distribution

The RPMJCN-1 and RPMJCN-2 obtained form the optimum formula were characterized by its particle's diameter using *Particle Size Analyzer* (PSA), and the size distribution in the solution (*Polydispersity Index/PI*). These two parameters are important since they have strong correlation with homogenicity affecting bulk characteristics, product performance, processability, stability and appearance of the finished product [11], [12]. The result of the particle size analyzer and its size distribution is shown in **Table 4**.

Table 4. Particle Size and Size Distribution of RMJCN-1 and RMJCN-2

Formula	Peak No	S.P. Area Ratio (%)	Diameter (nm)	PI	Z Avr (nm)
RMJCN-1	1	62	132.1 ± 7.9	0.49	259.7
	2	38	1072 ± 69.4		
RMJCN-2	1	8	30.1 ± 1.6	0.87	521.8
	2	92	739.8 ± 45.2		

It is shown that RMJCN-1 has relative low PI (0.486) with the average particle diameter size of 259.7 nm. Further analysis of RMJCN-2 has given results of the average diameter size of 521.8 nm, and PI value of 0.870. The high PI value comes from the big difference in the diameter of the smallest (30.1 nm; 8%) and the bigest (739.8 nm; 92%) particles. The small particle presumably came from the fall out fragmen of chitosan, and the average diameter of RMJCN-2 was detected as 739.8 nm, which still within the nanosize range and shows activity in the biological system [13].

3.1.5.2. Zetta Potential Analysis

The results of zetta potential measurement of RMJCN-1 and RMJCN-2 are shown in **Table 5** below.

Table 5. Zetta Potential Analysis Results of RMJCN-1 and RMJCN-2

Formula	Zeta Potential	Electrophoretic Mobility
RMJCN-1	+19.3 mV	0.000149 cm2/Vs
RMJCN-2	-14.4 mV	-0.000111 cm2/Vs

Zeta potensial analysis for RMJCN-1 gave a value of +19.3, which is sufficient to make repulsion force in nanoparticle system, and resulting a good stability of the nanoparticle. In the case of RMJCN-2, the zeta potenisal analysis gave value of -14.4 provides information about the existence of a negative charge on the surface of the nanoparticles.

3.2 Discussion

In the preparation of RMJCN-1, it was shown that the increase of pH environment led to gradual improvement on EE value. This might be related with the fact that electrostatic interaction is the determining factor for the association between protein - polysaccharide in nanoparticle [14], since pH will five influence on the degree of ionizaton. Formation of RMJCN-1 nanoparticles involved the protonated and deprotonated functional groups in both chitosan and RIP MJ-C. The use of medium with pH below the compound's pKa will induced the formation of protonated groups to interact with the negatively charged carboxyl group of RIP MJ-C, which also determines the strength of interaction in the nanoparticle system.

Apart from the correlation of pH environment to EE value, the effect of increasing concentration of chitosan on EE value is hardly concluded. The quantity of components involved in the reaction has an impact on the degree of interaction between each one as well as the charge or ionic group formation. The high density of chitosan fibre consequently may diminish the space for protein entrapment inside the system. Meanwhile, the low amount chitosan leads to a non-sufficient polymer quantity to encapsulate protein which contribute to the less compactness of nanoparticle system. In other words, the formation of RMJCN-1 with the best EE value could be accomplish by using a sufficient or optimum amount of chitosan theoretically. It can be concluded that the increase of pH environment gave significant effect rather than the chitosan concentration. All of the formula which is prepared in pH environment of 5.5 produced nanoparticles with good EE value (>50%), and the highest was found in the using of 0.5 % w/v chitosan.

The next step was the formulation of RMJCN-2, which was executed in pH of 5.5 as the optimum pH for RMJCN-1 preparation. RMJCN-2 is the form of double layered nanoprticles, where the second layer is constructed from alginate. In the appropriate pH medium, the carboxylate group (COO^-) of alginate will form electrostatic interaction with the protonated amine group ($-NH_3^+$) on the surface of RMJCN-1. This interaction resulted on the formation of second layer, which is strenghthened by Ca^{2+} ion from calcium chloride as the crosslinker. The predicted structure for the double layer nanoparticle RJMCN-2 is illustrated as in **Figure 2**.

From the preparation of RMJCN-2, it was found that the formulation with high concentration of calcium chloride (0.3%) resulting in nanoparticles with poor stability. This may relate to the fact found by Winarti (2011) that high concentration of crosslinker induces particles aggregation [15]. Alginate is known to be easily swell and forms gel in high temperature. These two characteristics can be mutually support the formation of aggregates in nanoparticle system.

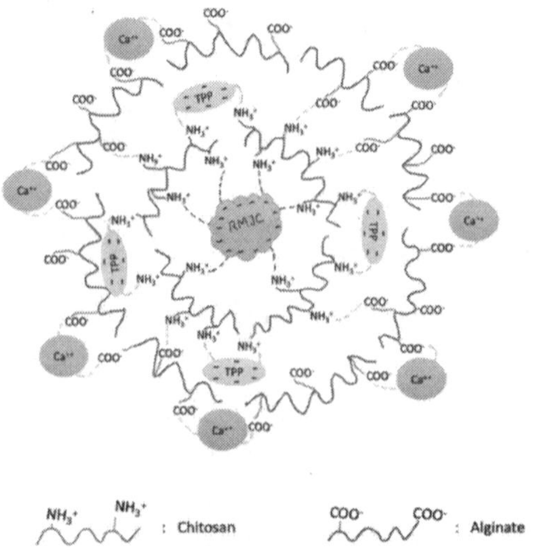

Fig. 2. Predicted complex structure for the double layer nanopartcles of RMJCN-2.

This may explain the results for EG 3, where the formation of the agreegate was likely triggered by the high concentration of calcium chloride. In lower concentration of calcium chloride (EG 1 with 0.1% and EG 2 with 0.2%), nanoparticles were more stable and showed no aggregation during 7 days of observation.

Furthermore, the tendency was more clearly seen in the transmitance tests. All of the RMJCN-2 show high %T value of >90%, indicating a low level of turbidity. Nevertheless, formula of EG3 (0.3% calcium chloride) has shown a big fluctuation ini %T value during the 7 days of observation. This phenomena may relate to the process of sedimentation in colloidal dispersion. The sedimentation began with the increasing turbidity on the surface and end with a clear appearance as the sediment went down. The transmittance value of EG1 and EG2

https://doi.org/10.1051/bioconf/20237504001

formula were more stable as almost no sedimentation occured. From all formula EG1 is the most stable nanoparticle, lead to conclusion that using 0.1% w/v calcium chloride for RMJCN-2 preparation a stable nanoparticle and is the optimum formula for the preparation of RMJCN-2.

In the analysis for the particle size of RMJCN-1 and RMJCN-2, it is shown that RMJCN-1 has the average particle diameter size of 259.7 nm, smaller than that of RMJCN-2 that was 521.8 nm. This bigger diameter of RMJCN-2 indicates the formation of second layer on the surface of RMJCN-1 by alginate and calcium chloride. Previous study by Ciptasari (2016) in preparing nanoparticle of RIP MJ-30 using alginate and calcium chloride gave the nanoparticle diameter range of 119-218 nm [16]. This proves that nanoparticle of RMJCN-2 represent a double coated form of nanoparticles.

Another proof of the formation of a double layer is from the results of zeta potential analysis. The results shows that the two nanoparticle systems provide different potential charges, where RMJCN-1 showed zeta potential of +19.3 mV, and for RMJCN-2 was -14.4 mV.

In a nanoparticle system with more than one component, the charges on the surface will formed with a certain value as the result of resultant ionic interaction between positive and negative charge. Chitosan is a cationic polymer with positive charge comes from ammonium group (NH_3^+). The negative charge of the protein RIP-MJC is predicted to be inside of the nanoparticle, interacts with the postive ammonium group of chitosan. Then the negative charge of STPP acts as a crosslinker to bind the positive ammonium group. Since RMJCN-1 covered with chitosan as the first layer, so the zetta potential of RMJCN-1 is postive. In the case of RMJCN-2, the zetta potential was found to be negative. This negative value is predicted caused by the carboxylate ion (COO^-) group from alginate on the outer layer. Hence, The change of zetta potential value from positive to negative in the formation of RMJCN-1 to RMJCN-2 give good indication for the success of the second layer formation by alginate and calcium chloride.

4 CONCLUSION

RIP MJ-C was successfully prepared in the form of double coated nanoparticle using chitosan-STPP dan alginate-calcium chloride. The optimum formula for first coat layer of chitosan was made from 0.5 % w/v chitosan and 0.3 % w/v TPP in medium pH of 5.5, resulting to RMJCN-1 core-particle. The optimum formula for second coat layer was made from 0.3 % v/v alginate and 0.1% w/v calcium chloride. The double-coated nanoparticles were proven to be stable in 7 days at room temperature, having zeta potential of -14.4 mV and particle size of 739.8 nm.

Further research needs to develop and to chalenge the potency of double-coated nanoparticles of RMJCN-2 as a promising drug model. Thus, the purpose of developing a stable drug delivery system for RIP MJ-C through oral administration could be accomplished.

REFERENCES

1. Xu, Modified natriuretic peptides and their potential roles in cancer treatment., *Biomed J*, pp. 118-131, (2022).
2. Chaddock, Major Structural Differences between Pokeweed Antiviral Protein and Ricin A-Chain do not Account for their Differing Ribosome Specificity, *European Journal of Biochemistry,* vol. **235**, pp. 159-166, (1996).
3. Sudjadi, Efek Sitotoksik suatu protein seperti Ribosome inactivating Proteins yang bersifat asam dari daun Mirabilis jalapa L. Pada sel kanker, *MFI,* vol. **18**, pp. 8-14, (2007).
4. Elzoghby, Albumin-based nanoparticles as potential controlled release drug delivery systems, *Journal of Controlled Release,* vol. **157**, pp. 168-182, (2012).
5. Peer, Nanocarriers as an emerging platform for cancer therapy., *Nature Nanotech,* vol. **2**, pp. 751-760, (2007).
6. Bowman, Chitosan nanoparticles for oral drug and gene delivery, *Int J Nanomedicin,* vol. **1**, pp. 117-128, (2006).
7. Kanjanabat, Preparation and Characterization of Nicotine–Magnesium Aluminum Silicate Complex-Loaded Sodium Alginate Matrix Tablets for Buccal Delivery, *AAPS PharmSciTech,* vol. **12**, pp. 683-692, (2011).
8. Rajesh, Formulation and evaluation of mucoadhesive microcapsules of aceclofenac using methyl cellulose and carbopol as mucoadhesive polymers, *Int. J. Pharm. Sci,* vol. **4**, pp. 362-366, (2012).
9. Bagre, Alginate coated chitosan core shell nanoparticles for oral delivery of enoxaparin: In vitro and in vivo assessment., *Int. J. Pharm,* vol. **456**, pp. 31-40, (2013).
10. F. Sekarningtyas, Formulasi dan Karakterisasi Nanopartikel Ribosome Inactivating Protein Mirabilis jalapa (RIP MJ) menggunakan Tripolifosfat Penaut Silang Kitosan Rantai Pendek (Skripsi), *Fakultas Farmasi Universitas Gadjah Mada, Indonesia (*2015).
11. . Danaei, Impact of Particle Size and Polydispersity Index on the Clinical Applications of Lipidic Nanocarrier Systems, *Pharmaceutics ,* vol. **10**, p. 57, (2018).
12. Rasmussen, Size and surface charge characterization of nanoparticles with a salt gradient, *Nat Commun,* vol. **11**, p. 2337, (2020).
13. How do Nanoparticles Enter Cells? [www document], (2014), *URL https://sustainable-nano.com/2014/08/19/how-do-nanoparticles-enter-cells/ (accessed 7.13.23).*
14. Ganzevles, Polysaccharide Charge Density Regulating Protein Adsorption to Air/Water Interfaces by Protein/Polysaccharide Complex

Formation, *J. Phys. Chem. B* , vol. **111**, pp. 12969-12976, (2007).

15. Winarti, Formulasi Nanopartikel Kitosan Rantai Pendek-TPP sebagai Sistem Penghantaran Gen Non Viral yang ditransfeksi pada Sel Kanker Payudara T47D., *Thesis, Fakultas Farmasi Universitas Gadjah Mada Indonesia,* (2011).
16. U. Ciptasari, Konjugasi Antibodi Anti-EpCAM pada Nanopartikel Ribosome Inactivating Protein Mirabilis jalapa L. Menggunakan Katalis EDAC (Skripsi), *Skripsi, Fakultas Farmasi Universitas Gadjah Mada, (*2016).

In Silico Pharmacokinetics Study of 2,5-Dibenzylidenecyclopentanone Analogs as Mono-Ketone Versions of Curcumin

Prajona Marbun[1], *Arief Rahman* Hakim[1], *Navista* Sri Octa Ujiantari[1], *Bambang* Sulistyo Ari Sudarmanto[1], *Agung* Endro Nugroho[1]*
[1]Faculty of Pharmacy, Universitas Gadjah Mada, Yogyakarta, Indonesia

Abstract. The absorption-distribution-metabolism-excretion (ADME) profile is a crucial parameter that indicates the pharmacokinetics of the drug. The pharmacokinetic properties of a drug represent the fate of the drug in the body. Curcumin is a main compound in turmeric produced by plants of the Curcuma longa species, and has several pharmacological effects in animal and human clinical studies. However, preclinical and clinical studies have shown that curcumin has pharmacokinetic limitations such as poor bioavailability and rapid metabolism which restrict its widespread use. Therefore, various modifications and synthesis of some analogs using curcumin as a lead compound with variations in the main structure and attached substituents have been carried out to explore the pharmacological effects as drug candidates. One of the widely developed methods is the modification of curcumin's main structure, specifically the conversion from diketone to mono-ketone.In 1997, 2,5-dibenzylidene cyclopentanone analogs were synthesized and their biological activity were performed. However, there is no further information related their pharmacokinetic properties. Therefore, those properties were predicted by performing ADME calculation in two online servers, ADMETsar 2.0 and ADMETlab 2.0.. By utilizing the online servers ADMETsar 2.0, and ADMETLab 2.0 for in-silico screening of pharmacokinetic properties, from the 17 compounds, it was found that the variation among pharmacokinetic aspects was observed, either decreasing or increasing drug likeness properties of 2,5-dibenzylidene cyclopentanone analogs compared to curcumin. In addition, the interaction those analogs with protein or enzymes involved during ADME process such as blood plasma protein (albumin), p-Glycoprotein, and CYP3A4 was evaluated by performing molecular docking.. The docking results showed a sufficiently positive correlation with ADME screening outcomes. Keywords: ADME, Curcumin, 2,5-dibenzylidenecyclopentanone, Pharmacokinetics, Molecular Docking

1 INTRODUCTION

The inability of a drug to enter the market is often associated with issues in the testing of drug efficacy and safety. The safety and efficacy of a drug are influenced by its absorption, distribution, metabolism, excretion (ADME), as well as its side effects/toxicity (T). Typically, methods used to assess the ADME/T properties of a drug involve animal testing (Phase I Clinical Trials), which can be costly and time-consuming. Phase I Clinical Trials also raise ethical concerns (animal rights) and economic considerations. One potential solution is the application of rational drug design methods using computational approaches [1].

Currently, a significant number of QSAR/SAR models have been developed, and specialized programs available on online servers have been created and developed based on these models. The advantages of these methods are the ability to accurately predict the ADME/T properties of a chemical compound and enable screening of most candidate compounds or their properties before conducting in vivo and in vitro testing. This approach helps to save resources, costs, and test animals [2].

Curcumin[1,7-bis-(4-hydroxy-3- methoxyphenyl) - 1,6-heptadiene- 3,5-dione] (figure 1) is a compound isolated from *Curcuma longa L.* It has been widely known

*Corresponding Author : nugroho_ae@ugm.ac.id

and used for centuries as a food pigment and spice. Curcumin has been found to have various traditional pharmaceutical applications in diseases, including external/internal wounds, liver diseases (especially jaundice), blood purification, antimicrobial effects, and inflammation in joints. In modern drug research, curcumin is still considered a highly promising compound for drug design and development based on its explicit bioactivity, minimal toxicity, and ease of synthesis.

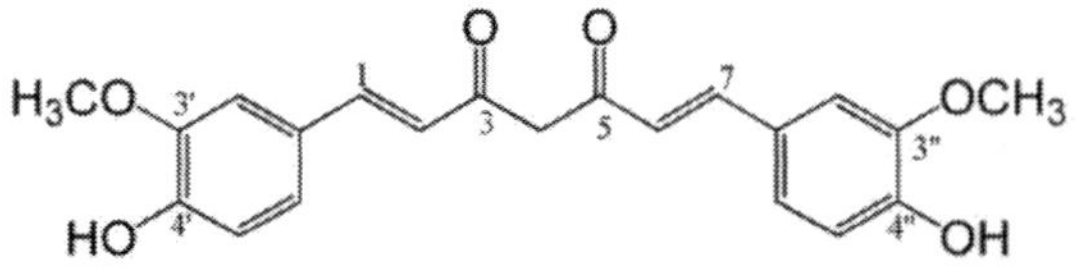

Figure 1. Curcumin [1,7-bis-(4-hydroxy-3-methoxyphenyl)-1,6-heptadiene-3,5-dione]

However, preclinical and clinical studies have shown that curcumin has pharmacokinetic limitations such as poor bioavailability, rapid metabolism, and the need for repeated oral doses, which restrict its widespread use. Curcumin is stable at pH below 6.5. Its instability above pH 6.5 is caused by the methylene group. By removing the methylene group and one carbonyl group, B.M. Markverich, M. Artico, and H.I. El-Subbagh synthesized a series of mono-carbonyl analogs of curcumin, 1,5-diaryl-1,4-pentadien-3-one, and evaluated their bioactivity. The results showed that the mono-carbonyl analogs exhibited stronger inhibitory effects in various cancer cells compared to curcumin [3].

One synthetic analog compound with a structure similar to curcumin, containing a mono-carbonyl group, is 2,5-Di2,5-dibenzylidenecyclopentanone, which was synthesized by Sardjiman in 1997 as part of a series of 17 analog compounds. The research conducted by Sardjiman in 1997 produced analog compounds of 2,5-dibenzylidenecyclopentanone with variations in substituents at carbon atoms 4, 5, and 6 that are symmetric in both benzylidene structures. In this study, a pharmacokinetic evaluation was performed by conducting ADME profiling of the analog compounds of 2,5-dibenzylidenecyclopentanone using a virtual screening approach with online servers ADMETLab 2.0, and ADMETsar 2.0 [4][5] .

2 METHOD

2.1 Instrumentation

Data processing was performed using a PC device with the following specifications: i5 processor, 8 GB RAM, 512 GB SSD, and Windows 10 operating system.

2.2 Data Source of Compounds

The studied compounds were the curcumin and 17 analog compounds of 2,5-dibenzylidenecyclopentanone that were synthesized and investigated by Sardjiman in 1997. The data on the structure of the analog compounds of 2,5-dibenzylidenecyclopentanone consisted of a table containing a list of 17 derivative compounds of curcumin that were synthesized in Sardjiman's research in 1997 (Figure 2) (table 1) [6]. The list of compounds represents a part of the 2,5-dibenzylidenecyclopentanone analogs with variations in functional groups attached to carbon atoms 3, 4, and 5 on the benzene ring attached to both sides of the cyclopentanone structure.

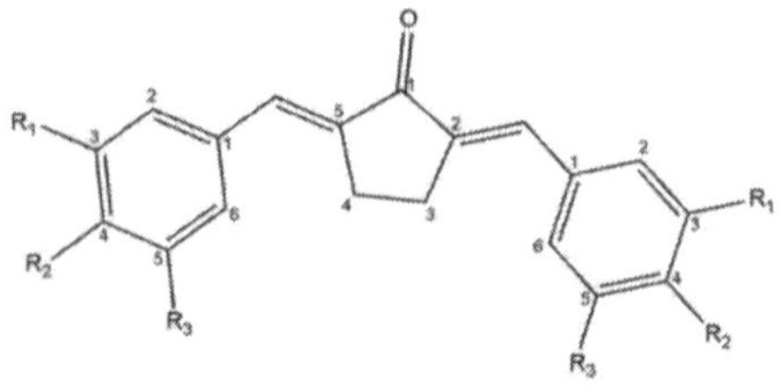

Figure 2 : Main Frame of 2,5-Di2,5-dibenzylidenecyclopentanone

Table 1. List of Chemical Group Variations and Derivatives of 2,5-Dibenzylidene Cyclopentanone [6]

Comp. Code	R1	R2	R3	Chemical Name	MW
B0	H	OH	H	2,5-Bis(4-hydroxy benzylidene)cy clopentanone	292.11
B1 (PGV-0)	OCH_3	OH	H	2,5-Bis(4-hydroxy-3-methoxy benzylidene)cy clopentanone	352.13

Comp. Code	R1	R2	R3	Chemical Name	MW
B2	H	H	H	2,5-Di2,5-dibenzylidenecyclopentanone	260.12
B3	H	Cl	H	2,5-Bis(4-chlorobenzylidene)cyclopentanone	328.04
B4	H	OCH_3	H	2,5-Bis(4-methoxy benzylidene)cyclopentanone	320.14
B5	H	CH_3	H	2,5-Bis(4-methylbenzylidene)cyclopentanone	288.15
B6	H	t-C_4H_9	H	2,5-Bis(4-tertier butylbenzylidene)cyclopentanone	372.25
B7	H	CF_3	H	2,5-Bis(4-methylfluoro benzylidene)cyclopentanone	396.33
B8	H	$N(CH_3)_2$	H	2,5-Bis(4-dimethyl aminobenzylidene)cyclopentanone	346.21
B9	Cl	Cl	H	2,5-Bis(3,4-dichloro benzylidene)cyclopentanone	395.96

Comp. Code	R1	R2	R3	Chemical Name	MW
B10	Cl	Cl	H	2,5-Bis(3,4-dichlorobenzylidene)cyclopentanone	329.23
B9	Cl	H	H	2,5-Bis(3-chloro benzylidene)cyclopentanone	328.04
B11 (PGV-1)	CH_3	OH	CH_3	2,5-Bis(4-hydroxy-3,5-dimethylbenzylidene)cyclopentanone	348.17
B12	C_2H_5	OH	C_2H_5	2,5-Bis(4-hydroxy-3,5-diethylbenzylidene)cyclopentanone	404.24
B13	i-C_3H_7	OH	i-C_3H_7	2,5-Bis(4-hydroxy-3,5-diisopropyl benzylidene)cyclopentanone	460.20
B14	i-C_4H_9	OH	i-C_4H_9	2,5-Bis(4-hydroxy-3,5-ditert.butylbenzylidene)cyclopentanone	516.36
B15	OCH_3	OH	OCH_3	2,5-bis(3,5-dichloro-4-hydroxy benzylidene)cyclopentanone	412.44
B16	Cl	OH	Cl	2,5-bis(4-hydroxy-3,5-	430.11

Comp. Code	R1	R2	R3	Chemical Name	MW
				dimethoxy benzylidene)cy clopentanone	

2.3 DATA PROCESSING

The compound structures were inputted into online servers that provide ADME profile screening, namely ADMETsar 2.0 and ADMETlab 2.0. The online servers were accessed through a web browser by visiting the screening pages at https://admetmesh.scbdd.com/ , and http://lmmd.ecust.edu.cn/admetsar2/ . Once the online servers were opened, the compound structures were entered into the provided input columns one by one.

The input of compound structures into the online servers for screening can be done in two ways: directly drawing the compound structure using the available tools or entering the SMILES (The simplified molecular-input line-entry system) code, which is a chemical notation that represents the molecular structure as a graph with optional chiral indications. After inputting the structure, the "submit" or "Run!" option was selected to start the screening process. If a manually drawn structure was entered, the SMILES code would appear first.

After the screening of the entered compounds, ADME/T data along with the physical-chemical properties of the compounds would be obtained. The screening results would produce positive/negative signs, probability values, or absolute numbers according to the parameters, indicating the capability or lack thereof, as well as the degree, of the compound's ability to undergo ADME processes in the body, based on the respective database used by each online server. The subsequent analysis involved interpreting and comparing the physical-chemical properties and ADME profiles of the same compounds screened using three different online servers and among the compounds in the analog list.

2.4 Data Analysis

The obtained ADME property data were then analyzed for their drug-likeness properties using various pharmacokinetic parameters, including physical-chemical properties, adsorption capacity, distribution ability, metabolic capacity, and elimination ability. The focus of this research was to compare the ADME properties of curcumin obtained with those of the analog compound 2,5-dibenzylidenecyclopentanone as its mono ketone version. The online servers, ADMETlab 2.0 and ADMETsar 2.0, have similarities in data presentation, where the majority of the data is qualitative, both in terms of absolute value parameters and parameters that generate probability data. However, there are differences between them. ADMETlab 2.0 for probabilistic parameters shows values representing the compound's usefulness (specifically referring to drug likeness of compounds) using symbols such as "+++" for the highest range of values (probability 0.9 - 1) and "---" for the lowest range of probability values (0-0.1). In this research, the probability values taken are the upper limits (endpoint probability) as the chosen values to indicate the level of probability. In contrast, ADMETsar provides comprehensive values ranging from 0 to 1 to indicate the probability results of the inputted compounds. Due to these differences, the correlation between the two cannot be determined, but it merely indicates the presence of two prediction values from two different servers. The research was conducted by inputting the SMILES code of each tested compound into all three types of online servers, followed by pharmacokinetic analys

https://doi.org/10.1051/bioconf/20237504002

3. RESULTS AND DISCUSSION

Table 2 . Physicohemical and ADME Parameters of 2,5-Benzylidenecyclopentanone Analog Compared to Curcumin Based On ADMETlab 2.0

PROPERTIES	[illegible]	B0	B1	B2	B3	B4	B5	B6	B7	B8	B9	B10	B11	B12	B13	B14	B15	B16	
Substient		p-OH	p-OCH2	H-H-H	p-Cl	p-OCH3	p-CH3	p-C4H9	p-CF3	p-N(CH3)2	m,p-2-Cl	m-Cl	m-2-CH3, p-OH	m-2-C2H5, p-OH	m-2-C3H7	m-2-C4H9	m-2-OCH3	m-2-Cl, p-OH	
Molecular Weight (g/mol)	368,39	292,33	352,39	260,34	329,23	320,39	288,39	372,55	396,33	346,47	398,12	329,23	348,44	404,54	460,66	516,77	412,44	430,11	
Hydrogen Bond Acceptor	6	2	2	0	0	0	0	0	0	0	0	0	2	2	2	2	2	2	
Hydrogen Bond Donors	2	3	5	1	1	3	1	1	1	3	1	1	3	3	3	3	7	3	
Rotation Bond	8	2	4	2	2	4	2	2	2	4	2	2	2	6	6	6	6	2	
Polar Surface Area (Å2)	93,06	57,53	75,99	17,07	17,07	35,53	17,02	17,02	17,02	23,55	17,02	17,07	57,53	57,53	57,53	57,53	94,45	57,53	
LogP	2,742	3,387	3,171	4,291	5,563	4,288	5,177	7,179	5,697	4,805	6,422	5,45	4,711	5,939	6,719	8,216	2,906	5,679	
Absorption																			
P-Glycoprotein Inhibitor (endpoint probability)	0,3	0,1	0,1	0,1	0,1	0,1	0,1	0,1	0,1	0,1	0,1	0,1	0,1	0,1	0,1	0,1	0,1	0,1	
P-Glycoprotein Substrat (endpoint probability)	0,1	0,5	1	1	1	1	1	1	1	1	1	1	1	1	1	1	1	1	
HIA (endpoint probability)	0,1	0,1	0,1	0,1	0,1	0,1	0,1	0,1	0,1	0,1	0,1	0,1	0,1	0,1	0,9	0,9	0,1	0,1	
Distribution																			
VD (L/Kg)	0,369	0,654	0,454	0,369	0,336	0,629	0,323	0,88	0,949	0,669	0,813	0,429	0,269	0,356	1,1	2,844	0,576	0,175	
PPB (%)	0,99799	0,98639	0,99643	0,98535	0,99605	0,97329	0,99373	1,00679	0,99809	0,96528	1,01418	1,00375	1,00825	0,99934	1,0177	1,01791	0,93573	1,02377	
BBB (endpoint probability)	0,3	0,1	0,1	0,1	0,1	0,1	0,1	0,1	0,1	0,3	0,1	0,1	0,1	0,1	0,1	0,1	0,1	0,1	
Metabolism (Endpoint Affnity Probability)																			
CYP1A2 substrat	0,9	0,3	1	0,3	0,3	1	0,9	0,9	0,7	0,9	0,3	0,5	1	1	1	1	1	0,1	
CYP1A2 Inhibitor	0,7	1	0,9	1	1	0,9	0,7	0,5	0,9	1	1	1	0,5	0,7	0,5	0,3	0,1	0,9	
CYP2C9 substrat	1	1	1	0,3	0,1	0,9	0,3	0,1	0,3	0,1	0,1	0,1	0,5	0,3	0,9	0,7	0,9	0,7	
CYP2C9 inhibitor	0,7	0,7	0,7	0,9	0,5	0,7	0,7	0,1	0,7	0,1	0,1	0,7	0,7	0,7	0,3	0,3	0,5	0,7	
CYP2C19 substrat	0,3	0,1	0,1	0,1	0,1	0,1	0,1	0,1	0,1	0,3	0,1	0,1	0,3	0,1	0,5	0,3	0,7	0,1	
CYP2C19 inhibitor	0,3	1	0,9	0,9	0,9	0,7	0,7	0,7	0,9	0,7	0,9	1	0,7	0,7	0,7	0,7	0,3	0,7	
CYP2D6 substrat	0,9	0,9	0,3	0,1	0,1	0,9	0,3	0,1	0,1	0,7	0,1	0,1	0,9	0,9	0,3	0,9	0,9	0,3	
CYP2D6 inhibitor	0,1	0,9	1	0,3	0,5	0,3	0,1	0,3	0,3	0,5	0,5	0,7	0,5	0,5	0,7	0,7	0,1	0,7	
CYP3A4 substrat	0,7	0,3	0,5	0,3	0,3	0,5	0,5	0,9	0,3	0,3	0,3	0,3	0,3	0,3	0,5	0,9	0,7	0,3	
CYP3A4 inhibitor	0,7	0,9	0,7	0,1	0,3	0,7	0,3	0,3	0,3	0,7	0,3	0,3	0,3	0,3	0,3	0,3	0,5	0,1	
Elimination																			
Cl (mL/min/Kg)	13,839	11,431	8,852	6,361	4,512	7,958	6,321	3,792	5,411	8,22	4,872	5,161	7,444	3,406	0,914	1,906	7,726	2,6	
T1/2 (hours)	0,948	0,877	0,897	0,24	0,054	0,119	0,097	0,021	0,007	0,177	0,027	0,089	0,597	0,52	0,067	0,048	0,915	0,153	**TOTAL**
Expected increase in pharmacokinetic properties		7	8	7	7	8	7	8	7	7	7	8	7	6	8	8	7	5	122
Expected decrease in pharmacokinetic properties		7	7	8	7	7	7	5	8	7	7	7	7	7	6	3	7	9	116
Similarity to curcumin		4	4	2	1	4	5	2	2	4	1	2	4	4	0	2	6	2	49
Unexpected increase in pharmacokinetic properties		2	2	3	4	2	3	4	2	3	3	2	3	2	4	4	4	2	49
Unexpected decrease in pharmacokinetic properties		4	3	4	5	3	2	5	5	3	6	5	3	5	6	7	0	6	72
Pharmacokinetic improvement		9	10	10	11	10	10	12	9	10	10	10	10	8	12	12	11	7	171
Pharmacokinetic deteriomtion		11	10	12	12	10	9	10	13	10	13	12	10	12	12	10	7	15	188

Color Legend of Heatmap

- Expected increase in pharmacokinetic properties (Good pharmacokinetic characteristics are already present in curcumin and are further enhanced in the analog 2,5-dibenzylidene cyclopentanone)
- Expected decrease in pharmacokinetic properties (The pharmacokinetic characteristics of the analog 2,5-dibenzylidene cyclopentanone are already good, but not as good as curcumin)
- Similarity to curcumin (the pharmacokinetic characteristics of the analog 2,5-dibenzylidene cyclopentanone that are identical to curcumin)
- Unexpected increase in pharmacokinetic properties (the pharmacokinetic characteristics of 2,5-dibenzylidene cyclopentanone that are poor, but not as poor as curcumin)
- Unexpected decrease in pharmacokinetic properties (The pharmacokinetic characteristics of 2,5-dibenzylidene cyclopentanone that are poor and worse than curcumin)

Pharmacokinetic improvement : Expected increase in pharmacokinetic properties + Unexpected increase in pharmacokinetic properties

Pharmacokinetic deterioration : Expected decrease in pharmacokinetic properties + Unexpected decrease in pharmacokinetic properties

Table 3. Physicohemical and ADME Parameters of 2,5-Benzylidenecyclopentanone Analog Compared to Curcumin Based On ADMETsar 2.0

PROPERTIES	Cur	B0	B1	B2	B3	B4	B5	B6	B7	B8	B9	B10	B11	B13	B14	B15	B16	
Substient		p-OH	p-OCH2	H-H-H	p-Cl	p-OCH3	p-CH3	p-C4H9	p-CF3	p-N(CH3)2	m,p-2-Cl	m-Cl	m-2-CH3, p-OH	m-2-C3H7	m-2-C4H9	2m-OCH3	m-2-Cl, p-OH	
Molecular Weight (g/mol)	368,39	292,33	352,39	260,34	329,23	320,39	288,39	372,55	396,33	346,47	398,12	329,23	348,44	460,66	516,77	412,44	430,11	
Hydrogen Bond Acceptor	6	2	2	0	0	0	0	0	0	0	0	0	2	2	2	2	2	
Hydrogen Bond Donors	2	3	5	1	1	3	1	1	1	3	1	1	3	3	3	7	3	
LogP	3,37	3,93	3,5	4,52	5,82	4,53	5,13	7,11	6,55	4,65	7,13	5,82	5,16	8,42	9,12	3,96	6,54	
Rotation Bond	8	2	4	2	2	4	2	2	2	4	2	2	2	6	6	6	2	
Polar Surface Area (Å2)	93,06	57,53	75,99	17,07	17,07	35,53	17,02	17,02	17,02	23,55	17,02	17,07	57,53	57,53	57,53	94,45	57,53	
Absorption																		
P-Glycoprotein Inhibitor (affinity probability)	0,5967	0,881	0,7524	0,8727	0,6927	0,7457	0,6375	0,6843	0,5992	0,6286	0,5895	0,6767	0,6	0,6147	0,4311	0,6466	0,6964	
P-Glycoprotein Substrat (affinity probability)	0,9362	0,9945	0,9903	0,9975	0,998	0,9955	0,9961	0,987	0,9913	0,9869	0,9964	0,9826	0,9916	0,9768	0,9802	0,9735	0,9945	
HIA (ability probability)	0.9803	0,9968	0,9914	0,9974	0,9968	1	1	0,9967	0,9966	0,9757	0,9974	1	1	1	1	0,9906	0,9971	
Distribution																		
BBB Permeability (probability)	0.6250	0,7	0,625	0,575	0,725	0,525	0,6	0,65	0,7	0,775	0,7	0,55	0,65	0,525	0,5	0,6	0,55	
Metabolism (Affinity Probability)																		
CYP1A2 Inhibitor	0.9106	0,8023	0,946	0,6615	0,5557	0,8885	0,6745	0,5647	0,5166	0,8758	0,8201	0,878	0,9558	0,9501	0,8243	0,8754	0,878	
CYP2C9 substrat	0.6120	0,8062	0,5948	0,6147	0,8062	0,594	0,5888	0,7907	0,6108	0,5969	0,8062	0,8054	0,5905	0,6107	0,7919	0,5948	0,8054	
CYP2C9 inhibitor	0.6796	0,5935	0,7715	0,6542	0,6253	0,6128	0,6543	0,5398	0,575	0,6202	0,7948	0,913	0,8383	0,8991	0,8814	0,519	0,913	
CYP2C19 inhibitor	0.8994	0,7073	0,9121	0,7049	0,8001	0,852	0,7614	0,7493	0,6937	0,5183	0,8864	0,8634	0,9168	0,9312	0,9076	0,8442	0,8634	
CYP2D6 substrat	0.7654	0,7194	0,687	0.7108	0,7409	0,6678	0,7515	0.7816	0,7224	0,3978	0,7409	0,7457	0,7516	0,7635	0,7775	0,687	0,7457	
CYP2D6 inhibitor	0.6715	0,8574	0,7373	0,7882	0,7174	0,8987	0,7607	0,8753	0,8364	0,7587	0,6802	0,7632	0,7018	0,6808	1	0,8477	0,7632	
CYP3A4 substrat	0.6442	0,6848	0,5904	0,7555	0,5928	0,6022	0,6908	0,6295	0,6652	0,6096	0,6019	0,5903	0,6327	0,5965	1	0,5774	0,5903	
CYP3A4 inhibitor	0,5392	0,5	0,5873	0,8217	0,6786	0,5371	0,6946	0,6829	0,5166	0,5296	0,6369	0,5921	0,6473	0,5867	0,5508	0,6226	0,5921	**TOTAL**
Expected increase in pharmacokinetic properties		9	6	9	9	6	8	9	8	6	8	8	6	5	7	4	6	114
Expected decrease in pharmacokinetic properties		3	5	3	2	6	3	2	3	6	3	3	5	6	3	8	5	66
Similarity to curcumin		0	1	0	0	0	0	0	0	0	0	0	0	0	0	0	0	1
Unexpected increase in pharmacokinetic properties		4	0	3	3	4	3	3	4	4	3	2	0	0	2	3	2	40
Unexpected decrease in pharmacokinetic properties		2	6	3	4	2	4	4	3	2	4	5	7	7	6	3	5	67
Pharmacokinetic improvement		13	6	12	12	10	11	12	12	10	11	10	6	5	9	7	8	154
Pharmacokinetic deterioration		5	11	6	6	8	7	6	6	8	7	8	12	13	9	11	10	133

3.1 Physicochemical Properties

In drug development, an important determinant of the suitability of a drug molecule is the Rule of Five (RO5) proposed by Lipinski. The use of this rule is to predict the drug-likeness of various chemical compounds with a specific biological activity designed for orally administered drugs. Drug likeness: According to the RO5, good physicochemical properties of a drug are indicated by a molecular weight (MW) <500 g/mol, Log P value <5 (representing hydrophobicity), hydrogen bond donors (HBD) <5 sites, hydrogen bond acceptors (HBA) <10 sites, and polar surface area ≤140 Å. In further studies, an additional parameter was found: the presence of rotatable bonds (RB) <10. Deviations from more than one RO5 rule indicate the possibility of low gastrointestinal absorption [7] [8].

The screening results of good physicochemical properties using the ADMETsar 2.0 (table 2), and ADMETLab 2.0 (table 3) servers show that the analogs of 2,5-Dibenzylidene cyclopentanone tend to have better physicochemical properties compared to curcumin, except for the hydrogen bond acceptor and donor parameters. All analogs of 2,5-Dibenzylidene cyclopentanone exhibit lower polar surface area compared to curcumin, except for compound B15, where the variation in its substitution is two methoxy groups on each side, which is twice the amount compared to curcumin. The addition of oxygen atoms is strongly suspected as the reason for the increase in the polar surface area value.

Both curcumin and all analogs of 2,5-Dibenzylidene cyclopentanone almost meet the RO5 rules, with deviations not exceeding 1, mostly due to the log P value exceeding 5 in 11 compounds of 2,5-dibenzylidenecyclopentanone analog. Among all the tested compounds, only curcumin and six analogs of 2,5-dibenzylidene cyclopentanone meet the criteria of having a logP value below 5. This is because of the presence of substituents that increase the non-polar properties such as C, F, and Cl, without being balanced by the presence of O. However, there is one compound that deviates from the 2 rules, making it not pass the RO5, namely compound B14. This is due to its substitution of C4H9 (figure 3), which is a relatively long hydrocarbon chain, resulting in a total molecular weight exceeding 500 g/mol and polarity value of more than 5.

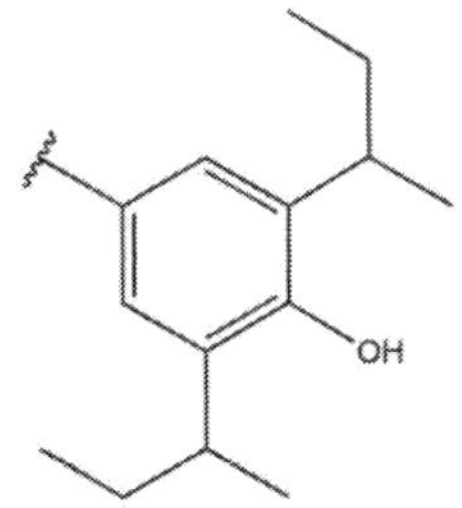

Figure 3. Compound with the code B14, 2,5-Bis(4-hydroxy-3,5-di tert-butyl benzylidene)cyclopentanone, which has bis-i-C4H9 substituents.

3.2 Absorption

Based on the screening results using ADMETLab 2.0, it was found that the ability of the 2,5-dibenzylidene cyclopentanone analogs to serve as P-Glycoprotein substrates is better than curcumin (table 2).

Conversely, the inhibitory effect on P-Glycoprotein is lower for all of compounds of 2,5-dibenzylidene cyclopentanone analogs compared to curcumin. The results of the prediction using ADMETlab 2.0 indicate that the tendency of the analogs to be P-glycoprotein inhibitors is higher than that of substrates, compared to curcumin. However, different from the screening results using ADMETsar 2.0, although the ability as a substrate is very high with probabilities close to 1 and better than curcumin, the inhibitory ability of the 2,5-dibenzylidene cyclopentanone analogs against P-Glycoprotein is almost always higher (table 3).

The difference in screening results for ligand-protein interactions regarding the parameter of interaction with P-glycoprotein between the two servers cannot be explained at the physicochemical level. This is because such interactions are specific ligand-receptor interactions that require further modeling and *in vivo* or *in vitro* research to validate them. However, both ADMETLab 2.0 and ADMETsar 2.0 servers show a high tendency for binding as substrates to P-Glycoprotein. This can be interpreted positively as the ability to enter cells to achieve therapeutic effects. However, excessive binding as a substrate can lead to negative effects, such as resistance to the efflux of substances from cells. P-gp is a transporter protein located on the cell membrane and plays a role in removing foreign compounds or drugs from within the cell. By strongly binding to P-gp, the compound can inhibit the normal function of P-gp in removing drugs from within the cell, leading to an increase in the drug concentration inside the cell and reducing its effectiveness. As a result, the compound can cause drug resistance, especially in cancer treatment cases.

Chemotherapy resistance is often associated with high P-gp activity, which can result in a decrease in the concentration of cytotoxic drugs inside cancer cells and reduce the effectiveness of treatment. Additionally, increased interaction with P-Glycoprotein can also affect the systemic bioavailability of the drug, potentially reducing the desired therapeutic effects.

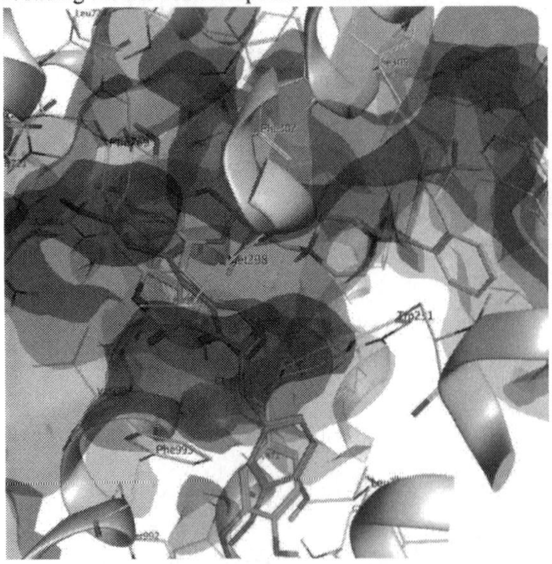

Figure 4. Interaction between P-Glycoprotein (6QEE) and the native ligand (green ribbon), curcumin (orange ribbons), and one of the analogs of 2,5-Dibenzylidene cyclopentanone (B1 /Pentagamavunons- 0 shown in purple ribbons)

The docking simulation results reveal interactions between curcumin and analogs of 2,5-dibenzylidene cyclopentanone, consistently demonstrating interactions with amino acid residues in the p-glycoprotein transport channel (figure 4). This indicates a favorable predictive affinity for both curcumin and the analog compounds of 2,5-dibenzylidene cyclopentanone.

In terms of the Human Intestinal Adsorption (HIA) parameter, the screening results using ADMETLab 2.0 showed that almost all of the 2,5-dibenzylidene cyclopentanone analogs have similarly poor intestinal absorption as curcumin, approaching 0, except for compounds B13 and B14, which are compounds with long hydrocarbon chain substituents (C_3H_7 and C_4H_9). This is thought to be due to an increase in solubility in the lipid layer of the intestinal membrane (increased non-polar properties that enhance solubility in the non-polar lipid phase). With these substituents, the solubility in the lipid phase in the intestine is increased, resulting in a higher ability to reach systemic circulation. Low HIA ability is associated with a large molecular size, which hinders penetration through the intestinal absorption barrier. Additionally, the presence of OH makes the compound sufficiently polar, reducing its solubility in the lipid layer of the intestinal membrane. However, the opposite prediction is shown by the ADMETsar 2.0 server, where all test compounds, including curcumin, are displayed as having very high intestinal absorption with probabilities close to 1. However, this prediction is less reliable considering various experimental studies that have shown very low intestinal absorption of curcumin and its derivatives.

3.3 Distribution

The distribution parameters examined in this study include the volume of distribution (VD), binding to Plasma Protein Binding (PPB), and blood-brain barrier (BBB) penetration ability. Based on the screening results with ADMETLab 2.0 (table 2), the VD values of the 2,5-dibenzylidenecyclopentanone analogs are generally higher than that of curcumin, except for compound B2, which has an OH group as a substituent. This may be due to the polar group experiencing an increase in polarity after passing through the metabolism phase. The OH group also has a strong binding ability to plasma proteins, reducing its distribution capability in systemic circulation. Additionally, the polar OH group can form hydrogen bonds with water, limiting its solubility to aqueous compartments such as blood and extracellular fluids, which have a relatively smaller volume compared to adipose tissues. The highest and most significant VD value is found in compound B14, which has a long hydrocarbon chain substituent (C_4H_9). This is likely due to its non-polar nature, which increases its solubility in adipose tissues, the largest distribution compartment in the body.

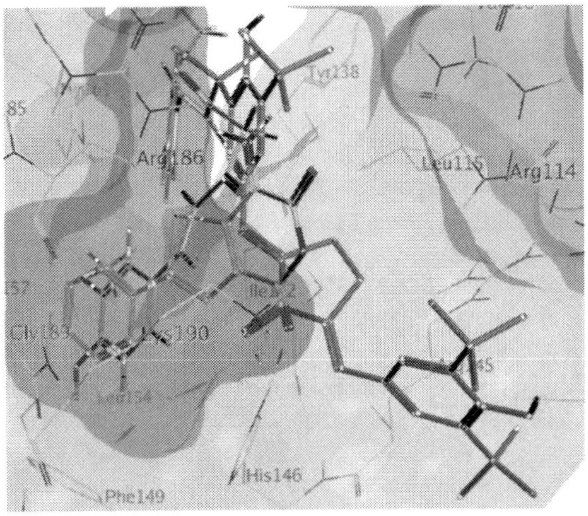

Figure 5 : The binding of Plasma Protein (PPB) Albumin with PDB 6WUW demonstrates interactions between albumin and the native ligand (green ribbon), curcumin (orange ribbons), and one of the analogs of 2,5-Dibenzylidene cyclopentanone (B14 shown in purple ribbon

In the docking simulation, an affinity is observed between curcumin and analogs of 2,5-dibenzylidene cyclopentanone, indicating specific interactions with the long-chain hydrocarbon substituent (B14) and His and Lys residues that form hydrophobic interactions with the hydroxyl group. This suggests a strong binding between these compounds (as shown in Figure 5 displaying PDB interactions with albumin) with a docking score representing a Gibbs free energy of -8.556 kcal/mol. Such interactions are not seen with curcumin, which demonstrates weak interactions with albumin.

Consistent with the theory, contrary to the value of VD, the screening results with ADMETLab 2.0 show that the binding ability of the test compounds to plasma proteins, both curcumin and analog 2,5-dibenzylidene cyclopentanone, is moderate to high.

Based on the theory in the equation of volume of distribution that uses the fraction of drug in the body as the multiplying factor and the fraction of free drug in the systemic circulation as the dividing factor, it can be concluded that drugs with strong binding to plasma proteins have low volume of distribution values. This is because the higher the binding between a compound and plasma proteins, the higher its fraction in the free form in the systemic circulation, resulting in lower volume of distribution values. This is consistent with the screening results from ADMETLab 2.0, where although all compounds fall within the optimal VD range (0.04-20 L/kg), these values are much closer to the lower limit than to the upper limit of the optimal VD range, in contrast to the close to 100% overall protein plasma binding. The strong binding of drugs to plasma proteins is often considered unfavorable as it reduces the drug's ability to distribute freely in the systemic circulation and reach its target. Drugs that bind strongly to plasma proteins tend to remain in a bound form and cannot easily transfer to target tissues, thereby potentially affecting the effectiveness of therapy. Drugs that bind strongly to plasma proteins are more likely to concentrate in the blood and other aqueous compartments, while distribution to non-aqueous tissues such as fat tissue or specific organs may be limited. Furthermore, if a drug is tightly bound to plasma proteins and difficult to release, it may have a longer half-life, leading to an increased risk of drug accumulation and undesirable side effects. However, for specific purposes, the binding ability to plasma proteins can be beneficial, such as helping to maintain a more stable therapeutic drug concentration in the blood, reducing the risk of rapid elimination through kidney filtration, and protecting the drug from rapid breakdown or elimination.

Based on the screening results with ADMETLab 2.0, the prediction of blood-brain barrier (BBB) penetration ability shows very low probabilities ranging from 0 to 0.3 (table 2), with a tendency for lower penetration compared to curcumin. In contrast, the prediction with ADMETsar 2.0 (table 3) shows that almost all compounds have BBB penetration probabilities above 0.5.

The ability of a drug to penetrate the BBB is not always a priority. Some drugs indeed need to penetrate the BBB to reach therapeutic targets in the brain, such as in the treatment of neurological disorders or mental disorders. Examples include drugs used in the treatment of epilepsy, depression, or Alzheimer's disease. For these drugs, the ability to cross the BBB becomes an important criterion for therapeutic effectiveness. However, not all drugs need to penetrate the BBB. Many drugs are designed to affect organs or tissues outside the brain, such as the heart, lungs, or other organs in the body. For these drugs, the BBB acts as a desired barrier, protecting the brain from side effects or potential damage caused by the drugs.

The physicochemical factors that influence the ability of a compound to penetrate the Blood-Brain Barrier (BBB) are molecular size, polarity, and lipophilicity. Compounds with small molecular size, non-polar (lipophilic) nature, and the ability to easily cross cell membranes tend to have better BBB penetration. On the other hand, compounds with large molecular size, polar nature, and low lipophilicity are more likely to face difficulties in crossing the BBB due to the barrier formed by the brain endothelial cells, which restricts the penetration of polar and large molecules from the bloodstream to the brain. Due to the majority of analogs of 2,5-Dibenzylidene Cyclopentanone being compounds with low polarity (logP>5), theoretically, the screening results would indicate that these compounds are more likely to easily penetrate the Blood-Brain Barrier (BBB). Therefore, based on these physicochemical properties, the predictive results from ADMETsar 2.0, which show high BBB penetration probability (probabilities>0.5), are more accurate compared to ADMETLab 2.0, which indicates probabilities not exceeding 0.3 for all test compounds.

3.4 Metabolism

Metabolism is the process of chemical modification of a drug within the body. It refers to the biotransformation of compounds in drugs so that they can be eliminated more easily. Most drug metabolism processes occur in the liver, as the enzymes that facilitate drug metabolism reactions are concentrated there [9].

Phase I metabolism is a series of reactions where functional groups on the drug are exposed for the first time, aiming to increase the compound's polarity. Phase I metabolism occurs in various tissues, with hepatic circulation being the main pathway. Other tissues involved include gastrointestinal epithelium, kidneys, skin, and lungs. Within the cells, most phase I enzymes are located in the endoplasmic reticulum, which is rich in microsomes [9] [10].

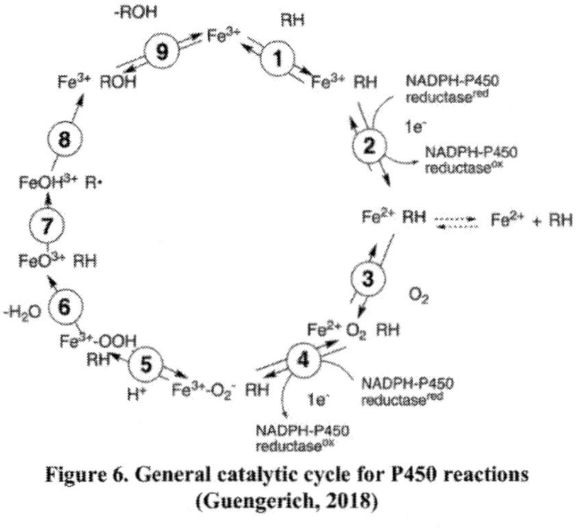

Figure 6. General catalytic cycle for P450 reactions (Guengerich, 2018)

The major catalyst in phase I metabolism reactions is the cytochrome P450 system (figure 6), a large family of membrane-bound enzymes found in the endoplasmic reticulum of hepatocytes. These enzymes are regulated by constitutive (relatively constant abundance regardless of substrate presence) and inducible (activated under specific conditions) regulation. The same gene sequence is used to classify cytochrome P450 enzymes, using family numbers (numeric notation) and subfamily letters (alphabetic notation) to distinguish between isoforms or individual enzymes. Subtypes of cytochrome P450 include CYP1A1/2, CYP1B1, CYP2A6, CYP2B6, CYP2C8, CYP2C9, CYP2C19, CYP2D6, CYP2E1, CYP3A4/5/7. The most common metabolic pathways involve CYP3A4/5, CYP2C9, CYP2D6, and CYP2C19. CYP3A4 is the most abundant isoform in the human liver and contributes to about 50% of all CYP450 activities [9] [11].

In the screening conducted, the ability of test compounds to bind as substrates to P450 is considered a desired pharmacokinetic property because the prediction indicates the likelihood of the compounds being metabolized and eliminated. Conversely, the ability to inhibit P450 is considered an undesired property because P450 inhibition can lead to various drug interaction issues, reduced drug effectiveness, and drug accumulation with potentially toxic properties due to the decreased ability of P450 to metabolize various types of drugs and other compounds [12].

Both the substrate and inhibitor abilities cannot be determined solely based on physicochemical properties. This is because the enzyme-compound complex is a ligand-receptor interaction that is specific to the chemical and stereochemical interactions of the compound with the active site of the enzyme[13]. Some screening results also indicate that the tested compounds can act as both inhibitors and substrates. This is strongly suspected to be related to the mechanism of competitive inhibition. However, further research is needed to determine the mechanism of action as a substrate or inhibitor.

The screening results with ADMETsar 2.0 and ADMETlab 2.0 show different results when considering each metabolic parameter, specifically isoforms of P450 (CYP1A1, CYP2C19, CYP2C9, CYP2D6, CYP3A4). For example, the screening results with ADMETlab 2.0 indicate that the ability of 2,5-dibenzylidenecyclopentanone analogs as CYP2C9 substrates is generally good but not better than curcumin. On the other hand, the screening results with ADMETsar

2.0 indicate that 8 compounds have better CYP2C9 substrate abilities compared to curcumin.
Another example is the prediction of CYP3A4 inhibition. The screening results with ADMETlab 2.0 indicate that 2,5-dibenzylidenecyclopentanone analogs tend to inhibit CYP3A4 but not are stronger than curcumin. However, the opposite results are found in the prediction results with ADMETsar 2.0, where almost all 2,5-dibenzylidenecyclopentanone analogs have stronger inhibitory abilities than curcumin.

Based on their structures, the differences in substrate-inhibition potency can be studied for substituents attached to the aromatic ring of analogs of 2,5-dibenzylidene cyclopentanone [14]. Isoform P450 specifically oxidizes aromatic groups through hydroxylation at the ortho, meta, and para positions of the aromatic ring (figure 7).

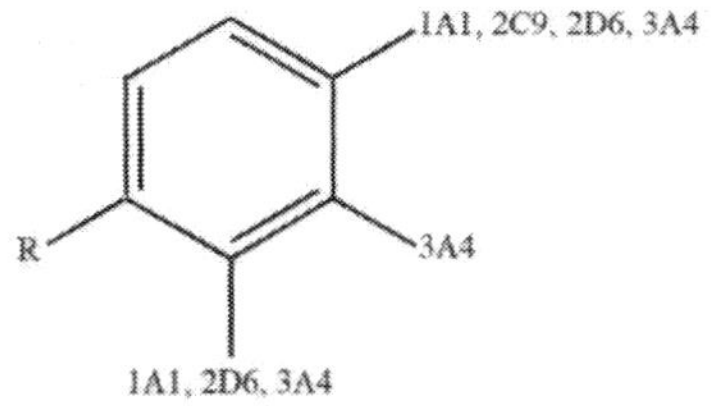

Figure 7 . The hydroxylation positions of P450 isoforms on the aromatic ring

Specifically, CYP1A1 and 1A2 tend to oxidize aromatic groups at the ortho and para positions. These isoforms are often involved in the metabolism of polycyclic aromatic compounds, such as benzo[a]pyrene and other polycyclic aromatics [15]. CYP2C9 tends to oxidize aromatic groups at the para position, as demonstrated by its ability to oxidize warfarin at the para position of its aromatic ring [16]. CYP2D6 tends to oxidize aromatic groups at the ortho and para positions. For example, CYP2D6 can oxidize substrates such as debrisoquine at the ortho and para positions of its aromatic ring [17]. Among the isoforms of P450, CYP3A4 has the broadest range of oxidation indices due to its ability to hydroxylate at the ortho, meta, and para positions [18].
Among the analogs of 2,5-dibenzylidene cyclopentanone, the compound most similar to curcumin is PGV-0 (Pentagamavunon-0) or in this study referred to as the compound with code B1. The similarity to curcumin is evident from the presence of one methoxy substituent at the meta position and one hydroxyl substituent at the para position. Theoretically, if there are no substituents on the aromatic positions of the aromatic ring and only hydrogen is present (as in the case of compound B2), its ability to act as a substrate would be higher, and its ability to act as an inhibitor would be lower compared to other compounds that have substituents on their aromatic rings. Screening results with ADMETLab 2.0 showed that compound B2, without any substituents on its aromatic ring, actually has a higher probability of inhibition compared to other compounds. In contrast, screening results using ADMETsar 2.0 showed that the inhibition ability of compound B2 is lower than other compounds. This indicates that the predictions for the metabolism parameter conducted by ADMETsar 2.0 are more accurate according to theory compared to ADMETLab 2.0. However, despite the differences between the two servers, there is one common observation: the tendency to be a substrate is lower than curcumin, and the inhibition ability is generally higher than curcumin. Curcumin is a natural compound, and natural compounds generally have high P450 inhibition potency. The screening results for analogs of 2,5-dibenzylidene cyclopentanone show unexpected outcomes, with a higher inhibition potency and lower substrate capability compared to curcumin.

Although there are differences between the two servers, there is a general trend of lower substrate ability compared to curcumin and higher inhibition ability compared to curcumin. Curcumin is a natural compound and natural compounds generally have relatively high P450 inhibition capabilities [19]. The screening results for 2,5-dibenzylidenecyclopentanone analogs, on the other hand, show unexpected results with a tendency for higher inhibitory potential and lower substrate ability.

In this study, molecular docking was performed using the MOE application version 2015 [20] between the ligand in the form of curcumin and the 17 analogs of 2,5-dibenzylidene cyclopentanone with CYP3A4, supporting the results of ADME screening for phase I metabolism parameters. CYP3A4 was employed as the receptor due to its role in catalyzing a significant portion of phase I metabolism in various drugs and being responsible for internal biotransformation, including hormones and various physiological glands. CYP3A4 is a P450 enzyme with the uniqueness of hydroxylating all positions of the aromatic carbons (ortho, meta, and para), thus rendering it widely catalytic. Through validation protocols, the visualization of interactions between the CYP3A4 receptor and the test ligands was obtained.

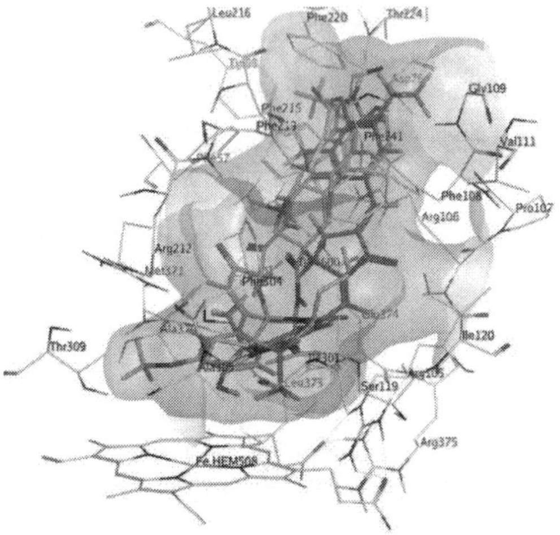

Figure 8. The docking results of the CYP3A4 receptor (3UA1) with curcumin (orange ribbon) and compound B14 (purple ribbon) in the best pose (visualized by MOE 2015)

By means of validated docking scores, predictive pIC50 values were obtained, indicating the inhibitory ability of CYP3A4 by curcumin and the 2,5-dibenzylidene cyclopentanone analogs (figure 8) (table 4).

Table 4. Docking scores of curcumin and 2,5-dibenzylidene cyclopentanone analogs along with the conversion to pIC50 values based on the validated GBV/WSA DG scoring equation.

Ligand	Dock. Score	pIC50	IC50 (μM)
Crc.	-8,2569	4,3230	47,5335
B0	-6,0997	1,7035	19794,5993
B1	-7,3219	3,1876	649,2194
B2	-6,1109	1,7170	19184,8774
B3	-6,1417	1,7545	17599,8370
B4	-6,8971	2,6717	2129,4751
B5	-6,8657	2,6337	2324,5849
B6	-8,1322	4,1716	67,3541
B7	-7,1156	2,9371	1155,7623
B8	-7,4862	3,3872	410,0467
B9	-6,7166	2,4526	3526,8799
B10	-6,5910	2,3001	5010,9835
B11	-7,4846	3,3852	411,9525
B12	-8,6074	4,7486	17,8407
B13	-9,2161	5,4878	3,2526
B14	-8,6550	4,8064	15,6162
B15	-8,5128	4,6338	23,2384
B16	-7,4121	3,2972	504,4404

Potency of Inhibitory Concentration at 50% (pIC50) is a logarithmic measure used to represent the potency or inhibitory activity of a substance, typically a drug or chemical compound, in inhibiting a specific biological target, such as an enzyme or receptor [21]. It quantifies the concentration of the substance required to achieve 50% inhibition of the target's activity. A lower pIC50 value indicates higher potency, meaning that a lower concentration of the substance is needed to achieve the desired inhibition. Out of the 18 test ligands, all compounds showed predictive pIC50 values. The highest predicted pIC50 values were observed for analogs compounds with codes B13, B14, B12, B15, followed by curcumin as the fifth compound with the highest pIC50 value.

Compounds with codes B13, B14, and B12 are analogs of 2,5-dibenzylidene cyclopentanone with alkyl hydrocarbon and hydroxyl substituents covering the ortho, meta, and para positions on the aromatic ring. The presence of these substituents, which exhibit the highest inhibition, is associated with the obstruction of CYP3A4 hydroxylation at those positions. On the other hand, the lowest pIC50 values were observed for compounds with codes B2 and B0, which only have hydrogen substituents and one hydroxyl group. This is associated with the availability of CYP3A4 hydroxylation space in the aromatic area of these compounds. The predictive pIC50 values obtained indicate that most of the 2,5-dibenzylidene cyclopentanone analog compounds are still inclined to inhibit CYP3A4. This result correlates positively with the screening results using both ADME online servers, which were used to predict the probability of inhibition of the test compounds against the P450 isoform. Therefore, further studies are needed regarding the inhibition potential and synthesis design of other mono-ketone analogs of curcumin, taking into consideration their pharmacokinetic factors.

Based on the research conducted by Kaur et al. in 2016 on the interaction of ritonavir with CYP3A4 as an inhibitor through pharmacophore modeling, the presence of Phe (phenylalanine) residue in the hydrophobic interaction near the heme as a key amino acid residue was identified as a marker of the inhibitory interaction by the ligand to the CYP3A4 receptor. Upon validation with a known ligand, in ligand X6V (tert-butyl [(2S)-1-(1H-indol-3-yl)-3- {[(2R)-1-oxo-3-phenyl-1- {[3-(pyridin-3-

yl) propyl] amino} propan-2- yl]sulfanyl} propan-2- yl]carbamate) (figure 9), a similar interaction with the CYP3A4 receptor 3UA1 was obtained [22].

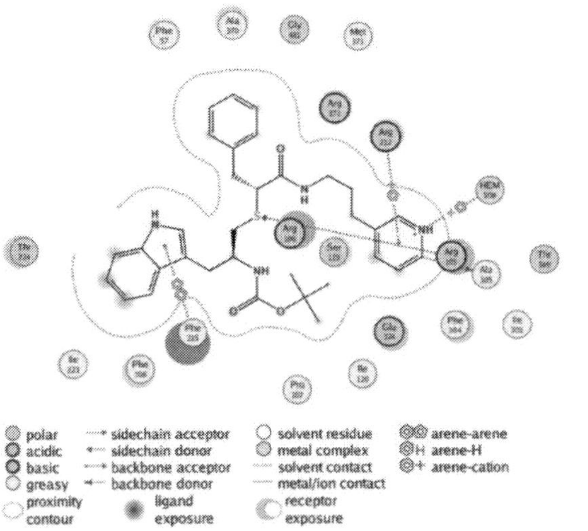

Figure 9. The known ligand X6V from PDB 7KVN interacts with CYP3A4 (3UA1) through aromatic interactions involving Phe (phenylalanine) and HEM-N (heme nitrogen) residues

In curcumin (figure 10) the observed inhibiton interaction involves the aromatic groups of kurkumin interacting with HEM (P450 cofactor).

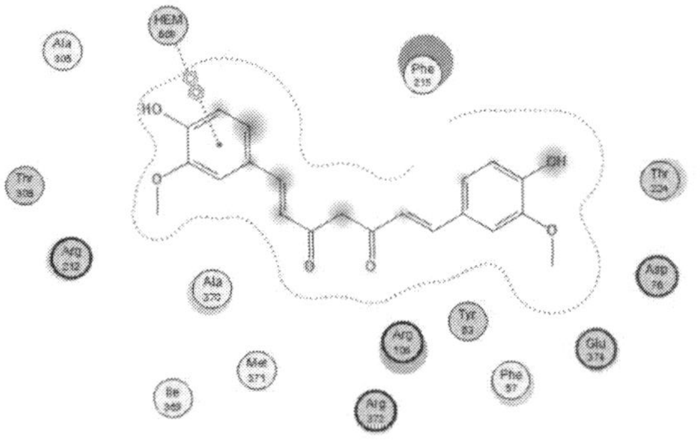

Figure 10. The interaction observed between CYP3A4 (3UA1) and kurkumin involves the HEM (P450 cofactor) interacting with the aromatic groups of kurkumin

In the compound B2 (figure 11), which lacks substituents, the observed inhibitory interaction involves the aromatic groups of B2 interacting with HEM (P450 cofactor).

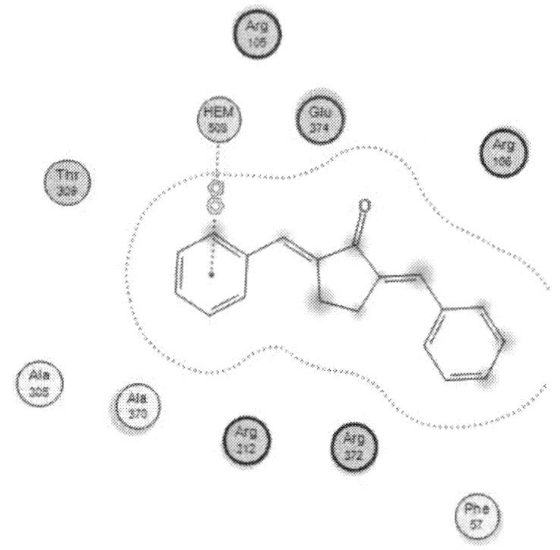

Figure 11. The interaction observed between CYP3A4 (3UA1) and B2 (analog without substituents) involves the HEM (P450 cofactor) interacting with the aromatic groups of B2

In the compound B1 (figure 12), commonly known as PGV-0 (Pentagamavunon-0) with the same substituents as kurkumin but in the form of a cyclic monocarbonyl, the observed inhibitory interaction involves HEM (P450 cofactor) interacting with the aromatic structure.

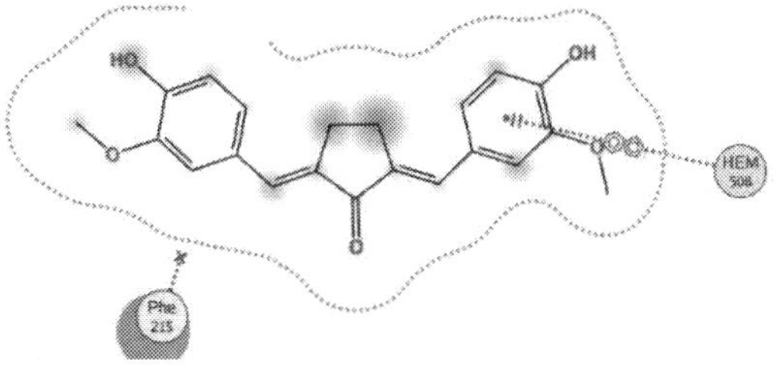

Figure 12. The interaction observed between CYP3A4 (3UA1) and B1 (PGV-0) involves the HEM (P450 cofactor) interacting with the aromatic structure of B1 (PGV-0).

In the compound B14, one of the compounds with the highest predictive pIC50 value, the interaction observed is similar to the known ligand X6V, involving the interaction between the Phe residue and the aromatic structure (figure 25).

3.5 Elimination Capability

The elimination parameters examined in this study are clearance (Cl) and half-life (T1/2). Both parameters are inversely related to the drug. Clearance is a pharmacological parameter that quantifies the irreversible release rate of a drug from the body[23]. Since drugs can be eliminated from the body by various organs (kidneys, liver, lungs, saliva, mammary glands, etc.), different types of clearance are distinguished based on these organs. Some of them include renal clearance (CLr) and hepatic clearance (CLh) [24]. On the other hand, the half-life is the time required for the drug concentration (usually in blood or plasma) to decrease by half from its initial value [25].

Among the three ADME screening online servers, only ADMETLab 2.0 provides data on clearance (klirens) and half-life (t1/2) as elimination parameters. The lower the T1/2 value, the shorter the elimination time. Conversely, the lower the clearance value, the longer the elimination time from the body. Screening results with ADMETLab 2.0 (table 2) show lower T1/2 values compared to curcumin. However, the clearance values of some 2,5-Dibenzylidene cyclopentanone analogs are lower than curcumin, including compounds B3, B6, B9, B12, B13, B14, and B16. These low clearance values are associated with the presence of hydrocarbon chain substituents in compounds such as B6 (t-C4H9 substituent), B12 (two C2H5 substituents), B13 (two i-C3H7 substituents), and B14 (two i-C4H9 substituents). It is strongly suspected that the lower solubility of these nonpolar substituents in water makes it difficult for them to dissolve during glomerular filtration. Additionally, the presence of long-chain hydrocarbon substituents results in a slower metabolism, higher binding to plasma proteins, and stronger binding to fatty tissues, leading to longer elimination times (clearance). However, contrary to the nonpolar-hydrophobic logic, the presence of halide substituents, such as Cl in compounds B3 and B16, actually shows low clearance values. This is strongly suspected to be associated with a series of pharmacological processes in the body that render these elements less polar and unable to have high clearance abilities, such as protonation (hydrogen bonding), electrostatic interactions with oppositely charged groups like carboxylate (-COO), and formation of double bonds. These chemical processes are believed to originate from enzymatic mechanisms and various other physiological processes in the body. These physiological processes can involve ion exchange and transport, which involve various specific transporters and ion channels present in cell membranes in various tissues and organs, such as the kidneys, inteFstines, and nervous system [26] [27]. One of the physiological processes strongly suspected is the first-phase metabolism by the P450 complex, which oxidizes drugs through hydroxylation of aromatic groups to increase the drug's polarity. The presence of halide as a substituent on the aromatic group in some analogs of 2,5-dibenzylidene cyclopentanone leads to the inhibition of hydroxylation by P450 [28]. This results in the inhibition of hydroxylation, causing the drug to become less polar and unable to be rapidly released from the hydrophilic fraction of the body. As a consequence, the clearance value of these compounds becomes low.

4 CONCLUSIONS

The results of pharmacokinetic character screening for 2,5-dibenzylidene cyclopentanone analogs indicate various predictive characteristics related to the physicochemical properties and ligand-receptor interaction abilities between the test compounds and proteins involved in drug pharmacokinetics in the body. Generally, the pharmacokinetic properties of 2,5-dibenzylidene cyclopentanone analogs that show improvements in drug-likeness compared to curcumin are parameters commonly related to physicochemical properties such as human intestine absorption (HIA), BBB permeability, Volume of Distribution (VD), total clearance value (Cl), and half-life (T1/2). On the other hand, the pharmacokinetic properties that tend to decrease are ligand-receptor interactions, such as an increasing trend in P-Glycoprotein inhibition, strong binding with plasma proteins (Plasma Protein Binding/PPB), and a higher probability of predictive P450 inhibition compared to curcumin. The interactions with BPP and P-glycoprotein are qualitatively demonstrated through interactions with key amino acid residues via molecular docking simulations. The docking process conducted is a validated process, including ligand and protein preparation, pose and scoring validation, and docking process following the validated results. he docking results of the CYP3A4 receptor with curcumin and 2,5-dibenzylidene cyclopentanone analogs indicate a predictive inhibitory ability, necessitating further in vitro or in vivo research to validate and provide a basis for the synthesis design of subsequent curcumin derivatives.

Although the in silico methods used in this study provide comprehensive pharmacokinetic properties and are reasonably correlated with the theoretical physicochemical properties of the drugs, these various parameters need to be validated in vitro and in vivo to prove their accuracy in the real physiological environment of the body. The in silico profiles developed in this study are expected to streamline future research, but further validation through experimental studies is still necessary.

REFERENCES

1. Zadorozhnii, P. V., Kiselev, V. V., & Kharchenko, A. V. (2022). In Silico ADME Profiling of Salubrinal and Its Analogues. Future Pharmacol, 2(2), 160-197. https://doi.org/10.3390/futurepharmacol2020013
2. Kiriiri, G.K., Njogu, P.M. & Mwangi, A.N. Exploring different approaches to improve the success of drug discovery and development projects: a review. Futur J Pharm Sci 6, 27 (2020). https://doi.org/10.1186/s43094-020-00047-9
3. Liang, R., Yuan, H., Xi, G., & Zhou, Q. (2009). Synthesis of wheat straw-gpoly (acrylic acid) superabsorbent composites and release of urea from it. Carbohydrate Polymers, 77(2), 181-187.
4. Xiong G, Wu Z, Yi J, Fu L, Yang Z, Hsieh C, Yin M, Zeng X, Wu C, Lu A, Chen X, Hou T, Cao D. ADMETlab 2.0: an integrated online platform for accurate and comprehensive predictions of ADMET properties. Nucleic Acids Res. 2021 Jul 2;49(W1):W5-W14. doi: 10.1093/nar/gkab255. PMID: 33893803; PMCID: PMC8262709.
5. Cheng F, Li W, Zhou Y, Shen J, Wu Z, Liu G, Lee PW, Tang Y. admetSAR: a comprehensive source and free tool for assessment of chemical ADMET properties. J Chem Inf Model. 2012 Nov 26;52(11):3099-105. doi: 10.1021/ci300367a. Epub 2012 Nov 1. Erratum in: J Chem Inf Model. 2019 Nov 25;59(11):4959. PMID: 23092397.
6. Sardjiman, S, MS Reksohadiprodjo., L Hakim., H Van der Goot, H Timmermen.1997.1,5-Diphenyl-1,4-pentadiene-3-ones and cyclic analogs as antioxidative agents.Synthesis and structure-activity relationship.Eur J Med Chem.Paris.
7. Doak, B. C., Over, B., Giordanetto, F., & Kihlberg, J. (2014). Oral Druggable Space beyond the Rule of 5: Insights from Drugs and Clinical Candidates. Chemical Biology, 21(9), 1115-1142. https://doi.org/10.1016/j.chembiol.2014.08.013
8. Veber, D. F., Johnson, S. R., Cheng, H. Y., Smith, B. R., Ward, K. W., & Kopple, K. D. (2002). Molecular properties that influence the oral bioavailability of drug candidates. Journal of Medicinal Chemistry, 45(12), 2615-2623. https://doi.org/10.1021/jm020017n
9. Phang-Lyn, S., & Llerena, V. A. (2023). Biochemistry, Biotransformation. In Dr. Kiran C. Patel College of Medicine & Joe DiMaggio Children's Hospital. Retrieved August 14, 2023.
10. Deodhar, M., Al Rihani, S. B., Arwood, M. J., Darakjian, L., Dow, P., Turgeon, J., & Michaud, V. (Tahun publikasi belum tersedia). Mechanisms of CYP450 Inhibition: Understanding Drug-Drug Interactions Due to Mechanism-Based Inhibition in Clinical Practice. Pharmaceutics, 12(9), 846. https://doi.org/10.3390/pharmaceutics12090846
11. Gilani, B., & Cassagnol, M. (2023). Biochemistry, Cytochrome P450. WakeMed Health & Hospitals & St. John's University. Last updated on April 24, 2023.
12. Gong, Y., Haque, S., Chowdhury, P., Cory, T. J., Kodidela, S., Yallapu, M. M., Norwood, J. M., & Kumar, S. (2019). Pharmacokinetics and pharmacodynamics of cytochrome P450 inhibitors for HIV treatment. Expert Opinion on Drug Metabolism & Toxicology, 15(5), 417-427. doi: 10.1080/17425255.2019.1604685
13. Selvaraj, C., Rudhra, O., Alothaim, A. S., Alkhanani, M., & Singh, S. K. (2022). Structure and chemistry of enzymatic active sites that play a role in the switch and conformation mechanism. In Advances in Protein Chemistry and Structural Biology (Vol. 130, pp. 59-83). https://doi.org/10.1016/bs.apcsb.2022.02.002
14. Bosak, A., Ramić, A., Šmidlehner, T., Hrenar, T., Primožič, I., & Kovarik, Z. (2018). Design and evaluation of selective butyrylcholinesterase inhibitors based on Cinchona alkaloid scaffold. PLOS ONE, 13(10), e0205193. https://doi.org/10.1371/journal.pone.0205193
15. Myers, S. R., Ali, M. Y., Wright, T., & Cunningham, C. (2007). BENZO(A)PYRENE METABOLISM: ROLE OF BIOALKYLATION. Journal Title, Volume(Issue), Pages. https://doi.org/10.1080/10406630701509104
16. Daly, A. K., Rettie, A. E., Fowler, D. M., & Miners, J. O. (2018). Pharmacogenomics of CYP2C9: Functional and Clinical Considerations. Journal of Personalized Medicine, 8(1), 1. https://doi.org/10.3390/jpm8010001
17. Llerena, A., Dorado, P., & Peñas-Lledó, E. M. (2009). Pharmacogenetics of debrisoquine and its use as a marker for CYP2D6 hydroxylation capacity. Pharmacogenomics, 10(1), 17-28. https://doi.org/10.2217/14622416.10.1.17

18. Sevrioukova, I. F., & Poulos, T. L. (2013). Understanding The Mechanism Of Cytochrome P450 3a4: Recent Advances And Remaining Problems. Dalton Transactions, 42(9), 3116-3126. https://doi.org/10.1039/c2dt31833d
19. Sharifi-Rad, J., El Rayess, Y., Abi Rizk, A., Sadaka, C., Zgheib, R., Zam, W., Sestito, S., Rapposelli, S., Neffe-Skocińska, K., Zielińska, D., Salehi, B., Setzer, W. N., Dosoky, N. S., Taheri, Y., El Beyrouthy, M., Martorell, M., Ostrander, E. A., Suleria, H. A. R., Cho, W. C., Maroyi, A., & Martins, N. (2020). Turmeric and Its Major Compound Curcumin on Health: Bioactive Effects and Safety Profiles for Food, Pharmaceutical, Biotechnological and Medicinal Applications. Frontiers in Pharmacology, 11, 01021. https://doi.org/10.3389/fphar.2020.01021
20. Vilar, S., Cozza, G., & Moro, S. (2008). Medicinal chemistry and the molecular operating environment (MOE): application of QSAR and molecular docking to drug discovery. Current Topics in Medicinal Chemistry, 8(18), 1555-1572. https://doi.org/10.2174/156802608786786624
21. Caldwell, G. W., Yan, Z., Lang, W., & Masucci, J. A. (2012). The IC(50) concept revisited. Current Topics in Medicinal Chemistry, 12(11), 1282-1290. https://doi.org/10.2174/156802612800672844
22. Sevrioukova, I. F., & Poulos, T. L. (2012). Structural and mechanistic insights into the interaction of cytochrome P4503A4 with bromoergocryptine, a type I ligand. Journal of Biological Chemistry, 287(5), 3510-3517. https://doi.org/10.1074/jbc.M111.317081
23. Horde, G. W., & Gupta, V. (2023). Drug Clearance. In StatPearls [Internet]. Treasure Island (FL): StatPearls Publishing. Retrieved June 20, 2023, from https://www.ncbi.nlm.nih.gov/books/NBK557758/
24. Miners, J. O., Yang, X., Knights, K. M., & Zhang, L. (2017). The Role of the Kidney in Drug Elimination: Transport, Metabolism, and the Impact of Kidney Disease on Drug Clearance. Clinical Pharmacology and Therapeutics, 102(3), 436-449. https://doi.org/10.1002/cpt.757
25. Smith, D. A., Beaumont, K., Maurer, T. S., & Di, L. (2015). Volume of Distribution in Drug Design. Journal of Medicinal Chemistry, 58(15), 5691-5698. https://doi.org/10.1021/acs.jmedchem.5b00201
26. Lu, X., Luo, C., Wu, J., Deng, Y., Mu, X., Zhang, T., Yang, X., Liu, Q., Li, Z., Tang, S., Hu, Y., Du, Q., Xu, J., & Xie, R. (2023). Ion channels and transporters regulate nutrient absorption in health and disease. Journal of Cellular and Molecular Medicine, 28 August 2023. https://doi.org/10.1111/jcmm.17853
27. Drozdzik, M., Drozdzik, M., & Oswald, S. (2021). Membrane Carriers and Transporters in Kidney Physiology and Disease. Biomedicines, 9(4), 426. https://doi.org/10.3390/biomedicines9040426
28. Grimm, S. W., Einolf, H. J., Hall, S. D., & He, K. (May 2009). The Conduct of in Vitro Studies to Address Time-Dependent Inhibition of Drug-Metabolizing Enzymes: A Perspective of the Pharmaceutical Research and Manufacturers of America. Drug Metabolism and Disposition: the Biological Fate of Chemicals, 37(7), 1355-1370. https://doi.org/10.1124/dmd.109.026716

A study of consumer behavior and willingness to pay towards cosmetic products of generation Z in Hochiminh city

Hien Thi Bich Tran [1,#,*], *Phuong Ngoc Duy Nguyen*[2,#], *Trung Quang* Vo[1], *Viet Nhu Nguyen*[1], *Thao Ho Dieu Nguyen*[1], *Susi Ari* Kristina[2], *Dwi* Endarti[2]

[1] Faculty of Pharmacy, Pham Ngoc Thach University of Medicine, Ho Chi Minh City 700000, Vietnam.
[2] Faculty of Pharmacy, Universitas Gadjah Mada, Yogyakarta 55281, Indonesia.
[#] These authors are the co-first authors and contributed equally to this work.

Abstract. The global cosmetics industry is experiencing robust growth and Generation Z (Gen Z) is a potential customer source of this market. This research examines customers' intentions to purchase cosmetic goods of Gen Z in Hochiminh city using perceived value factors as the antecedents of attitude in the model. A cross-sectional study, applying a convenient sampling method, was conducted to collect data from Gen Z people in January 2023. There were no specific cosmetics products included in the study. Descriptive analysis and Partial Least Square (PLS) method of Structural Equation Modeling (SEM) with SmartPLS 4.0.8.7 software were used to analyze the research data. The PLS-SEM analysis of 723 responses showed that attitudes toward purchasing cosmetic products are significantly positively impacted by perceived environmental value. A more positive attitude would result from increased brand credibility and product understanding of cosmetics. There was no evidence to support the impact of other perceived values (specifically, health, safety, social, spiritual, and ethical) on attitudes toward consumer behavior. Attitude was important in predicting willingness to pay (both direct and indirect). This study helps industry professionals to advance the qualities of cosmetic products by increasing and improving environmental value, product knowledge, and brand credibility. The eco-friendly pattern of production and marketing strategies focusing on product ingredients, manufacturing processes, and quality standards are necessary to enhance brand credibility and product understanding, align with consumer preferences and achieve greater success in the market.
Keywords: Consumer behavior, cosmetic products, generation Z, Hochiminh city, willingness to pay.

1 Introduction

Humans are more and more increasingly interested in aesthetic factors, always striving for perfection to express themselves as well as adapt to modern society [1-3]. Therefore, cosmetics were invented and quickly became an essential need of every person [4]. According to the Food and Drug Administration (FDA) of the United States, "a cosmetic is a substance intended to be applied to the human body for cleansing, beautifying, enhancing attractiveness, or changing appearance" (pure soap excluded) [5].

The cosmetics industry is a thriving global market. The cosmetics industry brought in $532 billion in total revenues in 2019, up 5.5% from the prediction in sales in 2018 [6]. Asia Pacific had the largest market share in terms of cosmetics in 2019, accounting for about 41% of the worldwide market. The leading and most influential countries in Asia are such as Japan, China and South Korea.

Vietnam is quickly catching up to the major markets in the region for cosmetics and beauty products. Vietnamese cosmetics generated $94.9 million in total revenue in 2018. Through 2019-2023, the market is anticipated to rise by an average of 6.1% per year [7]. A poll conducted in 2020 found that the majority of customers in the nation possessed lipsticks, with about 30% of Vietnamese women wearing makeup daily. Female customers utilized face care goods like face masks and face cleansers the most, and they made up the largest skincare category in terms of sales in the nation. Compare with nearby nations in the area like Singapore and Thailand. Vietnam is a developing country with potential for growth that is astounding [8].

Vietnamese customers like utilizing items from different Asian sources, including such as Japan, China and South Korea. The East Asian country has also been Vietnam's biggest cosmetics import partner, with imports valued at over $263 million in 2021. Local companies haven't yet captured a sizable portion of the market for these items because of the intense competition and strong consumer demand for multinational names. The sales of beauty goods are still growing thanks to consumer demand that is on the rise and major marketing efforts by cosmetics firms. This is strikingly and conspicuously

* Corresponding email: hienttb@pnt.edu.vn

apparent in major cities like Hanoi, Ho Chi Minh City, and Da Nang where the number of high-class consumers is rising [7].

The current youth generation is referred to as Generation Z (Gen Z) by numerous demographic researchers. The Pew Research Center refers to those who were born between 1997 and 2012 as Gen Z. They were born and reared in a rapidly evolving, technologically sophisticated society. Their lives have been significantly impacted by the fact that they are the first generation to have grown up using the internet and social media [9]. Gen Z has grown up in the era of social media and is highly connected online. Platforms like Instagram, TikTok, and YouTube have become influential spaces for beauty content. Gen Z's exposure to beauty influencers and online tutorials has led to increased interest and experimentation with cosmetics [10]. Gen Z is known to pay more attention to their appearance through routine skin care, so most of their spending is used to shop for cosmetic products [11]. Therefore, the increasing number of cosmetic marketers had started to target this generation. It was mentioned that Generation Z has $43 billion in spending power while also influencing an additional $600 billion of family spending. Their influence comes from new media, virtual friends and technology's power [12]. They account for 33% of the world population, and 21% of the population of Vietnam [13]. It is shown that Gen Z in Vietnam are potential customers of cosmetic brands.

In such a new and promising market, understanding the consumer behavior is the key for cosmetic companies to succeed in maintaining their brands. Consumer behavior refers to how individuals or groups select, obtain, utilize, and discard goods, services, concepts, or experiences in order to satiate their wants and requirements [14]. This means that the buying process not only begins with the purchase of the product, but even before the product is delivered. The buyer initiates the purchasing process, which then leads to the search for readily available substitutes with corresponding benefits and drawbacks. After that, the decision-making process for the purchase will start. A crucial component is post-purchase behavior, which informs marketers of the success or failure of the product. As a result, marketers will need to thoroughly investigate consumer demands and comprehend a variety of consumer needs [15]. In light of the fact that cosmetic products are typically perceived as promoting healthy and sustainable lifestyles, there are five consumer-perceived values related to health, safety, social values, hedonism, and the environment that can influence the attitude to purchase cosmetic products [16-18].

Together with desire to buy, willingness to pay is analyzed to determine whether customers are compelled to pay more for a good or service, in this case, natural and organic cosmetics. specific brand engines. The highest sum of money a consumer is ready to pay for a good or service is known as their "willingness to pay" (WTP). It is a typical indicator used in pricing research studies that enables companies to determine the best rates for their goods and services in order to draw in customers and increase revenues. The study's objective was to investigate the variables affecting Generation Z consumers' preferences for and willingness to pay for cosmetic items in Hochiminh City.

2 Materials and Methods

2.1 Study Design

A cross-sectional study was conducted among Gen Z in Hochiminh City in January 2023.

2.2 Study Subjects

Gen Z was the target demographic. This group was chosen on the basis that they are members of the public with discretionary income and are in charge of making decisions regarding the purchases of cosmetics.

The study included people who living in Hochiminh city; being born between 1997 and 2012; being able to read and understand Vietnamese; and not having any cognitive impairment. People refusing to participate in the study; not filling out the survey information completely; and choosing only one answer or playing zigzag were excluded.

The minimal sample size (N) was calculated using a single population proportion formula [19]. The following assumptions have been made: a 0.5 of positive attitudes toward cosmetic behavior (P), a 5% margin of error (d) and a 95% confidence interval ($Z_{\alpha/2}$=1.96).

$$N = \frac{\left(Z_{\frac{\alpha}{2}}\right)^2 \times P(1-P)}{d^2} = \frac{(1.96)^2 \times 0.5(1-0.5)}{0.05^2}$$
$$= 385$$

The researcher assumed a 10% non-response rate with a total of 423 respondents as the sample size. The researcher employed a stratified simple random probability sampling strategy for this study. The sample was obtained by stratifying Ho Chi Minh City's Gen Z population.

2.3 Survey Instrument

The questionnaire consisted of 70 questions and was broken down into seven main segments. Part 1 gathered personal characteristics such as gender, age, place to live, ethnicity, education category, marital status, occupation, average monthly personal income (million VND), purpose of cosmetic use, commonly used purchase channels, and purchase frequency. Part 2 collected consumer perceived value with 31 questions on the Likert-5 scale used to evaluate "consumer's perceived value" of cosmetic products. Part 3 assembled product knowledge including seven questions on the Likert-5 scale used to evaluate "consumer's product knowledge" of cosmetic products. Part 4 was about brand credibility with six questions on the Likert-5 scale used to evaluate "consumer's brand credibility" of cosmetic products. Part 5 assessed attitude towards consumer behavior with five questions on the Likert-5 scale used to evaluate

"Attitude towards consumer behavior" of cosmetic products. Part 6 focused on consumer purchase intention comprising seven questions on the Likert-5 scale used to evaluate "Consumer purchase intention" of cosmetic products. Part 7 considered willingness to pay including three questions on the Likert-5 scale used to evaluate "Willingness to pay" of cosmetic products. All answers were toward general cosmetic products, not specific.

Details are presented in **Appendix 01.** Each question was answered using a five-point Likert scale ranging from 1= strongly disagree, 2= disagree, 3= neutral, 4= agree, to 5= strongly agree [20]. The translation from English to Vietnamese followed the process recommended by the World Health Organization [21]. A modified Vietnamese version of a questionnaire served as the basis for the creation of a Google form (Google form). Before beginning to answer the major questions, participants must first sign a consent form that is located on the first page of the form. The main section's questions cannot be answered by those who select the "*I do not agree to participate in the study*" option; they will instead be taken to the Google form's final page. The key sections of the questionnaire's Vietnamese translation are represented on the form's subsequent pages.

2.4 Validation

A pilot test was conducted with a sample size of 60 participants to verify the questionnaire. Each person is asked about how they understand each question, explaining how to answer confusing or misleading words. All feedback about the questionnaire were acknowledged and questions were edited in terms of words to make it easier to understand and more suitable to Vietnamese culture.

The indicator loadings were examined in order to assess a reflective measurement model. In order to achieve appropriate item dependability, loadings above 0.708 should be sought after as they demonstrate that the construct accounts for more than 50% of the variation of the indicator [22]. Items such as CPI4, CPI5, CPI6, ENV5, ENV6, ENV7, SOV1, SPV4 were deleted because their outer loading was less than 0.708. All of the outside loadings, which range from 0.708 to 0.913, are shown in **Appendix 02**, demonstrating the accuracy of each measurement indicator.

Cronbach's alpha was calculated for each subscale and the overall score to measure internal consistency reliability. Reliability scores between 0.60 and 0.70 are considered "acceptable" in exploratory research, and reliability values between 0.70 and 0.90 are considered "sufficient to good." Values of 0.95 and above, however, present a problem because they demonstrate that the pieces are redundant and reduce construct validity [23, 24]. Cronbach's Alpha of 11 items is range from 0.864 to 0.915; with 4 items had Cronbach's Alpha value ≥ 0.900 (BCR - 0.908; ETC- 0.900; PKL - 0.913; SPV- 0.915) that are shown in **Appendix 03.** In addition, Composite Reliability value of 11 items range between 0.908 and 0.932; Average Variance Extracted value of 11 items is range from 0.851 to 0.915; and rho_a of 11 items is range from 0.851 to 0.915.

Convergent validity refers to how well a concept converges to take into account the variance of its constituent parts. The statistic used to evaluate the convergent validity of a concept is the average variance extracted (AVE) for all items on each construct. At least 0.50 must be the AVE. Every AVE score was higher than 0.50 (**Appendix 03**) that shows 50% or more of the variance of the construct's component parts.

Discriminant validity was assessed. The heterotrait-monotrait ratio (HTMT) of relationships was put out by Henseler et al. (2015) [25]. The HTMT is defined as the mean value of the item correlations across constructs, also known as the heterotrait-heteromethod correlations, as opposed to the (geometric) mean of the average correlations for the items measuring the same construct (i.e., the monotrait-heteromethod correlations). Problems with discriminant validity exist when HTMT values are high. Henseler et al. (2015) suggested a threshold value of 0.90 or 0.85 [25]. In addition to these recommendations, bootstrapping can be used to determine whether the HTMT value differs considerably from 1.00 or a lower threshold value like 0.85 or 0.90 [25]. As shown in **Appendix 04,** all HTMT between 0.417 and 0.815 demonstrated the study's ability to discriminate across items.

The Variance Inflation Factor (VIF) values were investigated to evaluate common method bias based on a thorough collinearity test of the model to complete the measurement model evaluation (**Appendix 05**). The collinearity among the constructs is not a significant problem in this study, according to the VIF values in this structural model, which was between 1.0 and 3.206. There was no significant multicollinearity that needed to be corrected.

2.5 Statistical Analysis

All collected data were entered into Microsoft Excel 2016 and analyzed using the Statistical Package for the Social Sciences (SPSS) version 20.0. Frequency and percentage were included in the descriptive analysis utilized in this study. The study was carried out using the Partial Least Square (PLS) method of Structural Equation Modeling (SEM), as the technique is appropriate for theoretical causal models [22]. Because standard regression does not distinguish between direct effect and indirect effect when assessing mediation relationships, PLS-SEM is notable for organizing the mediation effects using several indicators concurrently [26]. The duality between the academic research explanation and prediction for managerial consequences is resolved by the causal-predictive technique of PLS-SEM because the current investigation may lack a thorough foundation of established theories and proofs [22]. The study used SmartPLS software version 4.0.8.7 to analyze PLS-SEM.

To evaluate the structural model, it was proposed to use the Coefficient of Determination (R^2) Beta and corresponding t-values. The latent variable's R^2 value of 0.75 may be characterized as significant, 0.5 as moderate, and 0.25 as tiny. Also recommended to be added to the fundamental assessment were predictive relevance (Q^2) and impact size (f^2) [22]. The Q^2 prediction, which is comparable to the evaluation of Q^2 values acquired by the blindfolding process in earlier PLS-SEM value, needs to be assessed via PLS predict in addition to the size of the R^2 values as criteria for predictive accuracy. If the result is greater than 0, the external constructions have predictive value for the endogenous construct being studied [27]. Values of f^2 of 0.02, 0.15, and 0.35, respectively, reflect modest, medium, and substantial impacts of the exogenous latent variables [22]. In order to evaluate the hypothesized associations, the presented hypotheses were then evaluated using the bootstrap sampling method of 5,000 subsamples with and without the control variables.

3 Results

From the entire sample of 723 participants, the respondent profile (**Table 1**) included 55.3% female respondents and 44.7% male respondents. The majority of them were unemployed (82.6%), unmarried (96.1%), and had a low income. Only 26.6% have bought cosmetics less than 1 time/year. The remaining often shopped for beauty products, even more than 4 times/year (23.0%).

As shown in **Table 2**, the structural model explained 58.6 % of the variance in Attitude Toward Consumer Behavior (ATCB) and about 43% of the variance in Consumer Purchase Intention (CPI) and Willingness To Pay (WTP), which can be considered moderate. The Q^2 value was considerably above zero, with 0.574, 0.509,

Table 1. Demographic characteristics

Variables	N (%)	Variables	N (%)
Age		**Marital Status**	
≤22	484 (66.9)	Not Married	695 (96.1)
23-26	239 (33.1)	Married	28 (3.9)
Gender		**Occupation**	
Male	323 (44.7)	Not Working	597 (82.6)
Female	400 (55.3)	Working	126 (17.4)
Location		**Income (million VND per month)**	
Urban	593 (82.0)	No income	252 (34.9)
Rural	130 (18.0)	Under 4.5	257 (35.5)
Ethnicity		≥ 4.5	214 (29.6)
Kinh	696 (96.3)	**Purchase Frequency (time(s)/year)**	
Other	27 (3.7)	< 1	192 (26.6)
Education Level		1 - 3	263 (36.4)
High school or below	93 (12.9)	3 - 4	102 (14.1)
University	615 (85.1)	> 4	166 (23.0)
Postgraduate	15 (2.1)		

Table 2. Results of R^2 and Q^2

	R^2 adjusted	Q^2 predict	Interpretation
ATTITUDE TOWARD CONSUMER BEHAVIOR	0.586	0.574	Moderate
CONSUMER PURCHASE INTENTION	0.430	0.509	Moderate
WILLINGNESS TO PAY	0.422	0.321	Moderate

Table 3. Result of Effect Size

	f^2	Interpretation		f^2	Interpretation
ATCB → CPI	0.757	Large	HEV → ATCB	0.005	-
ATCB → WTP	0.039	Small	PKL → ATCB	0.068	Small
BCR → ATCB	0.156	Medium	SAV → ATCB	0.004	-
CPI → WTP	0.251	Medium	SOV → ATCB	0.001	-
ENV → ATCB	0.013	-	SPV → ATCB	0.001	-
ETC → ATCB	0	-			

ATCB Attitude Toward Consumer Behavior; BCR Brand Credibility; CPI Consumer Purchase Intention; ENV Environmental Value; ETC Ethical Concern; HEV Health Value; PKL Product Knowledge; SAV Safe Value; SOV Social Value; SPV Spiritual Value; WTP Willingness To Pay.

0.321 for ATCB, CPI and WTP, respectively, thus supporting the model's predictive relevance of ATCB, CPI and WTP.

Table 3 depicts the values of f^2 and its interpretation. It can be observed that one relationship has large effect size, two relationships have medium and two-relationships of small effect sizes, other relationships in that table were less than 0.02 indicate that there is no effect.

Figure 1 displays the PLS-SEM analysis outcome of the final structural model with control variables. All control variables were included in the model because they were, in some cases, statistically significant. As displayed in **Figure 1** and summarized in **Table 4**, the results indicated that hypotheses H5, H7, H8, H9a, H9b, H10 were accepted at $p < 0.05$ and $p < 0.001$. Meanwhile, H1, H2, H3, H4, H6 were not supported. Control variables also showed significant effect.

The results in **Table 5** shows that CPI positively and significantly mediated the relationship between ATCB (H11: $\beta = 0.327$, $p < 0.001$) and WTP, as both direct and indirect effects were also positive and significant, there was a complementary partial mediation. Additionally, CPI mediated the relationship between Gender ($\beta = 0.111$, $p < 0.001$), marital status (MAT) ($\beta = -0.207$, $p < 0.05$) and purchase frequency (PUF) ($\beta = 0.066$, $p < 0.001$) and WTP, when the direct effect was not significant whereas the indirect effect was significant this indicates an indirect-only mediation or only the indirect effect via the mediator exists. Therefore, Gender → CPI → WTP, MAT → CPI → WTP and PUF → CPI → WTP were fully mediation. Otherwise, CPI negatively and significantly mediated the relationship between place to live (PTL) ($\beta = -0.082$, $p < 0.05$) and WTP as direct effects are significantly positive and indirect effects are significantly negative, there was a competitive partial mediation.

Following the testing for mediation, the strength of the mediator can be further explained through the total effect (**Table 6**). The increased effect of ATCB, which in turn increases the effect on WTP can be explained via the partial mediation of CPI (H9b: $\beta = 0.516$, $p < 0.001$). In the contrary, although Gender, MAT, place to live (PTL) and PUF had significantly indirect effect to WTP mediated by CPI, the non-significant direct effect and opposite signs between indirect effect and direct effect led to total effect of these paths was not significant. It was an inconsistent mediation (suppression) effect which occurs when direct and indirect effects of similar magnitudes and opposite signs result in a nonzero but nonsignificant overall [28].

4 Discussion

The goal of this research was to see if there was a significant relationship between health value, product knowledge, brand credibility and consumer behavior, attitude toward consumer behavior and consumer purchase intention, attitude toward consumer behavior and willingness to pay, consumer purchase intention and willingness to pay. The function of ethical concern as a moderator in the relationship between perceived value and purchase intention was also investigated. The model was also evaluated, and demographic variables were controlled for. The current study's key results are presented below.

4.1 Demographic characteristics

Because Gen Z (aged 15 to 26) is the primary target population in this study, the majority of them were unemployed, not married, and had a low income. Furthermore, the high purchase frequency demonstrated that Gen Z was interested in cosmetics. grew up during the rapid development and widespread adoption of the Internet, Gen Z's interest in cosmetics has been influenced by the rise of social media platforms like Instagram, YouTube, and TikTok. These platforms have shaped beauty standards and trends, with influencers and content creators promoting cosmetics and skincare products through routines, tutorials, and reviews [29, 30]. This exposure to beauty content is maybe one of the major reasons sparking curiosity and interest in cosmetics among Gen Z.

4.2 Consumer behavior and willingness to pay towards cosmetic products

The findings show that attitude toward consumer behavior was found to have a significant influence on both willingness to pay and consumer purchase intention. This result was also in line with classical attitude-behavior theory [31] and previous studies which claimed that the desire to acquire cosmetics was directly, favorably, and reasonably strongly correlated with one's opinion toward organic products [32, 33]. In addition, the positive impact of attitude toward consumer behavior on willingness to pay was direct and partially mediated by consumer purchase intention. Thereby, hypothesis H11 was also supported. This indicated that attitude is a strong predictor for willingness to pay. Consuming cosmetics was essential and beneficial, so their positive ATCB will encourage them to pay more for cosmetics. Moreover, this study also showed that consumer purchase intention was found to have a significant influence on willingness to pay. This was consistent with study of Barber et al. (2012) showing that once customers had an intention for a certain cosmetic item, their payment will be more willing [34].

By defining the different types of consumers' perceived value on cosmetics based on prior experience, such as environmental value, the study's findings provided theoretical insight into consumer attitudes about cosmetic purchases. As a result, hypothesis H5 was confirmed. The results in this regard were in line with other research from both industrialized and developing nations about personal care products [35, 36]. Customers were worried about the effects of their purchases on the environment [37].

Table 4. Hypotheses testing

Hypothesis path coefficients	Beta	SE	t-value	*p*-value	Result
H1: HEV → ATCB	0.066	0.039	1.717	0.086	Not Supported
H2: SAV → ATCB	0.055	0.036	1.517	0.129	Not Supported
H3: SOV → ATCB	0.030	0.039	0.778	0.436	Not Supported
H4: SPV → ATCB	0.036	0.059	0.614	0.539	Not Supported
H5: ENV → ATCB	0.109	0.046	2.374	0.018*	Supported
H6: ETC → ATCB	-0.015	0.043	0.351	0.725	Not Supported
H7: PKL → ATCB	0.246	0.047	5.278	0*	Supported
H8: BCR → ATCB	0.397	0.050	7.905	0**	Supported
H9a: ATCB → CPI	0.618	0.030	20.843	0**	Supported
H9b: ATCB → WTP	0.189	0.046	4.106	0**	Supported
H10: CPI → WTP	0.529	0.046	11.457	0**	Supported
AGE → CPI	0.010	0.031	0.311	0.756	Not Supported
AGE → WTP	-0.048	0.035	1.399	0.162	Not Supported
EDU → CPI	0.045	0.029	1.561	0.119	Not Supported
EDU → WTP	0.002	0.028	0.057	0.954	Not Supported
ETH → CPI	0.240	0.170	1.409	0.159	Not Supported
ETH → WTP	0.086	0.212	0.404	0.686	Not Supported
Gender → CPI	0.210	0.057	3.694	0*	Supported
Gender → WTP	-0.071	0.061	1.154	0.248	Not Supported
INC → CPI	-0.056	0.036	1.531	0.126	Not Supported
INC → WTP	0.101	0.038	2.679	0.007*	Supported
MAT → CPI	-0.392	0.174	2.259	0.024*	Supported
MAT → WTP	0.133	0.129	1.026	0.305	Not Supported
OCP → CPI	0.167	0.105	1.596	0.110	Not Supported
OCP → WTP	0.002	0.101	0.016	0.987	Not Supported
PTL → CPI	-0.155	0.064	2.424	0.015*	Supported
PTL → WTP	0.211	0.066	3.203	0.001*	Supported
PUF → CPI	0.125	0.030	4.206	0**	Supported
PUF → WTP	-0.012	0.034	0.360	0.719	Not Supported

*** $p<0.001$; * $p<0.05$.*
ATCB Attitude Toward Consumer Behavior; BCR Brand Credibility; CPI Consumer Purchase Intention; EDU Education Level; ENV Environmental Value; ETC Ethical Concern; ETH Ethnicity; HEV Health Value; INC Income; MAT Marital Status; OCP Occupation; PKL Product Knowledge; PTL Place To Live; PUF Purchase Frequency; SAV Safe Value; SOV Social Value; SPV Spiritual Value; WTP Willingness To Pay

Other types of consumers' perceived value such as health value, ethical value, safe value, social value, spiritual value did not show any significant effect in predicting attitude in this study. In contrast, recent studies' constructs have a significant positive effect on consumer attitude [33, 38].

For health value, these consumers did not associate cosmetics as being healthy. Also noteworthy is the fact that the health awareness measurement items tended to lean more toward food and exercise. As a result, it was possible that the construct did not correctly reflect the participants' health value regarding the use of cosmetics. Surprisingly, safe value, spiritual value, ethical value did not show any significant effect in predicting attitude in this study. This could be due to cosmetics having a low degree of visibility as compared to clothing and/or foods.

Meanwhile, social value did not have a significant on attitude that was consistent with study of Ghazali et al., (2017) and Suphasomboon and Vassanadumrongdee (2022) [33, 38]. People had various skin types, responses to product components, and preferences for certain requirements, hence Ghazali et al. (2017) claimed that personal considerations rather than reference groups influenced product selections more (such as whitening products). The findings of this study thus implied that these individual aspects are taken into account in addition to societal impacts [33]. The results of this study, however, were at odds with those of a previous study, which discovered that social factors had a favorable effect on consumers' humanitarian intentions as well as their propensity to buy green cosmetics goods [39]. They concluded that reliable and convincing information on social media can increase customers' environmental awareness and green cosmetic purchase choices. Contrarily, consumers who placed a high value on individual gained over societal standards have lower inclinations to purchase cosmetics [40].

Table 5. Indirect effect (mediation)

Indirect Effect	Beta	SE	t-value	*p*-value	2.5 % CI	97.5 % CI	Meditation
H11: ATCB → CPI → WTP	0.327	0.032	10.309	0**	0.265	0.389	Partial [a]
AGE → CPI → WTP	0.005	0.016	0.310	0.757	-0.028	0.036	No
EDU → CPI → WTP	0.024	0.015	1.531	0.126	-0.005	0.056	No
ETH → CPI → WTP	0.127	0.091	1.394	0.164	-0.045	0.309	No
Gender → CPI → WTP	0.111	0.033	3.411	0.001*	0.049	0.180	Fully
INC → CPI → WTP	-0.029	0.020	1.505	0.132	-0.069	0.008	No
MAT → CPI → WTP	-0.207	0.094	2.217	0.027*	-0.398	-0.031	Fully
OCP → CPI → WTP	0.088	0.056	1.568	0.117	-0.012	0.212	No
PTL → CPI → WTP	-0.082	0.035	2.376	0.018*	-0.151	-0.016	Partial [b]
PUF → CPI → WTP	0.066	0.017	3.876	0**	0.036	0.103	Fully

*** p<0.001; * p<0.05*

[a] Complementary; [b] Competitive

ATCB Attitude Toward Consumer Behavior; BCR Brand Credibility; CPI Consumer Purchase Intention; EDU Education Level; ENV Environmental Value; ETC Ethical Concern; ETH Ethnicity; HEV Health Value; INC Income; MAT Marital Status; OCP Occupation; PKL Product Knowledge; PTL Place To Live; PUF Purchase Frequency; SAV Safe Value; SOV Social Value; SPV Spiritual Value; WTP Willingness To Pay

This study also incorporated product knowledge and brand credibility into the model, which demonstrated favorably influence attitude toward consumer behavior in addition to the six types of perceived values. Brand credibility was the most important factor in predicting attitude toward customer behavior, these results were consistent with those of Paul and Bhakar (2017) who conducted research in other contexts [41]. The findings clearly showed that when a company had high credibility, consumers had a positive attitude toward it. For product knowledge, these findings were consistent with those obtained by (Wang et al., 2019); (Nurhayati & Hendar, 2019); & (Sriminarti & Nora, 2018) in other settings [42-44]. The results clearly demonstrated that consumer awareness of cosmetic products was formed by knowledge of cosmetic products, and that awareness, in the long run, promotes cosmetic purchase intention.

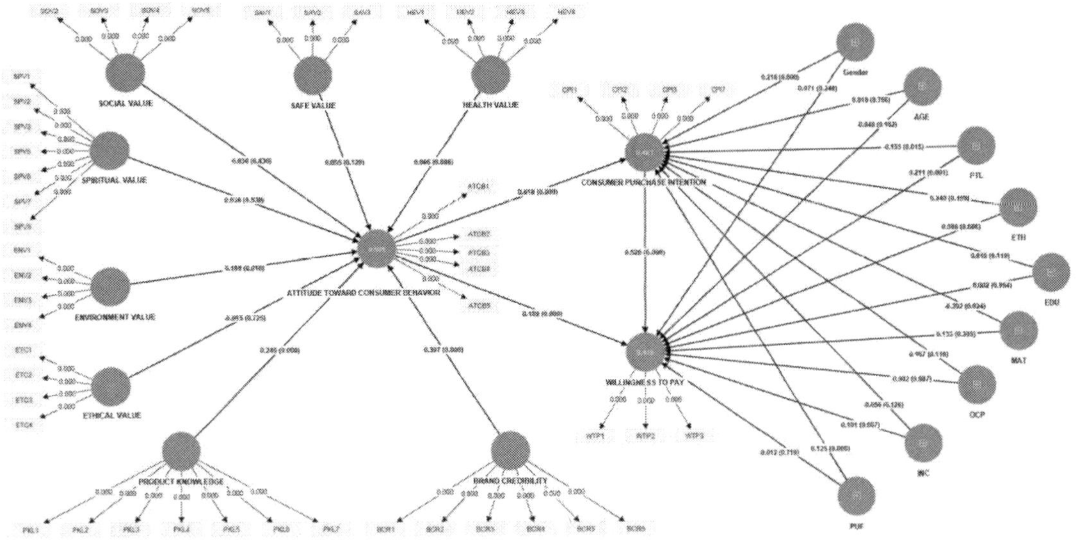

Fig. 1. PLS-SEM model result (with control variables)

Table 6. Total effects

Total effect	Beta	SE	t-value	*p*-value
H1: HEV → ATCB	0.066	0.039	1.717	0.086
H2: SAV → ATCB	0.055	0.036	1.517	0.129
H3: SOV → ATCB	0.030	0.039	0.778	0.436
H4: SPV → ATCB	0.036	0.059	0.614	0.539
H5: ENV → ATCB	0.109	0.046	2.374	0.018*
H6: ETC → ATCB	-0.015	0.043	0.351	0.725
H7: PKL → ATCB	0.246	0.047	5.278	0**
H8: BCR → ATCB	0.397	0.050	7.905	0**
H9a: ATCB → CPI	0.618	0.030	20.843	0**
H9b: ATCB → WTP	0.516	0.037	13.956	0**
H10: CPI → WTP	0.529	0.046	11.457	0**
AGE → CPI	0.010	0.031	0.311	0.756
AGE → WTP	-0.043	0.038	1.137	0.256
EDU → CPI	0.045	0.029	1.561	0.119
EDU → WTP	0.025	0.033	0.770	0.441
ETH → CPI	0.240	0.170	1.409	0.159
ETH → WTP	0.212	0.230	0.924	0.356
Gender → CPI	0.210	0.057	3.694	0**
Gender → WTP	0.040	0.068	0.586	0.558
INC → CPI	-0.056	0.036	1.531	0.126
INC → WTP	0.071	0.040	1.766	0.077
MAT → CPI	-0.392	0.174	2.259	0.024*
MAT → WTP	-0.075	0.179	0.418	0.676
OCP → CPI	0.167	0.105	1.596	0.110
OCP → WTP	0.090	0.120	0.746	0.456
PTL → CPI	-0.155	0.064	2.424	0.015*
PTL → WTP	0.128	0.074	1.736	0.083
PUF → CPI	0.125	0.030	4.206	0**
PUF → WTP	0.053	0.038	1.395	0.163

** $p<0.001$; * $p<0.05$

ATCB Attitude Toward Consumer Behavior; BCR Brand Credibility; CPI Consumer Purchase Intention; EDU Education Level; ENV Environmental Value; ETC Ethical Concern; ETH Ethnicity; HEV Health Value; INC Income; MAT Marital Status; OCP Occupation; PKL Product Knowledge; PTL Place To Live; PUF Purchase Frequency; SAV Safe Value; SOV Social Value; SPV Spiritual Value; WTP Willingness To Pay.

4.3 Limitations and recommendations

This research had several limitations. As the majority of users of cosmetic items lived mostly in metropolitan areas, this study focused on data gathered from Ho Chi Minh City in Vietnam. Nonetheless, it is advised that future studies gather responses from a larger swath of the nation. Furthermore, to test the model's ability to predict consumer behavior, additional variables like customer satisfaction and loyalty could be added. To enhance consumer appeal, prioritize eco-friendly practices and communicate sustainability efforts clearly. Increasing brand credibility and better product understanding contribute to more positive consumer attitudes. To capitalize on this, cosmetic brands should focus on transparent communication about product ingredients, manufacturing processes, and quality standards. Strengthening brand credibility through certifications and clear messaging can foster trust and enhance positive consumer attitudes. These strategies not only align with consumer values and foster favorable attitudes but also influence their willingness to invest financially. Consumer behavior and willingness to pay are dynamic and can evolve over time. Businesses should consistently monitor consumer preferences and conduct further research to stay updated on changing trends and preferences. Future research may also take into account deterrents like accessibility and cost and skepticism toward claims about cosmetics.

5 Conclusion

The study verified that attitude was the most significant component in predicting intention to purchase. Attitude and desire to acquire cosmetic products had both a favorable direct and indirect influence on willingness to pay. Brand reputation took precedence over product understanding and environmental value in influencing the attitude toward purchasing cosmetics. It is important for industry professionals to understand consumer purchasing habits. By understanding and addressing these influencing factors, it is crucial for the sector to build a reputable cosmetic brand for customers, to communicate intelligibly and honestly about the advantages of their cosmetic products, and to guarantee the veracity of their cosmetic products' environmental claims while meeting the demands and needs of their target market.

Supplemental Material

Supplementary data associated with this article can be found in the online version at https://s.pro.vn/qkFM

Conflict of Interest

There is no conflict of interest to declare.

Funding

No funding was received for this analysis.

Acknowledgements

The author would like to acknowledge the voluntary participants for data collection.

Authors Contribution

All authors contributed to the study conception and design. Material preparation, data collection and analysis were performed by HTTB, TVQ, VNN, THDN. All authors commented on all draughts of the manuscript. All authors read and approved the final version for submission.

References

1. T. L. Rochelle, *Health and help-seeking behavior among Chinese men in Hong Kong: The influence of culture.* Psychology of Men & Masculinities, 2019. **20**(1): p. 71.
2. M. Kowal and P. Sorokowski, *Sex Differences in Physical Attractiveness Investments: Overlooked Side of Masculinity.* International Journal of Environmental Research and Public Health, 2022. **19**(7): p. 3842.
3. M. d. l. L. C.-G. Francisca R, Nunes MA, et al, *12 - Cosmetics.* 2018. **393-427**.
4. T. Banerjee and A. Samanta, *Metal oxides for cosmetics and sunscreens*, in *Metal Oxides for Biomedical and Biosensor Applications*. 2022, Elsevier. p. 119-135.
5. U.S Food & Drug Administration. *Is It a Cosmetic, a Drug, or Both? (Or Is It Soap?).* 2022 [cited 2023 14 July]; Available from: https://www.fda.gov/cosmetics/cosmetics-laws-regulations/it-cosmetic-drug-or-both-or-it-soap.
6. R. Barreiras, *Consumer Behaviour changes in the Cosmetic Industry.* 2019.
7. *APAC: cosmetics market size 2015-2020*, in *Statista*. 2021.
8. *Developing Marketing Practices Based on Consumer Behavior. Case Company: Shiseido Group Vietnam*, in *Theseus*. 2019.
9. Herohunt.ai, *Generation Z definition, synonyms and explanation.*
10. D. Weinswig, *Gen Z: Get ready for the most self-conscious, demanding consumer segment.* Fung Global Retail & Technology, 2016: p. 1-19.
11. A. Fahira and M. D. Djamaludin, *The Influence of Brand Trust and Satisfaction towards Consumer Loyalty of a Local Cosmetic Products Brand X among Generation Z.* Journal of Consumer Sciences, 2023. **8**(1): p. 27-44.
12. K. C. Williams and R. A. Page, *Marketing to the generations.* Journal of behavioral studies in business, 2011. **3**(1): p. 37-53.

13. V. H. Nguyen, T. X. D. Truong, H. T. Pham, D. T. Tran, and P. H. Nguyen, *Travel intention to visit tourism destinations: A perspective of generation Z in Vietnam.* The Journal of Asian Finance, Economics and Business, 2021. **8**(2): p. 1043-1053.
14. *Consumer Behavior: Buying, Having, and Being (8th ed.).* Management Decision, 2009. **47**(5): p. 845-848.
15. A. Bhalerao and R. K. Pandey, *Consumer Behaviour: A literature review.* International Journal of Engineering Technology Science and Research, 2017. **4**(8): p. 1095-1103.
16. J. C. Sweeney and G. N. Soutar, *Consumer perceived value: The development of a multiple item scale.* Journal of retailing, 2001. **77**(2): p. 203-220.
17. J. N. Sheth, B. I. Newman, and B. L. Gross, *Why we buy what we buy: A theory of consumption values.* Journal of business research, 1991. **22**(2): p. 159-170.
18. M.-F. Chen, *Consumer attitudes and purchase intentions in relation to organic foods in Taiwan: Moderating effects of food-related personality traits.* Food Quality and Preference, 2007. **18**(7): p. 1008-1021.
19. R. V. Krejcie and D. W. Morgan, *Determining sample size for research activities.* Educational and psychological measurement, 1970. **30**(3): p. 607-610.
20. S. Easton and D. Van Laar, *User Manual for the Work-Related Quality of Life (WRQoL) Scale : A Measure of Quality of Working Life. 2nd ed.* 2018: University of Portsmouth.
21. World Health Organization. *Process of translation and adaptation of instruments.* 2019 [cited July 20, 2019; Available from: https://www.who.int/substance_abuse/research_tools/translation/en/.
22. M. Sarstedt, C. M. Ringle, and J. F. Hair, *Partial least squares structural equation modeling*, in *Handbook of market research.* 2021, Springer. p. 587-632.
23. A. Diamantopoulos, M. Sarstedt, C. Fuchs, P. Wilczynski, and S. Kaiser, *Guidelines for choosing between multi-item and single-item scales for construct measurement: a predictive validity perspective.* Journal of the Academy of Marketing Science, 2012. **40**: p. 434-449.
24. A. L. Drolet and D. G. Morrison, *Do we really need multiple-item measures in service research?* Journal of service research, 2001. **3**(3): p. 196-204.
25. J. Henseler, C. M. Ringle, and M. Sarstedt, *A new criterion for assessing discriminant validity in variance-based structural equation modeling.* Journal of the academy of marketing science, 2015. **43**: p. 115-135.
26. N. A. Ramli, H. Latan, and G. V. Nartea, *Why should PLS-SEM be used rather than regression? Evidence from the capital structure perspective.* Partial least squares structural equation modeling: Recent advances in banking and finance, 2018: p. 171-209.
27. G. Shmueli, M. Sarstedt, J. F. Hair, J.-H. Cheah, H. Ting, S. Vaithilingam, and C. M. Ringle, *Predictive model assessment in PLS-SEM: guidelines for using PLSpredict.* European journal of marketing, 2019. **53**(11): p. 2322-2347.
28. D. P. MacKinnon, J. L. Krull, and C. M. Lockwood, *Equivalence of the mediation, confounding and suppression effect.* Prevention science, 2000. **1**: p. 173-181.
29. J. Liu, *The Effectiveness of Influencer Marketing on Social Media: Facial Cosmetics Targeting Generation Y and Generation Z in Bangkok.* 2020.
30. G. Dao, *Tiktok influencer marketing as a marketing tool for cosmetic industry to attract generation Z.* 2022.
31. I. Ajzen, *The theory of planned behavior.* Organizational behavior and human decision processes, 1991. **50**(2): p. 179-211.
32. E. J. Van Loo, M. N. H. Diem, Z. Pieniak, and W. Verbeke, *Consumer attitudes, knowledge, and consumption of organic yogurt.* Journal of dairy science, 2013. **96**(4): p. 2118-2129.
33. E. Ghazali, P. C. Soon, D. S. Mutum, and B. Nguyen, *Health and cosmetics: Investigating consumers' values for buying organic personal care products.* Journal of Retailing and Consumer Services, 2017. **39**: p. 154-163.
34. N. Barber, P. J. Kuo, M. Bishop, and R. Goodman Jr, *Measuring psychographics to assess purchase intention and willingness to pay.* Journal of consumer marketing, 2012. **29**(4): p. 280-292.
35. J. Chin, B. C. Jiang, I. Mufidah, S. F. Persada, and B. A. Noer, *The investigation of consumers' behavior intention in using green skincare products: a pro-environmental behavior model approach.* Sustainability, 2018. **10**(11): p. 3922.
36. M. Sadiq, M. Adil, and J. Paul, *An innovation resistance theory perspective on purchase of eco-friendly cosmetics.* Journal of Retailing and Consumer Services, 2021. **59**: p. 102369.
37. A. Tewari, S. Srivastava, D. Gangwar, and V. C. Verma, *Young consumers' purchase intention toward organic food: exploring the role of mindfulness.* British Food Journal, 2022. **124**(1): p. 78-98.
38. T. Suphasomboon and S. Vassanadumrongdee, *Toward sustainable consumption of green cosmetics and personal care products: The role of perceived value and ethical concern.* Sustainable Production and Consumption, 2022. **33**: p. 230-243.
39. R.-A. Pop, Z. Săplăcan, and M.-A. Alt, *Social media goes green—The impact of social media on green cosmetics purchase motivation and intention.* Information, 2020. **11**(9): p. 447.
40. L.-C. Lu, H.-H. Chang, and A. Chang, *Consumer personality and green buying intention: The mediate role of consumer ethical beliefs.* Journal of Business Ethics, 2015. **127**: p. 205-219.
41. J. Paul and S. Bhakar, *Does celebrity image congruence influences brand attitude and purchase intention?* Journal of Promotion Management, 2018. **24**(2): p. 153-177.
42. E. S.-T. Wang and Y.-C. Chen, *Effects of perceived justice of fair trade organizations on consumers' purchase intention toward fair trade products.* Journal of Retailing and Consumer Services, 2019. **50**: p. 66-72.

43. T. Nurhayati and H. Hendar, *Personal intrinsic religiosity and product knowledge on halal product purchase intention: Role of halal product awareness.* Journal of Islamic Marketing, 2020. **11**(3): p. 603-620.

44. N. Sriminarti and L. Nora, *The role of product knowledge and attitudes toward purchasing intentions: the moderating effect of religiosity on halal cosmetics.* KnE Social Sciences, 2018: p. 518–529-518–529.

A Survey of Community Perceptions and Knowledge Toward Antibiotic Resistance: Case Study in Magelang, Indonesia

Erki Arfianto[1], *Indriyati* Hadi Sulistyaningrum[2], *Prasojo* Pribadi[3*], *Heni* Lutfiyati[3], *Vanny* Eka Septiana[3], *Amalia* Ratna Puspitadewi[3], *Rizka* Amalia[3], *Rayi* Citra Ayu Pangestuti[3], *Nufikha* Falyauma[3]

[1]Program Studi Pendidikan Profesi Apoteker, Faculty of Pharmacy, Universitas Islam Sultan Agung, Semarang, Indonesia.
[2]Department of Pharmacy Faculty of Pharmacy, Universitas Islam Sultan Agung, Semarang, Indonesia.
[3]Department of Pharmacy, Faculty of Health Sciences, Universitas Muhammadiyah Magelang, Indonesia

Abstract. One of the main causes of antibiotic resistance is the unprescription dispensing of antibiotics, which can result in the improper use of antibiotics. The purpose of a self-administered questionnaire-based study was to gauge participants' understanding and opinions regarding antibiotic resistance. In October 2019, a cross-sectional descriptive survey was self-administered to individuals. For this survey, Total of 455 respondents were used. Respondents who were residing in Magelang, spoke Indonesian, and were older than eighteen were the inclusion criteria. Six sections made up the questionnaire: demographic information; five questions about general antibiotic resistance information; five questions about the advantages of antibiotics; three questions about prescription antibiotics; four questions about using antibiotics personally; and five questions about antibiotic knowledge. Data analysis were used SPSS version 21, research data is treated as a descriptive analysis. The results show that majority of respondents have the perception that antibiotics are the best choice for treating fever (26.5%), 72.3% of participants are aware of antibiotic resistance. Furthermore, 53.8% of respondents said they completely trusted medical experts' advice. Health professionals have a greater responsibility to alert the public about antibiotic use and resistance. In conclusion, it is important to increase knowledge about antibiotic resistance because it concerns the general public's opinion regarding antibiotic resistance and their level of awareness of the risks associated with it. Ultimately, this can change people's attitudes and actions to use antibiotics wisely.
Keywords: Perception, Knowledge, Antibiotic, Resistance, Community

1 Introduction

Antibiotic resistance is one of the biggest health threats in the world, especially in developing countries including Indonesia [1]. The negative effects of antibiotic resistance are the increasing mortality rates, requiring longer time for therapy, decreasing functionality and the need for post-acute care [2]. Approximately 0.7 million deaths/year are caused by antibiotic problems and it is expected to exceed 10 million by 2050. This report shows the alarming spread of antibiotic resistance worldwide [3]. A research conducted in Indonesia showed that of 781 patients in the hospital, 81% of them had resistance with ampicillin, 73% resistance with cotrimoxazole and 56% resist against Escherichia coli [4]. This study was supported by research conducted by Huda in 2016, it was confirmed that Escherichia coli was the most resistant to Sulfamethoxazole (100%), Ampicillin (93.3%) and Cefadroxil (90.9%) [5] and another study conducted by Kibret in 2011 at northeast Ethiopia found that high levels of resistance to erythromycin (89.4%), amoxicillin (86.0%) [6].

Factors influencing antibiotic resistance are improper use of antibiotics, a low level of patient knowledge in the use of antibiotics such as using antibiotics for handling viruses like colds, cold coughs, and fevers. Patients with high income usually ask to be given antibiotic therapy at a high price, new and without a doctor's prescription even though it is not necessarily used, whereas patients with low income are often unable to continue the therapy with antibiotics. Massive sales and commercial promotions from pharmaceutical companies supported by the influence of globalization, would facilitate the exchange of goods and the number of antibiotics circulating in the wider community so it is easy to get antibiotics. The use of monotherapy is easier to cause resistance

* Corresponding email: prasojopribadi@unimma.ac.id

compared to the use of combination therapy. Based on research conducted by Bish et al (2009), it is stated that research in the discovery of new antibiotics is still small, the supervision carried out by the government in the distribution of antibiotic use is weak, and there is a lack of commitment from related agencies to improve drug quality and infection control [7].

Several studies in Europe showed that the increasing antibiotic resistance is related to high use in the community. It is caused by irrational use of drugs and lack of education about antibiotics for patients [8, 9], while several studies in developing countries showed

The high use of antibiotics without a prescription. Widayati on her research, stated that 324 respondents, namely 58.00%, out of 559 respondents do not consult with doctors and use antibiotics to treat symptoms of the disease and using antibiotics without a prescription for the past one month, they are about 7.30% [10, 11]. The global action plan outlines five strategic goals, including raising public awareness and understanding of antimicrobial resistance, strengthening knowledge via research and surveillance, lowering infection rates, maximizing the use of antimicrobial agents, and guaranteeing ongoing funding for the fight against antimicrobial resistance [12]. Moreover, the National Action Plan Antimicrobial Resistance (AMR NAP) Indonesia was created with five strategic goals in mind: raising awareness and comprehension of AMR through efficient communication, instruction, and training; fortifying the body of knowledge and evidence via research and supervision; decreasing the frequency of infections related to sanitation, hygiene, and preventive infections; optimizing the use of antimicrobial drugs in human and animal health; and creating business cases for sustainable investment and boosting spending on novel medications, medical equipment, immunizations, and other interventions [13].

Based on research conducted by Carter in the United State, it was found that people were aware of antibiotic abuse associated with antibiotic resistance. However, many people do not consider it is as an important issue [14]. Several studies related to the knowledge of antibiotics have been conducted in Putra Jaya, Malaysia. It showed a low level of knowledge of the community in the area of antibiotics [15]. Another study done in Malaysia showed that 43% respondents have good knowledge [16]. Similar research is limited in Indonesia. Research conducted by Khoirunnisa stated that the knowledge, perceptions and beliefs of the Indonesian people on antibiotics were still very low. Some efforts should be made to increase their knowledge and understanding of antibiotics [1]. There are many studies conducted in Indonesia investigating knowledge of antibiotic use, only a few of those links it with antibiotic resistance. Knowledge about how to use antibiotics is not enough if it is not balanced with awareness of the dangers of antibiotic resistance. Knowledge of antibiotic resistance can foster awareness to change people's behavior in using antibiotics properly. Therefore, this study aims to examine people's perceptions and knowledge about antibiotic resistance in Central Java, Indonesia.

2 METHOD

This study was carried out in October 2019 using a cross-sectional survey methodology. Descriptive analysis was employed in the research design to evaluate Central Javans' perceptions and level of knowledge regarding antibiotic resistance. In this study, 455 respondents were used. Respondents who were at least eighteen years old, spoke Indonesian, and lived in Central Java met the inclusion requirements. Data collection was done using questionnaire which consists of six parts. The first section contains demographic data including gender, age, education level and occupation. The second part contains five questions regarding general information about resistance. The third part contains five questions regarding the benefits of antibiotics. The fourth part contains three questions regarding antibiotic prescription. The fifth section deals with the personal use of antibiotics. The sixth part contains five questions related to antibiotic knowledge. The questionnaire was adapted from Carter [14]. The respondents were asked to rate on a 4-point Likert scale from 1 (fully disagree) to 4 (fully agree) the degree. A pilot study was conducted on 10 respondents to find out the simplicity and clarity of the questionnaire. Furthermore, it is done with the aim to detect the ambiguous language and the level of understanding of respondents. Research data is processed in the form of analysis descriptive using SPSS version 21 software.

3 RESULTS

Table 1 shows the demographic profile of the respondents. A total of 455 respondents' data was processed in this study. The majority of respondents in this study were women (69.9%), while male respondents were 30.1%. Respondents aged 18-29 years dominate in this study with a percentage of 91%. The most educational level of respondents were senior high school students (62.9%). The majority of respondents have an income level of less than 1,500,000 Indonesian Rupiah (IDR) with a percentage of 65.1%.

Table 1. Characteristics of Respondents

Characteristic	**Frequency (%)**
Gender	
Male	137 (30.1)
Female	318 (69.9)
Age (years old)	
18-29	414 (91)
30-39	25 (5.5)
40-49	13 (2.9)
50-64	3 (0.7)
Level of Education	
Elementary school	2 (0.4)
Junior high school	5 (1.1)
Senior high school	286 (62.9)
Diploma	63 (13.8)
Bachelor/ Master/ Ph.D	99 (21.8)
Occupation	
Students	266 (58.5)
Civil servant	31 (6.8)
Entrepreneur	40 (8.8)
Private employee	98 (21.5)
Farmer	7 (1.5)
Other	13 (2.9)
Monthly income (IDR)	
≤ 1,500,000	296 (65.1)
>1,500,000 – 2,500,000	74 (16.3)
>2,500,000 – 3,500,000	30 (6.6)
> 3,500,000	55 (12.1)

The knowledge about symptoms that can be relieved by antibiotic and knowledge of antibiotic resistance are shown in Table 2. More than 10% of respondents believed that antibiotics are the best choice to treat fatigue (5.5%), cough (8.8%) fever (26.5%) and runny nose and sore throat (25.5%). In the source section of the information obtained by respondents regarding antibiotic resistance including medical professionals are the most common source of information (38.24%). Furthermore, to hearing that information from medical professionals, there are several other sources that are often used is the internet (20.88%) and others (14.95%).

Table 2. Knowledge about Symptoms can be Relieved by Antibiotic, Knowledge of Antibiotic Resistence, and Source Information

Knowledge	Frequency (%)
Antibiotics are the best choice for treating the following symptoms	
Anxiety	12 (2.6)
Moodiness	13 (2.9)
Headache	13 (2.9)
Fatigue	25 (5.5)
Joint or muscle pain	28 (9.5)
Cough	40 (8.8)
Runny nose & sore throat	130 (25.5)
Fever	121 (26.5)
Ear pain	16 (3.3)
Buring with urination	57 (12.5)
Do you know about antibiotic resistance	
Yes	329 (72.3)
No	126 (27.7)
Type of Information Source	
Newspaper	2 (0.44)
Friend/ Family member	41 (9.01)
Medical professional	174 (38.24)
Internet	95 (20.88)
Television	2 (0.44)
Other	68 (14.95)
No answer	73 (16.04)

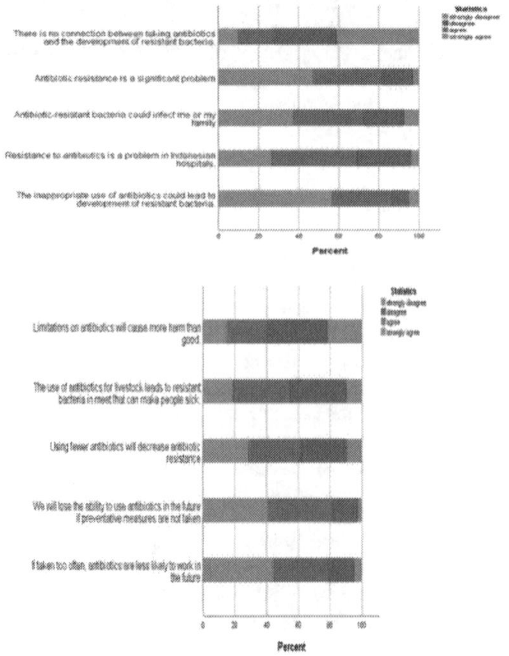

Fig. 1. Knowledge and Perception about Antibiotic Resistance and Preservation of Antibiotic Efficacy (n=455)

The results of the study on the assessment of people's knowledge and perceptions in Central Java are shown in Figure 1. It was found that a total of 256 (56.30%) of respondents acknowledged that inappropriate use of antibiotics could encounter the development of antibiotic resistance. The most of respondents agreed that antibiotic resistance was a problem in the hospital (43.1%). The majority of respondents agreed if resistant bacteria could infect them or their family members (37.1%). Most of respondents strongly agreed that antibiotic resistance was a significant problem (47%). A total of 33% of respondents agreed that the use of fewer antibiotics would reduce antibiotic resistance.

Similar with previous studies which stated that resistance to antibiotics can be prevented by stopping antibiotics as soon as they feel healed [17]. Moreover, a total of 36% respondents agreed that the use of antibiotics for livestock caused resistant bacteria in meat that could make people sick. However, 10.1% respondents strongly disagreed.

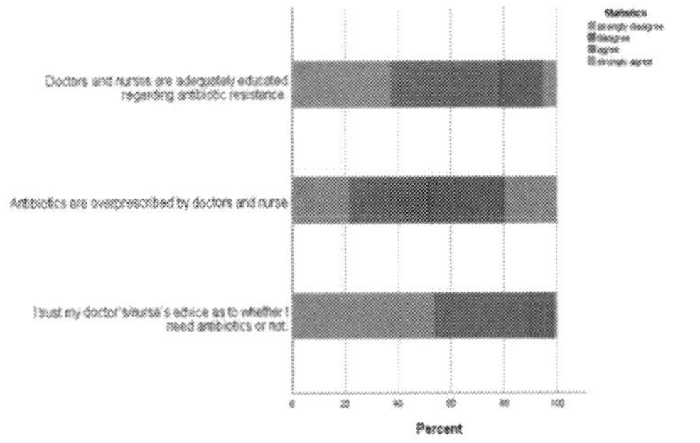

Fig. 2. Trust in Health Workers about Prescribing Antibiotic (n=455)

Shown in Figure 2 most of the respondents (53.8%) in this study trusted their doctors or nurses advice about the need for antibiotics to be consumed or not. This was reinforced by the community's assumptions in the next statement, there were 41.3% respondents who assumed that doctors or nurses have adequate knowledge about antibiotic resistance.

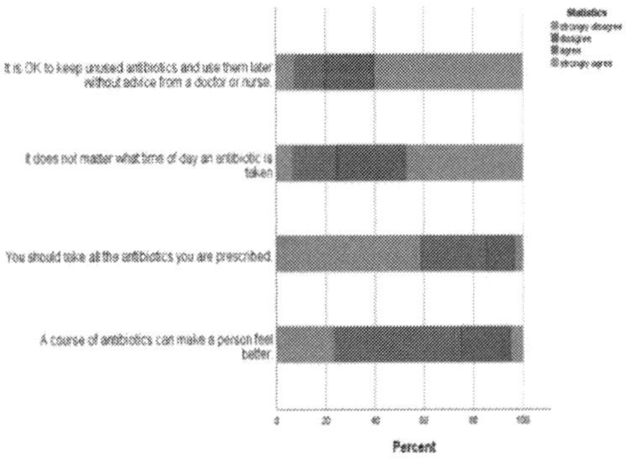

Fig. 3. Knowledge and Perception about Use of Antibiotic (n=455)

Figure 3 Shown in 44.5% of the respondents agreed that antibiotics can kill bacteria. This is supported by finding that 45.9% respondents agree if antibiotics can kill bacteria where they normally live on the skin and in the intestine. The majority of respondents (51.9%) agreed if they had to take antibiotics which had been prescribed. Total of 40.9% respondents disagreed if there was no relationship between taking antibiotics and developing bacterial resistance. Most respondents (47.3%) disagree if there is no problem on how long antibiotics are taken.

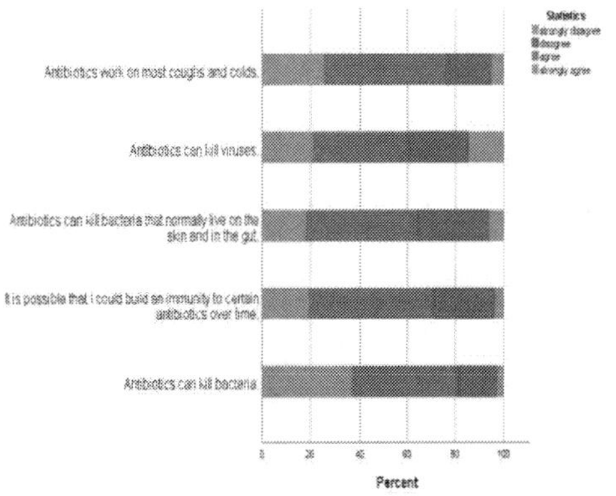

Fig. 4. Knowledge of Antibiotics (n=455)

The results of this study showed that 44.55% of respondents agreed with the statement that antibiotics can kill bacteria. In addition, 38.9% of respondents agreed that antibiotics can kill viruses (Figure 4).

4 DISCUSSION

Most of the respondents believed that several symptoms can be relieved by antibiotics. Only about 10% of respondents believed that antibiotics are the best choice to treat fatigue (5.5%), cough (8.8%), fever (26.5%), runny nose and sore throat (25.5%). The same results were mentioned in previous research conducted by Dadari which found that as many as 25.27% of respondents believed that antibiotics are used to treat fevers, colds, bacterial and viral infections [18]. Meanwhile, antibiotics are used to kill bacteria or prevent bacterial proliferation [19, 20]. Therefore, it can be seen that public knowledge about the use of antibiotics is sufficient. Public knowledge should be increased because inadequate knowledge can lead to misunderstandings and inappropriate use of antibiotics. This may contribute to antibiotic resistance [18]. The result show, there are 72.3% of respondents who know the term antibiotic resistance. Moreover, 43.7% of respondents said that they had never heard of the term before. The similar study conducted in Jordan stated that 70% of respondents were unaware of the term "antimicrobial resistance" [21]. In this study, most of respondents chose medical professionals as a source of information about antibiotic resistance. These results are supported by WHO, which reported that doctors or nurses are a great source of information on antibiotic resistance for respondents [22].

The assessment of people's knowledge and perceptions found that 56.30% of respondents acknowledged that inappropriate use of antibiotics could encourage the development of antibiotic resistance, whereas there were 5.1% of respondents stated disagree with the statement. One of the factors causing improper use of antibiotics is dispensing antibiotics without a prescription. According to Canton, the improper and excessive use of antimicrobials contributes to the persistence of resistant bacteria [23]. Approximately, 196 (43.1%) of our respondents also agreed that antibiotic resistance was a problem in the hospital and 19 (4.2%) our respondents disagreed about this. In previous studies there were also 69% of respondents considered it a problem in their hospital [24]. Out of 455 respondents, 35 (7.7%) respondents disagreed if resistant bacteria could infect them or their family members, however 169 (37.1%) respondents agreed with the statement. A total of 215 of our respondents strongly agreed that antibiotic resistance was a significant problem while as many as 13 (2.9%) respondents disagreed if antibiotic resistance was considered a significant problem. These findings are similar to survey reported in India which stated that 25% (125/500) of patients knew about the development of bacterial resistance to antibiotics & which could be fatal to themselves and even their family members and 47% (235/500) the patient is not aware about it [25].

This research found that 33% of respondents agreed that limiting antibiotic use could reduce antibiotic resistance and there were still 9.9% of respondents who disagreed that reducing antibiotic use could minimize antibiotic resistance. Similar with previous studies which stated that resistance to antibiotics can be prevented by stopping antibiotics as soon as they feel healed [17]. A total of 36% of the respondents agreed that the use of antibiotics for livestock caused resistant bacteria in meat that could make people sick. Referring to Hadi's research the spread of resistant bacteria comes from fisheries, hospitals, livestock and household waste [26]. According to Thapa (2014) and Usui (2014) reported that in the Indonesian traditional market found 16% of samples of chicken meat infected with listeria monocytogenes and most of these isolates showed multi-drug resistance especially against penicillin, ampicillin, and erythromycin [27, 28].

Most of respondents in this study trusted their doctors or nurses about the need for antibiotics to be consumed by them (53.8%). This was reinforced by the community's assumptions in the next statement, there were 41.3% respondents assumed that doctors or nurses have adequate knowledge about antibiotic resistance. The consistent finding was also conveyed by Chan's study, were found 84% of respondents agreed to take antibiotics on time according to doctor's instructions. This shows that adequate knowledge from health workers can lead to confidence for patients to be obedient in taking antibiotics [29]. The community considers that the knowledge and competence possessed by health workers can explain the rational use of antibiotics, because health workers are able to provide knowledge about side effects, pharmacological effects, drug interactions, and how to use them. In addition, information from doctors and pharmacists is also needed to support patient compliance in rational antibiotic use [30].

The majority of respondents (51.9%) agreed if they had to take antibiotics that had been prescribed. A total of 40.9% of the respondents disagreed and 9.9% agreed if there was no relationship between taking antibiotics and developing bacterial resistance. Most of respondents (47.3%) disagree if there is no problem how many days antibiotics are taken. These results are in accordance with previous studies, which stated that most of respondents from 5 hospital districts in Indonesia correctly answer that taking antibiotics should be regular. General guidelines to take antibiotics is every day until they run out; If the patient condition has improved, the dose of antibiotics taken should remain the same until all antibiotic tablets are used up [17].

The results in this study, there are 44.5% of respondents agreed with the statement about antibiotics can kill bacteria, and 38.9% of respondents agreed that antibiotics can kill viruses. Similar results were mentioned in a study conducted by Robert, there are 66% of patients answered correctly to the statement: "Antibiotics can be used to treat bacterial infections" and 31% of patients also believed antibiotics to be effective against viruses. Overall, 44% of patients knew that antibiotics were effective against bacteria and ineffective against viruses [31]. The antibiotics could kill viruses is a false statement because antibiotics are not effective against viruses [32]. A survey in Europe shows that 57% of Europeans and 35% of French people think that antibiotics are effective against viruses and they don't know if antibiotics are not effective against viral infections [33].The limitation of this study lies in the small number of samples, data obtained from one region, therefore this result cannot generalized.

5 CONCLUSION

This research shows that the majority of respondents have the perception that antibiotics are the best choice for treating fever (26.5%). As many as 72.3% of respondents understood the term antibiotic resistance. The majority of respondents also understand that inappropriate use of antibiotics can encourage the development of antibiotic resistance (56.39%) and 53.8% of respondents trusted doctors or nurses to handle the need for antibiotics to be consumed. In addition, a total 44.5% of the respondents agreed that antibiotics can kill bacteria.

ACKNOWLEDGEMENTS

This study was funded by LPPM Universitas Islam Sultan Agung.

REFERENCES

1. Khairunnisaa, Tanjunga HR, Sumantria IB. Penilaian Pengetahuan, Persepsi Dan Kepercayaan Masyarakat Kota Medan Terhadap Penggunaan Antibiotik. Trop Med. 2018;1(1):291–6.

2. Friedman ND, Temkin E, Carmeli Y. The negative impact of antibiotic resistance. Clin Microbiol Infect. 2016;22(5):416–22.

3. Singh R, Singh AP, Kumar S, Giri BS, Kim KH. Antibiotic resistance in major rivers in the world: A systematic review on occurrence, emergence, and management strategies. J Clean Prod. 2019;234:1484–505.

4. Akualing JS, Rejeki IPS. Antibiogram. Indones J Clin Pathol Med Lab. 2016;23(1):90–5.

5. Huda M. Resistensi Bakteri Gram Negatif Terhadap Antibiotik Di UPTD Balai Laboratorium Kesehatan Lampung Tahun 2012-2014 Resistance Gram Negative Bacteria Against Antibiotics in UPTD Health Laboratory Lampung Year 2012-2014. J Anal Kesehat. 2016;5(1).

6. Kibret M, Abera B. Antimicrobial susceptibility patterns of E. coli from clinical sources in northeast Ethiopia. Afr Health Sci. 2011;11(1).

7. Utami ER. Antibiotika, resistensi, dan rasionalitas terapi_Fakultas Saintek, Universitas Islam Negeri Maulana Malik Ibrahim Malang. Jalan Gajayana No 50 Malang. 2011;1(4):191–8.

8. Goossens H, Ferech M, Stichele R Vander, Elseviers M. Outpatient antibiotic use in Europe and association with resistance : a cross-national database study. lancet. 2005;365:579–87.

9. Franco BE, Martínez MA, Rodríguez MAS, Wertheimer AI. The determinants of the antibiotic resistance process. Dove Med Press Ltd. 2009;1–12.

10. Widayati A. Self medication with Antibiotics in Yogyakarta City Indonesia. 2013.

11. Djawaria DPA, Setiadi AP, Setiawan E. Analisis Perilaku dan Faktor Penyebab Perilaku Penggunaan Antibiotik Tanpa Resep di Surabaya. Media Kesehat Masy Indones. 2018;14(4):406.

12. WHO. Global Action Plan on Antimicrobial Resistance. Microbe Mag. 2015;10(9):354–5.

13. Kementrian Kesehatan RI. National Action Plan on Antimicrobial Resistance Indonesia 2017-2019.

14. Carter RR, Sun J, Jump RLP. A survey and analysis of the American public's perceptions and knowledge about antibiotic resistance. Open Forum Infect Dis. 2016;3(3):1–7.

15. Ka Keat L, Chew Charn T. A cross sectional study of public knowledge and attitude towards antibiotics in Putrajaya, Malaysia. South Med Rev. 2012;5(2):26–33.

16. Qamar M, Abdullah NHS, Khan J, Mahmud A, Ahmad A. Knowledge and Attitude towards Antibiotic Usage among General Public in Shah Alam, Malaysia. UK J Pharm Biosci. 2014;2(6):60.

17. Herawati F, Setiasih, Alhabsyi MM, Gunawan W, Palijama DE, Diah LF, Ardiansyah AB, Avanti RYC. A patient caregiver survey in Indonesia: Knowledge and perception of antibiotic use and microbial resistance. J Infect Public Health. 2019;4–8.

18. Dadari HIS. Antibiotics use, knowledge and practices on antibiotic resistance among breastfeeding mothers in Kaduna state (Nigeria). J Infect Public Health. 2019;8.

19. Nisak M, N AS, Y PSP, I AMK, Fatmawati L, Nilarosa AD,Fornia P, Pratiwi DWP, Apriliani D, Rosyidah S. Profil Penggunaan dan Pengetahuan Antibiotik pada Ibu-ibu. J Faramasi Komunitas. 2014;3(1):12–7.

20. Menkes RI. Peraturan Menteri Kesehatan Republik Indonesia. Peratur Menteri Kesehat No 2406 TAHUN 2011 Tentang Pedoman Umum Pengguna Antibiot. 2011;4.

21. Yusef D, Babaa AI, Bashaireh AZ, Al-Bawayeh HH, Al-Rijjal K, Nedal M, KAilani S. Knowledge, practices & attitude toward antibiotics use and bacterial resistance in Jordan: A cross-sectional study. Infect Dis Health. 2018;23(1):33–40.

22. WHO. Antibiotic Resistance: Multi-Country

Public Awareness Survey. WHO Press. 2015. 1–51 p.

23. Cantón R, Horcajada JP, Oliver A, Garbajosa PR, Vila J. Inappropriate use of antibiotics in hospitals: The complex relationship between antibiotic use and antimicrobial resistance. Enfermedades Infecc Microbiol Clínica. 2013;31:3–11.

24. Dyar OJ, Howard P, Nathwani D, Pulcini C. Knowledge, attitudes, and beliefs of French medical students about antibiotic prescribing and resistance. Med Mal Infect. 2013;43:423–30.

25. Desai AJ, Gayathri GV, Mehta DS. "Public's Perception, Knowledge, Attitude and Behaviour on Antibiotic Resistance-A survey in Davangere City, India." J Prev Med Holist Health. 2015;2(1):17.

26. Hadi MP, Fadlillah LN, Widasmara MY, Muziasari WI, Subaryono. Potensi sumber bakteri resisten antibiotik berdasarkan kondisi kualitas air dan penggunaan lahan di Sungai Code , Yogyakarta : suatu tinjauan metodologis hewan. Di Indonesia , permasalahan mengenai resistensi Health Organization (. J Pengelolaan Lingkung Berkelanjutan. 2018;2(1):88–100.

27. Thapa SP, Shrestha S, Anal AK. Addressing the antibiotic resistance and improving the food safety in food supply chain (farm-to-fork) in Southeast Asia. Food Control. 2020;108:1–7.

28. Usui M, Ozawa S, Onozato H, Kuge R, Obata Y, Uemae T, Ngoc PT, Heriyanto A, Chalemchaikit T, Makita K, Muramatsu Y, Tamura, Y. Antimicrobial Susceptibility of Indicator Bacteria Isolated from Chickens in Southeast Asian Countries (Vietnam, Indonesia and Thailand). Jpn Larg Platf Acad E-J. 2014;5(76):685–92.

29. Chan YH, Fan MM, Fok CM, Lok ZL, Ni M, Sin CF, WongM, Yeung R, Yeung, TR, Chow, WC, Lam TH, Schooling CM. Antibiotics nonadherence and knowledge in a community with the world's leading prevalence of antibiotics resistance: Implications for public health intervention. Am J Infect Control. 2012;40(2):113–7.

30. Tamayanti WD, D.M. Sari W, Dewi BDN. Penggunaan antibiotik di dua apotek di Surabaya: identifikasi faktor-faktor yang mempengaruhi kepatuhan pasien. Pharmaciana. 2016;6(2):151–61.

31. Robert A, Nguyen Y, Bajolet O, Vuillemin B, Defoin B, Vernet-Garnier V, Drame M, Bani-Sadr F. Knowledge of Antibiotics and Antibiotic Resistance in Patients Followed by Family Physicians. Med Mal Infect. 2016;1–10.

32. Fernandez B anna maria. Studi Penggunaan Antibiotik Tanpa Resep Di Kabupaten Manggarai dan Manggarai Barat – NTT. 2013;2:1–17.

33. Almohammed RA, Bird EL. Public knowledge and behaviours relating to antibiotic use in Gulf Cooperation Council countries: A systematic review. J Infect Public Health. 2019;12:159–66.

Analytical Data for Electronic Medical Records in Primary Health Care

Guardian Yoki Sanjaya[1*], *Wahyudi* Istiono[2], *Agustinus* Verry Ricki[1], *Devi* Emrianti R[1], *Miftah* Adiyaksa Luckyarno[1], *Mochammad* Arief Darmawan[1]and *Nela* Afirda Prastika[1]

[1]Department of Health Policy and Management, Faculty of Medicine, Public Health and Nursing, Gadjah Mada University.
[2]Department of Family Medicine, Community and Bioethics, Faculty of Medicine, Public Health and Nursing, Gadjah Mada University

Abstract. Digital health transformation encourages primary health facilities to implement electronic medical records (RME) that are interoperable according to standard medical classification and terminology. The standard RME also allows connecting to wearable devices for direct patient monitoring. An analytical approach to digital data has the potential to support clinical decision making for primary care physicians. This study aims to Strengthening primary care as a center for continuous patient care by using an analytical approach in the form of a dashboard.. This study uses a participatory action research approach in implementing RME in primary care. The 4 stages of action research were carried out by involving primary care physicians (dentists and general practitioners), medical records, nurses, pharmacists and electronic medical record developers. The trial implementation of RME and wearable devices was evaluated using the System Usability Scale (SUS). Structured RME data makes it easy to analyze and visualize in the form of a dashboard to support primary care management and monitor individual patient health status. The analytic features in RME that allow direct patient monitoring are perceived as useful for supporting continuous patient care. The use of data standards in clinical records such as ICPC, LOINC and SNOMED-CT makes it easier to achieve semantic interoperability including potential interoperability with portable medical devices.

Keyword: ICPC 3, Wearable Device, Dashboard Patient, Electronic Medical Record, Primary Health Care

1 Background

Primary health care facilities have an important role in organizing health service efforts, be it promotive, preventive, curative or rehabilitative. As well as being the first point of contact in the health care system, primary care is a liaison with other health services through an effective referral system [1]. Several studies show that strengthening primary services can increase the efficiency of health financing through its role as a gatekeeper [2]. In addition, primary services can support the quality of health services in the community, such as the ability to manage 144 diseases thoroughly, follow up on medical management of patients with chronic illnesses, select and identify appropriate health resources for patients with an integrated referral system.

However, the strengthening of primary services still faces several challenges such as the availability and distribution of primary care doctors, the gap in incentive models, the limits of authority and scope of services that can be provided, the availability of medical support facilities, as well as the satisfaction and culture of the community in obtaining health services which are still centered on hospitals and medical specialist. Research conducted by Werni, Nurlinawati, and Rosita in 2017 regarding the implementation of essential public health efforts (UKM) in remote and very remote Puskesmas showed that as many as 87% of 131 Puskesmas carried out 5 types of essential services (health promotion; environmental health; maternal health, children and family planning; nutrition services; and, disease prevention and control) and the rest carry out less than 5 types of essential services. Data from the Institute for the Evaluation of Health Metrics, Ministry of Health in 2019 showed that there were 96.8% of infant deaths, 76.4% of child deaths, 63.9% of adolescent deaths, 72.6% of productive age deaths and 73.5% of elderly deaths caused by preventable or partially preventable disease or condition. This shows that services that focus

* Corresponding email: gysanjaya@ugm.ac.id

on promotive and preventive efforts must be a priority even though curative and rehabilitative services are also available in primary health services [3]. Based on previous studies, not all health centers in 8 districts/cities have health human resources according to the Minister of Health of the Republic of Indonesia Number 75 of 2014, but general practitioners, midwives and nurses are available in all health centers, although the number is still lacking. Meanwhile, in all medical practices, the most widely available health human resources are general practitioners and nurses. There have been changes in planning for human resource procurement in the regions after JKN, increased workload and working hours, so that it is necessary to plan and procure human resources based on needs [4].

Some countries are pursuing some models of more equitable incentives for primary service delivery [5], job dissatisfaction from a person can increase improvement turnovers employees, and decreased employee productivity so that it can also increase staffing costs and medical costs [6] as an example, several studies of job satisfaction that have been studied in the field of health services concluded that the effect of employee job satisfaction on job quality, effectiveness and efficiency is very high. The better quality of work will be shown if the level of satisfaction with one's work is higher so as to produce effectiveness and efficiency in health services [7]. This is in line with the competence of primary care physicians through the primary care specialist program.

One of the first-level health facilities is a clinic, a clinic is a health service facility that organizes individual health services that provide basic and/or specialist medical services. A health service is required to provide adequate and satisfactory services. Therefore, as a health service provider, a health facility requires the support of electronic medical records to optimize recording in primary health services. Through the Republic of Indonesia Minister of Health Regulation number 24 of 2022 concerning the implementation of electronic medical records, every health service facility (Fasyankes) is required to run an electronic patient medical history recording system [8].

The solution for primary services that will be provided is related to interoperability issues and PMK 24 regulations, namely by adding several supporting features so that they can maximize existing services in primary health facilities, such as the electronic medical record (RME), which is a computerized health information system in provides detailed information records including patient demographic data, patient medical history, allergies, history of laboratory tests, and some decision support [9]. The challenges of electronic medical records are interoperability and integration [10]. The Ministry of Health of the Republic of Indonesia has inaugurated the Satu Sehat platform to achieve interoperability and integration of digital health services used by the public, health service facilities and other relevant stakeholders.

Electronic medical records contribute significantly to improving the quality of service, one example is the facilitation of primary health services, namely the family doctor clinic. In applying electronic medical records to family doctor clinics, it is ideal to use a special classification standard for primary health services, namely the International Classification for Primary Care (ICPC). So far, primary health services in Indonesia still use the International Statistical Classification of Disease (ICD). The International Classification of Primary Care (ICPC) is a classification system that aims to explain patient clinical information data in primary health care [11].

The provision of primary health services must be able to take advantage of the use of wearable devices to monitor patients. A wearable device is an electronic device that can determine physical conditions and patterns life of its users [12]. With the integration of wearable devices with electronic medical records will produce an interactive dashboard that allows users to easily and quickly view data visualizations, one of which is in the form of a patient's condition chart. The patient dashboard is useful for monitoring the continuity of patient care, where data comes from electronic medical records, wearable devices, or patient contributions through personal health records (data accessibility/medical information for patients).

2 Method

The method in this study used participatory action research, where researchers collected data through observation and interviews or focus group discussions involving research subjects. The research subjects involved were family doctors, nurses and pharmacists in primary care, as well as primary care patients.

Research Period January-November 2023, Data collection 10 patient respondents in primary care Family doctor, Nurse, Pharmacist, Medical recorder, Research Location UGM Korpagama Family Medicine Clinic.

2.1 Wearable Device and Dashboard System Design

The design of this Monitoring System and Dashboard begins with making flowcharts, entity relationship diagrams, data flow diagrams, use case diagrams, and activity diagrams. This Monitoring system will later be accessed via Mobile Apps which can be installed through the Play Store and consists of two Mobile Apps, one Web Based Application for Administration, while

https://doi.org/10.1051/bioconf/20237505003

the dashboard will later be accessed via a Web Based application.

2.2 Prototype

There are 2 types of prototypes for the Monitoring System, namely Mobile Apps for Patients and Web Apps for Administration, while for dashboards using Web Apps. This prototype is designed so that users/patients can interact with Monitoring and the dashboard that will be used and can provide feedback for further development. The Monitoring User Interface and dashboard were designed using figma and then entered into Bootstrap and displayed and used to get feedback from users to evaluate the usability of the Monitoring System.

3 Results

3.1 Stages of Data Needs Analysis

The stages of data analysis in this study include:

a. Action diagnosis stage, at this stage a literature review is carried out related to primary service electronic medical records with ICPC, use of wearable devices and presentation of patient information in the form of dashboards. Furthermore, a Focus Group Discussion (FGD) was carried out with family doctors and health staff at the family clinic to identify needs and problems. use of electronic medical records in primary care. The output of this stage is the need for electronic medical records and management of medical record data to support continuity of care.

b. In the action planning stage, at this stage a focus group discussion (FGD) was conducted to determine priorities for the use of electronic medical records in primary care. The output of this stage includes plans to implement electronic medical records connected to wearable devices to produce dashboards.

c. In the implementation phase, at this stage the development of electronic medical records based on the International Classification for Primary Care (ICPC) version 3 will be carried out, then testing the functions of electronic medical records, including reconciliation of data sources from wearable devices and personal health records. The output of this stage is an interoperable ICPC-based electronic medical record.

d. Evaluation Phase, at this stage an evaluation is carried out using the System Usability Scale (SUS) framework to evaluate the usability of electronic medical records. The output of this stage includes continuous improvement of electronic medical records in primary care.

3.2 Data Sources

Sources of data in this study are primary data and secondary data obtained from data collection through questionnaires, interviews and documentation.

3.3 Dashboard Design

The dashboard is designed according to the needs of primary services. Figure 1, Figure 2, and Figure 3 show the dashboard design that has been designed:

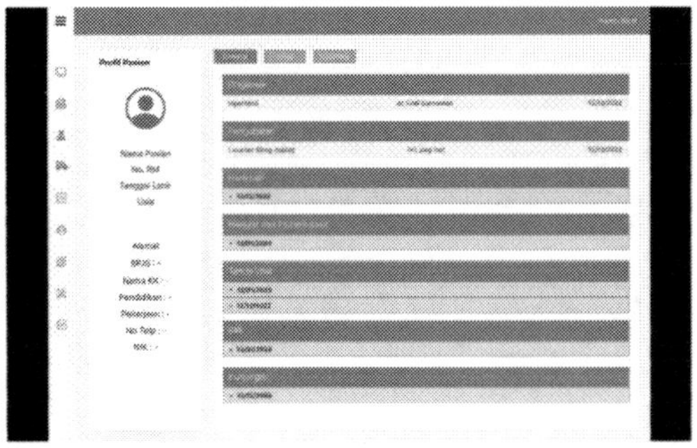

Fig. 1. General (Patient History) shows a dashboard related to the patient profile which diagnosis, treatment, lab results, history and examination, vital signs, etc.

Fig. 2. Trend (Patient Graph) shows a dashboard regarding patient trends containing blood pressure, BMI, lab results and oxygen saturation

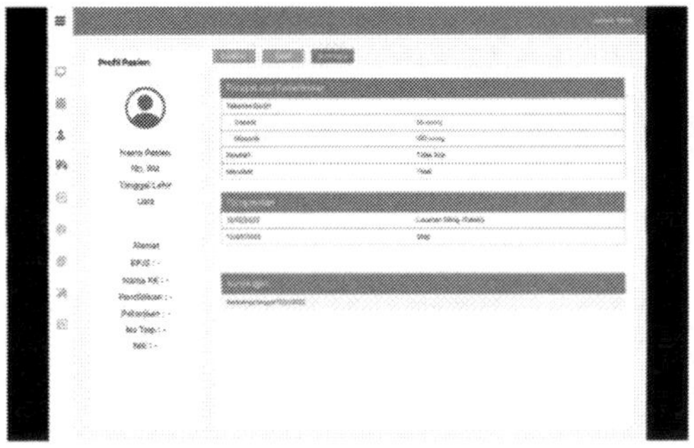

Fig. 3. Summary (Summary Patient) shows a summary containing the patient's history and examination, treatment that has been carried out, as well as history of visits to health facilities

https://doi.org/10.1051/bioconf/20237505003

3.4 Implementation of Electronic Medical Records based on ICPC-3

Figure 4 shows a general patient examination form with subjective indicators based on ICPC-3 and Figure 5 shows a general patient examination form with assessment indicators based on ICPC-3.

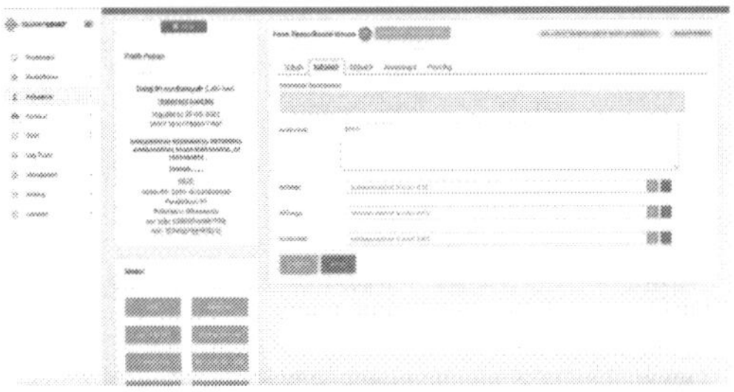

Fig. 4. Subjective with ICPC-3

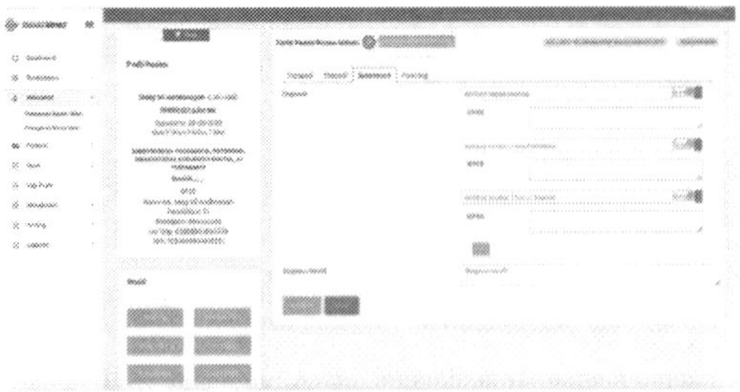

Fig. 5. Assessment with ICPC-3

3.5 Utilization of wearable devices in supporting electronic medical records

The following (Figure 6) is a prototype of a medical gamma application (Monitoring application) that utilizes wearable devices and is integrated with electronic medical records in primary care.

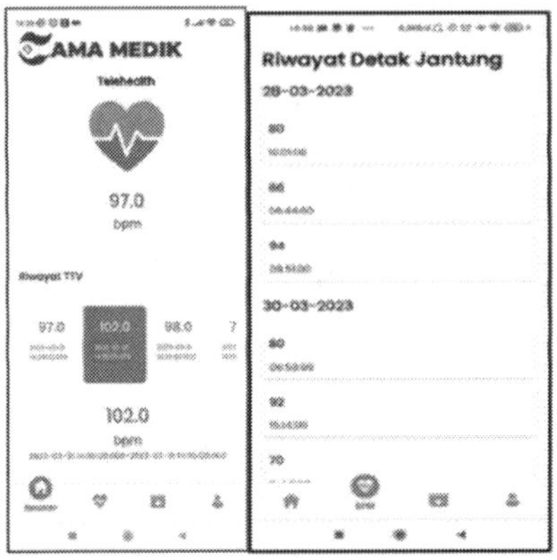

Fig. 6. Application of Gama Medik for a Wearable device

Evaluation of the Monitoring Application used the System Usability Scale (SUS) questionnaire and was carried out by 10 respondents. The questionnaire we sent resulted in the scores listed in Table 1.

Table 1. SUS Score Results for Monitoring Applications

Respondent Data	SUS Score/Score
1	62.5
2	67.5
3	70
4	77.5
5	52.5
6	52.5
7	57.5
8	62.5
9	55
10	62.5
Average SUS Score is 62	

4 Discussion

Electronic medical records contribute significantly to improving the quality of service, one example is the facilitation of primary health services, namely the family doctor clinic. In applying electronic medical records to family doctor clinics, it is ideal to use a special classification standard for primary health services, namely the International Classification for Primary Care (ICPC-3) where so far primary health services in Indonesia still use the International Statistical Classification of Disease (ICD). The International Classification of Primary Care (ICPC-3) is a classification system that aims to explain patient clinical information data in primary health care.

The provision of primary health services must be able to take advantage of the use of wearable devices to monitor patients. A wearable device is an electronic device that can determine the physical condition and lifestyle of its users. With the integration of wearable devices with electronic medical records, it will produce an interactive dashboard that allows users to easily and quickly view data visualizations, one of which is in the form of a patient's condition chart. The patient dashboard is useful for monitoring the continuity of patient care, where data comes from electronic medical records, wearable devices, or patient contributions through personal health records (accessibility of medical data/information for patients). Acceptance of patients in using wearable devices is quite acceptable, this is proven from the results of the SUS data showing that the SUS results with a value of 62 are classified as "Enough" which can be interpreted that the use of wearable devices can be well received by users and can be immediately implemented to support health services in the future because wearable devices are made to monitor health at this time generally equipped with movement sensors and heart rate. These two sensors have become a must-have feature for every smartwatch manufacturer today so that patients can see their condition while they are being treated. With this, as more and more sensor technologies can be embedded into smartwatches, the greater the opportunity for smartwatches to improve human health with various aspects [13].

5 Conclusion

In implementing electronic medical records in primary care to support continuity of care, the International Classification of Primary Care (ICPC-3) standard can be used. Patient telemonitoring can utilize wearable devices. As well as a patient dashboard that connects various useful data sources to support clinical decision making.

Reference

1. Hermawan, D. Nurcahyo, C. Afdal, A. "Quality Primary Services: A Literature Review The Qualified Primary Care: A Literature Review". Journal of National Health Insurance, vol 1, No. 1 (2021)
2. Rahma, A. Arso, SP Suparwati, A. "Implementation of Primary Health Center Functions as Gatekeepers in the JKN Program (Study at Juwana Health Center, Pati District)". Journal of Public Health, vol. 3 no. 3 (2015)
3. Situmorang, LF Citrawati, TN "Integration of Primary Health Services as an Effort to Transform Primary Services". Policy Briefs (2022)
4. Mujiati, M and Yuyun Yuniar. "Availability of health human resources at first-level health facilities in the era of National Health Insurance in eight districts-cities in Indonesia." Health Research and Development Media 26, no. 4 : 201-210.(2016)
5. Agustina, A. Budi H. and Pandujiwo N. "Systematic Study: Development of Risk Adjustment-Based Capitation Payment Systems in Various Countries." Journal of Indonesian Health Economics 6.2 (2022).
6. O'Leary, P., Wharton, N., Quinlan, T. "Job satisfaction of physicians in Russia". International Journal Of Health Care Quality Assurance, 22 (3), 221-31.(2009)
7. Pantouvakis, A., Mpogiatzidis, M. The impact of internal service quality and learning organization on clinical leaders job satisfaction in Hospital Care Services. Leadership in Health Services, 26(10), 34-49.(2013)
8. RI Minister of Health. Regulation of the Minister of Health of the Republic of Indonesia Number 24 of 2022 concerning medical records. Jakarta. (2022)
9. DA Ludwick and John Doucette. "Adopting electronic medical records in primary care: Lessons learned from health information systems implementation experience in seven countries," International Journal of Medical Informatics, vol. 78, pp. 22–31, (2009).
10. R. Hammami, H. Bellaaj, AH Kacem. "Interoperability for medical information systems: an overview," Health and Technology, vol. 4, pp. 261-272, (2014).
11. Ten Napel, H., van Boven, K., Olagundoye, O. A., van der Haring, E., Verbeke, M., Härkönen, M., Altbuis, Van Tjeerd., Agusto, K Daniel., Laurent, Letrilliart., Sebrans, Diego., Weel, Van Cbris ., Schers, H. (2022). Improving Primary Health Care Data With ICPC-3: From a Medical to a Person-Centered Perspective. The Annals of Family Medicine, 20(4), 358-361.
12. N. Wikansari and DB Santoso. "Diablock: Prototype of Mobile-Based Personal Health Records for Diabetics" Indonesian Journal of Health Information, Vol.8 No. 1, 19-27. (2022)
13. Siradj, Yahdi. "Smartwatch Potential for Health Smartwatch Potentials for Healthcare". Telekontran, vol.4, no.1, 35-41. (2016)

Assessing the relationship between work-related factors and the quality of working life among nurses: A cross-sectional study in Laos

Moukda Chanthavisouk [1,#], *Trung Quang* Vo[1,#,*], *Quang Vinh* Tran[1,#], *Bay Van* Vo[1], *Tuan Nguyen Anh* Ho[2], *Susi Ari* Kristina[3], *Dwi* Endarti[3], *and Shyamkumar* Sriram[4]

[1] Faculty of Pharmacy, Pham Ngoc Thach University of Medicine, Ho Chi Minh City 700000, Vietnam.
[2] Faculty of Medicine, Pham Ngoc Thach University of Medicine, Ho Chi Minh City 700000, Vietnam.
[3] Faculty of Pharmacy, Universitas Gadjah Mada, Yogyakarta 55281, Indonesia.
[4] College of Health Sciences and Professions, Ohio University, Athens OH 45701, Ohio, United States.
[#] These authors are the co-first authors and contributed equally to this work.

Abstract. The quality of working life is crucial for improving work productivity, particularly among nurses, who often experience high levels of stress. This study aims to evaluate the quality of working life among nurses in Laos and identify the factors that influence it. A cross-sectional study was conducted among Laos nurses from August 2021 to July 2022. Data collection was conducted using an anonymous questionnaire distributed via the Internet. The Quality of Working Life version 2 (WRQoL-2) questionnaire, comprising 32 items divided into seven subscales, was employed to assess the quality of working life. Statistical tests such as t-tests, ANOVA, and Spearman correlation were applied to examine differences and correlations. A total of 326 participants were included, with an average age of 32.62±8.21 years. Among the seven subscales, the highest score was observed in the Job Career Satisfaction subscale (3.72±0.56), while the lowest score was found in the Safety at Work subscale (3.22±0.71). The overall mean score was 3.49±0.54. Significant differences in the quality of working life were observed among different groups categorized by age, job position, salary, and working hours. The WRQoL-2 questionnaire was found to be suitable for assessing the quality of working life in this study.
Keywords: job satisfaction, healthcare staff, Laos, stress, WRQoL.

1 Introduction

Quality of working life (QoWL) has been mentioned since 1960 and is regarded as a key to augmenting worker productivity [1]. There were various definitions of QoWL proposed, with different relevant factors highlighted [2]. While some authors merely concentrated on job characteristic criteria, others emphasized numerous aspects, including personality, psychological wellbeing, relationships with managers and colleagues, life satisfaction, and happiness [2-9]. The QoWL reflects not only the employees' consciousness of physical and mental health relating to their work, but also their contentment based on their experiences in the organizations [10, 11].

It cannot be denied that healthcare staff usually have an extreme level of stress because of their heavy workload and the nature of their occupations. According to a report, occupations in the health sector ranked third in terms of depression [12]. Especially nurses have been admitted to usually cope with distress and stress-related burnout in high numbers of prevalence [13-15]. This negative psychology could lead to a decrease in nurses' work performance, which would have a detrimental effect on both them and their patients [16]. Therefore, it is necessary for governments to pay attention to the assessment of QoWL of nurses in order to mitigate some of the necessary preventions and interventions.

There are some tools to assess the QoWL under several language versions, such as the GHQ-12 General Health Questionnaire, Warr Job Satisfaction Scale (WJSAT), Warr Job Related Well-being Anxiety-Contentment Scale (WJRWB-AC) and the Work Locus of Control [2]. Among them, the Work-Related Quality of Life Scale (WRQoL) was acknowledged as a dependable indicator with strict generation and validation [2, 17]. It is likely to demonstrate a systematic view of the definitions of QoWL given [2]. Moreover, it is only presented on a single page, which has significant advantages for data collection. Many researchers applied this survey tool in their studies, addressing different types of subjects. There were some of them that could be mentioned, comprising higher education staff in the UK (The United Kingdom) (2009), nurses in China (2013),

* Corresponding email: trungvq@pnt.edu.vn

https://doi.org/10.1051/bioconf/20237505004

police in the UK (2013), train drivers in Iran (2014), nurses in Uganda (2014), nurses in Taiwan (2016), nurses in Turkey (2016) and General Surgery Residents in the US (The United State) [17-24]. Nowadays, the WRQoL scale version 2 was published with better psychometric properties to accomplish the objective of QoWL estimation when compared to the origin.

Laos has a large population, with more than 7.4 million residents in 2021 and the figures recorded over the last ten years showed a dramatic upward tendency [25]. This means the demand for health care in this nation will probably increase in the future. It is critical to establish reliable data on Lao healthcare workers' QoWL. However, there was still a limited amount of research carried out in this field. The objective of this study was to evaluate the quality of working life among Laos nurses and explore influencing factors by using the WRQoL questionnaire.

2 Materials and Methods

2.1 Study Design

A cross-sectional study was conducted among Laos nurses from August 2021 to July 2022. Data collection took place from January to March 2022 using an online questionnaire. An anonymous questionnaire was shared within nurse networks to invite nurses to participate in the study.

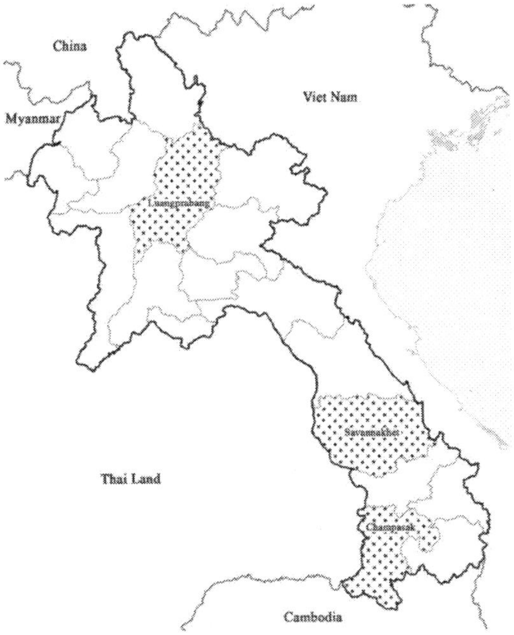

Fig. 1. Maps of Lao People's Democratic Republic with neighbor coutries

2.2 Study Subjects

The study included Laos' nurses who were working at a medical center and voluntarily chose to participate. The research objectives were explained to the participants before they answered the questionnaire.

2.3 Survey Instrument

The questionnaire consisted of two parts. Part one gathered demographic information such as age, gender, marital status, education level, work location, employer, job position, type of work, income, working hours, and extra work. Part two included the second edition of the Quality of Working Life scale (WRQoL-2), which was translated into the Laotian language. The WRQoL-2 comprised 32 questions divided into seven subscales: Job and Career Satisfaction (JCS - 6 items), General Well-Being (GWB - 6 items), Home-Work Interface (HWI - 4 items), Control at Work (CAW - 4 items), Working Conditions (WCS - 4 items), Stress at Work (SAW - 4 items), and Employee Engagement (EEN - 3 items). Details are presented in **Table 2** [26]. In addition, question 32, *"I am satisfied with the overall quality of my working life"*, was not included in any subscales but was used as a single measure of QoWL to validate the WRQoL scale. [26]. Each question was answered using a five-point Likert scale ranging from 1= strongly disagree, 2= disagree, 3= neutral, 4= agree, to 5= strongly agree [2]. The translation from English to Laotian followed the process recommended by the World Health Organization [27].

2.4 Validation

A pilot test was conducted with a sample size of 30 participants to verify the questionnaire. Cronbach's alpha was calculated for each subscale and the overall score. The obtained values for JCS, GWB, HWI, CAW, WCS, SAW, EEN, and overall were 0.721, 0.614, 0.800, 0.652, 0.750, 0.622, 0.829, and 0.906, respectively, indicating acceptable internal consistency.

2.5 Statistical Analysis

All collected data were entered into Microsoft Excel 2016 and analyzed using the Statistical Package for the Social Sciences (SPSS) version 20.0. Following the provided instructions, scores for the five dimensions and the overall quality of working life were calculated and presented as Mean±Standard Deviation (Mean±SD) [26]. The differences in WRQoL scores among groups categorized by demographic characteristics were assessed using t-tests or one-way analysis of variance (ANOVA). Spearman's correlation analysis was used to examine the correlation between two continuous variables based on their distribution.

3 Results

A total of 326 eligible nurses participated in the study, with an average age of 32.62±8.21. The study population had a distribution of 43.87% participants under the age of 30 and 17.18% participants over the age of 40. **Table 1** provides an overview of the demographic variables. The majority of participants were female (69.63%), married (60.43%), and working in urban areas (83.44%). Approximately 55.21% of the nurses had a bachelor's degree, and only a small minority worked for foreign medical centers. More than 90% of the nurses held staff positions and worked full-time. Figure 1 presents the distribution of nurses based on the number of disabled, elderly, and young family members. Nearly half of the nurses had an elderly family member (N=162) or children under 6 years old (N=157), while 53.07% (N=173) had children between the ages of 6 and 18. The majority of nurses did not have disabled family members.

As shown in **Table 1**, the income distribution of the participants indicated that 49.69% (N=162) earned less than 3 million Laotian Kip. Around 50% of the nurses reported having an extra income source or not having any other income source. Notably, 62.58% of the nurses expressed satisfaction with their income. In terms of working hours, Laos nurses reported spending an average of 39.82±16.79 hours on their tasks, and 213 out of 326 participants indicated that they occasionally or seldom had to work extra hours.

Table 1. Demographic characteristics

Variables	N (%)	Variables	N (%)
Age		**Type of work**	
Mean±SD	32.62±8.21	Full time	265 (81.29)
Gender		Part time	61 (18.71)
Male	99 (30.37)	**Extra income source**	
Female	227 (69.63)	Yes	148 (45.40)
Marital status		No	178 (54.60)
Single	121 (37.12)	**Monthly income (million Laotian Kip)**	
Married	197 (60.43)		
Divorced/Widow	8 (2.45)	< 1.5	162 (49.69)
Education level		1.5 - < 2.0	122 (37.42)
Short-time training	64 (19.63)	2.0 - <3.0	27 (8.28)
Bachelor	180 (55.21)	≥ 3	15 (4.60)
Post-graduate	82 (25.16)	**Satisfaction of income**	
Work location		Yes	204 (62.58)
Urban	272 (83.44)	No	122 (37.42)
Rural	54 (16.56)	**Working hour per week**	
Employer		Mean±SD	39.82±16.79
State	175 (53.68)	**Extra working hour**	
Private	143 (43.87)	Never	54 (16.56)
Foreign	8 (2.45)	Seldom	96 (29.45)
Job position		Sometimes	127 (38.96)
Manager/Head of department	21 (6.44)	Usually	39 (11.96)
Staff	305 (93.56)	Always	10 (3.07)

Table 2 displays the average scores for each subscale and the overall Quality of Working Life (QoWL) assessment for Laos nurses across multiple aspects. The highest score was observed in Job and Career Satisfaction (JCS) with a mean of 3.72±0.56, followed by Working Conditions (WCS) with 3.63±0.68, General Well-Being (GWB) with 3.60±0.68, and Employee Engagement (EEN) with 3.59±0.82. The lowest mean score was reported for Stress at Work (SAW) with 3.22±0.71. On average, Laos nurses scored 3.49±0.54 for QoWL.

Additionally, more than 34% of nurses responded "neutral" or "agree" to item 32, *"I am satisfied with the overall quality of my working life"*, which represented the highest proportion among the five response options.

Table 3 presents the mean QoWL scores for each group categorized based on demographic characteristics, along with the results of difference testing. Significant differences were observed in JCS scores when comparing age, education level, and job position groups. T-tests and ANOVA tests indicated statistically significant differences in GWB averages among groups classified by age, marital status, education level, and type of employer (p-value<0.05). This suggests a need to focus on improving QoWL for nurses in lower hierarchical positions. For HWI, age group and job position were

Table 2. Work-related quality of life of Laos nurse (N=326)

Aspect		Question	Mean±SD
JCS	Job and career satisfaction	q01, q03, q08, q11, q18, q20	3.72±0.56
GWB	General well-being	q04, q09, q10, q15, q17, q21	3.60±0.68
HWI	Home-work interface	q05, q06, q14, q25	3.36±0.80
CAW	Control at work	q02, q12, q23, q30	3.34±0.57
WCS	Working conditions	q12, q16, q22, q31	3.63±0.68
SAW	Stress at work	q07, q19, q24, q29	3.22±0.71
EEN	Employee engagement	q26, q27, q28	3.59±0.82
Overall	Average of all aspects above		3.49±0.54

identified as influential factors. In addition to these factors, gender influenced the mean score for Control at Work (CAW). Significant differences (P-value < 0.01) in WCS and overall scores were observed when analyzing age groups and job positions. No factors were found to be associated with SAW and EEN. **Figure 2** presents the average WRQoL overall score according to caregiving responsibilities. T-tests did not reveal any statistically significant differences between these groups (P-value > 0.05).

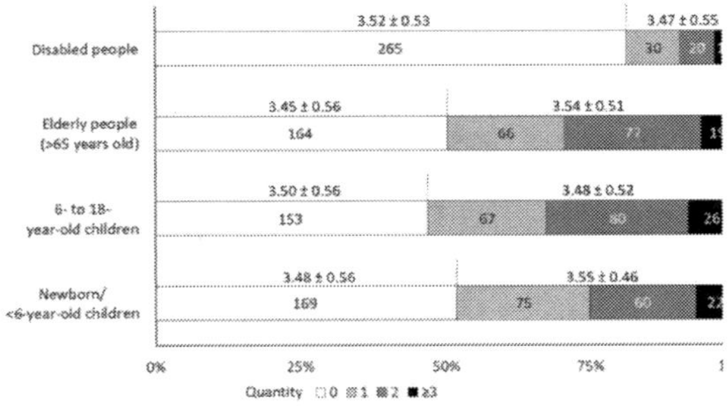

Fig. 2. WRQoL overall score followed by caregiving responsibilities (P-value > 0.05)

Table 3 also displays the WRQoL overall scores based on income and working hours. The importance of income in relation to QoWL was evident. Nurses who were satisfied with their income or had a monthly salary of less than 3 million Laotian Kip had higher scores in all aspects and overall QoWL. These results were statistically significant, except when comparing SAW between groups based on monthly income. Regarding working extra hours, ANOVA tests yielded $p<0.01$ for HWI and SAW, and $p=0.02$ for GWB. Having an extra income source was a significant factor that influenced all scores, except for WCS and SAW. Spearman's correlation analysis revealed a negative correlation between working hours and QoWL.

4 Discussion

4.1 Demographic characteristics

The average age of the participants in this study was 32.62±8.21 years, which was similar to a study conducted in Iran (33.1±8.00) but higher than a study in Turkey (29.5±7.1). The decrease in the number of participants with increasing age groups could be attributed to the methodology of data collection via the internet, which may have influenced the participation of older individuals. The gender distribution showed a higher proportion of females (69.63%) compared to males, which is consistent with the nature of the nursing profession. Similar results have been reported in studies conducted in Iran (61.4% females) and China (96.6% females) [19, 28].

4.2 Quality of working life

Among the subscales, Job and Career Satisfaction (JCS) had the highest average score of 3.72±0.56, indicating that Laos nurses felt content in their workplace and recognized their career development. This subscale plays a crucial role in the overall quality of working life [2]. When compared to other studies, the JCS score of Laos nurses was higher than that of nurses in Turkey (3.0-3.3) and China (3.48±0.58) [19, 23]. However, studies conducted in China and Uganda also identified JCS as the dimension with the highest score when evaluating the quality of working life of nurses [21, 28]. The subscale with the lowest score in this study was Stress at Work (SAW) with a mean of 3.22±0.71. The high work pressure in healthcare professions, particularly nursing, is well-known. Some studies have identified other subscales such as Working Conditions (WCS) or Home-Work Interface (HWI) as having the lowest scores [29]. In 2017, there were an estimated 2.1 nurses and midwives in Laos, which means Laos nurses usually have a heavy workload [30]. Abbasi *et al.* assessed the SAW score at 3.21±0.77 which was also the smallest value among the six dimensions [28]. Some studies have identified other subscales such as Working Conditions (WCS) or Home-Work Interface (HWI) as having the lowest scores [21, 23]. The overall score for quality of working life among

Laos nurses was 3.49±0.54, indicating that their quality of working life was above average. However, due to the lack of studies using the WRQoL version 2 to assess nurses' quality of working life, a direct comparison of the overall score is challenging.

4.3 Influencing factors

The analysis of demographic characteristics revealed that age and job position significantly influenced the quality of working life of nurses. Previous studies in Iran have also shown a significant relationship between age and WRQoL scores, while other studies have identified gender as an influencing factor [21, 28, 31]. Regarding job position, managers or heads of departments had significantly higher scores in JCS, HWI, Control at Work (CAW), and WCS compared to staff nurses. Similar findings have been reported by Shukla et al., except for WCS.

Factors related to income and working hours played an important role in the quality of working life scores. They significantly affected most of the subscales and the overall score. Nurses who were satisfied with their income had significantly higher scores in all dimensions compared to those who were not satisfied, and there was a significant difference between nurses with a salary of ≥ 3 million Laotian Kip and their counterparts. Therefore, providing reasonable remuneration and implementing wage increase policies are crucial for improving nurses' psychological well-being.

4.4 Limitations and recommendations

Several limitations should be considered in this study. Firstly, the data collection method relied on internet-based surveys, which may have limited the participation of certain individuals. Conducting direct surveys at medical centers would provide a more representative sample. Secondly, as a cross-sectional study, it cannot establish causal relationships. Future research should consider longitudinal designs to investigate the dynamic nature of the quality of working life. Lastly, the sample size was limited to 326 nurses, which may not fully represent the entire population. Future studies with larger sample sizes would provide a more comprehensive understanding of the quality of working life among nurses. Based on the limitations of this study, several recommendations can be made for future research. Direct surveys conducted at medical centers would ensure a more diverse and representative sample. Expanding the study to include healthcare staff beyond nurses would provide a broader perspective on the quality of working life in the health sector. Additionally, comparing job satisfaction between different occupations within the healthcare sector would be valuable for designing interventions to improve overall working productivity and well-being.

5 Conclusion

In conclusion, the WRQoL version 2 scale was found to be suitable for assessing the quality of working life (QoWL) of nurses. The average score for WRQoL was 3.49±0.54, indicating a relatively positive level of QoWL among nurses in Laos. Among the dimensions of WRQoL, job and career satisfaction had the highest score, while stress at work had the lowest score. It is important for managers and directors to consider factors such as working hours and salary to improve nurses' QoWL. By understanding and addressing these influencing factors, it is possible to enhance the quality of working life for nurses, leading to improved efficiency in community healthcare.

Table 3. Work-related quality of life by demographic characteristics

Variables		JCS	GWB	HWI	CAW
Age group	<30	3.64±0.57	3.48±0.65	3.23±0.81	3.28±0.54
	30-40	3.75±0.54	3.62±0.70	3.37±0.79	3.33±0.57
	>40	3.88±0.56	3.87±0.61	3.68±0.69	3.52±0.59
	p-value	**0.02***	**<0.01***	**<0.01***	**0.03***
Gender	Male	3.80±0.55	3.67±0.63	3.44±0.80	3.48±0.53
	Female	3.69±0.56	3.57±0.69	3.33±0.80	3.28±0.57
	p-value	*0.12*	*0.22*	*0.24*	***<0.01****
Marital status	Single	3.68±0.57	3.50±0.67	3.30±0.80	3.32±0.50
	Married	3.74±0.54	3.64±0.66	3.38±0.79	3.34±0.58
	Divorced/Widow	3.94±0.77	4.06±0.86	3.91±0.77	3.69±0.76
	p-value	*0.32*	***0.03****	*0.10*	*0.20*
Education level	Short-time training/Bachelor	*3.67±0.54*	3.55±0.66	3.32±0.78	3.33±0.55
	Master/Doctor	3.88±0.58	3.75±0.70	3.49±0.83	3.37±0.62
	p-value	***<0.01****	***0.02****	*0.10*	*0.59*
Work location	Urban	3.73±0.57	3.61±0.67	3.38±0.81	3.32±0.58
	Rural	3.69±0.49	3.56±0.69	3.25±0.73	3.43±0.50
	p-value	*0.61*	*0.67*	*0.28*	*0.20*
Employer	State	3.76±0.54	3.68±0.65	3.43±0.80	3.36±0.55

	Private/Foreign	3.68±0.57	3.50±0.70	3.29±0.79	3.31±0.59
	p-value	*0.21*	***0.02****	*0.12*	*0.40*
Job Position	Manager/Head of department	4.04±0.63	3.87±0.87	3.85±0.92	3.75±0.58
	Staff	3.70±0.55	3.58±0.66	3.33±0.78	3.31±0.55
	p-value	***0.01****	*0.06*	***<0.01****	***<0.01****
Type of work	Full time	3.72±0.57	3.59±0.69	3.35±0.78	3.32±0.58
	Part time	3.74±0.49	3.66±0.60	3.41±0.86	3.41±0.48
	p-value	*0.77*	*0.41*	*0.64*	*0.26*
Monthly income	< 3 million Laotian Kip	3.65±0.57	3.50±0.68	3.24±0.82	3.24±0.56
	≥ 3 million Laotian Kip	3.79±0.54	3.70±0.66	3.49±0.76	3.44±0.56
	p-value	***0.02****	***0.01****	***<0.01****	***<0.01****
Income satisfaction	Yes	3.85±0.54	3.81±0.64	3.61±0.72	3.44±0.55
	No	3.51±0.53	3.26±0.59	2.95±0.76	3.18±0.56
	p-value	***<0.01****	***<0.01****	***<0.01****	***<0.01****
Extra working hour	Never	3.73±0.55	3.59±0.77	3.52±0.80	3.38±0.50
	Seldom	3.76±0.61	3.71±0.63	3.47±0.76	3.28±0.62
	Sometimes	3.72±0.52	3.63±0.62	3.36±0.73	3.38±0.56
	Usually	3.68±0.63	3.36±0.74	3.06±0.95	3.33±0.53
	Always	3.57±0.39	3.20±0.71	2.63±0.80	3.15±0.63
	p-value	***0.86***	***0.02****	***<0.01****	***0.55***
Extra income source	Yes	3.82±0.56	3.69±0.68	3.49±0.74	3.42±0.52
	No	3.65±0.54	3.52±0.66	3.25±0.82	3.27±0.59
	p-value	***0.01****	***0.03****	***0.01****	***0.01****
Working hour	r	-	-0.185	-0.177	-
	p-value	0.08	***<0.01****	***<0.01****	0.27

Variables		**WCS**	**SAW**	**EEN**	**Overall**
Age group	<30	3.55±0.64	3.15±0.66	3.55±0.83	3.41±0.53
	30-40	3.63±0.71	3.27±0.76	3.56±0.83	3.50±0.54
	>40	3.83±0.69	3.27±0.72	3.74±0.79	3.69±0.52
	p-value	**0.03***	0.33	0.29	**0.01***
Gender	Male	3.70±0.69	3.13±0.74	3.61±0.90	3.55±0.52
	Female	3.60±0.68	3.25±0.70	3.58±0.79	3.47±0.55
	p-value	*0.23*	*0.17*	*0.77*	*0.24*
Marital status	Single	3.55±0.60	3.12±0.67	3.53±0.84	3.43±0.53
	Married	3.66±0.68	3.28±0.73	3.60±0.82	3.52±0.54
	Divorced/Widow	4.09±0.68	3.00±0.68	4.08±0.77	3.82±0.56
	p-value	*0.05*	*0.10*	*0.17*	*0.07*
Education level	Short-time training/Bachelor	3.61±0.63	3.23±0.68	3.58±0.79	3.47±0.52
	Master/Doctor	3.7±0.810	3.18±0.79	3.59±0.91	3.56±0.60
	p-value	*0.33*	*0.56*	*0.96*	*0.18*
Work location	Urban	3.65±0.69	3.24±0.72	3.59±0.84	3.50±0.55
	Rural	3.56±0.64	3.09±0.65	3.58±0.77	3.45±0.50
	p-value	*0.37*	*0.15*	*0.96*	*0.53*
Employer	State	3.66±0.69	3.24±0.69	3.64±0.83	3.54±0.53
	Private/Foreign	3.59±0.67	3.19±0.73	3.53±0.82	3.44±0.54
	p-value	*0.37*	*0.52*	*0.24*	*0.11*
Job Position	Manager/Head of department	4.13±0.70	3.24±0.83	3.76±1.05	3.80±0.67
	Staff	3.60±0.67	3.21±0.70	3.57±0.81	3.47±0.52
	p-value	***<0.01****	*0.88*	*0.31*	***<0.01****
Type of work	Full time	3.63±0.69	3.21±0.73	3.55±0.84	3.48±0.55
	Part time	3.65±0.62	3.26±0.59	3.75±0.76	3.56±0.51
	p-value	*0.79*	*0.61*	*0.08*	*0.32*
Monthly income	< 3 million Laotian Kip	3.55±0.67	3.20±0.72	3.48±0.88	3.41±0.55
	≥ 3 million Laotian Kip	3.71±0.69	3.23±0.71	3.69±0.76	3.58±0.52
	p-value	***0.03****	***0.69***	***0.03****	***<0.01****

Income satisfaction	Yes	3.79±0.65	3.34±0.71	3.72±0.78	3.65±0.50
	No	3.37±0.66	3.01±0.67	3.36±0.85	3.23±0.50
	p-value	***<0.01****	***<0.01****	***<0.01****	***<0.01****
Extra working hour	Never	3.69±0.65	3.44±0.63	3.48±0.89	3.54±0.56
	Seldom	3.63±0.75	3.32±0.75	3.56±0.85	3.53±0.56
	Sometimes	3.65±0.63	3.19±0.69	3.68±0.78	3.52±0.50
	Usually	3.53±0.71	2.83±0.69	3.51±0.88	3.33±0.57
	Always	3.40±0.71	2.95±0.39	3.50±0.61	3.20±0.50
	p-value	***0.64***	***<0.01****	***0.54***	***0.10***
Extra income source	Yes	3.71±0.69	3.27±0.71	3.73±0.79	3.59±0.52
	No	3.56±0.67	3.17±0.71	3.47±0.84	3.41±0.54
	p-value	***0.05***	***0.24***	***<0.01****	***<0.01****
Working hour	r	-	-	-	-0.132
	p-value	0.32	0.10	0.62	**0.02***

Note: - r was unnecessary to be calculated; *P-value<0.05; Test used: t-test, ANOVA and Spearman.

Conflict of Interest

There is no conflict of interest to declare.

Funding

No funding was received for this analysis.

Acknowledgements

The author would like to acknowledge the voluntary participants for data collection.

Authors Contributions

All authors contributed to the study conception and design. Material preparation, data collection and analysis were performed by Chanthavisouk Moukda, Trung Quang Vo, Quang Vinh Tran. All authors commented on all draughts of the manuscript. All authors read and approved the final version for submission.

References

1. E. Mayo, *The Human Problems of an Industrial Civilization.* Nature, 1934. **134**(3380): p. 201-201.
2. S. Easton and D. Van Laar, *User Manual for the Work-Related Quality of Life (WRQoL) Scale : A Measure of Quality of Working Life. 2nd ed.* 2018: University of Portsmouth.
3. V. V. Baba and M. Jamal, *Routinization of job context and job content as related to employees' quality of working life: A study of Canadian nurses.* Journal of Organizational Behavior, 1991. **12**(5): p. 379-386.
4. J. R. Hackman and G. R. Oldham, *Motivation through the design of work: test of a theory.* Organizational Behavior and Human Performance, 1976. **16**(2): p. 250-279.
5. R. A. Katzell and R. A. Guzzo, *Psychological approaches to productivity improvement.* American Psychologist, 1983. **38**(4): p. 468-472.
6. R. S. M. Lau and B. E. May, *A win-win paradigm for quality of work life and business performance.* Human Resource Development Quarterly, 1998. **9**(3): p. 211-226.
7. K. A. Loscocco and A. R. Roschelle, *Influences on the quality of work and nonwork life: Two decades in review.* Journal of Vocational Behavior, 1991. **39**(2): p. 182-225.
8. P. H. Mirvis and E. E. Lawler, *Accounting for the Quality of Work Life.* Journal of Occupational Behaviour, 1984. **5**(3): p. 197-212.
9. P. Warr, J. Cook, and T. Wall, *Scales for the measurement of some work attitudes and aspects of psychological wellbeing.* Journal of Occupational Psychology, 1979. **52**: p. 129-148.
10. H. Li, Z. Liu, R. Liu, L. Li, and A. Lin, *The relationship between work stress and work ability among power supply workers in Guangdong, China: a cross-sectional study.* BMC public health, 2016. **16**: p. 123.
11. K. Tuomi, J. Ilmarinen, L. Eskelinen, E. Jarvinen, J. Toikkanen, and M. Klockars, *Prevalence and incidence rates of diseases and work ability in different work categories of municipal occupations.* Scand J Work Environ Health, 1991. **17 Suppl 1**: p. 67-74.
12. Substance Abuse and Mental Health Services Administration, *Results From the 2006 National Survey on Drug Use and Health: National Findings.* 2007, Department of Health and Human services. Substance Abuse and Mental Health Services Administration. Office of Applied Studies: Rockville, MD.
13. L. P. Chou, C. Y. Li, and S. C. Hu, *Job stress and burnout in hospital employees: comparisons of different medical professions in a regional hospital in Taiwan.* BMJ Open, 2014. **4**(2): p. e004185.
14. N. Khamisa, K. Peltzer, and B. Oldenburg, *Burnout in relation to specific contributing factors and health*

outcomes among nurses: a systematic review. Int J Environ Res Public Health, 2013. **10**(6): p. 2214-40.
15. R. K. Roberts and P. L. Grubb, *The consequences of nursing stress and need for integrated solutions.* Rehabilitation nursing : the official journal of the Association of Rehabilitation Nurses, 2014. **39**(2): p. 62-69.
16. H. Sveinsdottir, P. Biering, and A. Ramel, *Occupational stress, job satisfaction, and working environment among Icelandic nurses: a cross-sectional questionnaire survey.* Int J Nurs Stud, 2006. **43**(7): p. 875-89.
17. J. A. Edwards, D. Van Laar, S. Easton, and G. Kinman, *The Work-related Quality of Life Scale for Higher Education Employees.* Quality in Higher Education, 2009. **15**(3): p. 207-219.
18. S. Easton, D. V. Laar, and R. Marlow-Vardy, *Quality of Working Life and the Police.* Management, 2013. **3**(3): p. 135-141.
19. X. Zhao, T. Sun, Q. Cao, C. Li, X. Duan, L. Fan, and Y. Liu, *The impact of quality of work life on job embeddedness and affective commitment and their co-effect on turnover intention of nurses.* J Clin Nurs, 2013. **22**(5-6): p. 780-8.
20. A. Mazloumi, Z. Kazemi, G. Nasl-Saraji, and S. Barideh, *Quality of Working Life Assessment among Train Drivers in Keshesh Section of Iran Railway.* International Journal of Occupational Hygiene, 2014. **6**(2): p. 50-55.
21. J. G. Opollo, J. Gray, and L. A. Spies, *Work-related quality of life of Ugandan healthcare workers.* Int Nurs Rev, 2014. **61**(1): p. 116-23.
22. H. D. Dai, F. I. Tang, I. J. Chen, and S. Yu, *Taiwanese Version of the Work-Related Quality of Life Scale for Nurses: Translation and Validation.* J Nurs Res, 2016. **24**(1): p. 58-67.
23. H. Kahyaoglu Sut and E. Mestogullari, *Effect of Premenstrual Syndrome on Work-Related Quality of Life in Turkish Nurses.* Saf Health Work, 2016. **7**(1): p. 78-82.
24. M. H. Zubair, L. R. Hussain, K. N. Williams, and K. J. Grannan, *Work-Related Quality of Life of US General Surgery Residents: Is It Really so Bad?* J Surg Educ, 2017. **74**(6): p. e138-e146.
25. The World Bank. *Population, total - Indonesia.* 2018 [cited Oct 15, 2019; Available from: https://data.worldbank.org/indicator/SP.POP.TOTL?locations=ID.
26. Quality of working life. *WRQoL-2 Scale.* 2015 [cited July 20, 2019; Available from: http://www.qowl.co.uk/researchers/qowl_downloads_qnr_overview.html#WRQoL2.
27. World Health Organization. *Process of translation and adaptation of instruments.* 2019 [cited July 20, 2019; Available from: https://www.who.int/substance_abuse/research_tools/translation/en/.
28. M. Abbasi, A. Zakerian, A. Akbarzade, N. Dinarvand, M. Ghaljahi, M. Poursadeghiyan, and M. H. Ebrahimi, *Investigation of the Relationship between Work Ability and Work-related Quality of Life in Nurses.* Iran J Public Health, 2017. **46**(10): p. 1404-1412.
29. K. Shukla, S. Shahane, and W. D'Souza, *Role of demographic and job-related variables in determining work-related quality of life of hospital employees.* J Postgrad Med, 2017. **63**(1): p. 4-10.
30. The World Bank. *Nurses and midwives (per 1,000 people).* 2019 [cited Oct 20, 2019; Available from: https://data.worldbank.org/indicator/SH.MED.NUMW.P3.
31. C. Carrillo-Garcia, C. Solano-Ruiz Mdel, M. E. Martinez-Roche, and C. I. Gomez-Garcia, *Job satisfaction among health care workers: the role of gender and age.* Rev Lat Am Enfermagem, 2013. **21**(6): p. 1314-20.

Association Between Religiosity, Spirituality, and Depression-Anxiety Among Pharmacist Students in Indonesia

Nisa Febrinasari[1*], *Chilmia* Nurul Fatiha[1] and *Risda* Fatin Fitria[1]
[1]Faculty of Pharmacy, Universitas Islam Sultan Agung, Semarang, Indonesia.

Abstract. Students in the Pharmacist Professional Study Program (PPSP) must pass the Indonesian Pharmacist Competency Exam (UKAI) to get the title of pharmacist. They feel the burden and fear of not passing the exam, which can trigger anxiety and depression, especially in the pandemic era. Religion also provides a perspective that individuals can use to reduce their distress when faced with many stressors. The study aims to determine the relationship between religiosity and spirituality and depression and anxiety in PPSP students in Indonesia. The method used in the study is a cross-sectional design with DUREL, DSES, SAS and SDS questionnaires as the data instruments. The study was carried out in all Association of Higher Education in Indonesian Pharmacy (APTFI) regions. A cluster random sampling technique was conducted and 362 students participated. The study found that Indonesian PPSP students had mild to moderate anxiety (21%). Fortunately, the relationship between the level of spirituality, depression, and anxiety represents negative values, with a correlation $R = -0.123$ ($p < 0.05$) and -0.115 ($p < 0.05$), which indicates that religiosity and spirituality in PSPP students are associated with lower levels of depression and anxiety. As a consequence, developing spirituality and religiosity for PPSP students and improving mental well-being is essential.

Keyword: mental illness, pharmacists, students, religion, spirituality

1 Introduction

Globally, the Covid-19 pandemic has had a significant impact on the mental health of the general population and groups at high risk [1]. Without the pandemic itself, students often experience anxiety or even melancholy as a result of their academic obligations. In addition, pharmacy education and pharmacy training are rigorous and arduous endeavors, it has been identified as one of the health disciplines that may contribute to student academic stress. Stress has a negative impact on the mental health of students, leading to stress-related disorders, low subjective life satisfaction, and poor academic performance [2].

Depression is one of the aspects of the comprehensive investigation of mental illness within its relationship with religiosity. At least 10 to 20 million people have ever experienced depression. The lack of spirituality has often been associated with the emergence of negative behavior and psychology, one of which is depression [3]. On the other hand, A Pew Research Center survey titled 'The Global God Divide' revealed that 96% of Indonesia's population stated that faith in God is very important, required for the instillation of morality and values [4]. Another study also reported that spirituality plays an important role in the life and health of most people [5].

Indonesia has a unique geographies, a large archipelago, with multicultural and multiracial society [6]. In some provinces or cities, religions other than Islam may be the majority. In Bali, Hinduism is the dominant religion ; in North Sulawesi, Papua, and West Papua, Christianity predominates; and in East Nusa Tenggara, Catholicism predominates. Officially, only Islam, Christianity, Catholicism, Hinduism, Buddhism, and Confucianism are recognized religions [4]. When contemplating religion in a psychological context, numerous factors come into play [7]. An individual may perform various rituals in a group setting in accordance with the tenets of religion, which consist of externally manifested motivated behavior with distinct goals. Spirituality, on the other hand, is an internal endeavor that is subjective in nature, has a unique relationship with a transcendent being, and may involve spiritual experiences [8].

Previous research on the topic of religiosity and mental health focused primarily on populations with a majority of a single religion [9]. Limited studies have been conducted to assess the relationship between religiosity and spirituality in multi-religion countries. Furthermore, Students in the Pharmacist Professional Study Program (PPSP) must pass the Indonesian Pharmacist Competency Exam (UKAI) to get the title of pharmacist. They feel the burden and fear of not passing the exam, which can trigger more anxiety and depression, especially in the era of covid-19 pandemic. Therefore, study about religiosity and spirituality and the prevalence of anxiety and depression among PPSP students need to be done in Indonesia.

* Corresponding author: nisafebrie@unissula.ac.id

The results are expected to provide an overview of the degree of relationship between the level of religiosity and spirituality with the level of anxiety and depression in students of the Pharmacist Professional Study Program in dealing with academic problems. The findings of this study can be useful to help students maintain mental health amid academic stressors and can be used as coping mechanisms for depression and anxiety in students.

2 Methods

2.1 Study design

An online survey, cross-sectional design with cluster random sampling is conducted in 40 universities that have PPSP that represent the five regions of APTFI in Indonesia. From 20 universities chosen, only twelve (12) universities gave permission to include in this study. Data was collected online using google forms that shared into university representative. Total of 367 respondents are participated in this study. The research was carried out from June until October 2020. The ethical clearance was obtained from Ethics Committee Faculty of Medicine Sultan Agung Islamic University with number 207/VII/2020/Komisi Bioetik.

2.2 Study instrument

The research uses four questionnaire instruments to measure each variable of religiosity, spirituality, anxiety, and depression. Religiosity was measured using Duke University Religion Index (DUREL) and translated into Indonesia languange by Primaningtyas in 2019 [10]. The item about Organized religious activity (ORA), non-organized religious activity (NORA) and intrinsic religiosity (IR) explanation were described in original study by Koenig in 2010[11]. In Spirituality is measured using DSES (*Daily Spiritual Experience Scale*) by Underwood in 2011 and translated into Indonesia language that called DSES-Ina by Qomarudin and Rahmah in 2019[12]. DSES is useful for measuring daily spiritual experiences that usually occur and can be aimed at someone who believes in a religion or do not. Anxiety was measured using the SAS (*Zung Self Rating Anxiety Scale*) by Zung in 1971 and translated into an Indonesian version by Setyowati in 2019 [13]. Depression was measured using the SDS (*Zung Self-Rating Depression Scale*) Indonesian version by Susanto et al in 2019 [14].

2.3 Data collection

The study made use of convenience sampling. The inclusion criteria were willingness to participate, ≥18 years of age and registration as an PSPP student, which as first taker or retaker in UKAI. The approximate time needed to complete the questionnaire estimated at around 15-20 minutes.

2.4 Data analysis

The collected data were analyzed using descriptive statistics. It summarized participant characteristics, reporting mean and standard deviation for continuous variables, while categorical variables were presented as the frequency with a corresponding percentage. Additionally, Spearman rank correlation analyses were used to examine potential correlations between religious factors and symptoms of anxiety or depression. Subsequently, result was consulted with the psychiatrist into conclusion drawing.

3 Results

In total 362 students participated in this study. Majority are female (84,8%), 17-25 years old (54,1%), have not taken UKAI or first taker (91,9%), single (95,9%), and Muslim. This study found that depression and anxiety are dominated by the young adult age group (17-25 years), woman, single and first taker of UKAI (**Table 1**).

Table 1. Demographic characteristics of the respondents

Table 1. Demographic of the Respondents				
Demographic Factor		**%**	**Depression N(%)**	**Anxiety N(%)**
Age				
17-25	233	64.4	20 (54.1)	46 (59)
26-35	123	34	16 (43.2)	32 (41)
36-45	6	1.6	1 (2.7)	0 (0)
Sex				
Male	55	15.2	6 (16.2)	9 (11.5)
Female	307	84.8	31 (83.8)	69 (88.5)
Marital Status				
Single	347	95.9	37 (100)	78 (100)
Married	15	4.1	0 (0)	0 (0)
UKAI Status				
First taker	339	93.6	34 (91.9)	70(89.7)
Retaker	23	6.4	3 (8.1)	8 (10.3)
University in 5 Region APTFI				
Andalas University	4	1,1		
STIFAR Riau	44	12.2		
UHAMKA	27	7.5		
Indonesia University	21	5,8		
Padjajaran University	50	13,8		
Jember University	33	9,1		
Hasanuddin Universit	26	7,2		
Halu Oleo University	12	3,3		
STIFI Perintis Padang (UPERTIS)	31	8,6		
Gadjah Mada Universi	37	10,2		
Unissula	11	3		
UMS	66	18,2		
Total	**362**	**100**	**37 (100)**	**78 (100)**

3.1 Religiousity of Pharmacist Students

This study presents that PSPP students in Indonesia have high index in religious based on DUREL scores (Mean : 23.48), with the highest score is in question "How often do you spend time (individually), example Shalat/learning holy book/Meditation (or another religious activity?)" (**Table 2**)

Table 2. DUREL Analysis

DUREL (*Duke University Religion Index*)				
No	**Question**	**M (SD)**	**Me**	**Mo**
1.	Pray/do religious activity to Mosque/Church/Monastery/ Pura/Temple (ORA)	4.38 (1.329)	4	6[a]
2.	Spend time (individually) to Shalat/learning holy book/Meditation (or another religious activity) (NORA)	5.25 (1.394)	6[b]	6
3.	In my life, I feel the presence of Allah/God/Jesus (IR1)	4.78 (0.468)	5	5
4.	My religious beliefs are what guide my daily life (IR2)	4.52 (0.675)	5	5
5.	I try my best to practice my religion in dealing with every problem in my life (IR3)	4.54 (0.609)	5	5
Total		23.48 (2.809)	24	24

3.2 Spirituality of Pharmacist Student

This study found that PSPP students in Indonesia have high index in spirituality based on DSES score. The highest score is on the statement of " I ask God for help in my daily activities", and high score in question "In general, how do you feel and how close are you to God?" with score 2,98 out of 3 (**Table 3**).

Table 3. DSES Analysis

DSES (*Daily Spiritual Experience Scale*)				
No	**Question**	**M (SD)**	**Me**	**Mo**
1.	I feel the presence of God or the Sacred Thing.	4.31 (1.4)	5	5
2.	I feel very close to life.	4.12 (1.421)	5	5
3.	During worship or other times of association with God, I feel joy and it frees me from everyday problems.	4.19 (1.46)	5	5
4.	I find strength in my religion and spirituality	4.3 (1.428)	5	5
5.	I feel comfortable in my religion and spirituality	4.43 (1,427)	5	5
6.	I feel peace and harmony inside	4.33 (1.405)	5	5
7.	I ask God for help in my daily activities.	4.56 (1.431)	5	5
8.	In my daily activities, I feel God's guidance.	4.44 (1.421)	5	5
9.	I immediately felt God's love for me (a).	4.41 (1.453)	5	5
10.	I immediately felt God's love for me (b)	4.22 (1.502)	5	5
11.	The beauty of creation touches me spiritually.	4.34 (1.420)	5	5
12.	I am thankful for my blessings/luck.	4.52 (1.407)	5	5
13.	I care about others selflessly.	4.15 (1.423)	5	5
14.	I accept other people even if they do what I think is wrong.	3.47 (1.487)	3	2
15.	I want to be close or united with God.	4.48 (1.438)	5	5
16.	In general, how do you feel and how close are you to God?	2.98 (0.795)	3[a]	3[a]
Total		67.28 (19.264)	75	78

3.3 Level of Anxiety of Pharmacist Students

The present study show that Indonesian PSPP students have mild into moderate score of anxiety, with the highest score in statement of "my hand feel dry and warm" (**Table 4).**

3.4 Level of Depression of Pharmacist Students

This study report that Indonesian PSPP students do not have - mild depression, with the highest score in "I don't want to think too long" (**Table 5**). Subsequently, this result was consulted with a psychiatrist, which conclude that Indonesian PSPP students do not have depression.

3.5 Correlation Religiousity-Spirituality with Anxiety and Depression of Pharmacist Students

This study indicates that there is a moderate positive correlation between depression and anxiety (r = 0.452**), which means that the higher the level of anxiety, the higher the level of depression. There is a weak negative correlation between depression and spirituality levels (r = -0123 *), which can be interpreted that the higher the spirituality level, the lower the level of depression. Then, there is a weak negative correlation between depression level and ORA, NORA, and IR religiosity level dimensions (r = -0.127 *, -0.166 ** and -0.216**). The results mean that the higher the religiosity level of the ORA, NORA, and IR dimensions, the lower the depression level.

Table 6 shows the correlation analysis between anxiety and spirituality shows a weak negative relationship (r = -0.115*), which can be interpreted that the higher the level of spirituality, the lower the level of anxiety. The relationship between anxiety and the dimensions level of religiosity of the ORA, NORA and IR dimensions is (r = -0.17**, 0.041, and 0.16**). It can be interpreted that anxiety is not correlated with NORA but is weakly and negatively correlated with ORA and IR. Therefore, it can be interpreted that the increasingly higher levels of religiosity (ORA and IR) make increasingly lower levels of anxiety to decrease. The correlation between spirituality and the religiosity dimensions of ORA, NORA and IR (r = 0.056, 0.167**, and 0.321**) indicates that spirituality is not correlated with ORA, but correlated positively weak with NORA and IR. Therefore, it can be interpreted that the higher the level of religiosity (NORA and IR), the higher the level of spirituality.

Table 4. SAS Analysis

SAS *(Zung Self-rating Anxiety Scale)*				
No	**Question**	**M (SD)**	**Me**	**Mo**
1.	I feel more nervous and anxious than usual	2.31 (0.702)	2	2
2.	I have no reason to be afraid	1.96 (0.773)	2	2
3.	I am temperamental and panic easily	2.16 (0.711)	2	2
4.	I feel lonely and depressed	1.78 (0.811)	2	2
5.	I think everything is fine, nothing bad will happen	2.29 (0.832)	2	2
6.	My arms and legs were shaking	1.47 (0.636)	1	1
7.	I am bothered by headaches, neck and back pain	1.88 (0.780)	2	2
8.	I feel weak and tired	2.04 (0.760)	2	2
9.	I feel calm and can sit still	2.21 (0.859)	2	2
10.	I feel my heart pounding	1.83 (0.698)	2	2
11.	I feel dizzy	1.64 (0.668)	2	2
12.	I suddenly wanted to pass out	1.19 (0.482)	1	1
13.	I can breathe easy	1.51 (0.781)	1	1
14.	I feel numbness and tingling in my fingers and toes	1.59 (0.694)	1	1
15.	I am bothered by stomachaches or indigestion	1.89 (0.828)	2	2
16.	I often urinate	2.27 (0.874)	2	2
17.	My hands are dry and warm	3.15 (0.934)	3	4
18.	My face gets hot easily	1.55 (0.740)	1	1
19.	I fall asleep easily and sleep well	2.26 (0.940)	2	3
20	I have nightmares	1.73 (0.619)	2	2
Total		37.83 (7.272)	38	40

Table 5. SDS analysis

SDS (*Zung Self-rating Depression Index*)				
No.	**Question**	**M(SD)**	**Me**	**Mo**
1.	I feel sad and grieve	1.88 (0.608)	2	2
2.	I feel fresher in the morning	2.13 (0.712)	2	2
3.	I often mourn (cry over, regret) myself	1.96 (0.79)	2	2
4.	I have difficulty sleeping at night	2.14 (0.916)	2	2
5.	My appetite is good as usual	1.82 (0.747)	2	2
6.	I have an interest in the opposite sex	1.71 (0.836)	2	1
7.	I'm getting thin	1.79 (0.819)	2	1
8.	I have difficulty defecating	1.79 (0.757)	2	2
9.	My heart rate beats faster	1.75 (0.688)	2	2
10.	I feel tired for no reason	1.9 (0.793)	2	2
11.	My mind is as clear (clear) as ever	2.22 (0.76)	2	2
12.	I can do the things I used to do well	1.90 (0.697)	2	2
13.	I am restless and unsettled	1.93 (0.739)	2	2
14.	I am hopeful (optimistic) about my future	1.8 (0.698)	2	2
15.	I get offended quickly	2.01 (0.812)	2	2
16.	I don't want to think too long	2.57 (0.772)	3	3
17.	I feel useful and needed	2.21 (0.694)	2	2
18.	My life is full of sufficiency	1.87 (0.748)	2	2
19.	I feel like everything will be better if I die	1.4 (0.792)	1	1
20.	I can still enjoy the things I used to do	1.73 (0.713)	2	2
Total		38.50 (7.669)	38	37

Table 6. Spearman Correlation Analysis

Spearman analysis						
	ORA	**NORA**	**IR**	**S**	**Anxiety**	**Depression**
S	0.056	0.167**	0.321**	1	-0.115*	-0.123*
A	-0.17**	0.041	-0.16**	-0.115*	1	0.452**
D	-0.127*	-0.166**	-0.216**	-0.123*	0.452**	1

4 Discussion

This study was first insight about correlation of religiousity and spirituality with anxiety-depression among pharmacists students in Indonesia. This study presents that depression and anxiety are majority found in women, similarly in Malaysia and United Kingdom [15], [16]. The experience makes them more concerned with themselves, does not focus on the learning process, has a much shorter memory range, and deteriorates the emotional intelligence that affects the optimal learning process[17]. It also majority occurred in single students, similarly in Malaysia [16]. Evidence shows that marriage produces positive mental health outcomes, such as decreasing levels of depression, anxiety, the desire to commit suicide, and drug abuse [18].

Furthermore, this study demonstrates depression and anxiety occur in most students who have not taken UKAI before. The result is similar in United States that pharmacy students who will take the NAPLEX (North American Pharmacist Licence Examination). Students can experience an increase in anxiety when they try to adjust learning habits and adjust preferences for intensity, demands, structure, and pace within the professional environment of the graduate level. Various changes and adjustments to the pharmacist's role, such as the apprentice in work activities ahead of UKAI, can result in confidence changes, an increase in insecurity, and stress within the pharmacy student [19]. However, this findings are different with some studies from Malaysia and Saudi Arabia, which reported that depression and anxiety increased as age increases. As students increase in age, they become more mature and aware of the responsibility and challenges to manage [15], [20].

This study found that the level of religiosity of pharmacy students in Indonesia is included in the high category. Based on the dimensions of ORA from DUREL, to the question "How often do you do worship/other religious activities at the mosque/church/monastery/temple ?" respondents answered "once a week" or "more than once a week" were 44.2%. In the IR dimension with the questions : "In my life, I feel the real presence of Allah/God/deity.", "My religious belief is guiding my everyday life" and "I tried as best as possible to practise the teachings of my religion in the face of every incident in my life.", respondents who answered "agree" or "very strongly agree" were amounted to 98.9%, 96.6%, and 96.9%, repectively. This findings are similar in previous research by Jacob et al in 2017 [21].

This study also indicating that the level of spirituality of pharmacy students in Indonesia is included in the category of high spirituality category even though the diversity is also high. On the final concluding question, "In general, according to your feelings, how close are you to God?" Respondents who answered "very close" or "as close as possible" are as much as 69.9%. However, this finding is contrastly with a study from United States [22]. In addition, this study interprets that most pharmacy students in Indonesia do not experience severe depression and anxiety. The highest prevalence level of depression and anxiety among PPSP students in Indonesia is 9.4% of mild depression and 21% of mild-moderate anxiety. Similarly in Malaysia [16]. However, the results is different with study in China[23].

This study demonstrates, it can be inferred that religious rituals can reduce the levels of depression and anxiety even though the association is low($r = -0.17^*$, -0.127^*). This may be due during this pandemic, all the limitations to conducting activities outside allow students have time at home and have many opportunities to practise worship and seek tranquility [24]. Then the weak negative correlation of the level of depression with religiosity (NORA) ($r = -0.166^{**}$) shows that in addition to obligatory worship activities/religious rituals, such as viewing religious lectures on social networks, giving can bring calm and reduce depression syndrome. Then religiosity (IR) in this investigation is weakly negatively correlated with anxiety and depression ($r = -0.16^{**}$ and -0.216^{**}). The results mean that the low levels of depression and anxiety of respondents could be caused by the high intrinsic religiosity/spirituality, such as practising religion as the highest value, with an orientation towards unity with a direction of good intentions, as well as trying to practise religion and obtain peace [21].

Based on these results, it is known that high religiosity and spirituality are associated with lower levels of depression and anxiety. ORA activities such as public or privately (dhikr, prayer) are formally arranged with teaching that a general group following may increase social support, healthy behaviour, and better lifestyle and happiness. Thus, religiosity and spirituality also help face stress, fear, sadness, misery, and anger. Individuals with a degree of spirituality and religiosity that are high tend to produce better mental health and quality of life because they develop internal and external mechanisms that can help overcome the difficulties in life's journey [25]. Religious involvement is thought to prevent the development of mental disorder (or ameliorate its course) and increase mental health resiliency [7].

The study results showed a weak negative correlation between religiosity and spirituality with levels of anxiety and depression because the research population was a heterogeneous sample composed of students from various cultures and religions. Given the pluralistic and multireligious nature of the study population, the concepts and interpretations of religiosity may vary and they may embrace the concept of spirituality rather than formal religion. Although spirituality and religiosity overlap, there is a significant difference between the two constructions. Spirituality refers to a more personal and individual interpretation in search of meaning, while dogmatic and institutionalised sacred interpretations characterize religiosity. Religiosity is a multidimensional construct; the measurement scale used may not fully reflect the strength and nature of a person's involvement in religious activities. Therefore, the scale alone may not be a measure of religiosity that is adequate in the real sense [8].

An explanation of the negative relationship between religiosity and anxiety depression can be found in study by Abdel-Khalek et al. in 2019 [26], which majority respondents in this study are muslim. Based on that study, Islam religion can be considered to be coping mechanism to face problems and difficulties as well as an effective anxiety release mechanism. In Islam, several practices are available to relieve anxiety and other negative emotions, including ablution and prayer five times a day, reciting the Qur'an, remembering Allah, praying and fasting for a month each year (Ramadan)[26]. In addition to spirituality and religiosity, other factors can be related to the level of depression and anxiety, such as smoking, parent marital status, GPA, family mental disorders/treatment history, health state, loss of someone valuable, age, ethnicity/race, religion, supportive social relationships, positive family environment and appropriate coping style to improve mental welfare[16].

Developing spirituality and religiosity for PPSP students and improving mental well-being is essential and will be helpful in work experience in the future. As a study in United Stated, which clinical pharmacist improving therapy medication through patient spiritual counseling therapy[27]. Pharmacy students can be encouraged to understand how spirituality affects patients' views, such as understanding its relationship with health care decision making. Pharmacy students will become sensitive and listen more actively to the patient when participating in the health care initiative, communicate persuasively about the specific medication intervention, give a lifetime of treatment that has implications associated with religious beliefs, and showing empathy when dealing with a varied patient population.

5 Conclusion

Presumably, this study has strength in the broad samples and first insight into religion-spirituality and mental health among pharmacists students in Indonesia. However, this study has limitations in a small sample. In conclusion, there is no depression detected in Indonesian PSPP students. Even though Indonesian PPSP students had mild to moderate anxiety, this study found a relationship between the level of spirituality, depression, and anxiety that represents negative values. Indicate that religiosity and spirituality in PSPP students are associated with lower levels of depression and anxiety. This finding suggests that improvements can be made through the competencies or curriculum of the pharmacist profession in Indonesian institutions that teach the importance of religiosity and spirituality and identify stressors and coping mechanisms that can be done to cope with anxiety and depression. Most importantly, students understand the importance of mental health and the impact that can be caused in the future.

Acknowledgment: The authors thank and appreciate the respondents who were willing to participate in this study.
Conflict interests: The authors declare no competing or potential conflicts of interest concerning the research and publication of this article

References

1. S. Pappa, J. Chen, J. Barnett, A. Chang, R. K. Dong, W. Xu, A. Yin, B. Z. Chen, A. Y. Delios, R. Z. Chen, S. Miller, X. Wan, and S. X. Zhang, "A systematic review and meta-analysis of the mental health symptoms during the Covid-19 pandemic in Southeast Asia," *Psychiatry Clin. Neurosci.*, vol. 76, no. 2, pp. 41–50, 2022, doi: 10.1111/pcn.13306.
2. O. Article, "Impact of Religiosity on Subjective Life Satisfaction and Perceived Academic Stress in Undergraduate Pharmacy Students," pp. 192–198, 2018, doi: 10.4103/jpbs.JPBS.
3. S. Kasen, P. Wickramaratne, M. J. Gameroff, and M. M. Weissman, "Religiosity and resilience in persons at high risk for major depression," *Psychol. Med.*, vol. 42, no. 3, pp. 509–519, 2012, doi: 10.1017/S0033291711001516.
4. R. Ardi, D. H. Tobing, G. N. Agustina, A. F. Iswahyudi, and D. Budiarti, "Religious schema and tolerance towards alienated groups in Indonesia," *Heliyon*, vol. 7, no. 7, p. e07603, 2021, doi: 10.1016/j.heliyon.2021.e07603.
5. P. Gavaza, B. M. Rawal, and E. Johnston, "Exploratory Research in Clinical and Social Pharmacy Pharmacy students ' perceived barriers to spiritual care : A qualitative study," *Explor. Res. Clin. Soc. Pharm.*, vol. 9, no. May 2022, p. 100246, 2023, doi: 10.1016/j.rcsop.2023.100246.
6. H. Surendra, D. Paramita, N. N. Arista, A. I. Putri, A. A. Siregar, E. Puspaningrum, L. Rosylin, D. Gardera, M. Girianna, and I. R. F. Elyazar, "Geographical variations and district-level factors associated with COVID-19 mortality in Indonesia: a nationwide ecological study," *BMC Public Health*, vol. 23, no. 1, pp. 1–12, 2023, doi: 10.1186/s12889-023-15015-0.
7. H. G. Koenig, F. Al-Zaben, and T. J. VanderWeele, "Religion and psychiatry: recent developments in research," *BJPsych Adv.*, vol. 26, no. 5, pp. 262–272, 2020, doi: 10.1192/bja.2019.81.
8. B. Francis, J. S. Gill, N. Y. Han, C. F. Petrus, F. L. Azhar, Z. A. Sabki, M. A. Said, K. O. Hui, N. C. Guan, and A. H. Sulaiman, "Religious Coping , Religiosity , Depression and Anxiety among Medical Students in a Multi-Religious Setting," pp. 1–13, 2019, doi: 10.3390/ijerph16020259.
9. N. Upadhayay, "Clinical training in medical students during preclinical years in the skill lab," *Adv. Med. Educ. Pract.*, vol. 8, 2017, doi: 10.2147/AMEP.S130367.
10. W. PRIMANINGTYAS, A. A. A. K. A. N. PUTRI, and H. HASTUTI, "Body Image and Religiosity in Adolescents: a Comparation Between Public and Private High School Students," *Smart Med. J.*, vol. 2, no. 2, p. 98, 2020, doi: 10.13057/smj.v2i2.35672.
11. R. M. Bonelli and H. G. Koenig, "Mental Disorders, Religion and Spirituality 1990 to 2010: A Systematic Evidence-Based Review," *J. Relig. Health*, vol. 52, no. 2, pp. 657–673, 2013, doi: 10.1007/s10943-013-9691-4.
12. M. bagus Qomarrudin;Rahmah Indawati, "Spiritual everyday experience of religious people," *J. Int. Dent. Med. Res.*, vol. 2, no. 2, pp. 827–834, 2019.
13. A. Setyowati, M. H. Chung, and A. Yusuf, "Development of self-report assessment tool

for anxiety among adolescents: Indonesian version of the zung self-rating anxiety scale," *J. Public Health Africa*, vol. 10, no. S1, pp. 3–6, 2019, doi: 10.4081/jphia.2019.1172.

14. T. D. Susanto, "Validity and Reliability of Indonesian Languages Version of Zung Self-Rating Depression Scale Questionnaire for Pulmonary Tuberculosis Patients," *Indian J. Public Heal. Res. Dev.*, vol. 10, no. 12, pp. 2023–2027, 2019, doi: https://doi.org/10.37506/v10%2Fi12%2F2019%2Fijphrd%2F192171.
15. U. Jaffer, C. M. Nasril, R. A. Osman, A. L. A. Razak, N. Allie, M. A. Ahmed, M. A. Jalaludin, and N. M. Kadri, "The Mediating Roles of Religious and Spiritual Coping between Religiosity, Spirituality, and Depression among Medical and Health Science Students," *Eur. J. Mol. Clin. Med.*, vol. 9, no. 8, pp. 1209–1223, 2022, [Online]. Available: https://www.embase.com/search/results?subaction=viewrecord&id=L2021958906&from=export
16. N. S. M. Yusof, Z. A. Zainal, H. Huri, Sabrina, A. Jacob, M. N. Alwi, Y. Hassan, R. Hashim, I. A. Wahab, A. Y. A. Nasiruddin, N. A. Jamludin, Amelah, M. A. Qader, S. A. Hisham, N. Jamil, C. Shin, Chee, and L. Ghazali, "Prevalence of depression among undergraduate pharmacy students in Malaysia," *Int. J. Pharm. Res.*, vol. 12, no. 3, pp. 2033–2042, 2020, doi: 10.31838/ijpr/2020.12.03.282.
17. S. Mirawdali, H. Morrissey, and P. Ball, "Academic anxiety and its effects on academic performance," *Int. J. Curr. Res.*, vol. 10, no. 6, pp. 70017–70026, 2018, [Online]. Available: https://www.journalcra.com/sites/default/files/issue-pdf/30653.pdf
18. G. G. Gan and Y. L. Hue, "Anxiety, depression and quality of life of medical students in Malaysia," *Med. J. Malaysia*, vol. 74, no. 1, pp. 57–61, 2019.
19. C. A. Spivey, M. A. Chisholm-Burns, and J. L. Johnson, "Factors associated with student pharmacists' academic progression and performance on the national licensure examination," *Am. J. Pharm. Educ.*, vol. 84, no. 2, pp. 269–276, 2020, doi: 10.5688/ajpe7561.
20. S. Samreen, N. A. Siddiqui, and R. A. Mothana, "Prevalence of anxiety and associated factors among pharmacy students in Saudi Arabia: A cross-sectional study," *Biomed Res. Int.*, vol. 2020, 2020, doi: 10.1155/2020/2436538.
21. B. Jacob, A. White, and A. Shogbon, "First-year student pharmacists' spirituality and perceptions regarding the role of spirituality in pharmacy education," *Am. J. Pharm. Educ.*, vol. 81, no. 6, pp. 1–7, 2017, doi: 10.5688/ajpe816108.
22. B. Jacob, T. Huynh, A. White, A. S. Nwaesei, R. Lorys, W. Barker, J. Hall, L. Bush, and W. L. Allen, "Pharmacy and nursing students' perceptions regarding the role of spirituality in professional education and practice," *Am. J. Pharm. Educ.*, vol. 84, no. 9, pp. 1218–1225, 2020, doi: 10.5688/ajpe7777.
23. R. Shao, P. He, B. Ling, L. Tan, L. Xu, Y. Hou, L. Kong, and Y. Yang, "Prevalence of depression and anxiety and correlations between depression, anxiety, family functioning, social support and coping styles among Chinese medical students," *BMC Psychol.*, vol. 8, no. 1, pp. 1–19, 2020, doi: 10.1186/s40359-020-00402-8.
24. I. Rammouz, L. Lahlou, Z. Salehddine, O. Eloumary, H. Laaraj, and M. Ouhamou, "symptoms among nursing and medical students during the middle stage of the COVID- pandemic : A cross-sectional study in Morocco".
25. L. M. Vitorino, G. Lucchetti, F. C. Leão, H. Vallada, and M. F. P. Peres, "The association between spirituality and religiousness and mental health," *Sci. Rep.*, vol. 8, no. 1, pp. 1–9, 2018, doi: 10.1038/s41598-018-35380-w.
26. A. M. Abdel-Khalek, L. Nuño, J. Gómez-Benito, and D. Lester, "The Relationship Between Religiosity and Anxiety: A Meta-analysis," *J. Relig. Health*, vol. 58, no. 5, pp. 1847–1856, 2019, doi: 10.1007/s10943-019-00881-z.
27. L. C. Ilechukwu, A. A. Sewagegn, and A. N. Amedu, "Intervention for depression among undergraduate religious education students," no. September, pp. 1–5, 2022.

Community Pharmacist' attitudes towards counterfeit medicines in Yogyakarta, Indonesia

Susi Ari Kristina[1*], *Kadek Hendra Darmawan*[2], *Rizka Prita Yuliani*[1], *Fathul Mu'in*[1], *Vo Quang Trung*[3]
[1]Faculty of Pharmacy, Universitas Gadjah Mada, Yogyakarta, Indonesia.
[2]Ministry of Health, Jakarta, Indonesia
[3]Pham Ngoc Thach Unibersity of Medicine, Ho Chi Minh, Vietnam

Abstract. Counterfeit medicines (CFMs) is a catastrophic public health issue that encompasses all therapeutic classes in Indonesia. It is a consensus that community pharmacists in many countries could contribute to preventing the distribution of CFMs. Information on community pharmacists' attitudes regarding CFMs is lacking. This study explores the attitudes of community pharmacists in Yogyakarta, Indonesia, regards to CFMs. The specific aim was to identify the perceived reason, the method used to identify CFMs, and complemented with recommendations on what aspects are lacking today. A cross-sectional study involving 350 community pharmacists in Yogyakarta province was conducted. A semi-structured questionnaire assessed the components of their attitude toward CFMs. The Likert-scale statements were implemented to best describe pharmacists' attitudes in responding to specific questions. Most community pharmacists perceived a business profit induced by exceeded market demand and a medicine shortage as significant reasons for CFMs in Indonesia. They are confident in identifying CFMs through the medicine's appearance, packaging, and noticeable price difference. This study captures the need for enforcing regulators to take more responsibility for specific CFMs regulations and build a centralized system for reporting the potential cases of CFMs, allowing for continuous country CFMs surveillance. Community pharmacists could be a crucial supporting function in preventing CFMs from reaching patients. To enhance their role, a robust reporting system is urged to be adopted and socialized in a massive way.
Keyword. Pharmacist, Attitudes, CFMs, Perceived reasons, counterfeit medicines

1 Introduction

Imagine a patient being prescribed chemotherapy to treat a life-threatening tumor. A responsible pharmacist dispenses the prescribed medication and directs the patient without realizing the medicine does not contain an active ingredient. In this case, the patient is not successfully achieving the targeted therapy, and the physician and pharmacist are biased in evaluating the treatment outcomes based on the patient's response to a placebo instead of an active pharmaceutical ingredient.
Counterfeit medicine (CFMs) is defined as illegal, generally lower priced, and often associated with lower qualities compared to its originator and poses a significant problem that is growing globally [1]. Contextually, CFMs are medical products that deliberately or fraudulently misrepresent their identity, composition, or source and are termed falsified [2].

CFMs encompass all therapeutic classes, from lifesaving to lifestyle products [3,4]. According to the World Health Organization (WHO) data, it is estimated that around 1% of prescribed medicines in developed nations and 10-50% in developing countries are counterfeit [5]. The European Commission predicts that 15% of the global medicine supply chain could be prone to counterfeit activity [6].

CFMs are a catastrophic issue that exposes significant risks to public health, taking the ineffectiveness and toxicity, and antibiotic resistance leading to an increased global morbidity and mortality rate, not to mention the economic consequences [7,8].

Indonesia, in 2016, experienced a widely publicized lawsuit against the counterfeited vaccine, which resulted in approximately 1500 children being injected with a fake vaccine [9]. This case got massive attention after a baby died later being vaccinated at one of the hospitals in Bekasi, West Java. This case revealed the distribution chain of counterfeit vaccines, which involved the falsifier, distributor, business owner, and healthcare professional (doctor and nurses) who were unresponsible and traded off the children's health with financial incentives from the counterfeit medicine. Police investigations later determined that counterfeiters had refilled used vaccine vials collected by an organized network of hospital cleaners [10].

The enforcement of regulations of CFMs differs among countries. In Indonesia, CFMs are not explicitly regulated in any government document to date. But the Ministry of Health's regulations number 284/2007[11] is mentioned as a facilitator of the high CFMs incidence; therefore, the newer MoH regulation (53/2016)[12] was enacted to prevent the circulation of CFMs in the drug store, known as *Apotek Rakyat* in local language [13].

* Corresponding author: susiari_k@ugm.ac.id

Providing an effective pharmaceutical service would not be adequate when the availability of CFMs may compromise the safety and efficacy of the medication. WHO good pharmacy practice guideline highlights the critical role of pharmacists in combating CFMs; ensure the dispensed medicines are effective and of quality [14,15].

The perception and attitudes of pharmacists about CFMs have been surveyed in several developing countries; its consensus that improvement in awareness through campaigns and pharmacy curriculum development, establishing strict control of medicine, building a CFMs reporting system, and enforcing the law will be needed to optimize pharmacists' contribution for preventing CFM reach patients [6,16,17]. Furthermore, community pharmacists are often the first point of access to affordable medicine in Indonesia [18]. However, there need to be more studies regarding CFMs in Indonesia, especially from the community pharmacy point of view. In addition, the current literature search did not reveal any studies related to community pharmacists and CFMs in Indonesia.

This study explores Indonesia's pharmacists' attitudes toward CFMs. We aimed to specifically outline the perceived reason for massively available CFMs in pharmacies, the method used to identify CFMs, and possible recommendations to address what aspects are lacking today.

2 Methods

2.1 Study design

This was a cross-sectional survey of 350 community pharmacists in five districts in Yogyakarta Province, and as concerned, this is the first study discussing the topic in Indonesia, conducted in March 2023. This study procedure was approved by Ethical Committee in Medical and Health Research, Gadjah Mada University no KE/FK/1035/EC/2023.

2.2 Study instrument

In order to collect the pertinent data, this study uses a questionnaire that was prepared after a prior literature review [6] and focus group discussion. In order to assess the content validity of the study, a consensus among professionals in the field of pharmacy practice was used. The current questionnaire was made up of Likert-scale statements and was divided into three main sections: (I) demographic information; (II) real-world experiences in identifying counterfeit medical products; and (III) professional attitudes of pharmacists toward CFMs.

The questionnaire was pre-tested, refined and finally administered to the target sample through personal contact by the researcher. Informed consent information was attached to each questionnaire. A total of three hundred and fifty questionnaires were continuously administered.

2.3 Data collection

The study made use of convenience sampling. The main inclusion criteria were practicing pharmacists willing to participate and could speak and read *Bahasa*. Prior to data collection, the researchers verbally explained the purpose of the study and the approximate time needed to complete the questionnaire estimated at around 15 minutes and returned it up to 30 minutes after completion.

2.4 Data analysis

The collected data were analyzed using descriptive statistics, expressed as numbers and percentages. For this Likert-type statement, participants' responses were evaluated after dichotomizing them to strongly agree/agree and strongly disagree/disagree. In addition, content analysis was employed for structuring the open-ended questions.

3 Results

A total of 350 pharmacies participated in the survey; of these, the majority worked in independent pharmacies. Demographic characteristics are given in **Table 1**. The dominant age of participants was between 31 and 40 years; 73.43% were female, geographically spreading at five sub-regions of Yogyakarta with an average of years of practicing between 5-10 years.

Table 1. Demographic characteristics of community pharmacists

Characteristics	Category	n	%
Age	21-30	95	27.14
	31-40	132	37.71
	41-50	98	28.00
	>50	25	7.14
Gender	Female	257	73.43
	Male	93	26.57
Location of pharmacies	Sleman	94	26.86
	Yogyakarta	78	22.29
	Kulonprogo	67	19.14
	Bantul	59	16.86
	Gunung Kidul	52	14.86
Years of practicing	>5 years	97	27.71
	5-10 years	134	38.29
	>10 years	119	34.00
Number of patients per day	<20	175	50.00
	20-50	90	25.71
	>50	85	24.29
Type of pharmacies	Independent pharmacy	312	89.14
	Chain pharmacy	38	10.86

3.1 Perceived reasons for the avaibility of counterfeit medicine by pharmacist

A series of questions with multiple responses allowed; of the 350 pharmacists, as shown in **Figure 1** reported that 81,71% perceived that a business and profit-oriented resulted in the availability of CFMs in the community pharmacies. In contrast, about 73,71% and 69,73%, respectively, believed the market factors, including the medicine shortage and high community demand for medicine, drive the sales of CFMs. In addition, 54% of participants criticized the lack of control from medicine regulators on CFMs, which successfully perceived in facilitating the high availability of CFMs in the market.

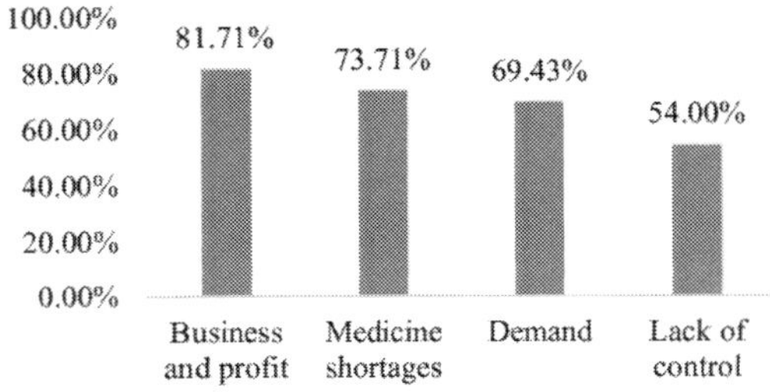

Fig. 1. The reason for the avaibility of CFMs, according to pharmacists (n=350)

3.2 Pharmacists' practice to identifying counterfeit medicines

A significant percentage of the 350 pharmacists (93,71%) identified CFMs through the medicine's physical appearance and 54% of them through their packaging, as reported in **Figure 2**. While cost is reported as one of the explicit determinants of CFMs that are recognizable, that could trigger the pharmacist's attention to the possibility of CFMs in their practice. In addition, the supplier, whether they are reputable is considered as the way to recognize the CFMs.

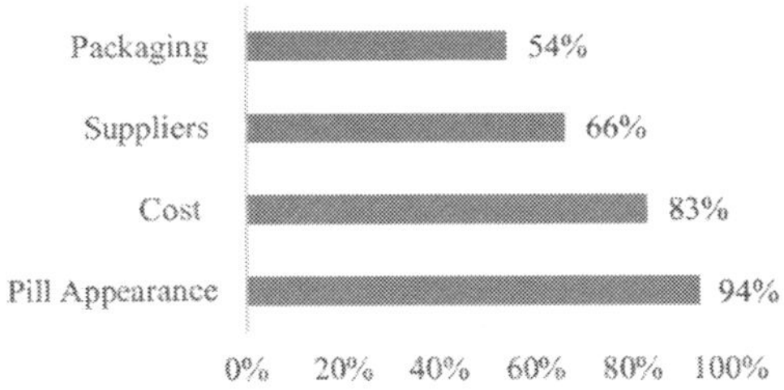

Fig. 2. Pratice in identifying CFMs among community pharmacist (n=350)

3.3 Pharmacists' attitude regarding counterfeit medicines

Figure 3 shows the respondents' attitudes regarding counterfeit medicines. Most pharmacists consider the Indonesian Ministry of Health (MoH) responsible for guaranteeing the safety and efficacy of medicine before it is distributed to the national drug market. About 70% of pharmacies perceive that CFMs are primarily sold through online platforms. The urgency of specific regulations to combat CFMs is raised by 89,14% of participants, including the law for selling medicine in e-commerce. In contrast, a significant number of pharmacists are still confused about how to report the CFMs even though they are confident enough in identifying the CFMs.

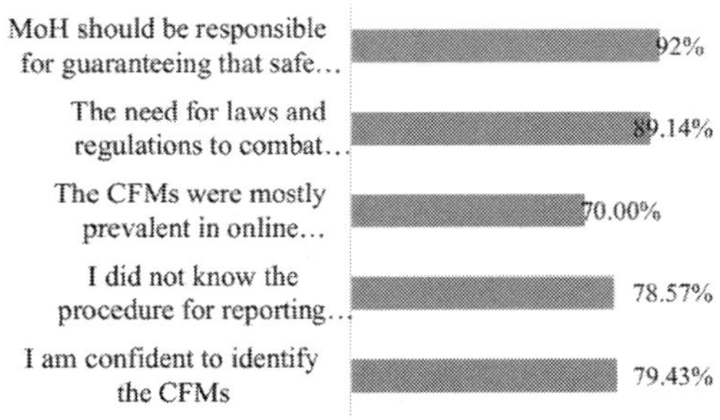

Fig. 3. Attitude regarding counterfeit medicines (n=350)

3.4 Recommendation for preventing counterfeit medicines

According to **Table 2**, most community pharmacists strongly agree on an action to report suspicious medicine to the CFMs reporting system if applicable, legalizing online pharmacies, and adopting continuous surveillance activities by regulators. Meanwhile, intrinsically, community pharmacists agree upon the consensus on avoiding purchasing medicine from non-reputable suppliers.

Table 2. Recommendation to prevent the CFMs from community pharmacists (n=350)

Recommendations from community pharmacists	Strongly agree/agree	
	(n)	(%)
Report suspicious medicines to the CFM reporting system	312	89.14
Adopt a Code of Ethics for pharmacists	280	80.00
Approved online pharmacies	256	73.14
Avoid purchasing drugs from unapproved suppliers	236	67.43
Continuous surveillance	213	60.86

4 Discussion

Community pharmacists' attitudes toward counterfeit medicine (CFMs) revealed deficiencies in knowledge of reporting procedures and the CFMs incidence, yet the respondents feel confident in identifying the potential cases of CFMs during their regular practice. It is an elsewhere consensus that community pharmacists hold a crucial role in preventing the distribution of CFMs, hence their function needs to be leveraged for better public health [19].

The demographic characteristics of respondents identified that female pharmacists were over-represented, which is aligned with data on the ratio of male to female pharmacists in Indonesia being 1:3. In contrast, the relevant study in Lebanon reported the opposite respondent's characteristics where male pharmacists were more interested and willing to share their practical experiences regarding CFMs [6]. Additionally, it is recognized that accumulated years of practicing impacted community pharmacists' attitudes toward CFMs [9].

Responses in the first section of the questionnaire showed various perceptions from Indonesian community pharmacists on how CFMs could circulate through a pharmacy chain. There is a market incentive when a high demand for medicine is not matched by the limited supply capacity from a legitimate supplier, which causes a medicine shortage. A massive counterfeit resulting from a medicine shortage was documented in the USA when Tamiflu (Oseltamivir) was prescribed during the Influenza season. Counterfeiters break into the medicine supply chain as a consequence of pharmacists' search for alternative medicine vendors due to unmet demand, resulting in sporadicities of fraudulent distributors [19].

Besides the market incentives, a more significant profit margin is another perceived reason for the availability of CFMs at pharmacies. The same circumstances were found in relevant studies involving community pharmacists in Egypt [20], Lebanon [6], and Ethiopia [21]; the pharmacy professional decided to carry out CFM transactions from uncertified sources for a significant profit. In the Indonesian context, this perceived activity has been pronounced since many pharmacies are not owned by the pharmacist but by the profit-oriented businessman; hence, a larger profit seems to be an option compared to the availability of safe products at the outlet.

Consistent with a study in Jordan [22], adequate legislation and lack of control from regulators were primary reasons perceived by community pharmacies that successfully incentives the pharmacist to make CFMs available at the pharmacy store. Even though the regulation is in place, the lack of in-house control made the pharmacist normalize counterfeiting activities, thus facilitating the spreading of CFMs in the community pharmacy.

Indonesian pharmacists' way of detecting CFMs was limited to standard practice. Firstly, through pill appearance and packaging. In Nigeria [23], pharmacists outlined the packaging checked, and included the product registration number and manufacturer code manually. Meanwhile, In Indonesia, the National Food and Drug Authority holds a website for everyone to check the originality components of drugs, herbal products, food, and cosmetics (batch number, manufacturer information, recall history, etc.). But the website does not fully recognize the counterfeiting aspect of medicine since the identity could be correct for CFMs. Other studies in Lebanon [6] reported the challenges of distinguishing CFMs from originators through visual examination, even for a pharmacist with more than ten years of service.

Along with drug appearance and packaging, in the U.S., advanced medicine labeling technologies were initiated that could primarily be utilized for product authentication since they are complex and/or expensive to copy. Pharmaceutical manufacturers currently use holograms, colour-shifting inks, embedded codes, images, and dyes on packaging to create an additional layer of protection. More recently, barcodes carrying medication products' identification for track-and-trace purposes were employed for advancing the medicine packaging complementing the anticounterfeiting technologies which can assist the community pharmacists in authenticating the products better [19].

The legitimate medicine supplier where counterfeiting incidence rarely occurs is another way perceived by the respondents to characterize the CFMs. Pharmacists were recommended to be selective when purchasing medicine only from authorized wholesalers or other reliable sources. Indonesian pharmacists are aware of the importance of choosing a trusted wholesaler to purchase medicine stock. Meanwhile, there is a rising concern in Nigeria since the pharmacists claimed that they sourced their medicine personally from wholesalers, which can be at high risk for counterfeiting medicine [23,24].

Aligned with other studies in other countries [17,20,25], pharmacists in Indonesia perceived that it is an urgent need for specific laws and regulations to combat CFMs. Even though some respondents may be aware of the existence of a chapter in Indonesia Health Law (36/2009) [26] about the penalty for medicine counterfeiters, it needs to regulate the CFMs comprehensively. Another center of discussion of implementing a medicine counterfeiter chapter in health law is the deviation of the penalty and the actual court ruling [27]. A court case in Medan, Indonesia, only ruled an antibiotic counterfeiter with one year of prison without any financial penalty. At the same time, it should be a maximum of five years of prison with financial consequences of IDR 2 billion [28]. In Ethiopia, weak regulation infringement is a driving factor for the high prevalence of medicine counterfeit cases [21].

Uniquely, responders in this study rely on the regulators to guarantee the distribution of safe and effective medicine through pharmacy circumstances. This is aligned with the perception of Lebanese pharmacists [6]. Meanwhile, in other countries like the U.S., collaborative activity between cross-function stakeholders is already initiated. The system calls MedWatch, a national system that involves pharma companies, Food and Drug Authority (regulators), community pharmacists, and business owners to ensure that only safe medicines circulate through the U.S. pharmacy channels [19].

A contradictive condition was revealed in this study; even though most responders feel confident in identifying the CFMs, almost 78% do not know the procedures for reporting the CFMs nor know the existing reporting mechanisms used by regulators. Currently, the National Food and Drug Authority in Indonesia operates a generic reporting or complaint system that the general Indonesian population could utilize to report any suspicious medicine, food, or cosmetics, which means not explicitly designated for clinicians or community pharmacists to report the potential counterfeit medicine. In other words, the current system does not specify supporting the continuous surveillance system for CFMs in the community.

To add to the attitude of community pharmacists in Indonesia toward CFMs, most of them believe that CFMs are prevalent in e-commerce (online commercial platforms). It is synergized with the data that approximately 60% of medicines that purchased online could be counterfeited or substandard [29]. In the U.S., thousands of online vendors sell unapproved and/or counterfeit medicines, including the sale of prospection regiment therapy without requiring a prescription [30].

This study seeks recommendations from the community pharmacy point of view regarding the CFMs. Since there is no reporting system to report suspicious medicine from health professionals, including pharmacists, this organized system needs to be prioritized by the medicine regulator in Indonesia. A relevant study in Lebanon [17] supports this recommendation; in the same circumstances where no CFMs reporting system exists in the country, establishing the integrated system would allow pharmacists to report any suspected CFMs in a more systematic order efficiently. In short, the centralized reporting system of CFMs will contribute to building a continuous country surveillance system for better response and track-and-trace for CFMs incidents.

Indonesian community pharmacists urged to strengthen the Code of Pharmacists implementation. A case study in Jordan [22] and Iran [16] reported several conditions where pharmacists could possibly name other pharmacists that deal with CFMs yet are still reluctant to inform authorities. As informing peers is highlighted in the current Code of Pharmacists, pharmacists are ideally able to remind peers not to deal with CFMs when considering the effect and consequences for themselves and the community.

Lastly, it is highly recommended that community pharmacists avoid purchasing from unapproved medicine suppliers. The same suggestion has campaigned in the U.S., where pharmacists urge to procure medications from known, reliable sources. This campaign is supported by wholesale distributors accreditation program called 'the Verified-Accredited Wholesale Distributor' by the National Association of Boards of Pharmacy (NABP) [19]. The same distributor system could be adopted in Indonesia to systematically help the pharmacist choose legitimate distributors, increasing trust and reliability of sources of safe medications.

This study potentially overlooked the representation of opinion and practice by Indonesian pharmacists as a whole nation since respondents were selected conveniently rather than by random sampling and it's limited to specific geographic areas; hence generalization is hard to achieve. However, since this is the first CFMs study using a pharmacist point of view in Indonesia, the result could serve as a preliminary overview of the attitude of community pharmacists regarding CFMs. Further study is needed to explore other aspects of CFMs, including a case study of Indonesian pharmacists on actual practice to identify CFMs and their consequences to the health system.

5 Conclusion

Community pharmacists hold a critical role as gatekeepers to prevent the distribution of CFMs in Indonesia before reaching the population. By understanding their attitude toward CFMs, a better recommendation could be formulated to strengthen medicine safety and prevent the public from the harmful consequences of CFMs' circulation throughout the legal supply chain.

Acknowledgment: The authors thank and appreciate the respondents who were willing to participate in this study.

Conflict interests: The authors declare no competing or potential conflicts of interest concerning the research and publication of this article

References

1. CAd. Matos, CT. Ituassu, CAV. Rossi. Consumer attitudes toward counterfeits: a review and extension. Journal of Consumer Marketing. **(24)**, 36-47, (2007).
2. Guidelines for the development of measures to combat counterfeit drugs. In. Geneva: World Health Organization; 1999.
3. DE. Baker, Medication Counterfeiting. Hospital Pharmacy.**(8)**, 683-684, (2014).
4. BJ. Venhuis, PH. Keizers, R. Klausmann, I. Hegger. Operation resistance: A snapshot of falsified antibiotics and biopharmaceutical injectables in Europe. Drug Test Anal. **(3-4)**:398-401, (2016).
5. PN. Newto, MD. Green, FM. Fernández, NP. Day, White NJ. Counterfeit anti-infective drugs. Lancet Infect Dis. **(9)**:602-613 (2006).
6. L. Sholy, P. Gard, S. Williams, A. MacAdam. Pharmacist awareness and views towards counterfeit medicine in Lebanon. Int J Pharm Pract.**(3)**:273-280, (2018).
7. TK. Mackey, BA. Liang. The global counterfeit drug trade: patient safety and public health risks. J Pharm Sci. **(11)**:4571-4579, (2011).
8. EA. Blackstone, JP. Fuhr, Jr., S. Pociask. The health and economic effects of counterfeit drugs. Am Health Drug Benefits. **(4)**:216-224, (2014).
9. A. Hasnida, MO. Kok, E. Pisani. Challenges in maintaining medicine quality while aiming for universal health coverage: a qualitative analysis from Indonesia. BMJ Global Health. **(6)**, (2021)
10. S. Putra. Kronologi Pengungkapan Kasus Vaksin Palsu di Bekasi dan Tengerang. (2016).
11. Peraturan Menteri Kesehatan Republik Indonesia Nomor 284/MENKES/PER/III/2007 Tentang Apotek Rakyat (2007).
12. Peraturan Menteri Kesehatan Republik Indonesia Nomor 53 Tahun 2016 Tentang Pencabutan Peraturan Menteri Kesehatan Nomor 284/MENKES/PER/III/2007 Tentang Apotek Rakyat, (2016).
13. Association IP. Jadi Biang Kerok Obat Palsu, Permenkes Apotek Rakyat Akan Dicabut. (2016).

14. Mdege ND, Chevo T, Toner P. Perceptions of current and potential public health involvement of pharmacists in developing nations: The case of Zimbabwe. Res Social Adm Pharm. 2016;12(6):876-884.
15. NN. Gowri Rajapandian, A. Maheswaran and P. Chanchal. Dangerous World Of Counterfeit Drugs – Pharmacist's Role And Its Prevention. International Journal of Pharmaceutical Science and Research (IJPSR). 1709-1713, (2013).
16. S. Shahverdi, M. Hajimiri, F. Pourmalek, H. Torkamandi, K. Gholami, S. Hanafi, NA. Shahmirzadi, M. Javadi. Iranian pharmacists' knowledge, attitude, and practice regarding counterfeit drugs. Iran J Pharm Res. **(3)**, 963-968, (2012).
17. LB. Sholy, PR. Gard, S. Williams, A. MacAdam, CE. Saliba CE. Public and pharmacist perceptions towards counterfeit medicine in Lebanon using focus groups. International Journal Of Community Medicine And Public Health. **(2)**, 489-499, (2018).
18. C. Brata, C. Fisher, B. Marjadi, CR. Schneider, Clifford RM. Factors influencing the current practice of self-medication consultations in Eastern Indonesian community pharmacies: a qualitative study. BMC Health Serv Res. **(16)**, 179, (2016).
19. WG. Chambliss, WA. Carroll, D. Kennedy, D. Levine, MA Mone, LD Ried, M Shepherd, M. Yelvigi.Role of the pharmacist in preventing distribution of counterfeit medications. J Am Pharm Assoc (2003). **(2)**:195-199, (2012).
20. A. Bashir, S. Galal S, A. Ramadan, A. Wahdan, L. El-Khordagui. Community pharmacists' perceptions, awareness and practices regarding counterfeit medicines: a cross-sectional survey in Alexandria, Egypt. East Mediterr Health J. **(5)**:556-564, (2020).
21. E. Abdunasir, T. Sosengo, F. Amare, M. Yimam, B. Hagos. Assessment of awareness and attitude towards counterfeit medicines among pharmacy professionals working in community drug retail outlets in Harar town, Ethiopia. (2021).
22. YA. Taleb, MR. Pharmacists' Awareness of Drug Counterfeiting in Jordan. Journal of The Royal Medical Services. **(2)**:57-70, (2013).
23. V U Odili SO, E U Eke, H A Okeri. Identification of Counterfeit Drugs by Community Pharmacists in Lagos State. Tropical Journal of Pharmaceutical Research. **(5)**, (2006).
24. OP. Adigwe, G. Onavbavba, DO. Wilson. Challenges Associated with Addressing Counterfeit Medicines in Nigeria: An Exploration of Pharmacists' Knowledge, Practices, and Perceptions. Integr Pharm Res Pract. **(11)**:177-186, (2022).
25. PN. Newton, MD. Green, FM. Fernández. Impact of poor-quality medicines in the 'developing' world. Trends Pharmacol Sci. **(3)**:99-101, (2010).
26. Undang-undang (UU) tentang Kesehatan, (2009).
27. Eni K. Vonis Rendah dari Hakim Membuat Pemalsu Obat tidak Jera. Media Indonesia, (2019).
28. AKHB. Lumbantobing, U. Utomo. Pertanggungjawaban Pidana Pelaku Yang Sengaja Menjual Obat-Obatan Yang Tidak Memenuhi Standar Mutu (Studi Putusan No: 1335/Pid.Sus/2018/Pn Medan). PATIK:Jurnal Hukum. **(9)**, 203 - 214, (2020).
29. A silent epidemic: protecting the safety and security of drugs. Pharmaceutical Outsourcing; (2011).
30. Counterfeit Pharmaceutical Inter-Agency Working Group report to the Vice President of the United States and to Congress., (2011).

Determining Priority Subgroup: Insight from a Health Literacy Survey in Universitas Gadjah Mada, Indonesia

Fatwa Sari Tetra Dewi[1*], *Yana* Yulyana[2], *Diana* Novianti[3] and *Ardhina* Ramania[3]

[1]Department of Health Behaviour, Environment and Social Medicine, Faculty of Medicine, Public Health and Nursing, Universitas Gadjah Mada, Yogyakarta, Indonesia
[2]Inahealth YouTube Channel, Faculty of Medicine, Public Health and Nursing, Universitas Gadjah Mada, Yogyakarta, Indonesia
[3]Health Behavior and Promotion Major, Public Health Postgraduate Program, Faculty of Medicine, Public Health and Nursing, Universitas Gadjah Mada, Yogyakarta, Indonesia

Abstract. Health literacy measurement and source of information are important to design health education programs. Aims: To measure the level of health literacy, its factors and pattern of source of information among faculty members in Universitas Gadjah Mada, Indonesia. Materials and methods: A cross-sectional study was done among lecturers, staffs and students with consecutive sampling methods. The HLS-EU-Q16 questionnaire was used to measure the health literacy. Furthermore, the respondents' characteristics of sex, age, faculty discipline, academic role, and education attainment were also collected. The data collection used E-HDSS online survey application distributed via the official university information system. Results: The respondent involved in this study was 1,036 people, with majority had adequate health literacy (74%). The proportion of adequate health literacy was significantly higher among staff, had bachelor and postgraduate education degree (OR 2.6, 1.4, and 1.8 respectively). Internet and Instagram were the most preferred source of information. Conclusions: Inadequate health literacy was more prevalent among students and those with lower education levels. This was the prioritized target subpopulation for health education programs, that should be delivered through Instagram and internet.
Keywords: HLS-EU-Q16; Health Promoting University; Health literacy; Information channel; Health education design; LMIC

1 Introduction

Adequate health literacy was higher among lecturer and staff, had bachelor and postgraduate education degree. Inadequate health literacy was higher among students and those with lower education levels, was the most prioritized target audience for health education programs. The main source of information was internet, and the most preferred social media was Instagram.

Affiliations of authors should be typed in 9-point Times. They should be preceded by a numerical superscript corresponding to the same superscript after the name of the author concerned. Please ensure that affiliations are as full and complete as possible and include the country.

Health literacy is the ability of individuals to seek, comprehend, appraise, and apply health information to maintain and enhance their health [1]. Poor health literacy tends to be associated with limited access to healthcare services and lower health outcomes [2]. Studies in the United States have shown that 36% of the population has low health literacy [3], and similarly in Europe, more than half of the population has low health literacy [4]. Person with low health literacy is often linked to disadvantage social determinants, which in turn result in health inequities. Health literacy is associated with several factors such as occupation type, income level, and educational level [5]. A higher education does not always guarantee higher health literacy, but cultural and linguistic facinterventions, thus enabling the development of better communication strategies to reach the target population (refetors also play a role in the status of health literacy [6].

The population health literacy needs to be known in order to identify the priority subpopulation for health education interventions, thus enabling the development of better communication strategies to reach the target population [7]. Health literacy is measured by defining and exploring its dimensions, which include an individual's ability to access, comprehend, appraise, and apply health information for the purpose of disease treatment, prevention, and improving health status [3]. The health education will be more effective if it is delivered through preferred sources of information by the target subpopulation. Therefore, understanding the sources of information is also important. This is essential for improving education and health promotion, healthcare services, reducing health inequalities, and enhancing the overall health status of the community [3]

*Corresponding email: fatwasari@mail.ugm.ac.id

https://doi.org/10.1051/bioconf/20237505007

Despite numerous studies conducted on health literacy, research on health literacy among academic communities remains limited. This study aims to 1) measure the level of health literacy and identify various factors associated with health literacy levels, and 2) determine the sources of health information among the academic community at Universitas Gadjah Mada in Yogyakarta, Indonesia. By measuring the health literacy of the academic community, specific subpopulations can be identified as priorities for health education initiatives.

2 Materials and Methods

This Health Literacy survey is part of the routine activities of the Health Literacy working group at UGM HPU, one activity under the health literacy working group [8]. A cross-sectional study was done, with the population of all lecturers, staffs and students. The HLS-EU-Q16 questionnaire was used to measure the health literacy, and demographic factors of sex, age, affiliated faculty, academic status, and education level were also measure. E-HDSS online survey application is distributed via the official university information system to collect the data. The sampling method was consecutive, based on the willingness of respondents to participate. We report this study following the STROBE guidelines for a cross-sectional study [9].

A cross-sectional study involving lecturers, staffs and students was done in year 2022. This study located in a University in Yogyakarta, Indonesia. The University has 18 faculties and has 4,336 faculty staffs and 61,440 students. Dependent variable, health literacy was measured using a health literacy behavior questionnaire adapted from the Health Literacy Study Short Form (HLS-EU-16Q-Indonesia) questionnaire developed by the Asian Health Literacy Association (AHLA) [10]. HLS-EU-16Q has been translated into Indonesian and has been tested for validity and reliability [11]. HLS-EU-16Q has also been used in routine survey activities in the population of Sleman Regency [12]. The determinant factors: sex, age, affiliated unit, academic status, and education level were measured using self-developed questions. Those questions were sought through a E-HDSS application online and distributed via the official informatics system of the University during 2 weeks periods in 2022. HLS-EU-Q16 questionnaire consist of 16 questions asking ability to search for health information, to understand the health information, to criticize the health information and to make an informed decision [10], the respondents were required to indicate the answers very difficult, difficult, easy, or very easy. We scored the very difficult and difficult as 0, and easy or very easy as 1. The health literacy data were categorized as adequate if the total score of the health literacy ≥ 13 and inadequate if < 13. Sex was determined as man or woman; age was categorized as ≥ 40 year and < 40 year; affiliated unit was categorized as natural science, social sciences and administrative units; academic status was grouped into lecturer, staff and student; and education was categorized into undergraduate, bachelor and postgraduate levels.

After the data was collected using Google Forms within a 2-week data collection period, the data were extracted from the system and cleaned for completeness. The data were analyzed using STATA version 14.2 software [13]. To determine the level of health literacy and provide a health literacy patterns, a descriptive analysis was conducted and presented in a table with percentages. Furthermore, to identify factors associated with health literacy, initial bivariate analysis was performed using bivariate logistic regression test. Variables that showed significant associations with health literacy were further analyzed using multivariate logistic regression test.

The most significant threat is sampling and selection bias, given the consecutive sampling method and online data collection. The sample may not represent the population adequately due to a higher likelihood of subgroup participation, considering the online data collection method might be easier for certain subgroup. Therefore, caution is required in interpreting the data. Despite the limitations of online data collection, it offers the advantage of broader reach.

Ethical approval was obtained from the Ethics Commission of the Faculty of Medicine, Public Health, and Nursing at UGM.

3 Results

The participating respondent was 1037 people, and after cleaning for data completion and outlier, 1036 respondents were considered for analysis. Out of all respondents were women (62 %), within the <40 years age group (72%), affiliated with natural science faculty (51%), currently active students (59%), and came from bachelor education background (43%) as shown in Table 1. In both men and women, majority of respondents were within the <40 years age group, affiliated with natural science faculty, were currently active students and came from bachelor education background.

The bivariate analysis showed that being aged >= 40 year, being staff, lecturer and had postgraduate degree were significantly had higher possibility for adequate health literacy (with OR 1.9, 2.7, 2.3, 1.9 respectively). Those determinants of age group, academic type, and education that had significant relationships with health literacy were then included into the model for a multivariate analysis using logistic regression. The results of multivariate analysis in Table 2 return to being staff, had bachelor and postgraduate education degree has more possibilities for adequate health literacy with OR 2.6, 1.4, and 1.8 respectively. The most priority group for a health education program was those aged < 40 year, student and with undergraduate education.

Table 1. Characteristics of respondents stratified by sex

Variable	Group	Men		Women		Total	
		n	%	n	%	n	%
N		397	38	639	62	1036	100
Health Literacy	Adequate	290	73	475	74	765	74
	Inadequate	107	27	164	26	271	26
Age in years, mean (SD)		35.1 (12.4)		30.8 (11.4)		32.5 (11.9)	
Age group	<40 years	245	62	505	79	750	72
	≥40 years	152	38	134	21	286	28
Affiliated Unit	Natural science	190	48	342	54	532	51
	Social science	105	26	175	27	280	27
	Administrative unit	102	26	122	19	224	22
Employment Status	Lecturer	38	10	71	11	109	11
	Staff	150	38	163	26	313	30
	Student	209	53	405	63	614	59
Education	Undergraduate	143	36	188	29	331	32
	Bachelor	146	37	292	46	438	42
	Postgraduate	108	27	159	25	267	26

Table 2. Bivariate and multivariate logistic regression analysis to predict health literacy adequacy according to sociodemographic characteristics

		Bivariate			Multivariate		
Variables	**Categories**		**CI**			**CI**	
		OR	**Lowest**	**Highest**	**OR**	**Lowest**	**Highest**
Sex	Man	1.0					
	Woman	1.1	0.8	1.4			
Age group	<40	1.0			1.0		
	≥40	1.9*	1.4	2.7	1.1	0.7	1.7
Affiliated Unit	Social	1.0					
	Administrative	1.0	0.7	1.5			
	Natural	1.2	0.9	1.7			
Academic type	Student	1.0			1.0		
	Staff	2.7*	1.9	3.8	2.6*	1.8	3.9
	Lecturer	2.3*	1.4	4.0	1.6	0.9	3.1
Education	Undergraduate	1.0			1.0		
	Bachelor	1.3	0.9	1.8	1.4*	1.0	1.9
	Postgraduate	1.9*	1.3	2.7	1.8*	1.1	2.8

Table 3 represents the means of health literacy score according to domain and activity. Among the three domains that were measured, disease prevention scored the lowest mean (2.89), followed by health promotion (3.14) and healthcare (3.18). While across the four activities, appraise and applying information both scored the lowest means (2.97 and 2.99, respectively).

Internet was the main source of health information for majority of the respondents (61.8%), followed by social media (23.9%) and conventional or other media (14.3%) (Figure 1). Out of the respondents with social media as their main source of health information, 63.7% were Instagram users, followed by Twitter (14.9%), Youtube (11.7%), Tiktok (4.84%), Facebook (2.0%) and others (2.8%) (Figure 2). Instagram is mostly popular among under 40-year-olds (91.8%), students (70,3%) and undergraduate and bachelor respondents (30.4% and 51.3%, respectively) (Table 4).

Table 3. Mean of Health Literacy Score by Domain and Activity

Domain	**Mean**	**Activity**	**Mean**
Healthcare	3.18	Access/finding information	3.15
Disease Prevention	2.89	Understanding information	3.23
Health Promotion	3.14	Appraise/judge/evaluate information	2.97
		Applying information	2.99

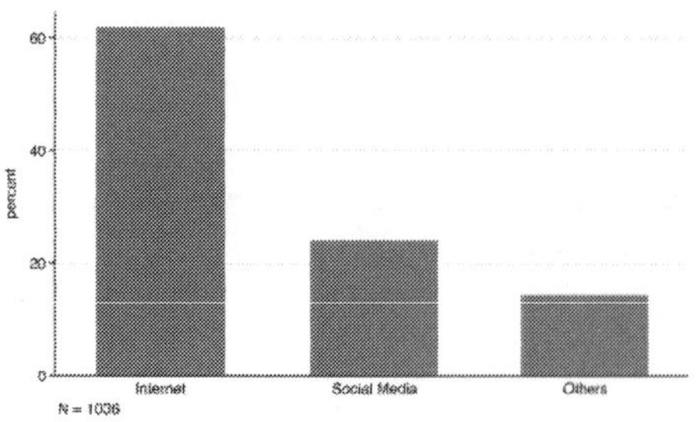

Fig. 1. Main source of health information

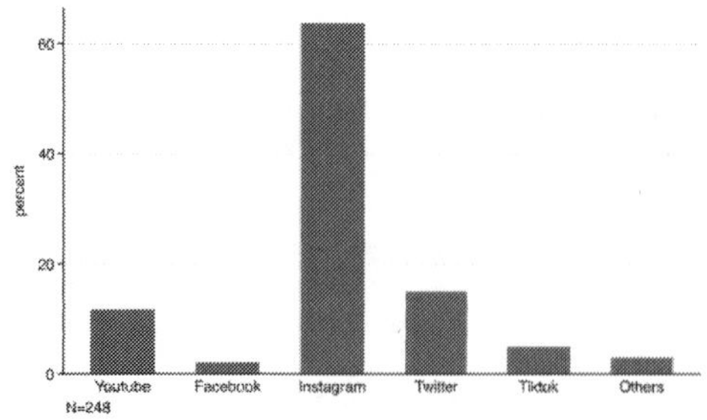

Fig. 2. Social media preference

4 Discussions

This study was conducted to measure the health literacy of the academic community in Universitas Gadjah Mada, Indonesia, and their source of information patterns. Our study show that the level of health literacy was majority adequate (74%). Our results found that being staff, had bachelor and postgraduate education degree has more possibilities for adequate health literacy. Internet was the main source of health information, followed by social media. The most preferred social media were Instagram, Twitter, YouTube, Tiktok, and Facebook respectively. Instagram is mostly popular among the under 40-year-olds, students and undergraduate respondents.

Previous studies have found that the relationship between health literacy and gender is not always significant. Some studies have shown significantly higher health literacy among women, while others have found higher health literacy among men [14]. In our study, we did not find a significant relationship between health literacy and gender. It is possible that health literacy is influenced by various factors that interact in complex ways, and gender is not the primary determinant of health literacy.

Our study found that adequate health literacy was higher among individuals aged ≥ 40 years and above. Health literacy tends to increase with age [15]. As individuals grow older, they accumulate more learning experiences and interactions with healthcare professionals, leading to an improvement in health literacy [15, 16]. However, health literacy tends to decline among the older age group (age 65 years and above) [17]. The lower health literacy among older age groups may be associated with factors such as cognitive decline and reduced exposure to health information as individuals age.

Table 4. Social media preference across respondents' sociodemographic characteristics

Variable		YouTube n (%)	Facebook n (%)	Instagram n (%)	Twitter n (%)	Tiktok n (%)	Others n (%)	P-value
N		29	5	158	37	12	7	
Age group	<40 years	14 (48)	3 (60)	145 (92)	34 (92)	8 (67)	2 (29)	<0.001
	≥40 years	15 (52)	2 (40)	13 (8)	3 (8)	4 (33)	5 (71)	
Employment status	Lecturer	2 (7)	0 (0)	6 (4)	0 (0)	0 (0)	0 (0)	0.007
	Staff	13 (45)	2 (40)	41 (26)	5 (14)	4 (33)	6 (86)	
	Student	14 (48	3 (60)	111 (70)	32 (87)	8 (67)	1 (14)	
Education	Undergrad	12 (41)	2 (40)	48 (30)	19 (51)	4 (33)	2 (29)	0.39
	Bachelor	13 (45)	2 (40)	81 (51)	15 (41)	6 (50)	2 (29)	
	Postgrad	4 (14)	1 (20)	29 (18)	3 (8)	2 (17)	3 (43)	

A study found that health literacy is associated with faculty affiliation. Those affiliated with healthcare units tend to have higher health literacy [18]. Individuals in a healthcare environment are more likely to be exposed to health information, resulting in higher health literacy. In contrast, our study found no relationship between health literacy and faculty affiliation. This may be due to the larger proportion of student participants (Table 1), who may have limited exposure to the specific health-related knowledge they are currently studying.

Our study revealed that adequate health literacy was higher among lecturers and staff compared to students. Previous research comparing health literacy based on academic roles is lacking. Research on health literacy among non-health faculty-affiliated lecturers showed lower levels of health literacy [19]. Another study among students indicated that 45% of students had inadequate health literacy [15, 20]. Conversely, research among staff members showed higher levels of health literacy regarding depression in Australia [21]. However, comparing the three academic roles of lecturers, staff, and students is difficult due to the different instruments and research contexts. Several studies have found a significant association between higher education levels and higher health literacy, although this significance disappears in multivariate analysis [17]. It is possible that lecturers have higher health literacy compared to students due to their higher educational attainment.

Health literacy is a multidimensional concept influenced by various factors that interact with each other in complex ways, ultimately impacting social determinants of health and health equity [22]. However, beyond that, based on these results, it can be concluded that students was the most important target group for interventions. Educational institutions have the opportunity to enhance the health literacy of their academic community and empower them to make informed decisions for themselves and their [15].

In addition to knowing the health literacy level of the population, this study also identified the main sources of information that were mostly used by the students and undergraduate education attainment, were the internet and social media. While the internet is the most popular way for students to access information, it is also associated with the worst health literacy scores, possibly due to the quality of information is often incorrect and hardly comprehendible [21].

The strength of this study lies in its sufficiently large sample size and its focus on developing appropriate health education strategies for priority targets based on their health literacy levels and preferred sources of information. Internet-based data collection has both advantages and disadvantages. On one hand, internet-based data collection offers the advantages of a wide reach to respondents, cost-effectiveness [22], and increased respondent confidentiality, thereby facilitating more honest responses [23, 24].

However, on the other hand, the biggest threat in internet-based data collection is non-response bias, which may limit the representativeness of the sample. Internet-based research and applications are still controversial. Studies comparing internet-based surveys with pen-and-paper surveys have shown no significant differences in the results of both types of surveys, suggesting that internet-based surveys can be considered as an additional research approach alongside conventional surveys [2, 27]. However, it is necessary to design questionnaires and ensure better data collection validity in internet-based surveys [28].

5 Conclusions and Recommendations

Adequate health literacy was higher among lecturer and staff, had bachelor and postgraduate education degree. Inadequate health literacy was higher among students and those with lower education levels, was the most prioritized target audience for health education programs. The main source of information was internet, and the most preferred social media was Instagram.

Acknowledgements

We would like to acknowledge the financial support for the accomplishment of this study by the Health Promoting University (HPU), UGM.

References

1. D. Nutbeam. Health Promotion Glossary. *Health Promotion International [Internet]*. (1998). **13**(4):349-64. Available from: https://doi.org/10.1093/heapro/13.4.349
2. N. D. Berkman, S. L. Sheridan, K. E. Donahue, D. J. Halpern, & K. Crotty, *Low health literacy and health outcomes: An updated systematic review*. AOIM. (2011). **155**(2):97-107. doi:10.7326/0003-4819-155-2-201107190-00005.
3. K. Sørensen, S. Van den Broucke, J. Fullam, G. Doyle, J. Pelikan, Z. Slonska, & H. Brand, *Health literacy and public health: a systematic review and integration of definitions and models*. BMC Public Health. (2012). **12**(1):80. doi: 10.1186/1471-2458-12-80
4. World Health Organization (WHO) Health literacy, The solid facts. 2013. U.S. Department of Health and Human Services, Office of Disease Prevention and Health Promotion. Washington, DC: National Action Plan to improve health literacy; (2010). health.gov/communication/HLActionPlan/pdf/Health_Literacy_Action_Plan.pd
5. ARPH 42:159-173 (Volume publication date April 2021) First published as a Review in Advance on October 9, 2020 https://doi.org/10.1146/annurev-publhealth-090419-102529

6. A. Noblin, , M. H. Gabriel, K. Cortelyou-Ward, & K. Holmes, *Health literacy among visiting college students in the U.S.: A pilot study*. JOACH. (2020). 1–9. doi:10.1080/07448481.2020.1758114 10.1080/07448481.2020.175811
7. D. Nutbeam, and J. Lloyd, E, *Understanding and Responding to Health Literacy as a Social Determinant of Health*. ARPH. (2021). **42**:159–73. https://doi.org/10.1146/annurev-publhealth-090419-102529
8. https://hpu.ugm.ac.id/
9. https://www.equator-network.org/wp-content/uploads/2015/10/STROBE_checklist_v4_cross-sectional.pdf
10. https://www.researchgate.net/publication/327039694_Health_literacy_scale-European_union-Q16_a_validity_and_reliability_study_in_Turkey
11. https://ahla-indonesia.dinus.ac.id/2022/11/30/sicerdik-udinus/
12. https://hdss.fk.ugm.ac.id/katalog-data-hdss-sleman/
13. StataCorp. Stata Statistical Software: Release 14. (2015). College Station, TX: StataCorp LP.
14. N. Aljassim & R. Ostini, *Health literacy in rural and urban populations: A systematic review, Patient Education and Counseling*. (2020). **103**(10):2142-2154, https://doi.org/10.1016/j.pec.2020.06.007.
15. L. Kühn,, P. Bachert, C. Hildebrand, J. Kunkel, J. Reitermayer, H. Wäsche, & A. Woll, *Health Literacy Among University Students: A Systematic Review of Cross-Sectional Studies*. FPH. (2022). 9:680999. doi: 10.3389/fpubh.2021.680999
16. S. Ayaz-Alkaya, & H. Terzi, *Investigation of health literacy and affecting factors of nursing students, Nurse Education in Practice.* (2019). (34):31-35, doi.org/10.1016/j.nepr.2018.10.009
17. EM. Berens, D. Vogt, M. Messer, K. Hurrelmann, & D. Schaeffer, *Health literacy among different age groups in Germany: results of a cross-sectional survey*. BMC Public Health. (2016). **16**, 1151. https://doi.org/10.1186/s12889-016-3810-6
18. C. Lauber, V. Ajdacic-Gross, N. Fritschi, N. Stulz, & W. Rossler, *Mental health literacy in an educational elite – an online survey among university students*. BMC Public Health. (2005). **5**, 44. https://doi.org/10.1186/1471-2458-5-44
19. *Evaluation of Health Literacy in Academics at a University of Turkey Evaluation of Health Literacy in Academics at a University of Turkey* https://literacy.mums.ac.ir/article_19795.html
20. I. van der Heide, J. Wang, M. Droomers, P. Spreeuwenberg, J. Rademakers & E. Uiters, *The Relationship Between Health, Education, and Health Literacy: Results From the Dutch Adult Literacy and Life Skills Survey*. JOHC. (2013). **18**:sup1, 172-184. doi: 10.1080/10810730.2013.825668
21. A. Gulliver, L. Farrer, K. Bennett & K. M Griffiths, *University staff mental health literacy, stigma and their experience of students with mental health problems, Journal of Further and Higher Education*. (2019). **43**:3:434-442, doi: 10.1080/0309877X.2017.1367370
22. D. Schillinger, *Social Determinants, Health Literacy, and Disparities: Intersections and Controversies*. HLRP. (2021). **5**(3):e234–e243. https://doi.org/10.3928/24748307-20210712-01
23. P. Santos, L. Sa, L. Couto, & A. P. Hespanhol, *Sources of information in health education: a cross-sectional study in Portuguese university students*. MSJA. (2018). **11**. doi: 10.21767/AMJ.2018.3435
24. A. Zeleke, T. Naziyok, F. Fritz, L. Christianson, & R. Röhrig *Data Quality and Cost-effectiveness Analyses of Electronic and Paper-Based Interviewer-Administered Public Health Surveys: Systematic Review.* JMIR. (2021). **23**(1):e21382. doi: 10.2196/21382
25. G. Eysenbach, & J. Wyatt. *Using the Internet for surveys and health research*. JMIR. (2002). **4**(2):E13. doi: 10.2196/jmir.4.2.e13. PMID: 12554560; PMCID: PMC1761932.
26. M. L. Remillard, K. M. Mazor, S. L. Cutrona, J. H. Gurwitz, & J. Tjia. *Systematic review of the use of online questionnaires of older adults*. JAGS. (2014). **62**(4):696-705. doi: 10.1111/jgs.12747. Epub 2014 Mar 17. PMID: 24635138; PMCID: PMC4098903.
27. S. D. Gosling, S. Vazire, S. Srivastava, & O. P. John. *Should we trust web-based studies? A comparative analysis of six*

preconceptions about internet questionnaires. AP. (2004). **59**(2):93-104. doi:10.1037/0003-066x.59.2.93
28. J. S. M. Belisario, J. Jamsek, K. Huckvale, J. O'Donoghue, C. P. Morrison, & J. Car. *Comparison of self-administered survey questionnaire responses collected using mobile apps versus other methods*. CDOSR (2015). Issue 7. Art. No.: MR000042. doi: 10.1002/14651858.MR000042.pub2.

Educational Curriculum for Diabetes Mellitus (DM) Patients at the Public Health Center in Salatiga City, Central Java Province

Anita Kumala Hati[1,2], *Susi* Ari Kristina[3*] , *Nanang* Munif Yasin[3], and *Lutfan* Lazuardi[4]

[1]Doctoral Program in Pharmacy, Faculty of Pharmacy, Gadjah Mada University, Yogyakarta, Indonesia, 55281
[2]Pharmacy Study Program, Faculty of Health, Ngudi Waluyo University, Semarang, Indonesia, 50512
[3]Department of Pharmaceutics, Faculty of Pharmacy, Gadjah Mada University, Yogyakarta, Indonesia, 55281
[4]Department of Public Health, Faculty of Medicine, Public Health, and Nursing, Gadjah Mada University, Yogyakarta, Indonesia, 55281

Abstract. The efficacy of diabetes mellitus (DM) therapy is not only contingent on the pharmacological agents but also significantly influenced by the depth of patient comprehension regarding therapeutic management. Therefore, designed educational interventions pertinent to individual requirements must be administered, including the intricacies of DM therapy management. This qualitative study was carried out in January - February 2023 at 6 Public Health Center in Salatiga City. The investigative method employed was semi-structured interviews, conducted with a cohort of DM patients between the ages of 55 to 65 years, who possessed proficient communicative skills in the Indonesian language, and voluntarily answered each question. Meanwhile, individuals concurrently engaged in healthcare vocations were excluded and the interview process was documented in the form of an audio recording. The results showed that patients still need education regarding proper diet for diabetics including the type, amount, and schedule. Guidance relating to appropriate physical activities catering to DM patients, in conjunction with the discernment of medication indications, potential side effects, storage requisites, and administration protocols, emerged as critical topics. This study stated that DM patients need educational curriculum related to diet/nutrition, physical activity, medicines, and DM complications.
Keywords: diabetes mellitus, education, curriculum, outcomes, Indonesia

1 Background

International Diabetes Federation (IDF) [1] in 2017 placed Indonesia as the 7th country out of the top 10 with the largest diabetes mellitus (DM) population in the world. However, there has been a*n increase in 2021 IDF placing Indonesia as the 5th country with the largest population in the world [2]. DM patients based on the results of Basic Health Research (RISKESDAS) in 2013 amounted to 6.9% and increased to 8.5% in 2018. The disease is a prominent health problem related to the increased cost of therapy. The Health Social Security Administering Body (BPJS) claims that the amount of funds disbursed continues to increase annually. In 2018, 2019, and 2022, the amount increased to Rp. 6.5 trillion, Rp. 7.1 trillion, and Rp. 7.5 trillion, respectively [3].

DM patients need to carry out secondary prevention through behavioral changes such as physical activity, changes in diet, and medication adherence. Behavior change requires understanding, persistence, and education in self-management to optimize health outcomes [4–7]. Randomized controlled trials and national observational studies show that education in DM management results in cost-effective [8, 9], clinical outcome [10–13], and quality of life [14–16]. Education is the basis of treatment for DM patients to achieve controlled results [17].

The American diabetes association (ADA) formulated a self-management education curriculum in the National Standards for Diabetes Self-Management Education and Support. The content includes a description of the process of developing DM disease and its treatment options, regulation of nutrition and physical activity into lifestyle, safe and effective use of medicines, monitoring, and interpretation of blood sugar levels and other parameters for self-decision making, detection, prevention, and treatment of acute and chronic complications, psychosocial development strategies, and Health Promotion and attitude change [17]. Education programs must be specific to certain populations, such as the type of DM and ethnic, social, language, cognitive, literacy, and cultural factors [18]. Based on this background, this study explores the need for educational programs and curriculum for patients in Public Health Center in Salatiga City.

2 Materials and Methods

2.1 Study Design

This study used a qualitative design and semi-structured interviews were conducted to identify the forms of educational media and curriculum needed by

*Corresponding author: susiari_k@ugm.ac.id

DM patients. A total of 10 respondents were DM patients participating in PROLANIS at Public Health Center in Salatiga City and the officer who conducted the interviews was Anita Kumala Hati. The interview process was documented in the form of an audio recording. The study was approved by the Ethics Committee of the Faculty of Medicine, Nursing and Public Health, Universitas Gadjah Mada, with number KE-FK-1595-EC-2022.

2.2 Study Sample

Respondents were obtained through the snowball sampling method [19]. The purpose of this study was explained and the willingness to participate was also reported. Inclusion criteria were DM patients participating in PROLANIS at the Public Health Center in Salatiga City aged over 18 years to 65 years, can communicate well with the Indonesian language, and voluntarily answered each question in the interview. Meanwhile, the exclusion criteria were DM patients who work as health workers and the recruitment of respondents continued until thematic saturation was reached.

2.3 Interview Guide

Respondents willing to take part filled out a consent form and received an explanation of the purpose and stages of this study. The following related questions were given:

1. What educational media do DM patients like and why?
2. How is the educational curriculum needed by DM patients?

2.4 Data Collection

In semi-structured interviews, respondents were encouraged to express issues related to questions through open-ended questions [20].

2.5 Data Analysis

Thematic analysis was conducted to obtain qualitative data [21]. Thematic analysis was applied to various theoretical approaches, but the underlying assumptions of recent studies were towards a postpositivist paradigm, including systematic efforts, direct interpretation, and transparency of data [22]. During the analysis, the themes guide the questions and are used to sort the data. The data obtained were coded deductively, to produce the initial code using the theoretical framework and objectives as a starting point. Furthermore, an inductive approach was used to code interesting additional features of the data that formed subthemes. The reconsideration of the theme involved a comprehensive review of the dataset. The main theme was refined and elaborated on through a process of identifying the various elements. The analysis was summarized in five distinct stages, namely transcription, initial coding, generating themes, reviewing themes, and defining themes [21]. Furthermore, the COREQ checklist was used in data analysis to report this study [20].

3 Result and Discussion

This study involved 10 DM patients participating in PROLANIS at the Public Health Center in Salatiga City. There were 8 females and 2 males aged between 57 to 65 years. The average length of time required to interview 1 respondent was 14 minutes.

3.1 Educational Media

The results of opinion exploration showed 2 themes regarding the selection of educational media. A total of 7 out of 10 respondents preferred educational media in the form of videos due to the following reasons:

"It is a bit difficult for old eyes to read."
(Respondent 1)
"Video will make us more able to understand the meaning of education because there are audio and visual."
(Respondent 6)

Respondents prefer educational media in the form of videos due to decreased ability to read, hence, movement and sound can be interpreted. It was also easier to understand educational curriculum by seeing the visual movement. Meanwhile, 3 respondents selected reading texts as educational media for the following reasons:

"I prefer reading scripts because videos on a cellphone are easily shifted and lost, I do not understand technology."
(Respondent 4)
"I like to read because my eyes are still clear to read."
(Respondent 10)

Respondents select education in text media because videos shared through smartphones are easily lost due to weak digital literacy. This study shows that DM patients prefer education given in video form due to several advantages of transferring information in the form of changing attitudes, increasing knowledge, or improving control of disease conditions. Video interventions have proven potential to promote and empower patients with chronic conditions [23, 24], as well as optimize self-management in adolescents with Type 1 DM [25]. The concepts have the potential to empower, transfer information and provide positive changes to adolescents with chronic conditions to carry out self-management routines [26, 27]. The perception and willingness to comply with therapy have increased with the existence of Intervention education-based videos [28]. The addition of a video-based lifestyle education program to conventional therapy is effective in increasing the promotion of glycemic control in DM patients. According to Gupta, almost 40% of respondents experienced a 1% decrease in HbA1c, where a simple technology-based educational program consisting of four videos were added to the routine treatment of DM patients [29]. The study conducted by Ratri, (2020) proved that type 2 DM patients at the Airlangga University Hospital in Surabaya who received educational video interventions on insulin therapy reported a significant increase in knowledge. Furthermore, there is an improvement in patient attitudes toward the practice of using insulin injections independently and managing DM conditions [30]. Leong, (2022) conducted interventions in DM patients using an educational program of 51 videos through social media consisting of 10 videos each on the definition of DM and daily care, as well as 6, 21, and 4 videos on nutritional care, DM medicine, and quizzes, respectively. This study was conducted for 3 months and proved to be effective in increasing the knowledge, attitudes, and self-care activities of DM patients [31]. Educational programs through video can overcome problems related to the low health literacy and answer the challenges of traditional face-to-face education [32]. A systematic literature study based on 28 studies with a total of 12,703 subjects concluded that video interventions were very effective for modifying health behavior depending on the target behavior to be influenced. The modeling also facilitated the learning of new behaviors as an important consideration in future video interventions [27]. Abrar (2020) suggested developing educational videos using a trans-cultural approach to traditional language in overcoming communication barriers in the knowledge transfer process. Educational videos in traditional languages increased patient knowledge of disease management [33]. Another analysis to determine the effectiveness of video compared to text as a medium of health education in chronic disease patients by comparing 4 treatment groups, group 1 - text (4000 characters), group 2 - video clip (5 minutes), group 3 - text followed by video, and group 4 - video followed by text, proved that the video followed by text intervention was more effective than the others [34].

3.2 Educational Curriculum

This study obtained 5 thematic educational curriculum needed by respondents, namely DM treatment management, diet, physical activity, drug use, and complications. The initial theme pertained to the management of DM therapy and a significant proportion of respondents require education concerning the therapy management. This necessity arose due to their perception of unwavering adherence to medication, contrasted against the backdrop of consistently elevated blood sugar levels. Furthermore, respondents exhibited limited awareness of the factors contributing to fluctuating blood sugar levels, lack knowledge regarding strategies to mitigate hyperglycemia, and expressed a fervent aspiration for complete recuperation.

"I have taken the medicine 3 times, in the morning before eating glimepiride, after eating metformin, then metformin in the evening?"
(Respondent 2)
"What I want to know is, how can I fully recover from DM and hypertension?"
(Respondent 6)

The first material in the National Diabetes Self-Management Education and Support (DSME) Standard Curriculum is to describe DM, which includes an explanation of the disease process and its treatment options [17]. The second theme is related to DM medications, where 4 out of 10 respondents need education regarding their use, such as dosage and how to store drugs, drug names, and indications, rules for use, and medication adherence.

"I do not know the uses of the drug (pointing to candesartan)."
(Respondent 5)
"I do not know the name of the drug, but what I know is, this is a blood pressure and sugar medicine. I just think of the little blue ones."
(Respondent 10)

DM patients with good adherence give positive results on treatment and reduce mortality [35–37]. Compliance with taking medication is influenced by the level of knowledge of the disease. Patients with a low level of knowledge will also have reduced adherence to taking medication [15, 38–40]. Educational curriculum on DM medications can include the importance of adhering to hypoglycemic medication, the correct technique for using independent insulin [41–43], explaining the indications, rules for use, and side effects of drugs [41, 43], as well as the right time and frequency of taking the medication [44]. The third theme is related to physical activity, where 2 out of 10 respondents state that education is needed regarding the physical activity suitable for DM patients and the time or sports schedule.

"What type of exercise is suitable for DM patients."
(Respondent 3)
"Exercising how much time every day."
(Respondent 8)

Cardiorespiratory physical activity with moderate or high intensity has been shown to reduce cardiovascular events and mortality [45–47], improve fitness in type 1 and Type 2 DM [48], and slow the development of peripheral neuropathy [49]. The recommended physical activity for DM patients is aerobic exercise and resistance. A total of 150 minutes per week of aerobic exercise is needed with at least two sessions per week of resistance training. Recommended types of aerobic exercise include walking, cycling, moderate brisk walking, jogging, basketball, and swimming. The fourth theme is related to a healthy diet, where 7 out of 10 respondents state that education on diet for DM patients is still needed, including the type, amount, and schedule.

"Concerning food, yes, the menu and the amount, say white rice weighing 2.5 ounces, how many scoops or spoons."

(Respondent 1)

"Good and bad foods eaten by DM patients."

(Respondent 5)

"DM patients need to know that you cannot eat too much sweet and fried foods with sachet drinks."

(Respondent 8)

Nutrition intake education can increase knowledge, attitudes, and healthy eating patterns proven to improve DM disease control [50]. Educational curriculum related to healthy eating patterns delivered through role-playing, group discussions, watching a short video, and listening to a lecture has been shown to significantly improve HbA1c control, blood glucose levels, lipid profiles, BMI, as well as blood pressure both systolic and diastolic [41]. Dietary education has been shown to improve the control of clinical parameters in patients [42, 44, 51, 52]. Healthy diet education in the Bukhsh study (2018) which includes meal planning, Eating tips for weight loss for obese/overweight patients, and types of healthy and unhealthy foods for T2DM sufferers are proven to significantly reduce HbA1c [42]. Ukrainian patient education makes guidelines for healthy eating including a meal plan calculated for calories, as well as foods to be consumed and avoided [53]. The fifth theme is related to DM disease, where 6 out of 10 respondents state that DM patients should be educated on the complications.

"Do not know about the complications yet, it seems important to know."

(Respondent 2)

"Want to know about complications to prevent than cure."

(Respondent 3)

"You should know the complications for early treatment and prevention."

(Respondent 8)

According to Bukhsh (2018), educational curriculum pertaining to DM plays a crucial role in imparting to patients the knowledge needed to mitigate the risks associated with DM-related complications. Identification and prevention of common complications such as hypoglycemia, diabetic foot ulcer, diabetic retinopathy, Cardiovascular issues, and diabetic nephropathy, are crucial [42]. Educational curriculum related to cardiovascular complications may include quitting smoking, maintaining blood pressure within the normal range, and controlling cholesterol and triglycerides within the normal range [54]. Diabetic foot complications can be prevented by regular checking and daily foot care. In the event of discovering any instances of compromised skin integrity such as abrasions, ulcers, blisters, regions displaying heightened warmth or reddening, and the formation of calluses, it is recommended to promptly seek consultation with a healthcare professional [54].

Previous studies also showed some of the need for a comprehensive educational curriculum covering a variety of materials. Sanaeinazab (2021) provided an educational curriculum with content including etiologic descriptions and risk factors related to DM, available treatments and their effectiveness, lifestyle modifications with healthy eating patterns and exercise, skills to measure blood sugar with a glucometer, complications of DM and how to assess and control these complications, and control to prevent anxiety, stress, or depression. The study proved to significantly improve HbA1c control, blood glucose levels, lipid profiles, BMI, and systolic and diastolic blood pressure [55]. Sabo (2021) conducted an analysis with educational curriculum for DM patients on the material covering the disease process and treatment, nutritional management, exercise, medication, blood glucose monitoring, acute complications, chronic complications, mental health, and goal setting. A significant decrease in HbA1c and blood pressure in patients compared to the control group was reported [56]. The results of a systematic literature review showed that 4 out of 5 educational curriculum succeeded in improving the quality of life [57]. Educational curriculum on DM and its complications, physical activity, self-care and behavioral self-efficacy, healthy food nutrition, medication adherence, and self-monitoring of blood glucose levels with an intervention period of 3 months is proven to improve quality of life, level of knowledge, and confidence in therapy [58].

4 Conclusion

In conclusion, video-based education was preferred by DM patients compared to text-based. This study recommended an educational curriculum suitable for DM patients including management of therapy, healthy eating patterns, physical activity, medicines, and complications of the disease.

Acknowledgment

The authors are grateful to respondents, who are very cooperative and willing to share experiences and ideas on DM education forms and curriculum. This study was supported by the Faculty of Pharmacy, Gadjah Mada University, and the Pharmacy Study Program, Faculty of Health, Ngudi Waluyo University.

References

1. Ogurtsova K, da Rocha Fernandes JD, Huang Y, et al. IDF Diabetes Atlas: Global estimates for the prevalence of diabetes for 2015 and 2040. *Diabetes Res Clin Pract* 2017; 128: 40–50.
2. Sun H, Saeedi P, Karuranga S, et al. IDF Diabetes Atlas: Global, regional and country-level diabetes prevalence estimates for 2021 and projections for 2045. *Diabetes Res Clin Pract* 2022; 183: 109119.

3. BPJS kesehatan. *Laporan Pengelolaan Program dan Keuangan BPJS Kesehatan 2022*. Jakarta, 2023.
4. Ivers NM, Jiang M, Alloo J, Singer A, Ngui D, Casey CG YC. Diabetes Canada 2018 clinical practice guidelines: Key messages for family physicians caring for patients living with type 2 diabetes. *Can Fam Physician* 2019; 65: 14–24.
5. *National Institute for Health and Care Excellence, Type 2 diabetes in adults: management. NICE Guidelines*. 2015.
6. Beck J, Greenwood DA, Blanton L, et al. 2017 National Standards for Diabetes Self-Management Education and Support. *Diabetes Educ* 2017; 43: 449–464.
7. Powers MA, Bardsley J, Cypress M, et al. Diabetes Self-management Education and Support in Type 2 Diabetes: A Joint Position Statement of the American Diabetes Association, the American Association of Diabetes Educators, and the Academy of Nutrition and Dietetics. *Clin Diabetes* 2016; 34: 70–80.
8. Urbanski P, Wolf A, Herman WH. Cost-Effectiveness of Diabetes Education. *J Am Diet Assoc* 2008; 108: S6–S11.
9. American Diabetes Association. Third-Party Reimbursement for Diabetes Care, Self-Management Education, and Supplies. *Diabetes Care* 2014; 37: S118–S119.
10. Deakin TA, McShane CE, Cade JE, et al. Group based training for self-management strategies in people with type 2 diabetes mellitus. In: Steinsbekk A (ed) *Cochrane Database of Systematic Reviews*. Chichester, UK: John Wiley & Sons, Ltd. Epub ahead of print 20 April 2005. DOI: 10.1002/14651858.CD003417.pub2.
11. Ryan JG, Jennings T, Vittoria I, et al. Short and Long-Term Outcomes from a Multisession Diabetes Education Program Targeting Low-Income Minority Patients: A Six-Month Follow Up. *Clin Ther* 2013; 35: A43–A53.
12. Brown SA, Garcia AA, Kouzekanani K, et al. Culturally Competent Diabetes Self-Management Education for Mexican Americans. *Diabetes Care* 2002; 25: 259–268.
13. A Steinsbekk, LO Rygg, M Lisulo, MB Rise and AF. Group based diabetes self-management education compared to routine treatment for people with type 2 diabetes mellitus. A systematic review with meta-analysis. In: *Database of Abstracts of Reviews of Effects (DARE): Quality-assessed Reviews*. New York, http://www.crd.york.ac.uk/CRDWeb/ (2012).
14. Goodarzi M, Ebrahimzadeh I, Rabi A, et al. Impact of distance education via mobile phone text messaging on knowledge, attitude, practice and self efficacy of patients with type 2 diabetes mellitus in Iran. *J Diabetes Metab Disord* 2012; 11: 10.
15. Lerman I. Adherence to Treatment: The Key for Avoiding Long-Term Complications of Diabetes. *Arch Med Res* 2005; 36: 300–306.
16. Pace AE, Ochoa-Vigo K, Caliri MHL, et al. Knowledge on diabetes mellitus in the self care process. *Rev Lat Am Enfermagem* 2006; 14: 728–734.
17. Haas L, Maryniuk M, Beck J, et al. National standards for diabetes self-management education and support. *Diabetes Care* 2014; 37 Suppl 1: 1630–1637.
18. Chatterjee S, Davies MJ, Heller S, et al. Diabetes structured self-management education programmes: a narrative review and current innovations. *Lancet Diabetes Endocrinol* 2018; 6: 130–142.
19. Noy C. Sampling Knowledge: The Hermeneutics of Snowball Sampling in Qualitative Research. *Int J Soc Res Methodol* 2008; 11: 327–344.
20. Tong A, Sainsbury P, Craig J. Consolidated criteria for reporting qualitative research (COREQ): a 32-item checklist for interviews and focus groups. *Int J Qual Heal Care* 2007; 19: 349–357.
21. Braun, V., Clarke, V. *Successful Qualitative Research: A Practical Guide for Beginners*. London: SAGE Publications Inc, 2013.
22. Denzin, N.K., Lincoln YS (Eds). *Handbook of Qualitative Research*. London: SAGE Publications Inc, 1994.
23. Sleath B, Carpenter DM, Lee C, et al. The development of an educational video to motivate teens with asthma to be more involved during medical visits and to improve medication adherence. *J Asthma* 2016; 53: 714–719.
24. RICH M, LAMOLA S, WOODS E. Effects of creating visual illness narratives on quality of life with asthma: A pilot intervention study. *J Adolesc Heal* 2006; 38: 748–752.
25. C. Webster, A. Greene, S. Greene. Development of "VIG-Diabetes", a video intervention to optimise self-management in young people with type 1 diabetes. *J Diabetes Nurs* 2015; 19: 22–26.
26. Krouse HJ. Video modelling to educate patients. *J Adv Nurs* 2001; 33: 748–757.
27. Tuong W, Larsen ER, Armstrong AW. Videos to influence: a systematic review of effectiveness of video-based education in modifying health behaviors. *J Behav Med* 2014; 37: 218–233.
28. Juo Y-Y, Freeby MJ, Arguello V, et al. Efficacy of video-based education program in improving metabolic surgery perception among patients with obesity and diabetes. *Surg Obes Relat Dis* 2018; 14: 1246–1253.
29. Gupta U, Gupta Y, Jose D, et al. Effectiveness of a Video-Based Lifestyle Education Program Compared to Usual Care in Improving HbA1c and Other Metabolic Parameters in Individuals with Type 2 Diabetes: An Open-Label Parallel Arm

Randomized Control Trial (RCT). *Diabetes Ther* 2020; 11: 667–679.

30. Ratri DMN, Hamidah KF, Puspitasari AD, et al. Video-Based Health Education to Support Insulin Therapy in Diabetes Mellitus Patients. *J Public health Res* 2020; 9: jphr.2020.1849.
31. Leong CM, Lee T-I, Chien Y-M, et al. Social Media–Delivered Patient Education to Enhance Self-management and Attitudes of Patients with Type 2 Diabetes During the COVID-19 Pandemic: Randomized Controlled Trial. *J Med Internet Res* 2022; 24: e31449.
32. Kim SH, Utz S. Effectiveness of a Social Media–Based, Health Literacy–Sensitive Diabetes Self-Management Intervention: A Randomized Controlled Trial. *J Nurs Scholarsh* 2019; 51: 661–669.
33. Abrar EA, Yusuf S, Sjattar EL, et al. Development and evaluation educational videos of diabetic foot care in traditional languages to enhance knowledge of patients diagnosed with diabetes and risk for diabetic foot ulcers. *Prim Care Diabetes* 2020; 14: 104–110.
34. Bezzubtseva MV, Demkina AE, Lipilina MN, et al. Video or text? Education through a social media website in hypertension. *Int J Cardiol Cardiovasc Risk Prev* 2022; 14: 200139.
35. Krapek K, King K, Warren SS, et al. Medication Adherence and Associated Hemoglobin A 1c in Type 2 Diabetes. *Ann Pharmacother* 2004; 38: 1357–1362.
36. Rhee MK, Slocum W, Ziemer DC, et al. Patient Adherence Improves Glycemic Control. *Diabetes Educ* 2005; 31: 240–250.
37. Simpson SH, Eurich DT, Majumdar SR, et al. A meta-analysis of the association between adherence to drug therapy and mortality. *BMJ* 2006; 333: 15.
38. Al-Qazaz HK, Sulaiman SA, Hassali MA, et al. Diabetes knowledge, medication adherence and glycemic control among patients with type 2 diabetes. *Int J Clin Pharm* 2011; 33: 1028–1035.
39. Vlasnik JJ, Aliotta SL, DeLor B. Medication adherence: Factors influencing compliance with prescribed medication plans. *Case Manager* 2005; 16: 47–51.
40. Albright TL, Parchman M, Burge SK, et al. Predictors of self-care behavior in adults with type 2 diabetes: an RRNeST study. *Fam Med* 2001; 33: 354–60.
41. Bell AM, Fonda SJ, Walker MS, et al. Mobile Phone-Based Video Messages for Diabetes Self-Care Support. *J Diabetes Sci Technol* 2012; 6: 310–319.
42. Bukhsh A, Nawaz MS, Ahmed HS, et al. A randomized controlled study to evaluate the effect of pharmacist-led educational intervention on glycemic control, self-care activities and disease knowledge among type 2 diabetes patients. *Medicine (Baltimore)* 2018; 97: e9847.
43. Chong M. Clinical outcomes of a diabetes education program for patients with diabetes mellitus in the Micronesian community in Hawaii. *J Res Pharm Pract* 2016; 5: 205.
44. Rusdiana, Savira M, Amelia R. The Effect of Diabetes Self-Management Education on Hba1c Level and Fasting Blood Sugar in Type 2 Diabetes Mellitus Patients in Primary Health Care in Binjai City of North Sumatera, Indonesia. *Open Access Maced J Med Sci* 2018; 6: 715–718.
45. Church TS, LaMonte MJ, Barlow CE, et al. Cardiorespiratory Fitness and Body Mass Index as Predictors of Cardiovascular Disease Mortality Among Men With Diabetes. *Arch Intern Med* 2005; 165: 2114.
46. Gregg EW, Gerzoff RB, Caspersen CJ, et al. Relationship of Walking to Mortality Among US Adults With Diabetes. *Arch Intern Med* 2003; 163: 1440.
47. Hu FB, Stampfer MJ, Solomon C, et al. Physical Activity and Risk for Cardiovascular Events in Diabetic Women. *Ann Intern Med* 2001; 134: 96.
48. Nielsen PJ, Hafdahl AR, Conn VS, et al. Meta-analysis of the effect of exercise interventions on fitness outcomes among adults with type 1 and type 2 diabetes. *Diabetes Res Clin Pract* 2006; 74: 111–120.
49. Balducci S, Iacobellis G, Parisi L, et al. Exercise training can modify the natural history of diabetic peripheral neuropathy. *J Diabetes Complications* 2006; 20: 216–223.
50. Sami W, Ansari T, Butt NS, et al. Effect of diet on type 2 diabetes mellitus: A review. *Int J Health Sci (Qassim)* 2017; 11: 65–71.
51. Azami G, Soh KL, Sazlina SG, et al. Effect of a Nurse-Led Diabetes Self-Management Education Program on Glycosylated Hemoglobin among Adults with Type 2 Diabetes. *J Diabetes Res* 2018; 2018: 1–12.
52. Dong Y, Wang P, Dai Z, et al. Increased self-care activities and glycemic control rate in relation to health education via Wechat among diabetes patients. *Medicine (Baltimore)* 2018; 97: e13632.
53. Ukrainian Patient Education. *DIABETES MELLITUS AND NUTRITION, ADULT*, https://elsevier.health/en-US/preview/diabetes-mellitus-and-nutrition-adult (2021).
54. Deborah J Wexler, MD Ms. Patient education: Preventing complications from diabetes (Beyond the Basics), https://www.uptodate.com/contents/preventing-complications-from-diabetes-beyond-the-basics (2023).
55. Sanaeinasab H, Saffari M, Yazdanparast D, et al. Effects of a health education program to promote healthy lifestyle and glycemic control in patients

 https://doi.org/10.1051/bioconf/20237505008

with type 2 diabetes: A randomized controlled trial. *Prim Care Diabetes* 2021; 15: 275–282.

56. Sabo R, Robins J, Lutz S, et al. Diabetes Engagement and Activation Platform for Implementation and Effectiveness of Automated Virtual Type 2 Diabetes Self-Management Education: Randomized Controlled Trial. *JMIR Diabetes* 2021; 6: e26621.
57. Cunningham AT, Crittendon DR, White N, et al. The effect of diabetes self-management education on HbA1c and quality of life in African-Americans: a systematic review and meta-analysis. *BMC Health Serv Res* 2018; 18: 367.
58. Mohammadi S, Karim NA, Talib RA, et al. The impact of self-efficacy education based on the health belief model in Iranian patients with type 2 diabetes: a randomised controlled intervention study. *Asia Pac J Clin Nutr* 2018; 27: 546–555.

Factors affecting customers' decisions to purchase medicines in Ho Chi Minh City: a quantitative study

Thao Le Thi Quy[1*], *Vinh* Huynh The[1], Susi Ari Kristina[2], and *Vy* Nguyen Huynh Thao[1]

[1] Faculty of Pharmacy, Pham Ngoc Thach University of Medicine, Hochiminh City 700000, Vietnam.
[2] Faculty of Pharmacy, Universitas Gadjah Mada, Yogyakarta 55281, Indonesia.

Abstract. Understanding customer behavior and the factors influencing decisions to purchase medicines is crucial for businesses. This study aimed to examine the factors influencing customers' decisions to purchase medicines in Ho Chi Minh City (HCM). This quantitative study used a self-administered questionnaire. A convenience quota sample of 599 participants was recruited online and offline, and data were collected from July to October 2022. The questionnaire included 8 items on personal characteristics, 6 on customers' medicine use characteristics, 13 on factors determining where to buy medicine, and 17 on factors influencing customers' decision to buy medicine. A two-step cluster analysis method was used to identify three distinct customer clusters based on common characteristics. Statistical significance was set at $p<0.05$, with a 95% confidence interval. Of the 632 distributed questionnaires, 599 were returned and analyzed, resulting in a response rate of 94.78%. When making decisions about purchasing medicines, customers expressed the highest level of concern regarding the therapeutic effects of medicines (mean = 4.40 ± 0.77) and the least interest in advertisements (mean = 3.29 ± 1.19). Offering a wide range of products from different brands provides customers with more choices, which attracts them to drugstores. It is recommended that companies minimize their efforts to advertise medicines in Vietnam.
Keywords: two-step cluster, customer behavior, survey, questionnaires, Vietnam

1 Introduction

The COVID-19 pandemic, declared a worldwide health emergency by the World Health Organization (WHO) in January 2020, has profoundly impacted global public health [1]. Following the WHO's characterization of COVID-19 as a pandemic in March 2020, the focus on disease prevention and health awareness has increased among individuals as healthcare systems recover [2, 3]. Studies indicate that the COVID-19 outbreak has significantly influenced consumer purchasing decisions [4-6].

Consumer behavior has long been a hot issue in marketing. Understanding how and why consumers behave a specific way and make confident purchasing decisions enables businesses to enhance their marketing tactics and increase their market share. Thus, one of the challenges facing marketers today is persuading people to buy their goods or services. Therefore, understanding consumer purchasing behavior sheds light on the psychology of how consumers reason, feel, and choose among available options (such as brands, products, and retailers), as well as how the consumer's environment (such as culture, family, and the media) influences them. It also clarifies how consumer motivation and decision-making differ between products. This helps us understand how marketers might enhance their marketing initiatives to better connect with consumers. Understanding customer decisions to purchase medicine, marketing campaigns from companies will become more effective, and public health campaigns can use this to implement actively.

Consumer purchasing decisions encompass various aspects, including shopping habits, buying behavior, brand preferences, and purchase locations. These decisions are influenced by customer characteristics and a range of cultural, social, familial, personality, psychological, and environmental factors [7]. Understanding these influencing factors provides companies with valuable insights for developing effective strategies, marketing messages, and advertising campaigns that align with the needs and preferences of their target consumers, ultimately enhancing customer satisfaction and driving sales.

Vietnam's pharmaceutical industry has experienced substantial growth over the past decade owing to changes in domestic regulations on imported drugs [8]. The Vietnamese healthcare market, valued at USD 17.4 billion, offers numerous opportunities for the pharmaceutical supply sector. Per-capita spending on healthcare is predicted to quadruple from USD 170 in 2017 to USD 400 by 2027. Furthermore, Vietnam's pharmaceutical market is projected to witness double-

* Corresponding author: thaoltq@pnt.edu.vn

https://doi.org/10.1051/bioconf/20237505009

digit growth over the next five years, with an estimated turnover of USD 5.9 billion in 2018, representing an 11.7% annual increase. Vietnam relies heavily on pharmaceutical imports, indicating the significance of this market [9].

While acquiring new customers is important for any business, retaining existing ones is crucial and challenging. Retaining current customers is more cost-effective than acquiring new ones and adds substantial value to a company [10]. Therefore, this study aimed to investigate the factors influencing consumers' purchase decisions regarding medicines in Ho Chi Minh (HCM), Vietnam. The specific objectives were to examine the general characteristics of customers purchasing medicines in HCM in 2022, investigate the factors influencing consumers' purchase decisions for medicines in the community, and explore the correlation between common characteristics and the factors affecting customers' decisions to buy medicines in HCM in 2022.

2 Materials and methods

2.1 Study design

This cross-sectional study was conducted in HCM, Vietnam, from July to October 2022.

2.2 Sample size and data collection

Sample size calculation was performed using the following formula:

$$n = \frac{z^2 \times P \times (100 - P)}{d^2}$$

where P is the anticipated awareness score percentage, d is the desired precision, z is the appropriate value from the normal distribution for the desired confidence interval, and n is the required sample size [11]. With a desired precision level of 5% at the 95% confidence level, a minimum of 385 cases was required at $p = 0.5$.

Convenience sampling was used to collect data from individuals who met the following inclusion criteria: (1) at the age of 18 or over, (2) nationality of the People's Republic of Vietnam, (3) the ability to complete the online questionnaire independently or with the assistance of investigators, (4) understanding of the questionnaire items, and (5) having purchased drugs in HCM in 2022. Individuals with cognitive impairment or mental disorders and those who did not complete the survey were excluded.

The survey was launched through social networks, utilizing Google Forms, with the questionnaire link included in articles posted on platforms such as Facebook and Zalo. The Google Forms made the data meet the sampling criteria by excommunicating participants who were not from Ho Chi Minh City or did not agree to participate in the research.

To mitigate the limitations associated with online surveys, face-to-face interviews were conducted to increase the sample size and reduce potential bias towards specific age groups. The participants will be asked in 10 minutes with four parts of the questionnaire, including demographic characteristics of customers (8 questions), customers' medicine use characteristics (6 questions), factors affecting pharmacy selection (13 questions), and factors affecting customer's purchase decision of medicines (17 questions).

2.3 Survey instrument

Data collection involved a structured questionnaire comprising closed-ended, open-ended, and multiple-choice questions. The questionnaire development process comprised several steps. First, a set of questions was prepared based on previous studies through a comprehensive literature review. Second, an expert panel comprising a specialist pharmacist II, a university pharmacist, and a pharmacy pharmacist evaluated the content and developed the questionnaire, resulting in an initial version with 53 questions. Third, the questionnaire was tested on 59 individuals aged 18 years or older [12], and the consistency of the questions was assessed using Cronbach's alpha. An exploratory factor analysis (EFA) was conducted to identify the underlying structures influencing responses. Based on the test results, the questionnaire was revised to produce the final version containing 44 questions. The questionnaire was originally developed in Vietnamese, the participants' native language.

2.4 Data analysis

Data from the questionnaire were coded, entered, and stored in Microsoft Excel. Statistical Package for the Social Sciences (SPSS) software was used for data analysis. Descriptive statistics, including frequencies and percentages, were used to analyze the demographic characteristics of the respondents. The means and standard deviations were calculated to examine the variables and questionnaire items. The reliability of the dataset was assessed using Cronbach's alpha with a threshold set at 0.7. An EFA was performed to identify the underlying structure influencing the responses. Additionally, a two-step cluster analysis was conducted to identify customer segments.

3 Results

The sample consisted of 599 participants, with women accounting for 51.8%. The average age of the participants was 36 ± 15 years. Notably, fewer respondents belonged to the income group of VND 2-4 million than that of less than VND 2 million because of the inclusion of unemployed individuals (18%). Further details on the socio-demographic characteristics of the study population are presented in Table 1.

Table 1. Socio-demographic characteristics

Variable	n = 599	%
Gender		
Male	289	48.2
Female	310	51.8
Age (years)		
18-25	195	32.6
26-35	131	21.9
36-45	87	14.5
46-55	83	13.9
≥ 56	103	17.2
Marital status		
Others (single, divorced, etc.)	323	53.9
Married	276	46.1
Education		
Illiteracy	1	0.2
Primary school	14	2.3
Secondary school	65	10.9
High school	91	15.2
Elementary level/intermediate level	42	7.0
College/ university	386	64.4
Occupation		
Student	148	24.7
Retired/unemployed	67	11.2
Healthcare Assistants	58	9.7
Office staff	164	27.4
Worker	49	8.2
Other	113	18.9
Income in one month* (VND)		
Under 2,000,000	108	18.0
2,000,000- 4,000,000	55	9.2
4,000,000- 6,000,000	103	17.2
6,000,000- 9,000,000	147	24.5
Over 9,000,000	186	31.1
Self-reported health status		
Very poor	3	0.5
Poor	20	3.3
Good or fair	516	86.1
Excellent or very good	60	10.0

Of the 599 participants, 468 (78.1%) reported purchasing medications six months prior to the survey. Among those who made purchases, the majority (40.4%) bought medicine only 1 or 2 times during that period. Participants indicated buying medicines from various categories, with the highest percentages observed for pain and palliative care medicines (58.1%); vitamins and minerals (44.0%); allergy medications (25.9%); medicines for eye, ear, nose, and throat infections (23.9%); and medicines for digestive problems (20.5%). The distribution of the medication groups is shown in Figure 1.

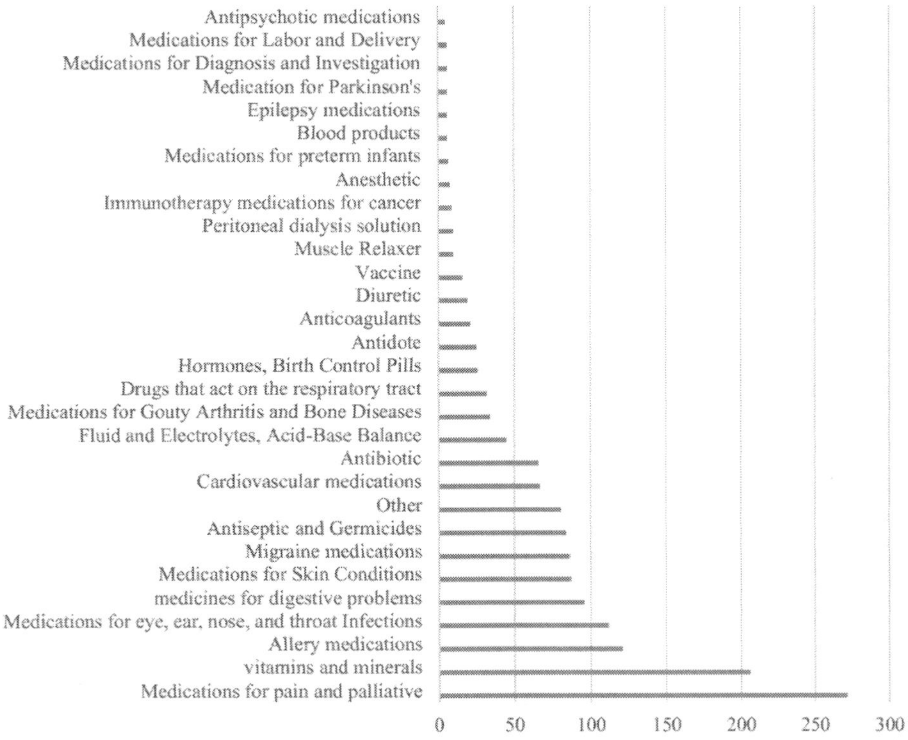

Fig. 1. Medication groups customers purchased within 6 months prior to the survey

Table 2. Factors affecting pharmacy selection

	SD	D	N	A	SA	Mean ± SD
Pharmacy's staff	1	2	3	4	5	4.18 ± 0.85
Anonymity/ confidentiality	1	2	3	4	5	3.89 ± 1.13
Store's atmosphere	1	2	3	4	5	3.63 ± 1.05
Pharmacy in long-term activity	1	2	3	4	5	3.90 ± 0.98
Reputation of pharmacy	1	2	3	4	5	3.92 ± 1.00
Product range	1	2	3	4	5	4.19 ± 0.89
Product quality	1	2	3	4	5	4.43 ± 0.80
SD: Strongly disagree, D: Disagree, N: Neutral, A: Agree, SA: Strongly agree						

Table 2 displays the factors influencing the participants' choice of pharmacy. All factors received mean values above 3, indicating that the participants considered these criteria when deciding where to buy medicines. The factor with the highest average score was product quality (4.43 ± 0.80), while the lowest average score was observed for advertisements (3.29 ± 1.19).

Table 3. Factors affecting customer's purchase decision of medicines

	SD	D	N	A	SA	Mean ± SD
Previous experience	1	2	3	4	5	3.75 ± 1.02
Pharmacist's advice	1	2	3	4	5	4.00 ± 0.95
Country of origin	1	2	3	4	5	3.89 ± 1.00
Advertisement	1	2	3	4	5	3.29 ± 1.19
Drug information	1	2	3	4	5	4.00 ± 1.03
Brand of product	1	2	3	4	5	3.93 ± 0.98
Therapeutic efficacy of medicine	1	2	3	4	5	4.40 ± 0.77
Product composition	1	2	3	4	5	4.13 ± 0.97
SD: Strongly disagree, D: Disagree, N: Neutral, A: Agree, SA: Strongly agree						

Table 3 presents the factors influencing customer decisions to purchase medicines. As in the previous section, all factors had mean scores above 3, indicating their consideration by the participants. The factor with the highest average score was the therapeutic efficacy of the medicine (4.40 ± 0.77), while the advertising factor received the lowest average score (3.29 ± 1.19).

Three distinct customer clusters were identified using a two-step cluster analysis method based on common characteristics. Table 4 illustrates the correlation between common characteristics and factors influencing the selection of pharmacies for each cluster. The same analysis was conducted for the factors influencing customer purchase decisions regarding medicines. The results are presented in Table 5. A summary of the three customer profiles is provided in Table 6, which highlights the demographic characteristics and main influencing factors for each cluster.

Table 4. Correlation between common characteristics and factors affecting selection of pharmacy

	Cluster 1 (n = 268)	Cluster 2 (n = 173)	Cluster 3 (n = 158)
		Mean	
Pharmacy's staff	4.28	3.98	4.23
Anonymity/ confidentiality	4.06	3.33	4.22
Store's atmosphere	3.77	3.39	3.66
Pharmacy in long-term activity	4.03	3.70	3.89
Reputation of pharmacy	4.06	3.69	3.92
Product range	4.37	3.91	4.17
Product quality	4.51	4.32	4.41

Table 5. Correlation between common characteristics and factors affecting customer's purchase decision of medicines

	Cluster 1 (n = 268)	Cluster 2 (n = 173)	Cluster 3 (n = 158)
		Mean	
Previous experience	3.94	3.41	3.79
Pharmacist's advice	4.12	3.76	4.06
Country of origin	4.10	3.73	3.73
Advertisement	3.47	2.78	3.53
Medicine's information	4.22	3.62	4.04
Brand of product	4.08	3.63	3.99
Therapeutic efficacy of medicine	4.50	4.24	4.39
Product composition	4.37	3.70	4.18

Table 6. Comparative summary of the three customer profiles

	Cluster 1	Cluster 2	Cluster 3
Demographic characteristics	High level of education, office staff, high income, average age	Average level of education, either office staff or worker, average income, older age	High level of education, student, low income, low age
Factors affecting pharmacy selection	The main influencing factors include Product quality, pharmacy's staff, and product range. The store's atmosphere factor is the least influential	Factors are less positive feedback than Clusters 1 and 3. The main influencing factors include product quality, pharmacy's staff, and product range. The anonymity/ confidentiality factor is the least influential	The main influencing factors include product quality, pharmacy's staff, and anonymity. Unlike Cluster 2, Anonymity/ confidentiality has a much greater influence in Cluster 3. The store's atmosphere factor is the least influential
Factors affecting customer's purchase decision	The main influencing factors include drug composition, drug information, and the therapeutic effect of the medicine	The main influencing factors include the therapeutic effect of the medicine, the pharmacist's advice, and country of origin. The advertising factor has no effect on this cluster	The main influencing factors include the therapeutic effect of the medicine, drug composition, and the pharmacist's advice

4 Discussion

4.1 Selection of pharmacy

All factors influenced customers' decisions to purchase medicines, with product quality being the most influential factor and store atmosphere being the least influential. Customers prefer pharmacies that offer high-quality products, which positively impact treatment outcomes. This, in turn, leads to positive reviews and electronic word-of-mouth recommendations. Ehsani (2015) defined product quality as a customer's perception of overall quality or superiority over alternatives [13]. However, pharmacy owners should also consider the cost of medicines when deciding which brands to import because high-quality products often come at higher prices. It is important to balance offering high-quality products and ensuring customer affordability.

Compared with similar studies conducted in other countries, Vietnamese customers placed less emphasis on services provided by pharmacies. The mean values for pharmacy staff, store atmosphere, and anonymity/confidentiality were lower in this study than in comparative studies [14]. This finding can be attributed to differences in customers' knowledge of drug information and services. In developed countries, customers are better educated about pharmaceuticals, leading them to consider pharmacy services' quality. In contrast, Vietnamese customers rely more on the perception of product quality, assuming that all pharmacies offer products of similar quality. Additionally, pharmacies in Vietnam often sell a wide range of products in addition to medicines that are convenient for customers.

4.2 Customer's purchase decision of medicines

The therapeutic efficacy of medicines has emerged as the most influential factor in customer purchase decisions. Customers are more inclined to purchase medicines that are perceived to be more effective. This finding aligns with the importance of product quality in the selection of pharmacies, emphasizing the overall impact of effective treatment outcomes on customer decision-making. Compared to previous studies, factors such as pharmacists' advice, product brand, and country of origin had higher mean values in this study [15].

Vietnamese customers have limited knowledge of medicines, leading them to prefer medicines from developed countries to those manufactured domestically. Medicines from developed countries are considered more effective. Consequently, customers rely heavily on pharmacists' advice as they lack the knowledge to make informed decisions about which medicines to purchase. This finding underscores the importance of knowledgeable and experienced pharmacists, who can provide accurate advice to customers. Additionally, positive customer perceptions of pharmacists' services contribute to customer engagement and loyalty [10]. Therefore, pharmacies should prioritize providing competent and socially skilled pharmacists to enhance customer satisfaction and loyalty.

Advertising was found to have minimal influence on customers' purchasing decisions, which is consistent with previous research [16, 17]. Customers perceive advertising as having little impact on their decisions, especially regarding over-the-counter (OTC) drugs. Trust in advertising is lower than trust in physicians' prescriptions. Physicians and the general public have negative views of pharmaceutical advertising, potentially affecting the physician-patient relationship [18].

4.3 Correlation between common characteristics and factors affecting customers' decision to buy medicines

Cluster analysis was used to identify three distinct customer groups based on their common characteristics. The largest group (Cluster 1) consists of customers with a high level of education, predominantly office staff with higher incomes, and an average age of 37.12. The selection of pharmacies was influenced by factors such as product quality, pharmacy staff, and product range. Regarding the decision to purchase medicines, factors such as drug composition, drug information, and the therapeutic effect of the medicine played a significant role.

The second largest group (Cluster 2) primarily comprised older customers with an average level of education, consisting of office staff and workers with average incomes and an average age of 47.10. They shared influencing factors in the selection of pharmacies with Cluster 1, including product quality, pharmacy staff, and product ranges. However, their purchase decisions regarding medicines were influenced by different factors, namely the therapeutic effect of the medicine, the pharmacist's advice, and the country of origin.

The smallest group (Cluster 3) mainly consisted of young students with low incomes but high education levels. This group differed from Clusters 1 and 2. The key factors influencing this group include product quality, pharmacy staff, and anonymity. Factors such as the therapeutic effect of the medicine, drug composition, and pharmacists' advice played a significant role in purchase decisions.

Notably, store atmosphere had a lesser impact on customers' selection of pharmacies in Clusters 1 and 3, while anonymity/confidentiality had a stronger influence in Cluster 3. This finding indicates that young people are more concerned about information security when choosing where to buy medicines, whereas middle-aged or older individuals with average incomes are less interested in customer information protection. Level of education and occupation also influenced customers' perspectives. Customers with higher educational levels were more attentive to drug composition and information, whereas those with lower educational levels and older adults placed less importance on these factors. Pharmacists' advice emerged as the third influencing factor in the student group's decisions to purchase medicines, indicating that pharmacists' advice holds greater weight than drug information for this particular demographic [19].

5 Conclusion

In conclusion, this study revealed that Vietnamese customers prioritize product quality when selecting a pharmacy, whereas factors such as store atmosphere, pharmacy staff, and anonymity have less influence on their decisions. Vietnamese customers prefer foreign pharmaceuticals, particularly those from developed countries, and factors such as the country of origin and product brand play a significant role in their purchase decisions. The therapeutic efficacy of medicines is highly valued by customers when making purchase decisions. Students in Vietnam prioritized information security when purchasing medicines. Therefore, pharmacies targeting this demographic group should consider this factor. Advertising has the least impact on customer purchase decisions, indicating that companies should not overly rely on advertising strategies. Segmentation analysis demonstrates variations in customer preferences, selection criteria, and demographic characteristics among the three customer clusters.

6 Limitation

The study's limitations are evident as it solely targeted the inner-city population of Ho Chi Minh City. This limitation arose due to the considerable geographical distance separating the researcher from the suburban areas, resulting in the omission of suburban subjects. Furthermore, the study exclusively engaged retail customers purchasing drugs, while leaders within health agencies remained unexplored.

To address these shortcomings, it is recommended that future studies consider these limitations. However, it is worth noting that we currently lack any rationale to anticipate that rectifying these weaknesses would substantially alter the primary outcomes we have attained.

7 Recommendation

To enhance the comprehensiveness of the study, it is recommended to extend the research beyond the inner-city population of Ho Chi Minh City and incorporate suburban areas. This would provide a more holistic understanding of the phenomenon under investigation and capture diverse perspectives and experiences. Efforts should be made to overcome geographical challenges and ensure representation from suburban subjects. Address the geographical distance issue by implementing strategies such as collaborating with local research partners or utilizing advanced communication technologies. By doing so, you can overcome the constraints posed by the researcher's location and ensure that the suburban population is adequately represented in the study.

To achieve a comprehensive understanding of the subject matter, extend the scope of the study to include leaders in health agencies. Their insights and perspectives are invaluable, as they possess a broader overview of health-related trends and policies. Engaging these stakeholders through interviews, surveys, or focus groups can provide a well-rounded perspective on the topic.

Consequently, more valuable insights and recommendations can be generated for the entire sector.

8 Theoretical Implications

Based on research, certain factors distinctly influence customer purchasing behavior. These factors collectively shape the pharmaceutical market in Ho Chi Minh City, with their impact varying in strength depending on the country and region. While the study effectively uncovers elements with significant or weak influences on customers' decisions to purchase pharmaceuticals in Ho Chi Minh City, its significance is contingent upon the timing of examination. It's important to acknowledge that these factors could evolve in the future.

9 Practical Implications

In light of the factors that influence medication purchasing decisions, businesses can fortify their weaker aspects or promote their strengths during business transactions, thereby leading to increased consumer satisfaction and loyalty. Furthermore, these factors facilitate the segmentation of the pharmaceutical usage market, equipping firms with a more comprehensive understanding of the pharmaceutical market in Ho Chi Minh City.

10 Supplementary Information

10.1 Funding statement

This research was not funded by any specific grants from the public, commercial, or not-for-profit sectors.

10.2 Acknowledgements

The authors express their gratitude to all survey respondents for their participation and willingness to share sensitive information for research purposes. They also acknowledge the valuable comments and suggestions of the reviewers.

Competing interests: **The authors declare no conflicts of interest in this work.**

REFERENCES

1. Velavan, T.P. and C.G. Meyer, *The COVID-19 epidemic.* Tropical medicine & international health. **25**(3): p. 278 (2020).
2. Betz, C.L., *COVID-19 Special Collection.* Journal of Pediatric Nursing. **62**: p. A9 (2022).
3. Barber, S.J. and H. Kim, *COVID-19 worries and behavior changes in older and younger men and*

https://doi.org/10.1051/bioconf/20237505009

women. The Journals of Gerontology: Series B. **76**(2): p. e17-e23 (2021).
4. Hesham, F., H. Riadh, and N.K. Sihem, *What have we learned about the effects of the COVID-19 pandemic on consumer behavior?* Sustainability. **13**(8): p. 4304 (2021).
5. Patil, B. and N. Patil, *Impact of COVID-19 pandemic on consumer behaviour.* Mukt Shabd Journal. **9**(5): p. 3074-3085 (2020).
6. Verma, M. and B. Naveen, *COVID-19 impact on buying behaviour.* Vikalpa. **46**(1): p. 27-40 (2021).
7. Ramya, N. and S.M. Ali, *Factors affecting consumer buying behavior.* International journal of applied research. **2**(10): p. 76-80 (2016).
8. Nguyen, T., A. Vitry, and E. Roughead, *Pharmaceutical policy in vietnam In: Pharmaceutical policy in countries with developing healthcare systems.* 2017, Springer Switzerland.
9. Quan, N. H. K., Adegoke, A. O., Wankasi, M. M., & Nnatuanya, I. N, *International Integration of Pharmaceutical Supply Chains in Vietnam: An Overview of Challenges and Opportunities at Hospitals in Ho Chi Minh City.* International Journal of Medical and Health Sciences Research. **7**(1): p. 37-48 (2020).
10. Nitadpakorn, S., K.B. Farris, and T. Kittisopee, *Factors affecting pharmacy engagement and pharmacy customer devotion in community pharmacy: A structural equation modeling approach.* Pharmacy Practice (Granada). **15**(3) (2017).
11. Bootsumran, L., S. Siripipatthanakul, and B. Phayaphrom, *Factors influencing consumers' purchase intention at pharmacies in Thailand.* Journal of Management in Business, Healthcare and Education. **1**(1): p. 1-16 (2021).
12. Viechtbauer, W., Smits, L., Kotz, D., Budé, L., Spigt, M., Serroyen, J., & Crutzen, R., *A simple formula for the calculation of sample size in pilot studies.* Journal of clinical epidemiology. **68**(11): p. 1375-1379 (2015).
13. Razak, I., N. Nirwanto, and B. Triatmanto, *The impact of product quality and price on customer satisfaction with the mediator of customer value.* IISTE: Journal of Marketing and Consumer Research. **30**: p. 59-68 (2016).
14. Kevrekidis, D. P., Minarikova, D., Markos, A., Malovecka, I., & Minarik, P., *Community pharmacy customer segmentation based on factors influencing their selection of pharmacy and over-the-counter medicines.* Saudi pharmaceutical journal. **26**(1): p. 33-43 (2018).
15. Temechewu, M.W. and M. Gebremedhin, *Factors affecting consumers' purchase decision of over-the-counter (OTC) medicines: Empirical evidences from community pharmacies in Ethiopia.* Journal of Medicine, Physiology and Biophysics. **65**: p. 8-25 (2020).
16. Villako, P., D. Volmer, and A. Raal, *Factors influencing purchase of and counselling about prescription and OTC medicines at community pharmacies in Tallinn, Estonia.* Acta Pol Pharm. **69**(2): p. 335-340 (2012).
17. Haramiova, Z., Z. Kobliskova, and J. Soltysova, *Purchase of prescription and OTC medicines in Slovakia: factors influencing patients' expectations and satisfaction.* Brazilian Journal of Pharmaceutical Sciences. **53** (2017).
18. Robinson, A. R., Hohmann, K. B., Rifkin, J. I., Topp, D., Gilroy, C. M., Pickard, J. A., & Anderson, R. J., *Direct-to-consumer pharmaceutical advertising: physician and public opinion and potential effects on the physician-patient relationship.* Archives of internal medicine. **164**(4): p. 427-432 (2004).
19. Leonard, S.T. and M. Droege, *The uses and benefits of cluster analysis in pharmacy research.* Research in Social and Administrative Pharmacy. **4**(1): p. 1-11 (2008).

Knowledge, Acceptance, and Willingness to Pay for Human Papillomavirus (HPV) Vaccine: A Systematic Review

Fatimah Endriyanti[1], *Dwi* Endarti[2*], and *Tuangrat* Phodha[3,4]

[1]Master Program of Pharmacy Management, Faculty of Pharmacy, Universitas Gadjah Mada, Yogyakarta, Indonesia
[2]Department of Pharmaceutics, Faculty of Pharmacy, Universitas Gadjah Mada, Yogyakarta, Indonesia
[3]Drug Information and Consumer Protection Center, Faculty of Pharmacy, Thammasat University, Rangsit Campus, Pathum Thani, Thailand
[4]Center of Excellence in Pharmacy Practice and Management Research Unit, Faculty of Pharmacy, Thammasat University, Rangsit Campus, Pathum Thani, Thailand

Abstract. Cervical cancer is the fourth most common type of cancer incidence in women. It has been recognized that Human Papillomavirus (HPV) is causative agent in the pathogenesis of cervical cancer. In 2022, 117 countries have included HPV vaccine in their national immunization program. Understanding participants' decision regarding their children and themselves being vaccinated is important to ensure high coverage of the vaccine. This study aimed to conduct a systematic review of knowledge, acceptance, and willingness to pay for HPV vaccine. The required information was collected by searching with relevant keywords during October-December 2022 for articles published from 2013 – 2022 in PubMed, Scopus, ScienceDirect, and SpringerLink. The data were analyzed in Excel and reported descriptively. Finally, 22 studies were included to review. Most of the participants were female (96.44%) and the majority had health workers background (50.09%). The knowledge of mostly results was poor (40.9%). The acceptance of HPV vaccine was high, the range vary from 39–98.8%, especially for the vaccine with higher degree, longer duration of protection, lower out-of-pocket, and lower risk of side effects. Overall, the respondents' attitude was positive toward vaccination. The most frequent method used for measuring WTP was CVM (54.54%). The average WTP range from 0.1 – 17.51% to GDP per capita (9.9–745.25 USD). The cost was the primary reason that impact to the willingness to pay and acceptance. Results showed that the acceptance rate of HPV vaccination and WTP were relatively high when the vaccine was offered for free or reasonable price, even though their knowledge was poor. It is recommended to reduce the cost of vaccination program and to increase knowledge, awareness, and attitude of people.
Keywords: willingness to pay, Human Papillomavirus (HPV), vaccine, knowledge, acceptance.

1 Introduction

Cervical cancer is the fourth most common cancer among women globally, with an estimated new cases and deaths were more than 600,000 and 300,000 in 2020 respectively. Furthermore, 90% of them occured in low- and middle-income countries[1]. Human Papillomavirus (HPV), especially variants 16 and 18, has been identified as the causative agent in the pathogenesis of cervical cancer[2]. In 2006, the United States granted a license for the first HPV vaccine[3]. As of March 2022, 117 countries (corresponding to approximately one-third of the global target population) have introduced HPV vaccine into their national routine immunization schedules, with 10 new introductions planned by the end of 2022[4].

Clinical trials and post-marketing surveillance have shown that HPV vaccines are safe and effective in preventing infections with HPV infections, high grade precancerous lesions and invasive cancer[5]. SAGE advised the following updates to HPV dosage schedules: 1-2 dose schedules for the major target population of girls aged 9 to 14, 1-2 dose schedules for young women aged 15-20, and 2 doses separated by 6 months for women over the age of 21[6]. Since the HPV vaccine should be administered to young adolescents, understanding parents' decision-making processes regarding their children being vaccinated is important to ensure high coverage of the vaccine[7]. Beside the vaccine prices, there were some other obstacles. Barriers include low risk

* Corresponding author: endarti_apt@ugm.ac.id

perception, parents' beliefs that their daughters do not require vaccinations because they are not sexually active, worries about safety and efficacy, ignorance of the necessity for vaccination, and a lack of information[8].

To help address this gap, this study aimed to conduct a systematic review of knowledge, acceptance, and willingness to pay for HPV vaccine. The analysis was conducted to synthesize evidence of participants' information needs, views and preferences regarding HPV vaccination. The availability of such information is critical for decision-making and planning to implement the most suitable interventions to develop HPV vaccination as one of the most effective cancer prevention programs.

2 Methods

Methods are reported according to the Preferred Reporting Items for Systematic Reviews and Meta-Analysis (PRISMA) checklist[9].

2.1 Data Source and Search Strategy

A search strategy was developed in PubMed, Scopus, ScienceDirect, and SpringerLink. The timeframe selected for searching the articles is the ones which have published during 2013 – 2022. The strategy included an extensive list of keywords and related subject headings to broaden the search by finding articles related to HPV vaccine. The following key words were used in search strategy using operator Boolean construction: "Willingness to Pay" AND "HPV vaccine" AND "knowledge, acceptance, attitude", and all articles were retrieved from 23rd October to 17th December 2022.

2.2 Eligibility Criteria

Studies were included if they were reporting: knowledge, acceptance, attitude, Willingness to Pay and factors associated of participants concerning HPV vaccine. Studies were excluded if they were reviews, abstracts, editorials, conference reports, and not using English language. Furthermore, other economic evaluations such as cost benefit analysis, cost effective analysis, and cost utility analisis were also excluded.

2.3 Study Quality Assessment

The quality of included studies was evaluated by reviewer using the Strengthening the Reporting of Observational Studies in Epidemiology (STROBE) checklist[10]. The checklist options include "Yes", "No", "Unclear", and 'Not Applicable". It is also to identify bias in each study.

2.4 Data Extraction

Two forms of data extraction were designed. The 1st form was for extracting general characteristics of the included studies, while the 2nd was for extracting the studies' results. Key data for general characteristics extracted included author, year, country, aim of study, setting, study design, duration of observation, number of respondents, instrument, and method for measuring WTP. Key data to extract the study results included demographics, % of positive WTP, WTP, % of GDP per capita, knowledge, attitude and acceptance, factors influenced in WTP, reason for not WTP, and other results.

2.5 Data Analysis Method

Extracted data were analysed in Excel. The countries' currency value reported in included studies was presented in original currency and USD[11]. In some studies, the value was reported as a percentage of the participants (e.g: 45.5% participants were willing to pay for the HPV vaccine). To calculate the percentage of WTP from GDP per capita, the amounts of WTP counted by currency in USD were converted to 2022 USD using free online website US Inflasion Calculator then were divided by GDP per capita (counting the highest amount if there were some WTP measurements) in the year of study (2022)[12]. The World Bank data were used as a basis for calculating GDP per capita and country segmentation based on economic status[13].

3 Results

3.1 Details of Included Studies

Out of 487 articles found from databases and other sources, 43 were excluded due to duplication between databases. In the title and abstract screening phase, 117 were also excluded. A total of 97 studies were excluded from full-text review. Additionally, 2 studies were found by hand searching. Finally, 22 articles were included in this study, as presented in Figure 1. All of the studies included in the final set were published in the last ten years (2013 - 2022). The characteristics and results of the reviewed articles were presented in Table 1 and 2.

3.2 Characteristic of Included Studies

As shown in Table 1, all of the included studies were conducted in 10 different countries (China, Vietnam, Iran, Jordan, Nigeria, Ethiopia, Hongkong, Malaysia, Thailand, Argentina). Three studies were conducted in 2 low-income economies countries[14]–[16], 5 studies in 2 lower and-middle-income economies countries[17]–[21], 12 studies in 5 upper-middle-income economies countries[22]–[33], and 2 studies in high-income-economies country[34], [35]. The country classification was based on New World Bank Country Classification by Income Level 2022-2023[13].

Regarding their aims, the articles fell into 4 categories. The majority (19) focused on willingness to pay; 14 on knowledge, awareness and acceptance; 8 on attitude; and 3 on factors associated with WTP.

Regarding the study population, 8 studies focused on mothers, 2 on workers, 8 on students, 2 on parents, 2 on outpatients of hospital/clinic, and 1 on teachers. Total participants of 22 studies were 32,722 and most of them were female (96.44%). The participants background consisted of: 9,125 mothers (27.89%); 1,099 parents (3.36%, both mother and father); 16,392 health workers (50.09%); 377 teachers (1.15%); 606 married women (1.85%); 911 outpatients of hospital/clinic (2.78%), and 4,212 students (12.87%).

All included studies were cross-sectional and used questionnaires as the instrument to data collecting method. Twelve of them used a self-administered questionnaire, 5 used structured questionnaire, 4 with online survey, and 4 prefer to face-to-face interview.

3.3 Knowledge, Attitude and Acceptance of HPV and HPV Vaccine

Mostly studies assessed knowledge of HPV or knowledge of its vaccination. Of the 22 studies included, awareness and/or knowledge were assessed with various method of

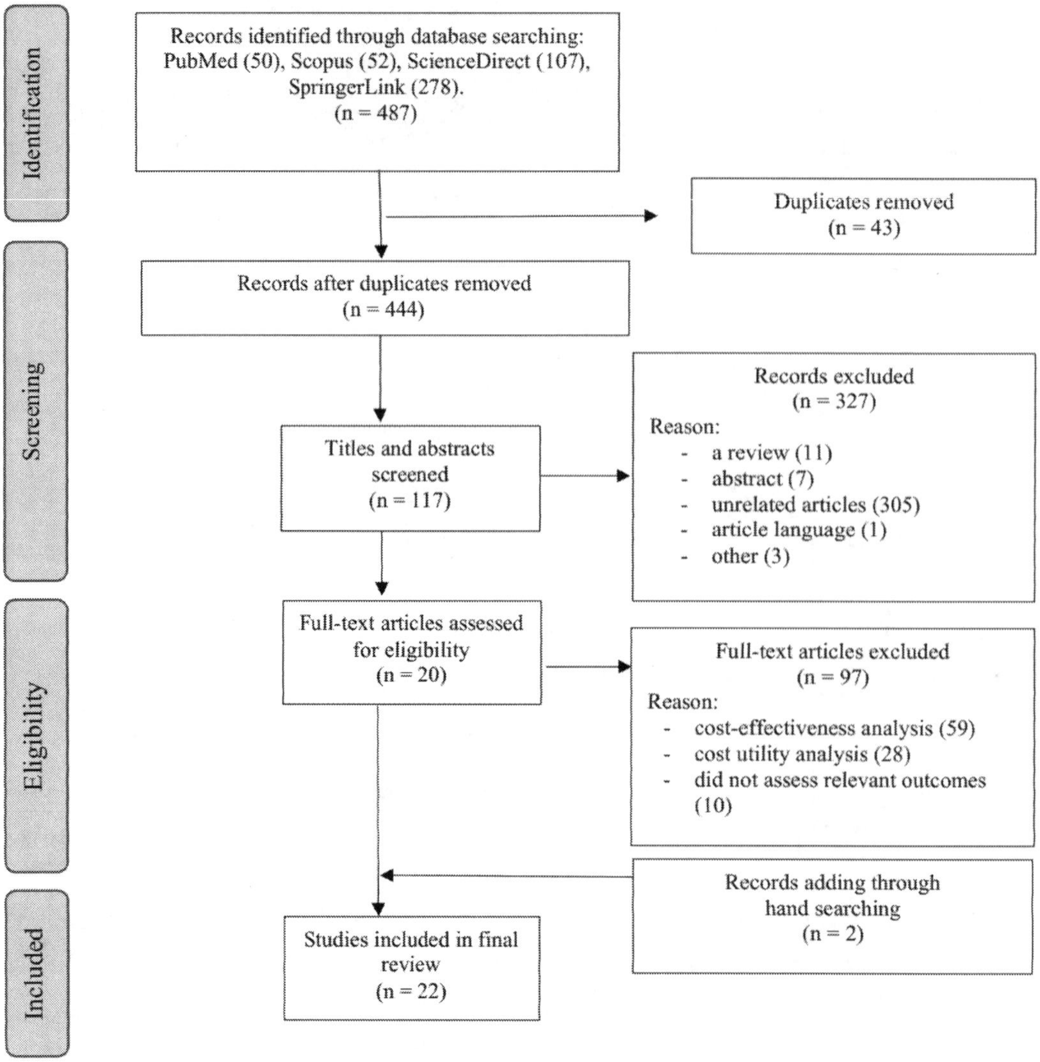

Fig. 1. PRISMA Flowchart

Table 1. Characteristics of the 22 studies

	n	%
Study location		
Asian		
Malaysia	3	14
Thailand	2	9
Vietnam	4	18
China	5	23
Iran	1	5
Hongkong	2	9
Jordan	1	5
Non-Asian		
Nigeria	2	9
Ethiopia	1	5
Argentina	1	5
Study design		
Cross-sectional	22	100
Study setting*		
Healthcare setting	4	18
School setting	6	27
College setting	8	36
General population setting	5	23
Data collection method		
Self-administered survey	10	45
Face-to-face interview	6	27
Online survey	6	27
Sample*		
Adult women		
Teachers, workers	3	13
Outpatients of hospital/clinic	2	9
Mothers	8	35
Mothers and fathers	2	9
Students	8	35
Sample size		
<100	0	0
100 - 500	12	54
501 - 1000	7	32
>1000	3	14
Methods for eliciting WTP**		
CVM		
Bidding game/Double dichotomous choice	6	27
Payment card	2	9
Open-ended questions	5	23
DCE	4	18
Market price offered	6	27

Note: *some studies were conducted in multiple settings; **some studies used >1 combination methods in eliciting WTP
Abbr: WTP = Willingness to Pay, DCE = Discrete Choice Experiment.

measurements. Eight studies assessed awareness with an initial question 'Have you heard/are you aware of HPV?" and rows of other "Yes or No" questions related. Other 6 studies assessed awareness and/or knowledge of HPV and HPV vaccination by answering "True or False" questions.

The attitude toward HPV vaccine were positive. Most of them would accept the vaccine for their children or themselves. Acceptance rates in the studies range 39 – 98.8%. Significant factors associated with acceptance are location of high school, study year, paternal educational level, annual household income, monthly disposable income, perceived self-confidence in taking the HPV vaccine in the near future, having no barriers to taking time off to take the HPV vaccination, and regular exposure to HPV vaccination information in the mass media. Despite positive attitude and high acceptance, the awareness and knowledge about cervical cancer and HPV vaccine were poor[15], [17], [20], [22], [24], [26], [27], [29], [31], moderate[30], [32], [33], [35], to high[16], [21], [25], [28]. Lack of knowledge were caused by rare information and difficulty to access[20].

Misconceptions and suspicions related to cervical cancer and the HPV vaccine were also common. Some participants were afraid of the safety and possible side effects due to the vaccine[18], [22], [30]. A small number of studies examined the religious beliefs regarding the HPV vaccine[18]. Religious norms also influenced how parents saw the need for the vaccine. Some parents believed they had raised their children properly and they would not engage in premarital sex, and therefore a vaccine to prevent sexually transmitted infection was not needed at that stage of their lives[7]. Surprisingly, a study mentioned that 70.7 % of its respondents have thought that vaccination will encourage the young population to become sexually active[33].

3.4 Willingness to Pay for HPV Vaccine

Out of the 22 articles reviewed in the study, 12 studies used Contingent Valuation Method (CVM) to elicit WTP, 4 studies used Discrete Choice Experiment (DCE), and 6 studies used market price. In general, cost of the vaccine was viewed as an important factor. Most participants thought that the HPV vaccine was expensive (10 studies). and cost was important factor when deciding whether or not to give the vaccine to their children. Many parents reported a high intention to vaccinate their children if the vaccine was going to be provided for free. Even if it was not free, many stated they were willing to pay for the vaccine so long as the price was reasonable[23].

The WTP were also counted for the proportion based on GDP per capita. The result showed that the percentage of WTP (converted to USD rate in 2022) compared to 2022 GDP per capita range from 0.91–1.34% (9.9–14.49 USD) in low-income economics, 0.1–17.34% (55–417.5 USD) in lower-middle income economics, 0.32–17.51% (41.91–745.25 USD) in upper-middle income economics, and 1.83–1.87% (241–247 USD) in high income economics. Based on economic status of the countries, the highest proportion was in Thailand (UMICs)[29].

 https://doi.org/10.1051/bioconf/20237505010

s of the included studies

)R, ₹, 'RY	% OF POSITIF WTP	WTP (CURRENCY)	WTP AS % OF GDP PER CAPITA	KNOWLEDGE	ATTITUDE AND ACCEPTANCE	FACTORS INFLUENCED IN WTP		REASON FOR NOT WTP	
						SIGNIFICANT	NOT SIGNIFI-CANT		
.,	92.54	mean values (in RMB) of 1,689.80 (±926.13) or 242.49 USD for imported bivalent, 2,216.61 (±1190.62) or 318.08 USD quadrivalent, 3,252.43 (±2064.71) or 466.72 USD 9-valent vaccines. 910.63 (±647.03) or 130.68 USD, 1,861.69 (±1147.80) or 267.15 USD, and 2,866.96 (±1784.41) or 411,41 USD for their domestic counterparts.	3.53 - 10.97	751 (92.83%) had heard of HPV, 728 (89.99%) had heard of HPV vaccine. The average cognitive score: 13.05 (±5.09) points. The respondents maintained a high level of cognition of HPV infection, transmission, and vaccination population but reported insufficient awareness of postvaccination.	60.82% and 88.01% wished to be vaccinated and support the partners to be vaccinated. 30.90% did not have the willingness to be vaccinated at this stage, and a total of 8.28% were unwilling to be vaccinated.	- educational background - perception of imported vaccine	-	-vaccine is unnecessary -not once and for all -cost -painful injection -prefer to use condom -Lack of authoritative evidence for long-term side effects	\ v l r - c v l f v
t al.,	46.87	WTP amount for two doses of HPV vaccine was 3,053,005 VND (137.5 USD) ranging from 200,000 VND (9 USD) to 4,180,000 VND (188.3 USD). Among the parents who accepted to get their sons vaccinated, 63.7% of them were willing to pay <3,580,000 VND (161.2 USD) for two doses of Gardasil	4.43 - 17.34	2/3 of participants had never heard of HPV. Only 18.9% of parents achieved good level of knowledge. only a few of parents (<8%) knew ideal age to receive HPV vaccination, dose of HPV vaccine for children under 15 years old, and asymptomatic HPV infection.	49.2% of the parents agreed to have their son vaccinated. The parents with good knowledge were more likely to accept HPV vaccine compared to those with worse knowledge. The parents who attained college degree or higher were more likely to get HPV vaccine for their sons than those with a lower level of education	-	knowledge	cost	

OR, R, RY	% OF POSITIF WTP	WTP (CURRENCY)	WTP AS % OF GDP PER CAPITA	KNOWLEDGE	ATTITUDE AND ACCEPTANCE	FACTORS INFLUENCED IN WTP		REASON FOR NOT WTP
						SIGNIFICANT	NOT SIGNIFI-CANT	
	Initial amount of General WTP: 800 CNY (66.2), 1600 CNY (57.1) , 2400 CNY (51.1), 3200 CNY (40.8), 4000 CNY (35.0)	The median general WTP was 2000 CNY (303 USD) (interquartile range, 1000–3200 CNY), The median WTP out-of-pocket was 1250 CNY (189 USD) (540–2000 CNY)	2.29 - 7.12	-	majority of respondents did not change their attitude towards HPV vaccination between two payment scenarios; those with higher price HPV vaccines (51.1%) had higher WTP out-of-pocket (1400 CNY; 560–2250 CNY) than those with cheaper vaccines (39.0%) (1120 CNY; 490–1960 CNY) ($P < 0.001$).	-cost (payment method) -younger age -unmarried status -higher monthly income -fewer children -more positive vaccination behavior -working in tertiary hospital -higher local GDP and HDI (each $P < 0.05$).	-educational level -professional title	-
al.,		- bivalent: US $ – 432 - quadrivalent US $ 380	0.10 - 0.38	-	-low incidence of cervical cancer cause lack of awareness and weak attitude towards the importance of prevention	-fewer serious side effects -protection against genital warts -protection duration -protection against cervical cancer -cost	-	-serious side effect -religious and cultural barriers

)R, ₹, 'RY	% OF POSITIF WTP	WTP (CURRENCY)	WTP AS % OF GDP PER CAPITA	KNOWLEDGE	ATTITUDE AND ACCEPTANCE	FACTORS INFLUENCED IN WTP		REASON FOR NOT WTP	
						SIGNIFICANT	NOT SIGNIFI-CANT		
l.,	-	-making an appointent through school, by phone 291.4 CNY (42.3 USD) - by the internet 241.7 CNY (35.0 USD) -on-site 203.6 CNY (29.5 USD) - WTP more to have the service on both weekdays and weekends 121.9 CNY (17.7 USD). -WTP 25.0 CNY (3.6 USD) but if they had to wait an additional week for immunization after the visit, they would prefer to pay less amount -WTP 171.1 CNY (24.8 USD) but less if they must wait for 30-60 minutes -WTP 297.9 CNY (43.2 USD) but less if they must wait for >60 minutes.	0.50 - 1.55	Average score of HPV knowledge * (SD): 4.30 (0.08)	-Class 1 (86.1% of the respondents) chose a vaccine with a greater level of protection, a longer period of time it would last, a lower risk of major side effects, and a lower price. -Class 2 (13.9% of the respondents) seemed unconcerned about the characteristics of the vaccine. 98.8% (of 742 respondents) still chose to receive the vaccine in service-related scenes	-	-	-	v a a a r c e a i c
al.,	16.0	reported only among 84 students (16.03%)	-	-prior knowledge (62.7) -aware of the existence of HPV vaccines (48.66) -The lowest knowledge level was in nursing students	willingness to receive HPV vaccination if provided freely (75.0%)	cost	-	the perceived low risk to get HPV infection	c e (c t v r
l.,	74.3	for three doses of bivalent, out of pocket was 15.01USD (N5,568.0), while among all participants, the average WTP was 11.47 USD (N4, 242.84)	1.16	-	-	-social factors (household) -economic factors (husband's monthly income).	-	-vaccine high price	V t p

OR, R, TRY	% OF POSITIF WTP	WTP (CURRENCY)	WTP AS % OF GDP PER CAPITA	KNOWLEDGE	ATTITUDE AND ACCEPTANCE	FACTORS INFLUENCED IN WTP		REASON FOR NOT WTP
						SIGNIFICANT	NOT SIGNIFI-CANT	
	2vHPV (81.2) 4vHPV (75.9) 9vHPV (67.7)	2vHPV: RMB580 (USD 88.27) per shot 4vHPV: RMB798 (USD 121.39) per shot 9vHPV: RMB1298 (USD 197.44) per shot	1.72 – 5.34	high score, aware of HPV and its vaccine (49.1%)	confident in getting the HPV vaccine (84.1%). Intent to obtain HPV vaccine (58.3%).	-Household income -mass media exposure to HPV vaccination, -perceived self-efficacy in HPV vaccination, -spouse/partner approval	-	-
	-metrople : 38.2 -non metrople (rural) : 4.4 -mean : 20.4	N/R	-	Participants who knew more about HPV vaccinations and cervical cancer had considerably more intention to be vaccinated.	Positive towards vaccination. Before learning the vaccine price -metropole: 59.3 -non metropole: 90.9 -mean: 75.9	monthly household income	-	vaccine price
	-2vHPV: 78.6 -4vHPV: 68.0 -9vHPV: 49.3	-2vHPV RMB580/ 82.38 USD per shot -4vHPV RMB798/ 113.34 USD per shot -9vHPV RMB 1298/184.36 per shot (1 USD: 7.04 RMB)	1.63 – 5.04	-aware that HPV infection can occur without symptoms: 42.5% -aware that HPV can cause oral cancer: 34.5% -have the misconception that HPV can be cured by antibiotics: 70.7% -hev the misconception if there is a cure for HPV infection: 81.5% The mean total knowledge score was 12.8 (SD ± 6.0) out of a possible score of 22. The median was 14 (inter quartile range, IQR, 9–17)	high perception of women's susceptibility to contracting HPV (81.1%), having cervical or vulvar cancer (75.4%), considered that women have a high risk of contracting genital warts (58.8%).	-	-	-

 https://doi.org/10.1051/bioconf/20237505010

OR, R, TRY	% OF POSITIF WTP	WTP (CURRENCY)	WTP AS % OF GDP PER CAPITA	KNOWLEDGE	ATTITUDE AND ACCEPTANCE	FACTORS INFLUENCED IN WTP		REASON FOR NOT WTP	
						SIGNIFICANT	NOT SIGNIFI-CANT		
9	85.97	231.34 ETB (US$8.50) per service	0.91	97.96% had heard about human papilloma virus vaccine. 68.62% had excellent knowledge about cervical cancer danger signs and its risk factors	-	-Age -educational status -knowledge -monthly income	-	-	
8	53.1	US$23 to US$46.	1.28 – 5.02	Only 33% of participants had adequate knowledge. Nearly all mothers, whether WTP or not, would like additional knowledge on the pertinent topic (97%).	81.5% WTP mothers and 77.8% not WTP mothers had positive attitude toward the vaccination (mean average: 79.7%)	cost	-	Vaccination high cost	T s i v b h c s
.,	86.6	US$49.3 (1.119 million VND)	1.37 – 5.38	Typical methods for finding out about HPV vaccine: -Social networks and internet (56.8%), -asking doctors, nurses, and other health professionals (41.9%). The percentages of-subjects correctly answering questions regarding -the best age for HPV vaccination: 67.9%, -the benefits of HPV vaccination: 94.6%, -the target subject for HPV vaccination: 12.3%.	-believed that vaccine was safe (92.8%) and effective (90.8%) -desired to get vaccinated (71.1%),	-high household income -had children > 6 years old -vaccine service user: adult male -age 20-29 -white-collar worker -getting information except from health pro -Believes that hPV vaccine is effective -has ever examined reproductive health	-Believes that hPV vaccine is safe -has family member who ever had STI	-being male (33.3%), -high cost (38.2%) -vaccine being seen as unnecessary (34.5%) -unsuitable age (22.6%)	

OR, R, TRY	% OF POSITIF WTP	WTP (CURRENCY)	WTP AS % OF GDP PER CAPITA	KNOWLEDGE	ATTITUDE AND ACCEPTANCE	FACTORS INFLUENCED IN WTP		REASON FOR NOT WTP
						SIGNIFICANT	NOT SIGNIFI-CANT	
al., 3	-	Maximum WTP for ideal vaccines (i.e., maximum protection, lifetime protection duration, and 0% adverse effects) = HK$8976 (US $1129). The estimated WTP for vaccines currently available was HK$1620 (US $208)	1.87	-80% mothers were concerned about their daughters' risk of cervical cancer and HPV infection -More than 50% mothers believed that the vaccines are unsafe and some of them refused the vaccines to be administered to their daughters.	88.4% of mothers made a more sensible decision by opting for more effective protection, lasted longer, cost less out-of-pocket, and had fewer side effects	- high household income -education level except the out-of-pocket cost attribute for primary education level	-	-
al.,	91.6	The average WTP was US$ 11.68.	1.34	19.1% ever heard of HPV infection	-7.5 % rejected HPV vaccination of their daughters. -Demands for HPV vaccinations for daughters were 18.8 times more likely to come from mothers who had previously been diagnosed with HPV infection.	-	-	-
1 5	12.25	up to RM 500 / 118.15 USD (1 USD; 4.2319 MYR	1.12 – 3.48	The mean score for knowledge of HPV, cervical cancer and Pap smear test were 12.2 out of a maximum score of 17±2.54.	overall attitude of participants toward HPV vaccination was positive	knowledge	-	vaccine cost

)R, R, 'RY	% OF POSITIF WTP	WTP (CURRENCY)	WTP AS % OF GDP PER CAPITA	KNOWLEDGE	ATTITUDE AND ACCEPTANCE	FACTORS INFLUENCED IN WTP		REASON FOR NOT WTP	
						SIGNIFICANT	NOT SIGNIFI-CANT		
				90.0% students have a high level of knowledge					
:hes 5	-	quadrivalent 21,189.9 Baht (593.32 USD), bivalent 10,479.9 Baht (293.44 USD) (1 USD = 35.72 THB)	5.64 – 17.51	fathers might have poor knowledge or awareness of HPV vaccine. Heard about HPV vaccine: 44.0 % fathers, 51.2 % mothers.	-Positive results for the risk factors for genital warts and cervical cancer suggested that parents chose immunizations that lower these risks. -They favored less side effects and paying less for vaccines, as seen by the negative indicators of the typical adverse effects and cost criteria.	-	-	-	r v 5 2 E f t 2 v r
l.,	59.8	mean price: 23.20 euro (24,66 USD), with a range of 0.70-128.60 euro/dose (0,74-136,71 USD).	1.30 – 4.03	-52.2%: considered that HPV vaccine should be given to age <14 -45.6%: considered their daughter to be protected against cervical cancer after HPV vaccination. 73.9%: thought it was also necessary to vaccinate boys. 83.3%: aware of cervical cancer	90.1% accept vaccination for their daughter	-gainful employment -household income -awareness of cervical cancer	-education level	being unsure about vaccine safety	l t g v v

)R, ł, 'RY	% OF POSITIF WTP	WTP (CURRENCY)	WTP AS % OF GDP PER CAPITA	KNOWLEDGE	ATTITUDE AND ACCEPTANCE	FACTORS INFLUENCED IN WTP		REASON FOR NOT WTP	
						SIGNIFICANT	NOT SIGNIFI-CANT		
groj 4	68.9: bivalent 67.3: quadrivalent	33.33% participants would copay less than 500 baht or approx US$16.67 (30 baht = approx US$1) while other 30% would pay 500-1,000 baht (approx US$33.33) for three doses of bivalent vaccine, respectively.	0.32 - 0.98	Knowledge regarding the HPV vaccine among parents was quite low. 49.2% knew about the link between HPV and cervical cancer while 28.5% knew that the vaccine should be administered to the children before they become sexually active.	Acceptance if vaccine was offered for free: 76.9% for the 2-v and 74.4% for 4-v vaccine. if it was not totally free: 68.9% for 2-v to 67.3% for 4-v vaccine.	-	-	financial limitations (39%-43%)	T r a c t v e
., g	27.5: mothers in 2008. 37.6: mother in 2012 27.1: shoolgirls	for full-course vaccination among mothers had a median of US$128/HK$1000 (50% central range = US$64–192/HK$500–1500), ie: substantially lower than the current market price	1.83	-Had heard of HPV: 28.8% schoolgirls, 40.5% mothers2008, 68.5% mothers2012. -Had heard of HPV vaccines: 40.3% schoolgirls, 68.3% mothers2008, 43.7% mothers2012.	one third of mother 2008, mothers 2012 and shoolgirls were willing to be vaccinated at market price. regardless of vaccine price, the overall acceptability was 44.6% (mother 2008) and 66.7% (mothers 2012) and 54.8% (the unvaccinated schoolgirls 2008).	-knowledge -higher monthly income -younger age of the daughter -perception on the health of daughter	-risk factors -education level -history of cervical screening	Vaccine price	r N v s r h c v a I i
al,	-	Most of the participants:RM 500 (122,25 USD) or less for themselves, and the number of students kept declining as the price went up. The total mean of WTP for themselves: RM 1304 (318.83 USD) In future aspect of vaccination for their children: mean amount of RM 1477.7 (361.30 USD)	3.32 – 10.31	Knowledge on HPV and cervical cancer did not impact on attitudes towards vaccines. Mean knowledge score was 10.43. The majority of students had moderate knowledge (57.7%). Knowledge of students predicts 23% ($R2$ = 0.232, F = 60.55, P< 0.01) of their WTP.	The mean attitude score for students was 3.86 (SD = 0.37, median = 3.83). The majority of students (66.2%) had a positive attitude toward HPV	-knowledge	-	-	S p t r b v v v c l

)R, ₹, 'RY	% OF POSITIF WTP	WTP (CURRENCY)	WTP AS % OF GDP PER CAPITA	KNOWLEDGE	ATTITUDE AND ACCEPTANCE	FACTORS INFLUENCED IN WTP		REASON FOR NOT WTP
						SIGNIFICANT	NOT SIGNIFI-CANT	
al,	-	The average amounts: -USD 108.66 for themselves. -<50% respondents want to spend around USD 200 for their children	1.90 – 5.90	-Mean knowledge score: 9.3 (moderate) out of 17 - -there was a significant difference in knowledge between genders (12 ± 4.23 males vs. 14 ± 2.24 females; p = 0.003)	-positive, mean attitude score: 34.8 -there were significant differences in attitude by the respondents' living condition.	-	-	-The government should cover the cost of vaccination

Some studies cited reasons for unwillingness to pay related to unwillingness to accept. In 10 out of 22, the cost was the main reason. Other reasons were afraid of safety and side effects of the vaccine[18], [22], [30], religious belief[18], felt unnecessary to be vaccinated[21], [22], [25], being male[21], unsuitable age[21], and presumption that vaccinating young will encourage sexual activity[33].

3.5 Factors Associated with Willingness to Pay

Sixteen studies assessed some factors that influenced WTP for HPV vaccine. The significant factors influenced WTP were: educational background[16], [22], [34], knowledge[28], [30], [32], [35], perception of imported vaccine[22], younger age[16], [21], [23], [35], unmarried status[23], household income[14], [16], [19], [21], [23], [26], [30], [33], [35], fewer children[23], behavior/ perception of health[23], [35], working site[23], gainful employment[30], higher local GDP[23], fewer serious effects[18], protection duration[18], efficacy/higher degree of protection[18], [21], [26], has examined reproductive health[21], cost[18], [20], [23], [25], mass media exposure[21], [26], and partner/spouse approval[26]. Other factors such as knowledge[17], educational level[23], professional title[23], vaccine safety[21], and history of family's sexual transmission infection[21] had no significant influences.

4 Discussion

Most of the reviewed studies were conducted in Asian, especially South East Asian countries, others were in Africa and South America. These setting areas were aligned with the Globocan Data Report 2020 that mentioned cervical cancer caused by HPV as the most commonly diagnosed cancer and the leading cause of cancer death in those regions[1]. Epidemiological studies of cancers, especially cervical cancer, have shown that this type of cancer is one of the most common cancers in these regions. In addition, cervical cancer-related deaths in South East Asia are among the highest in the world[36]. For the past few decades, mortality and incidence rates of cervical cancer actually have decreased in the majority of the world's regions. The decreases are attributed to elements that are either associated with rising socioeconomic averages or a declining risk of persistent infection with high-risk HPV as a result of advances in genital cleanliness, decreased parity, and a declining prevalence of sexually transmitted diseases. Despite the observations of rising risk among younger generations of women in some countries which may in part reflect changing sexual behavior and increased transmission of HPV that is insufficiently compensated by screening approach, cervical cancer screening programs have accelerated declines in many countries in Europe, Oceania, and Northern America. Although incidence rates are still high, rates have also fallen in the 2000s throughout the Caribbean and in Central and South America (such as Argentina, Chile, Costa Rica, Brazil, and Colombia). Furthermore, premature cervical cancer mortality has increased dramatically in recent generations in regions lacking efficient screening, such as Eastern Europe and Asia[1].

One of the studies focused on vaccination for boys. An interesting finding is that no participant mentioned their sons' prior HPV immunization. This is understandable as many countries still targeting girls aged 9-14 as priorities. WHO recommends that where possible and practical, vaccination of secondary targets, such as boys and older females, is advised, after the primary target girls were highly achieved[37]. Some of high-income countries that have achieved high coverage of girls vaccinated start planning to move from a girls-only HPV vaccination strategy to universal or gender-neutral strategy[38]. This may affect the availability and adequacy of the HPV vaccine worldwide.

According to some studies, people considered HPV vaccination was only for female, and male did not need it[21]. The same circumstances regarding knowledge level of people about HPV vaccine and related disease also happened in other countries. In Indonesia, a study conducted a structured-educational intervention to parents because of their low level of knowledge about HPV infection and the vaccines. Only 48.8% of parents and 49.2% of parents, respectively, had heard of the vaccine and HPV infection before receiving the intervention[39]. Meanwhile in Vietnam, only 18.9% of the parents achieved good level of knowledge about HPV and HPV vaccine[17]. Along with the knowledge level findings, it is recommended to policy makers to create innovative approaches to raise the people knowledge level widely. However, our findings showed that the HPV vaccine acceptance was relatively high. Significant factors associated with acceptance are location of high school, study year, paternal educational level, annual household income, monthly disposable income, perceived self-confidence in taking the HPV vaccine in the near future, having no barriers to taking time off to take the HPV vaccination, and regular exposure to HPV vaccination information in the mass media, similar with the previous systematic review in 2018[40].

The result also showed a relatively high WTP if vaccine was offered free or at reasonable price. These results were in line with previous systematic review of ASEAN countries[7]. This may also due to high prevalence of this type of cancer and its association with high-risk sexual behavior, also the growth of educational level and higher household income. As WHO recommends the HPV vaccine be included in countries vaccination program since had been introduced in 2006, about one third of global population in 117 countries has been introduced HPV vaccination as national immunization program today, making it become one of the most common types of cervical cancer prevention. This action was taken early based on many previous studies stated that HPV vaccination was very cost effective, measured with every disability-adjusted life-year averted costing less than the gross domestic product per head in 87% or 156 of 179 countries[41].

As the WTP was high before learning the vaccine price then dropped after knowing the high price offered, it is important to pay more attention to the policies and strategies provided by many organizations including WHO and Government in each country as policymakers, in order to gain better control and to increase cost-effectiveness of prevention methods. Reducing the cost, simplifying the access of the vaccination program, maintaining the vaccine availability and its supported cold chain facilities, followed by developing user-friendly and all over information system will be beneficial. HPV vaccination should be part of a multifaceted public health strategy entailing screening, condoms, and education of all stakeholders to reduce the significant burden of sexual transmitted diseases[42].

On the other hand, the results study also showed that people's knowledge and awareness of HPV infection and HPV vaccination were poor. In this regard, developing effective interventions to increase those things that related to acceptance and WTP is critical.

This systematic review provided summaries of the concerns in all included studies. However, this review had several limitations. The author could not conduct a meta-analysis due to the type of data reporting. In this study, only published articles in English version that were reviewed due to author's language literation. In addition, although it is easier to identify published articles with the use of online databases in this study, evidence selection biases from missed studies continue to be a problem when systematically reviewed. Moreover, unpublished studies that have related topic but cannot be included here potentially will be publication bias. These conditions can lead to reporting bias toward its findings. Therefore, findings from this review should be interpreted cautiously.

5 Conclusion

Results showed that the acceptance rate of HPV vaccination and WTP were relatively high among individuals when the vaccine was offered for free or reasonable price, even though their knowledge was poor. According to the results of this study, it is recommended for the government to reduce the cost of vaccination program and to increase knowledge, awareness, and attitude of people through better healthcare interventions and suitable approaches.

Conflict of Interest

The author declares that she has no conflict of interests or personal relationships that could have appeared to influence the work reported in this study.

Acknowledgments

This study was a part of thesis' requirement in Master in Pharmacy Management Program, Faculty of Pharmacy at Universitas Gadjah Mada.

References

1. H. Sung, J. Ferlay, R. L. Siegel, M. Laversanne, I. Soerjomataram, A. Jemal, and F. Bray, 'Global Cancer Statistics 2020: GLOBOCAN Estimates of Incidence and Mortality Worldwide for 36 Cancers in 185 Countries', *CA. Cancer J. Clin.*, vol. 71, no. 3, pp. 209–249, May 2021, doi: 10.3322/caac.21660.
2. E. J. Crosbie, M. H. Einstein, S. Franceschi, and H. C. Kitchener, 'Human papillomavirus and cervical cancer', *The Lancet*, vol. 382, no. 9895, pp. 889–899, Sep. 2013, doi: 10.1016/S0140-6736(13)60022-7.
3. F. Cutts, 'Human papillomavirus and HPV vaccines: a review', *Bull. World Health Organ.*, vol. 85, no. 09, pp. 719–726, Sep. 2007, doi: 10.2471/BLT.06.038414.
4. WHO, 'WHO HPV Vaccine Global Market Study, April 2022'. Accessed: Dec. 10, 2022. [Online]. Available: https://www.who.int/publications/m/item/who-hpv-vaccine-global-market-study-april-2022
5. J. Lei, A. Ploner, K. M. Elfström, J. Wang, A. Roth, F. Fang, K. Sundström, J. Dillner, and P. Sparén, 'HPV Vaccination and the Risk of Invasive Cervical Cancer', *N. Engl. J. Med.*, vol. 383, no. 14, pp. 1340–1348, Oct. 2020, doi: 10.1056/NEJMoa1917338.
6. WHO, 'One-dose Human Papillomavirus (HPV) vaccine offers solid protection against cervical cancer'. Accessed: Dec. 10, 2022. [Online]. Available: https://www.who.int/news/item/11-04-2022-one-dose-human-papillomavirus-(hpv)-vaccine-offers-solid-protection-against-cervical-cancer
7. K. E. Wijayanti, H. Schütze, C. MacPhail, and A. Braunack-Mayer, 'Parents' knowledge, beliefs, acceptance and uptake of the HPV vaccine in members of The Association of Southeast Asian Nations (ASEAN): A systematic review of quantitative and qualitative studies', *Vaccine*, vol. 39, no. 17, pp. 2335–2343, Apr. 2021, doi: 10.1016/j.vaccine.2021.03.049.
8. S. Marshall, A. Fleming, A. C. Moore, and L. J. Sahm, 'Views of parents regarding human papillomavirus vaccination: A systematic review and meta-ethnographic synthesis of qualitative literature', *Res. Soc. Adm. Pharm.*, vol. 15, no. 4, pp. 331–337, Apr. 2019, doi: 10.1016/j.sapharm.2018.05.013.
9. M. J. Page, J. E. McKenzie, P. M. Bossuyt, I. Boutron, T. C. Hoffman, C. D. Mulrow, L. Shamseer, J. M. Tetzlaff, E. A. Akl, S. E. Brennan, R. Chou, J. Glanville, J. M. Grimshaw, A. Hrobjartsson, M. M. Lalu, T. Li, E. W. Loder, E. Mayo-Wilson, S. McDonald, L. A. McGuinness, L. A. Stewart, J. Thomas, A. C. Tricco, V. A. Welch, P. Whiting, and D. Moher, 'The PRISMA 2020 statement: an updated guideline for reporting systematic reviews', *BMJ*, p. n71, Mar. 2021, doi: 10.1136/bmj.n71.
10. J. P. Vandenbroucke, C. D. Mulrow, and M. Egger, 'Strengthening the Reporting of Observational Studies in Epidemiology (STROBE)', *Epidemiology*, vol. 18, no. 6, 2007.
11. 'Konversi Mata Uang - Investing.com'. Accessed: Dec. 15, 2022. [Online]. Available: https://id.investing.com/currency-converter/
12. 'Inflation Calculator | Find US Dollar's Value from 1913-2022'. Accessed: Dec. 16, 2022. [Online]. Available: https://www.usinflationcalculator.com/
13. 'World Bank Country and Lending Groups – World Bank Data Help Desk'. Accessed: Dec. 15, 2022. [Online]. Available: https://datahelpdesk.worldbank.org/knowledgebase/articles/906519-world-bank-country-and-lending-groups
14. J. T. Enebe, 'Willingness to Pay for Cervical Cancer Vaccines Among Female Secondary School Teachers in Enugu, Nigeria', *TEXILA Int. J. PUBLIC Health*, vol. 9, no. 2, pp. 196–209, Jun. 2021, doi: 10.21522/TIJPH.2013.09.02.Art018.
15. I. B. Umeh, S. O. Nduka, and O. I. Ekwunife, 'Mothers' willingness to pay for HPV vaccines in Anambra state, Nigeria: a cross sectional contingent

valuation study', *Cost Eff. Resour. Alloc.*, vol. 14, no. 1, p. 8, Dec. 2016, doi: 10.1186/s12962-016-0057-0.
16. A. A. Tarekegn and A. E. Yismaw, 'Health professionals' willingness to pay and associated factors for human papilloma virus vaccination to prevent cervical cancer at College of Medicine and Health Sciences University of Gondar, Northwest Ethiopia', *BMC Res. Notes*, vol. 12, no. 1, p. 58, Dec. 2019, doi: 10.1186/s13104-019-4085-7.
17. L. H. Nguyen, T. B. T. Le, N. Q. N. Le, and N. T. T. Tran, 'Acceptance and Willingness to Pay for Vaccine Against Human Papilloma Virus (HPV) Among Parents of Boys in Central Vietnam', *Front. Public Health*, vol. 10, p. 801984, Mar. 2022, doi: 10.3389/fpubh.2022.801984.
18. N. Sargazi, A. Takian, M. Yaseri, R. Daroudi, A. Ghanbari Motlagh, A. Nahvijou, and K. Zendehdel, 'Mothers' preferences and willingness-to-pay for human papillomavirus vaccines in Iran: A discrete choice experiment study', *Prev. Med. Rep.*, vol. 23, p. 101438, Sep. 2021, doi: 10.1016/j.pmedr.2021.101438.
19. X. T. T. Le, P. T. N. Nguyen, T. T. T. Do, T. H. Nguyen, H. T. Le, C. T. Nguyen, G. H. Ha, C. L. Hoang, B. X. Tran, C. A. Latkin, R. C. M. Ho, and C. S. H. Ho, 'Intention to Pay for HPV Vaccination among Women of Childbearing Age in Vietnam', *Int. J. Environ. Res. Public. Health*, vol. 17, no. 9, p. 3144, Apr. 2020, doi: 10.3390/ijerph17093144.
20. H. Dinh Thu, H. Nguyen Thanh, T. Hua Thanh, L. Nguyen Hai, V. Tran Thi, T. Nguyen Manh, and A. Buvé, 'Mothers' willingness to pay for daughters' HPV vaccine in northern Vietnam', *Health Care Women Int.*, vol. 39, no. 4, pp. 450–462, Apr. 2018, doi: 10.1080/07399332.2017.1411914.
21. B. Tran, P. Than, T. Doan, H. Nguyen, H. Thi Mai, T. Nguyen, H. Le, C. Latkin, M. Zhang, and R. Ho, 'Knowledge, attitude, and practice on and willingness to pay for human papillomavirus vaccine: a cross-sectional study in Hanoi, Vietnam', *Patient Prefer. Adherence*, vol. Volume 12, pp. 945–954, May 2018, doi: 10.2147/PPA.S165357.
22. L. Zhou, B. Gu, X. Xu, Y. Li, P. Cheng, Y. Huo, G. Liu, and X. Zhang, 'On Imported and Domestic Human Papillomavirus Vaccines: Cognition, Attitude, and Willingness to Pay in Chinese Medical Students', *Front. Public Health*, vol. 10, p. 863748, May 2022, doi: 10.3389/fpubh.2022.863748.
23. X. Lu, M. Ji, A. L. Wagner, W. Huang, X. Shao, W. Zhou, and Y. Lu, 'Willingness to pay for HPV vaccine among female health care workers in a Chinese nationwide survey', *BMC Health Serv. Res.*, vol. 22, no. 1, p. 1324, Nov. 2022, doi: 10.1186/s12913-022-08716-6.
24. Y. Wang, Y. Hu, Y. Chen, and H. Liang, 'Preference and willingness to pay of female college students for human papillomavirus vaccination in Zhejiang Province, China: A discrete choice experiment', *Hum. Vaccines Immunother.*, vol. 17, no. 10, pp. 3595–3602, Oct. 2021, doi: 10.1080/21645515.2021.1932215.
25. M. Sallam, K. Al-Mahzoum, H. Eid, A. M. Assaf, M. Abdaljaleel, M. Al-Abbadi, and A. Mahafzah, 'Attitude towards HPV Vaccination and the Intention to Get Vaccinated among Female University Students in Health Schools in Jordan', *Vaccines*, vol. 9, no. 12, p. 1432, Dec. 2021, doi: 10.3390/vaccines9121432.
26. Y. Lin, Z. Lin, F. He, H. Chen, X. Lin, G. D. Zimet, H. Alias, S. He, Z. Hu, and L. P. Wong, 'HPV vaccination intent and willingness to pay for 2-,4-, and 9-valent HPV vaccines: A study of adult women aged 27–45 years in China', *Vaccine*, vol. 38, no. 14, pp. 3021–3030, Mar. 2020, doi: 10.1016/j.vaccine.2020.02.042.
27. Y. Lin, Z. Lin, F. He, Z. Hu, G. D. Zimet, H. Alias, and L. P. Wong, 'Factors influencing intention to obtain the HPV vaccine and acceptability of 2-, 4- and 9-valent HPV vaccines: A study of undergraduate female health sciences students in Fujian, China', *Vaccine*, vol. 37, no. 44, pp. 6714–6723, Oct. 2019, doi: 10.1016/j.vaccine.2019.09.026.
28. M. K. Maharajan, K. Rajiah, K. S. F. Num, and N. J. Yong, 'Knowledge of Human Papillomavirus Infection, Cervical Cancer and Willingness to pay for Cervical Cancer Vaccination among Ethnically Diverse Medical Students in Malaysia', *Asian Pac. J. Cancer Prev.*, vol. 16, no. 14, pp. 5733–5739, Sep. 2015, doi: 10.7314/APJCP.2015.16.14.5733.
29. S. Ngorsuraches, K. Nawanukool, K. Petcharamanee, and U. Poopantrakool, 'Parents' preferences and willingness-to-pay for human papilloma virus vaccines in Thailand', *J. Pharm. Policy Pract.*, vol. 8, no. 1, p. 20, Dec. 2015, doi: 10.1186/s40545-015-0040-8.
30. S. Alder, S. Gustafsson, C. Perinetti, M. Mints, K. Sundström, and S. Andersson, 'Mothers' acceptance of human papillomavirus (HPV) vaccination for daughters in a country with a high prevalence of HPV', *Oncol. Rep.*, vol. 33, no. 5, pp. 2521–2528, May 2015, doi: 10.3892/or.2015.3817.
31. S. Kruiroongroj, U. Chaikledkaew, and M. Thavorncharoensap, 'Knowledge, Acceptance, and Willingness to Pay for Human Papilloma Virus (HPV) Vaccination among Female Parents in Thailand', *Asian Pac. J. Cancer Prev.*, vol. 15, no. 13, pp. 5469–5474, Jul. 2014, doi: 10.7314/APJCP.2014.15.13.5469.
32. K. Rajiah, M. Maharajan, K. S. Fang Num, and R. C. How Koh, 'Knowledge about Human Papillomavirus and Cervical Cancer: Predictors of HPV Vaccination among Dental Students', *Asian Pac. J. Cancer Prev.*, vol. 18, no. 6, Jun. 2017, doi: 10.22034/APJCP.2017.18.6.1573.
33. K. Rajiah, M. K. Maharajan, N. S. Chin, and K. S. F. Num, 'Awareness and acceptance of human papillomavirus vaccination among health sciences students in Malaysia', *VirusDisease*, vol. 26, no. 4, pp. 297–303, Dec. 2015, doi: 10.1007/s13337-015-0287-3.
34. C. K. H. Wong, K. K. C. Man, P. Ip, M. Kwan, and S. M. McGhee, 'Mothers' Preferences and Willingness to Pay for Human Papillomavirus Vaccination for Their Daughters: A Discrete Choice Experiment in Hong Kong', *Value Health*, vol. 21, no. 5, pp. 622–629, May 2018, doi: 10.1016/j.jval.2017.10.012.
35. H. C. W. Choi, G. M. Leung, P. P. S. Woo, M. Jit, and J. T. Wu, 'Acceptability and uptake of female adolescent HPV vaccination in Hong Kong: A survey of mothers and adolescents', *Vaccine*, vol. 32, no. 1, pp. 78–84, Dec. 2013, doi: 10.1016/j.vaccine.2013.10.068.

36. J. S.-Y. Ng and I. Ismail-Pratt, 'Cervical cancer prevention training in South East Asian LMICs', *Gynecol. Oncol. Rep.*, vol. 19, pp. 13–17, Feb. 2017, doi: 10.1016/j.gore.2016.11.008.
37. WHO, 'WHO updates recommendations on HPV vaccination schedule'. Accessed: Jul. 14, 2023. [Online]. Available: https://www.who.int/news/item/20-12-2022-WHO-updates-recommendations-on-HPV-vaccination-schedule
38. E. Colzani, K. Johansen, H. Johnson, and L. Pastore Celentano, 'Human papillomavirus vaccination in the European Union/European Economic Area and globally: a moral dilemma', *Eurosurveillance*, vol. 26, no. 50, Dec. 2021, doi: 10.2807/1560-7917.ES.2021.26.50.2001659.
39. M. N. Sitaresmi, N. M. Rozanti, L. B. Simangunsong, and A. Wahab, 'Improvement of Parent's awareness, knowledge, perception, and acceptability of human papillomavirus vaccination after a structured-educational intervention', *BMC Public Health*, vol. 20, no. 1, p. 1836, Dec. 2020, doi: 10.1186/s12889-020-09962-1.
40. D. Santhanes, C. P. Yong, Y. Y. Yap, P. S. Saw, N. Chaiyakunapruk, and T. M. Khan, 'Factors influencing intention to obtain the HPV vaccine in South East Asian and Western Pacific regions: A systematic review and meta-analysis', *Sci. Rep.*, vol. 8, no. 1, p. 3640, Dec. 2018, doi: 10.1038/s41598-018-21912-x.
41. M. Jit, M. Brisson, A. Portnoy, and R. Hutubessy, 'Cost-effectiveness of female human papillomavirus vaccination in 179 countries: a PRIME modelling study', *Lancet Glob. Health*, vol. 2, no. 7, pp. e406–e414, Jul. 2014, doi: 10.1016/S2214-109X(14)70237-2.
42. D. T. Michaeli, S. Stoycheva, S. M. Marcus, W. Zhang, J. C. Michaeli, and T. Michaeli, 'Cost-Effectiveness of Bivalent, Quadrivalent, and Nonavalent HPV Vaccination in South Africa', *Clin. Drug Investig.*, vol. 42, no. 4, pp. 333–343, Apr. 2022, doi: 10.1007/s40261-022-01138-6.

Motivational Interviewing Effect on Medication Adherence and Other Outcomes in People with Schizophrenia (PwS): A Review

Noor Cahaya[1,2], *Susi Ari Kristina*[3*], *Anna Wahyuni Widayanti*[3], *James Green*[4]

[1]Doctoral Programme in Pharmaceutical Science, Faculty of Pharmacy, Universitas Gadjah Mada, Yogyakarta, Indonesia
[2]Department of Pharmacy, Faculty of Mathematics and Science, Universitas Lambung Mangkurat, Banjarmasin, Indonesia
[3]Department of Pharmaceutics, Faculty of Pharmacy, Universitas Gadjah Mada, Yogyakarta, Indonesia
[4]School of Allied Health and Physical Activity for Health, Health Research Institute (HRI), University of Limerick, Limerick, Ireland

Abstract. Adherence will have an impact on therapy because schizophrenia is a chronic mental disorder that requires long-term treatment. One strategy to improve adherence to medications is motivational interviewing (MI), although more study is needed to see how well it works and whether it has any other effects on schizophrenia. The study aimed to assess the effectiveness of motivational interviews in improving adherence to medications and other positive impacts on PwS. A literature review using PubMed, Science Direct, Springerlink, and google scholar databases from 2010-2023 focused on keywords adherence, schizophrenia, and motivational interviewing. The results showed that MI has inconsistencies in their effect on improving medication adherence in PwS, but some studies found evidence of an association between MI and other outcomes, such as improvement in psychotic symptoms and decreased hospitalisation rates. Differences in patient characteristics and MI interventions in each study, the to perform MI techniques, and the trusting relationship built by the counsellor with the patient will affect the impact of MI on adherence. MI showed inconsistencies in improving medication adherence in people with schizophrenia. Several factors will affect the effectiveness of MI. However, MI also has the potential to improve psychosis symptoms and reduce hospitalisation rates, although more research is needed.
Keywords: schizophrenia, motivational interviewing (MI), adherence, effectiveness, psychosis

1 Introduction

Schizophrenia is a chronic mental illness that is influenced not only by genetic, but also by neurobiological factors. It manifests itself as a combination of psychotic symptoms such as hallucinations, delusions, and disorganisation, as well as motivational and cognitive dysfunction [1]. Schizophrenia affects a person's thoughts, feelings, and behaviours, inhibits their ability to assess reality, and isolates them from social interactions [2]. Some symptoms of schizophrenia can make it difficult for the patient to cooperate during the treatment process [3].

People with Schizophrenia (PwS) have been reported to have a poor understanding of how they assess the symptoms of the disease and what they expect from treatment [4]. The lack of motivation is one of the problems faced by PwS [5]. Several studies have identified motivation as a primary negative symptom in schizophrenia that is associated with a poor functional outcome [6, 7]. Motivational deficiencies are common in schizophrenia patients, even in the early phases of the illness, and these deficiencies are one of the most major barriers to persons with schizophrenia achieving functional recovery [6].

Schizophrenia is a chronic and often disabling disease, requiring long-term antipsychotic treatment. Nonadherence to antipsychotic medications is one of the problems in PwS [8]. About 75% of PwS discontinue their antipsychotic drug treatment within 18 months [9]. Difficulties in maintaining a medication regimen may arise from a lack of motivation, in addition to any of the other complicated factors that are often intertwined with the disease process of schizophrenia [10]. This is called a disease-related factor where this factor comes from the symptoms of schizophrenia itself. These are called disease-related factors, where these factors come from the symptoms of schizophrenia itself which affect the adherence to treatment [11].

Non-adherence caused higher readmission rates, more aggressive incidents, more suicides, a significant emotional and social burden for PwS and their families, and higher financial costs [3, 12]. Treatment with antipsychotic drugs reduces the risk of relapse and the risk of readmission [13]. In a systematic review, Higashi et al. [3] found that lack of knowledge of the disease, beliefs about the effectiveness of medications, substance abuse, and the quality of the therapeutic relationship were important factors influencing the relationship. Enhancing patient motivation, taking these factors into account, may be the key to encouraging drug adherence [14].

* Corresponding email : susiari_k@ugm.ac.id

Several studies have shown that patient adherence to treatment can be improved by helping them understand and accept their disease and treatment, and dealing with problems related to the drug used and its side effects. Schizophrenics with a better understanding of their disease can show better adherence to treatment [15] .

Motivational interviews were found to be particularly useful for people with addiction or high resistance to treatment or aversion to change their behaviour [16]. This therapeutic approach to behavioural intervention was adopted to improve adherence to treatment in schizophrenia, with early positive evidence of the reduction of psychotic symptoms and relapse rates in patients [5]. The positive results obtained indicate the need for more research on the effects of this motivational and self-empowerment approach (both cognitive and emotional focus) on different patient outcomes in terms of not only the severity of symptoms and recurrence of the disease, but also patient medication adherence, knowledge about the disease. and/or medication, and psychosocial functioning in those with poor adherence to antipsychotic medication [17].

This review evaluates the effectiveness of motivational interviewing on medication adherence and the other effect on clinical outcomes in people with schizophrenia.

2 Methods

We searched all studies published in 2010-2023 reporting on motivational interview-based adherence therapy as an intervention in PwS. Search literature using several databases such as PubMed, Science Direct, Scopus, and Google Scholar. A comprehensive search combined key terms using Boolean operators (e.g. AND, OR). The search terms were schizophrenia AND motivational interviewing OR motivational interviewing OR compliance therapy OR adherence therapy AND adherence OR compliance.

The first author (NC) independently extracted the following information: study question or aims; characteristics of participants, and details of interventions. Two authors (SAK and AWW) checked all the extracted data, and discussion between the two authors resolved any disagreement; with the help of the fourth author (JG) when necessary.

3 Results

3.1 Study Selection

The total of studies found in the database was 1,900 studies. The extraction articles were done by reading the title of the article and the abstract. From the selection process, the articles will be decided according to the purpose of this review. In the screening stage, 26 studies were obtained and entered the eligibility stage. At this stage, 26 studies were read in full text to assess their suitability to inclusion and exclusion criteria. Of the 26 studies, several were excluded for the following reasons: Not measure adherence (n = 11), not English language (n = 3), not MI (n = 3), protocol (n = 2) and no complete article found (n=3). The number of articles reviewed and discussed was 4 articles.

3.2 Study Characteristics

The characteristics of the included studies are described in Table 1. The four studies were conducted between 2013 and 2019 and involved a total of 466 participants who had been diagnosed with schizophrenia spectrum disorders such as schizophrenia or schizoaffective disorders. Of the four included studies, three of the studies were carried out in Asia [18–20] and the rest were carried out in Europe [21]. Participants in two of the studies were in community setting [18, 20]. One study in a hospital and community setting [21], and the rest in a hospital setting [19].

Table 1. Characteristics of included studies

Authors (year)	**Country**	Study design	Setting
Barkhof et al, (2013) [21]	Netherland	RCT	Hospital and Community
Chien et al., (2015) [18]	China	RCT	Community
Ertem & Duman, (2019) [19]	Turkey	RCT	Hospital
Chien et al., (2019) [20]	China	RCT	Community

3.3 Characteristics of the Participants

All patients have a diagnosis of schizophrenia. The age of the participants ranged from 18 to 65 years old. Three studies established inclusion criteria for patients to be recruited as participants, one of them being patients who were poor adherence [18, 20, 21]. The characteristics of the participants are described in Table 2.

The determination of adherence or nonadherence to taking medication is based on measurements using the Drug Attitude Inventory or having a history of previous nonadherence to taking medication as assessed by adherence to coming to see a psychiatrist or a self-reported method carried out by the patient himself.

However, for the rest there was no information on the inclusion of patients with non-adherent conditions (19). One study has exclusion criteria for participants who were dependent on substances or drugs (19), one study excludes patients who had experienced relapse (21), three studies exclude patients with organic disturbances (18,20,21) and all studies exclude patients with intellectual dysfunction or cognitive impairment.

3.4 Characteristics of the MI intervention

The MI interventions used in these studies (Table 3) each have their own characteristics. MI intervention was carried out in several sessions with a range of 5-8 sessions and a different duration for each session. In a study conducted by Barkhof et al. (2013) [21], the MI intervention was carried out in 5-8 sessions with the duration of the intervention sessions varying between

Table 2. Characteristics of the Participants

Characteristics	Study Author (year)			
	Barkhof et al. (2013) [21]	Chien et al. (2015) [18]	Ertem and Duman (2015) [19]	Chien et al. (2019) [20]
Age (years)	18-65	18-60	18-65	18-65
Diagnosis of schizophrenia	Yes,	Yes,	Yes,	Yes,
Poor adherence	Yes,	Yes,	NA	Yes,
Drug or substance use	NA	NA	No	NA
Had experience psychotic relapse	Yes,	NA	NA	NA
Had an organic disturbance	No	No	NA	No
Cognitive impairment	No	No	No	No

over a period of 26 weeks. Chien et al. (2019) [20] study administered the intervention in 6 sessions in 24 weeks. Each session lasted 2 hours and was held once every 2 weeks. Therefore, the total duration of the intervention was 12 weeks, or approximately 3 months. Unlike other studies, the intervention was carried out with patients in the intervention group throughout the study period. Each interview lasted 40-60 minutes and the process was completed in a total of six interviews held every two weeks [19]. Another study that was also conducted by Chien et al. (2015) [18], the intervention was administered once every two weeks for four months. Each session lasted two hours.

The professionals who delivered MI and act as therapists include psychologists, psychiatrists, mental health nurses, and also individually by researchers who have been trained in MI techniques.

Table 3. Characteristics of MI in studies

Characteristics	Study Author (year)			
	Barkhof et al. (2013) [21]	Chien et al. (2015) [18]	Ertem and Duman (2015) [19]	Chien et al. (2019) [20]
Ways of delivery:				
Individual interview	Yes,	Yes,	Yes,	Yes,
Number of sessions	5-8	8	6	6
Professional(s) who delivered the intervention:				
Psychologist	Yes,			Yes,
Psychiatrist	Yes,	Yes,		Yes,
Nurse	Yes,	Yes,		Yes,
Researcher			Yes,	
Duration of the session (minutes)	20-45	120	40-60	120
Frequency of interview	NA	Every 2 weeks	Weekly	NA
Duration of intervention	26 weeks	16 weeks	24 weeks	48 weeks

3.5 Effect of MI intervention on adherence to medications, rehospitalisation, and severity of symptoms

Intervention with a motivational interview approach has an effect on several outcome parameters for people with schizophrenia who are undergoing antipsychotic therapy. The effects of MI intervention are described in Table 4.

Based on the results of a study conducted by Barkhof et al. [21], the MI intervention was shown not to be effective in improving therapy adherence in patients with multiepisode schizophrenia who were not adherent to medication, while the study results obtained from Chien et al. [18, 20], MI-based adherence therapy (AT) can improve the adherence of schizophrenia patients to their antipsychotic drugs and improve their treatment results. AT can also improve psychosocial functioning and quality of life in patients with schizophrenia. AT can improve patients' understanding of the disease and treatment and reduce rehospitalisation rates. Motivational interview-based adhesion therapy (AT) can improve patient adherence to their antipsychotic medications and reduce psychotic symptoms. In addition, this approach can also help reduce treatment-resistant negative symptoms, such as amotivation, anhedonia, and social withdrawal [18]. The study also showed that adherence therapy can provide benefits at a low incremental cost, especially in improving outcomes at 12 months of follow-up [20].

The study by Ertem et al. [19] also concluded that the use of motivational interviewing (MI) is effective in improving the level of adherence to medication and understanding of patients with schizophrenia. Patients who received the MI intervention showed a significant improvement in the level of adherence to medications and insight compared to the control group. Furthermore, follow-up after the MI programme also improved the mean scores for the insight and medication adherence levels of the intervention group.

4 Discussion

Motivational interviewing is defined as "a collaborative conversational style for enhancing an individual's motivation and commitment to change". It involves four essential interactive elements consisting of participation, goal setting, motivation, and planning, each of which must be reviewed and adapted as the patient's condition in a treatment process [22]. Motivational interviewing (MI) is an effective intervention to increase motivation for behavioural change [23–25].

The results obtained from studies reviewed, MI interventions still show inconsistencies in their effect on improving medication adherence in people with schizophrenia. Similarly, a systematic review study conducted by Palacio et al. (2016) [29] also found that

Table 4. MI Effects

Authors (year)	Adherence to medication	Re-hospitalisation	severity of the symptoms
Barkhof et al. (2013) [21]	MI did not improve medication adherence in people with multiepisode schizophrenia who did not adhere to therapy.	MI did not have a significant effect on rehospitalisation in patients who were previously nonadherent and experienced a psychotic relapse.	MI did not affect overall symptom severity in previously non-adherent patients who experienced psychotic relapse
Chien et al. (2015) [18]	Motivational interview-based adherence therapy interventions had a positive effect on medication adherence in patients with schizophrenia spectrum disorders.	Motivational interviews-based adherence therapy interventions had a positive effect on rehospitalisation rates in patients with schizophrenia spectrum disorders.	There was a significant improvement in symptom severity as measured by PANSS scores after the motivational interviewing-based adherence therapy intervention.
Ertem & Duman (2019) [19]	There was a significant increase in the mean adherence score before and after the MI intervention.	NA	NA
Chien et al. (2019) [20]	Adherence therapy that involves motivational interviewing techniques can improve patient adherence to antipsychotic medications.	Adherence therapy can reduce rehospitalisation rates in patients with schizophrenia	The positive and negative syndrome scale was significantly reduced between the intervention and routine care group

MI improved medication adherence only in some populations, thus showing inconsistent results on improving medication adherence. MI had no statistically significant impact on chronic disease, eating disorder behaviour, marijuana abstinence, medication adherence, and others [23]. A review study mentioned that most studies did not support a direct relationship between motivational interviewing and adherence to medications [26].

The inconsistent effect of MI on medication adherence appears to be influenced by various factors. Based on the studies in this review, there are differences in the characteristics of the MI intervention such as the number of sessions, as well as the duration of each session, the frequency of interviews, the intervention period, and the professionals who provide the intervention. An interesting finding relates to the duration of MI exposure; it was stated that the number of MI sessions was not related to the outcome, but the total amount of time participants received MI interventions showed a relationship with the outcome, a longer time in one MI visit could give better results [27].

As also described in a systematic review and meta-analysis, MI is stated to improve adherence to medications at different delivery modes, exposure times and level of therapist education. Not only frequency or duration of exposure to MI, but also the interactions between the therapist and the patient influence the mechanism of change in MI for the adherence to medications [28].

As found in the study conducted by Dobber et al. (2018), there are three factors that influence the success of MI in improving treatment motivation in schizophrenia patients. The first factor is a trusting relationship that will be promoted in the engagement process. The second factor is the therapist's ability to adapt MI strategies to the patient's process. Finally, the third factor is the explicit conversation delivered by the therapist about the patient's values or goals in relation to adherence to treatment, which will hopefully increase the intrinsic motivation of the patient [5].

A mixed-method study conducted by Dobber et al. [29] found that the trusting and empathic relationship between the therapist and the patient, the ability of the therapist to perform MI techniques adapted to the patient's condition, and should be factors that will also affect the effectiveness of MI [5]. Health professionals face several obstacles in performing MI, which are identified as barriers. These barriers include practitioner-related factors, client-related factors, lack of continuous training and supervision, and workplace-related factors. Factors related to work were identified as management and time as barriers.

Motivational interviewing has the potential to be administered by healthcare practitioners with diverse backgrounds, including mental health, medicine, nursing, pharmacy, psychology, and allied health disciplines [23, 30]. Therefore, to provide MI, healthcare professionals need to develop the skills essential to guide the individual through the process. This is achievable through training, practise, and experience. Practitioners with negative attitudes towards MI did not practise MI correctly [31, 32]. When administered by trained professionals in addition to routine care, MI can help patients have a positive attitude toward adherence to medication [33].

However, some studies found evidence of an association between motivational interviewing and other results, such as improvement in psychotic symptoms and decreased hospitalisation rates [26].

5 Conclusions

Lack of motivation, insight and poor treatment adherence are common in PwS that have a major impact on treatment, course and outcome. MI is an effective therapeutic approach to explore the personal goals of

PwS and allows them to develop their willingness to participate more actively in the treatment.

The results of the study show inconsistent MI interventions in improving medication adherence in people with schizophrenia. However, MI also has the potential to improve psychosis symptoms and reduce hospitalisation rates, although further research is needed.

Several factors will affect the effectiveness of MI and these factors must be modified appropriately according to the specific needs of the patient, so that additional alternative therapies are needed as additional therapy with MI for people with schizophrenia.

Acknowledgment: The authors thank and appreciate the Universitas Gadjah Mada that facilitate the publication

Conflict of interest: The authors declare that they have no competing or potential conflicts of interest with respect to the publication of this article.

References

1. Farah FH. Schizophrenia: An Overview. Asian J Pharm. 2018;12(2):77–87.
2. Gaebel W, Zielasek J. Schizophrenia in 2020: Trends in diagnosis and therapy. Psychiatry Clin Neurosci. 2015;1–13.
3. Higashi K, Medic G, Littlewood KJ, Diez T, Granström O, de Hert M. Medication adherence in schizophrenia: Factors influencing adherence and consequences of nonadherence, a systematic literature review. Ther Adv Psychopharmacol. 2013;3(4):200–18.
4. Joseph B, Narayanaswamy JC, Venkatasubramanian G. Insight in Schizophrenia : Relationship to Positive , Negative and Neurocognitive Dimensions. Indian J Psychol Med. 2015;37(1):5–11.
5. Dobber J, Latour C, de Haan L, Scholte op Reimer W, Peters R, Barkhof E, van Meijel B. Medication adherence in patients with schizophrenia: A qualitative study of the patient process in motivational interviewing. BMC Psychiatry. 2018 May 18;18(1).
6. Fervaha G, Foussias G, Agid O, Remington G. Motivational deficits in early schizophrenia : Prevalent , persistent , and key determinants of functional outcome. Schizophr Res [Internet]. 2015;166:9–16. Available from: http://dx.doi.org/10.1016/j.schres.2015.04.040
7. Foussias G, Siddiqui I, Fervaha G, Mann S, Mcdonald K, Agid O. Motivated to do well : An examination of the relationships between motivation , effort , and cognitive performance in schizophrenia. Schizophr Res [Internet]. 2015;166:276–82. Available from: http://dx.doi.org/10.1016/j.schres.2015.05.019
8. Phan S V. Medication adherence in patients with schizophrenia. Int J Psychiatry Med. 2016 Feb 1;51(2):211–9.
9. Lieberman JA, Stroup TS, McEvoy JP, Swartz MS, Rosenheck RA, Perkins DO, Keefe RSE, Davis SM, Davis CE, Lebowitz BD, Severe J, Hsiao JK. Effectiveness of antipsychotic drugs in patients with chronic schizophrenia. N Engl J Med. 2005 Sep;353(12):1209–23.
10. Najas-garcia A, Carmona VR, Gomez-Benito J. Trends in the Study of Motivation in Schizophrenia : A Bibliometric Analysis of Six Decades of Research (1956-2017). Front Psychol. 2018;9(February):1–16.
11. Roberts DL, Velligan DI. Medication adherence in schizophrenia. Drug Discov Today Ther Strateg [Internet]. 2011;8(1):11–5. Available from: https://www.sciencedirect.com/science/article/pii/S1740677311000362
12. Chapman SCE, Horne R. Medication nonadherence and psychiatry. Curr Opin Psychiatry. 2013;26(5):446–52.
13. Leucht S, Tardy M, Komossa K, Heres S, Kissling W, Salanti G, Davis JM. Antipsychotic drugs versus placebo for relapse prevention in schizophrenia: a systematic review and meta-analysis. Lancet (London, England). 2012 Jun;379(9831):2063–71.
14. Fiszdon JM, Kurtz MM, Choi J, Bell MD, Martino S. Motivational interviewing to increase cognitive rehabilitation adherence in schizophrenia. Schizophr Bull. 2016;42(2):327–34.
15. Gray R, Wykes T, Gournay K. From compliance to concordance : a review of the literature on interventions to enhance compliance with antipsychotic medication. J Psychiatr Ment Health Nurs. 2002;9:277–84.
16. Gray R, White J, Schulz M, Abderhalden C. Enhancing medication adherence in people with schizophrenia: an international programme of research. Int J Ment Health Nurs. 2010 Feb;19(1):36–44.
17. Harmanci P, Budak FK. The Effect of Psychoeducation Based on Motivational Interview Techniques on Medication Adherence, Hope, and Psychological Well-Being in Schizophrenia Patients. Clin Nurs Res [Internet]. 2022 Oct 1;31(2):202–16. Available from: https://doi.org/10.1177/10547738211046438
18. Chien WT, Mui JHC, Cheung EFC, Gray R. Effects of motivational interviewing-based adherence therapy for schizophrenia spectrum disorders: A randomized controlled trial. Trials. 2015;16(1).
19. Ertem MY, Duman ZÇ. The effect of motivational interviews on treatment adherence and insight levels of patients with schizophrenia: A randomized controlled study. Perspect Psychiatr Care. 2019;55(1):75–86.
20. Chien WT, Cheung EFC, Mui JHC, Gray R, Ip G. Adherence therapy for schizophrenia: a randomised controlled trial. Hong Kong Med J.

2019 Feb;25 Suppl 2(1):4–9.
21. Barkhof E, Meijer CJ, De Sonneville LMJ, Linszen DH, De Haan L. The effect of motivational interviewing on medication adherence and hospitalization rates in nonadherent patients with multi-episode schizophrenia. Schizophr Bull. 2013;39(6):1242–51.
22. Miller WR, Rollnick S. Motivational Interviewing Helping People Change. Third Edit. New York: The Guilford Press; 2013. 1–615 p.
23. Lundahl B, Moleni T, Burke BL, Butters R, Tollefson D, Butler C, Rollnick S. Motivational interviewing in medical care settings : A systematic review and meta-analysis of randomized controlled trials. Patient Educ Couns [Internet]. 2013;93:157–68. Available from: http://dx.doi.org/10.1016/j.pec.2013.07.012
24. Szczekala K, Wiktor K, Kanadys K, Wiktor H. Benefits of Motivational Interviewing application for patients and healthcare professionals. Pol J Public Heal. 2018;128(4):170–3.
25. Bischof G, Bischof A, Rumpf H. Motivational Interviewing : An Evidence-Based Approach for Use in Medical Practice. Dtsch Artebl Int. 2021;118:109–15.
26. Vanderwaal FM. Impact of Motivational Interviewing on Medication Adherence in Schizophrenia. Issues Ment Health Nurs. 2015;36(11):900–4.
27. Keeley RD, Burke BL, Brody D, Dimidjian S. Training to Use Motivational Interviewing Techniques for Depression : A Cluster Randomized Trial. 2014;000156:621–36.
28. Palacio A, Garay D, Langer B, Taylor J, Wood BA, Tamariz L. Motivational Interviewing Improves Medication Adherence : a Systematic Review and Meta-analysis. J Gen Intern Med [Internet]. 2016;31(8):929–40. Available from: http://dx.doi.org/10.1007/s11606-016-3685-3
29. Dobber J, Latour C, van Meijel B, ter Riet G, Barkhof E. Active Ingredients and Mechanisms of Change in Motivational Interviewing for Medication Adherence . A Mixed Methods Study of Patient-Therapist Interaction in Patients With Schizophrenia. Front Psychiatry. 2020;11:1–12.
30. Madson MB, Loignon AC, Lane C. Training in motivational interviewing : A systematic review. J Subst Abuse Treat [Internet]. 2009;36:101–9. Available from: http://dx.doi.org/10.1016/j.jsat.2008.05.005
31. Miller WR, Rollnick S. Ten Things that Motivational Interviewing Is Not. Behav Cogn Psychother. 2009;37:129–40.
32. Bell DL, Roomaney R. Exploring the barriers that prevent practitioners from implementing motivational interviewing in their work with clients. Soc Work. 2020;56(4):416–29.
33. Gray R, Bressington D, Ivanecka A, Hardy S, Jones M, Schulz M, von Bormann S, White J, Anderson KH, Chien W. Is adherence therapy an effective adjunct treatment for patients with schizophrenia spectrum disorders? A systematic review and meta-analysis. BMC Psychiatry. 2016;16(90):1–12.

Optimization of the Procurement Process of Pharmaceutical Supplies at Dr. M Goenawan Partowidigdo Pulmonary Hospital with a Lean Six Sigma Approach

Ulfia Mutiara Supatmanto[1*], *Amal* Chalik Sjaaf [2], *Helen* Andriani[2], *Wellya* Hartati[3], and *Lukluk* Utaminingsih[3]

[1]Magister of Hospital Administration Studies, Faculty of Public Health, University of Indonesia
[2]Department of Health Policy Administration, Faculty of Public Health, University of Indonesia
[3]Hospital Pharmacy Installation, DR. Goenawan Cisarua Pulmonary Hospital

Abstract. Internal barriers in the procurement process and pharmaceutical supply planning are barriers that can be controlled through interventions. The research focuses on proposing changes to the pharmaceutical supply management system, particularly in the aspects of planning and procurement within the internal team of the RSPG Cisarua Bogor. The intervention in this research applying lean six sigma extends only up to the "improve" phase. Data collection involves in-depth interviews with informants involved in the pharmaceutical supply planning and procurement process, as well as observation and document analysis. The research ends in group discussions to reach an agreement. The research findings reveal that a lack of established procedures in the pharmaceutical supply procurement process contributes to prolonged procedures and the absence of efficiency benchmarks within the system. Moreover, an unavailability of tools to facilitate the pharmaceutical supply procurement process causes low internal team communication. The "improve" phase of lean six sigma leads to proposed changes to the standards of operational procedures for routine pharmaceutical procurement, indicator applications of procurement efficiency and the utilization of the ABC VEN functioning as a tool within the pharmaceutical supply procurement process. The proposed improvement to tackle the matter includes the utilization of ABC VEN and efficiency indicators to overcome waste overproduction identified during the pharmaceutical supply planning process. In addition, it also involves the establishment of a new standard of operating procedure covering timelines, tools for pharmaceutical supply categorization based on ABC VEN and procurement efficiency indicator to handle waste waiting during the pharmaceutical supply procurement process.

Keywords: Pharmaceutical procurement; Lean Six Sigma; ABC VEN; Efficiency indicator; Hospital Procurement Process

1 Introduction

Hospital pharmacy installations are a few of the contributors to hospital performance, reflecting service performance or financial performance. The hospital pharmacy installations are also closely related to other functional units such as medical units and financial units. It emphasizes the importance of effective and efficient management of the pharmaceutical installation aiming to provide optimal service performance and sound financial performance leading to seamless inter-unit coordination. The efficiency of the hospital pharmacy installation focuses on service workflows, pharmaceutical inventory control processes, as well as the capacity and the credibility of human resources in pharmacies and associated units [1–5].

The complexity of the procurement process involves multiple individuals with diverse educational backgrounds so that a necessary tool is required to ensure the efficient and effective execution of the procurement processes. The hospital has already established buffer stock notifications and ABC VEN reports with limited access only for the pharmaceutical installations. However, at present, procurement efficiency indicators are not available causing insufficient evaluations of the procurement process. This situation recalls the necessity for a collective agreement on how the procurement system flows and whether additional tools are needed in the decision-making process with both existing *buffer stock* notification and ABC VEN reports that can be accessed by all individuals involved in the procurement process. Furthermore, it also underscores whether the assessment of efficiency indicators can function as references in evaluating the efficiency of the pharmaceutical supply procurement processes. This research aims to propose solutions for tackling the challenges at RSPG Cisarua Bogor involving planning management and procurement management of pharmaceuticals by formulating proposals for improvement efforts in the procurement process system at the hospital.

Hospital pharmacy installation have gather a wealth data about pharmaceutical usage, tren and requiretment. The data can be used in the pharmaceutical procurement

* Corresponding author: ulfia.mutiara@gmail.com

 https://doi.org/10.1051/bioconf/20237505012

planning process as a basis information. However, the existing data has not been properly utilized. This has an effect on the management of pharmaceutical supplies. Besides that, it can also influence the management of the pharmaceutical supplies procurement budget. Data processed within a certain period of time will provide accurate information to predict the need for hospital pharmaceutical supplies. However, it it not properly analyze and utilize, it leads to missed opportunity for efficient procurement and budget management.

2 Theoretical Review

The logistics management cycle commences with planning, budgeting, procurement, storage, distribution, reporting and destruction or elimination [6, 7]. Each part of the cycle must be completed clearly and continuously between one another for achieving a cyclical equilibrium. This procurement process must be centralized within the hospital pharmacy installation. The procurement cycle is reflected in [6] *MDS-2: The Selection, Procurement, Distribution and Use of Pharmaceuticals* issued by the *Managing Drugs Supply* department, a division of *Management Science for Health* and a non-profit institution focusing on global health management.

The procurement flow at each hospital has different stages [8]. Procurement involves pharmaceutical elements as management of pharmaceutical supplies and financial elements as a part of budget management [1]. The medical department is required as a consideration from the user side of pharmaceutical supplies. All of these elements have the same goal, namely providing optimal health services to patients in accordance with the existing budget capacity [9, 10].

Pharmaceutical supply management efficiency indicators published by Pudjastuti in 2006 are designed based on understanding, experience and studies of applicable regulations. Accordingly, these indicators possess a more comprehensive ability to describe the management of pharmaceutical supplies. There are four categories in these indicators namely planning, procurement, storage, and distribution. The efficient planning indicators mainly deal with the similarities between the formulated plan based on the necessities and required budget. Meanwhile, the procurement efficiency indicator emphasizes the procurement patterns and the payment patterns of pharmaceutical supplies. This also acts as a tool for measuring effectiveness and efficiency of the procurement process. Moreover, the Lean Six Sigma is utilized as an approach to optimize the procurement process,

The application of Lean Six Sigma approach is intended to delineate waste and variation within each phase of the procurement process. In addition, it is also a way of continuous improvement. The impact of Lean Six Sigma interventions in hospital business is expected to be improvements in service as well as effective and efficient care to patients. In this way, this can elevate customer value as the main focus of healthcare services. Lean Six Sigma is based on DMAIC concepts such as Defining, Measuring, Analyzing, Improving and Controlling. These concepts enable us to clarify problems as a whole, identify weak points, analyze the roots of the problems and provide innovative suggestions to address the issues. The next process is to evaluate the suggested innovations based on their effectiveness according to the results of the interventions. Thus, the overall process offers extensive understanding regarding the core problem of the discussed issue.

3 Research Methods

This study follows the Lean Six Sigma approach with a qualitative design supported by quantitative data. The initial step of the study is to examine the ABC VEN data that has been compiled by the pharmacy installation of RSPG Cisarua Bogor during the periods of January to March 2023. Subsequently, in-depth interviews were conducted with predetermined informants who were personnels at the implementation level. The final step is Focus Group Discussion with the board of directors, relevant managers related to the procurement and the operational staff. In this way, implementing the consensus generates advantages namely to involve existing stakeholders in the decision-making process and to mitigate potential researcher bias. The research timeline spanned from March 2023 to May 2023. The research has passed ethical clearance issued by the Ethics Review Team of the Faculty of Public Health, University of Indonesia. In addition, permissions from the Chief Executive Officer of RSPG Cisarua Bogor were obtained. The informants also had signed consent forms as they had agreed to participate in the research. The researcher maintains the confidentiality of data and informant identities.

4 Research Results

The informants in this study are at the managerial level and the operational staff. There is a new appointed director during the research due to a change in the board of directors.Thus, the new director refused to be interviewed and the previous one also objected to being interviewed.

4.1 Flow of Pharmaceutical Supplies Planning and Procurement System at RSPG Cisarua Bogor

4.1.1 Notification of buffer stock in the process of planning the procurement of pharmaceutical supplies

Within the Hospital Information System, there is a module designed to generate *buffer stock* notification systems named the *stock warning system* consisting of warehouse warning stock, outpatient depot warning stock and inpatient depot *warning stock*. During field observation, data buffer stock is presented on the working documents as the foundation for pharmaceutical supply planning. The working documents are refined with the evaluation from the head of the installation before being passed to the head of supporting service coordinator. Both data buffer stock and warning stock are only accessible within the scope of pharmacy. In the permanent procedure issued by the hospital No. OT.02.02/XXXIV.1/06285/2022, it is stated that the calculation of future usage plus buffer stock (d + % *buffer stock* = e) is conducted at the planning stage. The results of collaborative discussion confirms in depth-interviews stating that the utilization of warning stock as a basis for consideration in the preparation of pharmaceutical supply planning has been implemented. The formulation of buffer stock is conducted in layers from warehouse buffer stock and buffer stock of each pharmaceutical depot. The arrangement of buffer stock refers to the average of consumption patterns within the hospital over a period of two months. Referring to the triangulated results, there is a planning process based on buffer stock notification. This practice is completed within the scope of pharmacy representing a fundamental competency of pharmaceutical work.

4.1.2 The use of ABC VEN in the pharmaceutical supply procurement planning process

The preparation of ABC VEN is also carried out by the head of IFRS. The ABC VEN data is not comprehended by other staff regardless of their medical and non-medical educational backgrounds. It is reflected in the following excerpts of the interview with an informant possessing non-medical educational background:

"What's that? I don't understand that at all. But looking at it again, it seems to be helpful in facilitating the grouping of procurements so that it is clear what the purchase plan is." (I1)

Non-medical informants heavily rely on their experiences in deciding or simply following the recommendations proposed by the technical unit. According to hospital regulation No. OT.02.02/XXXIV.1/06285/2022, the formative planning refers to the study of disease development, treatment patterns, priority scales and development plan. The discussion process in the collaborative forum reveals that ABC VEN has been applied in the pharmaceutical scope even though other divisions cannot comprehend its implementation. The truth is that grouping using ABC VEN method can be the solutions of budget allocation needed by the budget planning division of the hospital.

The triangulated results show that the use of ABC VEN in planning has been carried out in accordance with the direction from the highest leadership at RSPG Cisarua Bogor. In the planning process, this concept has been applied and supported by other analysis. Nevertheless, this concept is still widely unknown to components outside the pharmacy so that it becomes the factor in determining the perception of insufficient planning proposals.

4.1.3 Flow of procurement of pharmaceutical supplies

Based on all informants' reports, the procurement flow at RSPG Cisarua Bogor is similar since the submissions are made by the technical unit, in this case the Hospital Pharmacy Installation. Submissions are centralized through the pharmaceutical installations. Another important information reveals that there is informal information sharing regarding the procurement plan. This signifies a problem related to Human Resources since PPBJ functional position does not exist causing the procurement process to be regarded as supplementary activity.

The flow of pharmaceutical procurement contained in the standard procedure No. OT.02.02/XXXIV.2/00073/2023 for urgent procurement states that procurement is proposed by the head of IFRS either verbally or in writing to the coordinator of support services, who then communicates to the technical director and KDP. The procurement workflow follows the existing working relation procedures at RSPG Cisarua Bogor. In limited discussions at the hospital, it was concluded that there is an urgency to form certain guidelines in the process of procuring pharmaceutical supplies. This is related to the absence of timelines in written reference documents. At present, the proposals of pharmaceutical supplies are still in manual forms causing difficulties in monitoring the pharmaceutical supply procurement process. The triangulated findings indicate that the procurement flow is not written in the fixed standard procedures as per the written regulations. This impacts unclear and non-accountable procurement processes in forms of the workflow and timeline.

4.2 VA, NVA and Waste Activities in the Procurement Process of Pharmaceutical Supplies at RSPG Cisarua Bogor

4.2.1 Identify value activities based on value stream maps

The results of informant interviews show that the VA process looks more than the NVA. NVA-shaped processes are *necessary*. The document also reflects that both planning and urgent procurement activities must be acknowledged by the head of the Pharmaceutical Installation.

The President Director as the Power of Attorney for Budget Users must also know during the urgent procurement process. The results of observations show that working papers are a means of evaluating pharmaceutical supply planning. This is included in value-added activities. Meanwhile, observations during the procurement process by ULP show that there is a lack of value-adding activities, including them as non-value added. The results of collaborative discussion showed that value added and *non-value added* activities were in accordance with the results of interviews, document tracing and observation. Non-value added activities are included activities involving the board of directors and regarded as exclusive activities. This particular activity can be categorized as *necessary but Non value added* (NNVA). These activities can be seen in the value stream map in Figure 1:

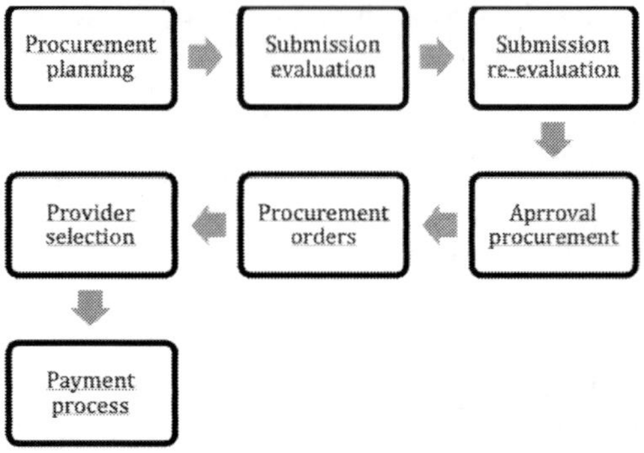

Fig. 1. Value Stream Map (Current Map)

4.2.2 Waste identification

The results of in-depth interviews conducted by researchers show that there are activities included in the categories of I *overproduction and waste waiting*. The result of observing the document No. OT.02.02/XXXIV.1/06285/2023 on pharmaceutical procurement planning reveals the identification of excessive activity like waiting for the approval of the technical director without clarity of timeline in fixed procedure. In the group discussion process, the waste waiting process happens because the submission of pharmaceutical supplies procurement was manual. In contrast, overproduction occurs due to the cross-test process that does not use the similar data source causing repeated calculations. This increases the process time of procuring pharmaceutical supplies at RSPG Cisarua Bogor. Based on the triangulated method carried out, it can be concluded that *waste* activities can be described in the table 1 as follows:

Table 1. VA and NVA identification

Activities	VA/ NVA	Kinds
Planning		
Planning the procurement of pharmaceutical supplies with working papers	VA	
Evaluation of pharmaceutical supply procurement plan with ABC VEN	VA	
Re-evaluation of the procurement of pharmaceutical supplies by looking at medical record files and casemix	NVA	Over Production
Placement of budget allocation for procurement of pharmaceutical supplies	VA	
Length of time for approval of the application for the procurement plan of pharmaceutical supplies by KDP	NVA	Waiting
Procurement		
Length of time for approval of the procurement of pharmaceutical supplies by the power of budget users	NVA	Waiting
The duration of the process of selecting pharmaceutical supplies in the e-catalog by ULP	NVA	Waiting
Process payments according to the predetermined budget currency	VA	

4.3 The Root of the Problem in the Pharmaceutical Supplies Procurement System at RSPG Cisarua Bogor

In the planning process, the source of the problem is the distrust of the presented data. In the procurement process, external factors are regarded as the cause of the obstruction of the procurement process. At the time of observation, problems do not exist both in the work process and the variable compositions on the working paper. Meanwhile, during the procurement process, the existing problem is informal reference used in the process since formal reference does not have a clear timeline for the procurement process. The result of document observation reflects that the planning process has already contained procedures in accordance with pharmaceutical standards, namely by using *buffer stock* calculations and estimates of future needs. Meanwhile, fixed procedures for the procurement process of pharmaceutical supplies have not been established. The documents found only discuss the procurement of urgent pharmaceutical supplies. In the directed discussion process, it is concluded that the problem in the pharmaceutical supply procurement system in the internal RSPG Cisarua Bogor is a manual process so that the data for the verification process cannot be accessed by the procurement department functioning as a decision maker. In addition, the grouping of pharmaceutical supplies is still unclear, causing confusion in the budget section of RSPG Cisarua Bogor. The three methods carried out in this study show that there are internal and external factors that influence the procurement process. Internal factors indicate that there are indications of ineffective internal communication in the process of procuring pharmaceutical supplies. Constraints in internal communication are reflected in confusion in the grouping of pharmaceutical supplies from the perspective of budget compilers. In addition, manual processes lead to an inability to know real time information. This obstacle can be overcome by referring to the fixed procedures that exist in the hospital. Nevertheless, in tracing the documents, these fixed procedures have not been obtained. The lack of human resources due to the regulation of exclusion of the position of BLU in regulations related to PPBJ causes the procurement process to be constrained. In-depth analysis using fishbone analysis can be shown in the Figure 2:

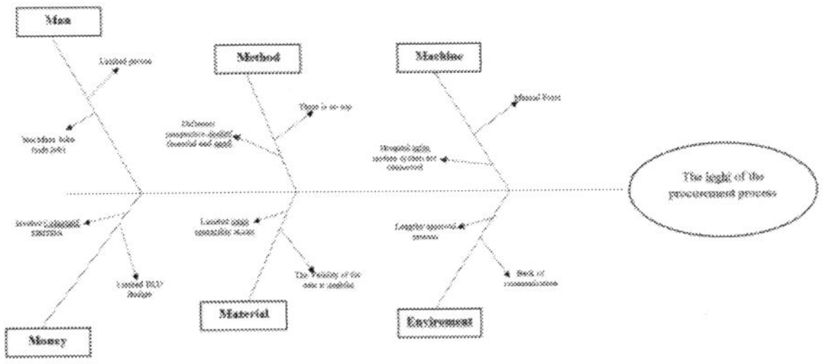

Fig. 2. Fishbone Diagram

4.4 Proposed Efforts to Improve the Procurement Process at RSPG Cisarua Bogor

4.4.1 Proposed improvement of SPO in the process of procuring pharmaceutical supplies

In an in-depth interview, one informant supports improvements in hospital procedures. Based on the document tracing carried out, there is a lack of standard procedure for procurement of pharmaceutical supplies containing the flow of procurement of pharmaceutical supplies. Based on field observations, it indicates that the absence of SPO in the procurement process causes the inexistence of well-documented standard procedures in the formal procurement process. Moreover, the results of the group discussion shows that the process flow of procurement of pharmaceutical supplies does not have a fixed standard procedure. This leads to the absence of standard procedures for the length of time to procure pharmaceutical supplies. Accordingly, in the newly created written procedure document *the timeline* of procurement. In addition, another alternative is making a digital procurement process with the purpose of integrating the hospital's internal system with the hospital's information system. The results of the triangulated method clarify that the absence of written standard operating procedures for the procurement of pharmaceutical supplies causes chaotic flow in the process of procurement of pharmaceutical supplies. Eventually, the creation or change of standard operating procedures is urgently required as a part of continuous improvement efforts for hospitals.

4.4.2 Proposed use of planning and procurement indicators in the process of procurement of pharmaceutical supplies

Pharmaceutical supply procurement indicators are unidentified so they do not serve as part of pharmaceutical supply procurement evaluation. In tracing the documents, indicators of pharmaceutical supply procurement either from annual reports, annual performance reports or work unit reports were not found. The applied indicators are only those of hospitals in general. The results of the focused discussion reveals that the urgency of applying indicators in the procurement processes is assumed to be unnecessary considering that the hospital only uses mandatory indicators issued by the Ministry of Health. The implementation of procurement efficiency indicators can only be applied once it is required by the hospital. The comprehensive results of the research methods conducted indicate that the utilization of efficiency indicators in the procurement process has not been fully understood within RSPG Cisarua Bogor. The urgency of incorporating efficiency indicators into the procurement process is still considered unnecessary because it is not mandatory based on the Ministry of Health. The application of the indicators in addition to being a benchmark for efficiency remains to be a consideration of a comprehensive system evaluation for hospitals.

4.4.3 ABC VEN proposal as a tool in the procurement process

In the planning process carried out by pharmacists, the use of ABC VEN as a tool has been carried out. The document search carried out shows that the use of ABC VEN in the planning process has been written in the permanent procedure No. OT.02.02/ XXXIV.1/06285/2022, namely in the planning procedure for the procurement of pharmaceutical supplies. During focused discussions, the proposed use of ABC VEN in the pharmaceutical supply procurement process can be used in budget allocations. Grouping with ABC VEN can help in budget allocations. In addition, the ABC VEN allocations can function as a reference in the planning process as well as a priority of pharmaceutical supply procurement at RSPG Cisarua Bogor. The application of ABC VEN has been used in the planning process, but grouping in the procurement process is not conducted. Accordingly, it can be suggested as a proposal for the use of grouping pharmaceutical supplies in the procurement process. The ABC VEN proposal as a grouping of pharmaceutical supplies becomes more applicable when it is used not only by the planning department but also by the budgeting or finance department. The ABC VEN grouping is an applicable tool in grouping pharmaceutical supply items for the budget department. Furthermore, it also simplifies identification for pharmaceutical supply procurement activities.

5 Discussion

5.1 Flow of Pharmaceutical Supplies Planning and Procurement System at RSPG Cisarua Bogor

5.1.1 Notification of buffer stock in the process of planning the procurement of pharmaceutical supplies

Notification of *buffer* stock already existed in the pharmacy module in the hospital information system in the form of a warning stock function as a tool in the planning process. This is supported by the results of research which shows that in the pharmaceutical scope, the utilization of this system as a tool in compiling a list of required pharmaceutical supply plans has been carried out. However, this system is not accessible to decision makers in the pharmaceutical supply planning process.

Planning the procurement of pharmaceutical supplies must be executed in a measured and well-organized manner so that it ensures a streamlined system that operates efficiently with minimal obstacles [11]. The flow of the planning system involves notification of *buffer* stock, in this case the *warning stock* in the Hospital Information System, including the composition from the demand and financial perspectives. The workflow of the procurement system is carried out effectively and efficiently by prioritizing a balance between demands and financial capabilities [12–14]. The application of accurate data in the planning process results in streamlined hospital business processes [1, 15).

Accurate and up to date data must always support the decision making process [10, 16]. When decision makers do not have data or cannot access data, bias can arise in the results of their decisions. Data processed properly can produce information that can be accounted for with the purpose of ensuring that the decision can be right on target [3, 15]. Therefore, decision makers must be able to access data used as a basis for decision making in order to provide proper and correct decisions.

5.1.2 The use of ABC VEN in the pharmaceutical supply procurement planning process

The results reveal that the preparation of pharmaceutical supply planning has used ABC VEN as one of the preparation tools since it is significantly proven based on in-depth interviews, document reviews, observations and FGDs. Nevertheless, the concept is not widely known by other divisions to improve the effectiveness of the procurement process.

The use of ABC VEN as part of the grouping of pharmaceutical supplies procurement has been applied in pharmaceutical work [13, 14, 17]. Grouping with ABC VEN is completed because it is classified based on hospital needs while considering the costs incurred by each item of pharmaceutical supplies [14, 17]. The composition results will certainly be unique and different in each hospital. The ABC VEN method is a combination of ABC analysis and VEN classification [13, 17].

Traditionally, each element in the procurement system will arrange procurement based on a priority scale. However, individuals do not have the same view on this matter. The use of ABC VEN grouping which is widely known in pharmacy certainly cannot stand alone. [13, 14]. These characteristics in the ABC VEN grouping will make it easier for team members to understand the position of each item of pharmaceutical supplies in the inventory stock turnover.

The application of the ABC VEN evaluation model in the planning process also helps sharpen the results of pharmaceutical supply procurement planning [1, 18]. The ABC VEN evaluation capturing pharmaceutical supplies in terms of needs and financial value empower this tool to optimize the pharmaceutical supply planning process [10, 12]. The ABC VEN grouping can be a solution to the problem of how to make pharmaceutical supplies groups on budget because it can be easily interpreted by divisions that do not have any health educational background with the purpose of facilitating the evaluation of a pharmaceutical supply procurement application.

The problems faced by this planning department can be assisted by the pharmacy department by making groups according to the pharmaceutical supply group in ABC VEN. This grouping is more practical because it focuses on the nature of supplies in managing pharmaceutical supply logistics instead of focusing on the types of supplies. Pharmacists working in IFRS can provide an understanding of the ABC VEN grouping to the budget planning department. It is reflected in the discussion process at FGD that the grouping can be a solution of crucial problems in the budget planning department.

5.1.3 Flow of procurement of pharmaceutical supplies

The flow of pharmaceutical procurement has a crucial problem where the hospital's fixed procedures regarding procurement do not exist. A procurement process without a recognized fixed procedure will make it a customary process without any legal reference. A fixed procedure is only understood as supporting documentation for accreditation instead of a basis for action. There are seven stages in the planning and procurement flow at RSPG Cisarua Bogor. A manual procurement process with a flow involving two boards of directors will create inefficiencies in the pharmaceutical supply procurement system.

The efficiency of the procurement flow of pharmaceutical supplies also concentrates on regulations [1, 4]. Standard operating procedures or fixed procedures in hospitals can be evaluated periodically both in terms of the procurement process flow and from the procurement process timeline. The purpose of this evaluation is to simplify the procurement process so that it is maintained well concerning the bureaucracy of procurement documents and the speed of meeting procurement demands. A well-documented procurement system that has fixed standard procedures confirmed by the leadership, in this case the president director, can facilitate the evaluation and conclusion of information obtained from pharmaceutical supply procurement data [1, 4].

Without a procedure in the internal policy order, the procedure proceeded is only an informal procedure that can potentially cause interdivisional problems. This certainly must be avoided in the implementation of a business process especially in hospitals involving many professions and educational backgrounds in their business processes.

The process flow of planning and procuring pharmaceutical supplies conducted manually and through long stages has the possibility of inefficiencies. Regulations issued by the Ministry of Finance regarding the procurement of goods and services exclude BLU. This is intended for the purpose of developing concepts in accordance with the existing ecosystem in the hospital. The aspiration is for hospitals to be managed with a fine business framework, ensuring effective and efficient business operations. Acting in the capacity of the leader of the hospital, the President Director possesses the autonomy to improve the hospital optimally.

5.2 VA, NVA and Waste Activities in the Procurement Process of Pharmaceutical Supplies at RSPG Cisarua Bogor

5.2.1 Identify value activities based on value stream maps

Internal constraints are the most controllable in a logistics management cycle. Internal barriers involve resources within the hospital itself [6]. When the hospital can identify internal obstacles, it can be more *agile* in changing following external situations and conditions. In this way, it is crucial to understand the weak points in the business processes prevailing in the hospital [19, 20].

Evaluation in an existing business process is a necessity for every business unit [18, 21]. The evaluation is mainly related to the unit that generates revenue and costs for the company [21, 22]. The example is the pharmaceutical unit with its important planning and procurement processes.

The results signifies that the activity points in pharmaceutical supply planning included in non-value added occurs due to doubts in the data presented in the pharmaceutical supply planning application. This leads to the re-evaluation process using other methods and it can be regarded as one of the obstacles in the pharmaceutical supply planning process. In addition, trust in the competence and integrity of personnel in pharmaceutical installations is also one of the factors for this re-evaluation.

This issue constitutes a crucial concern because of the lack of understanding of each other's points of view in the procurement process. Divergent points of views cause different justifications. Thus, the variations of justifications can be reconciled through the definite guidelines within the planning system and tools in the pharmaceutical supply planning system.

5.2.2 Waste identification

The identification of waste in this study was spotted in the ranks of decision makers in the process of procuring pharmaceutical supplies. Waste at the planning point happens due to revaluation applying different methods in calculating pharmaceutical supply needs. *Waste* that occurs on the board of directors includes Necessary but Non Value Added (NNVA).

Waste at the planning point can be optimized by opening access to supporting service coordinators enabling them to access data for conducting the same basic calculation standards. In this way, it can minimize calculations that are not in accordance with needs.

Waste on the board of directors exists because the process is carried out manually. This affects the procurement process negatively once the directors are not in place. Utilizing information technology is the best solution to tackle the issue [23]. The use of information technology is also one of the pillars in increasing the speed and ease of leadership in monitoring the performance of the subordinate tiers [3, 15]. Although this activity is considered non-value added, it remains important in the process of procuring pharmaceutical supplies. The role of leadership decisions is paramount to manage a balanced and compliant business process aligned with the applicable regulations [23].

5.2.3 The Root of the Problem in the Pharmaceutical Supplies Procurement System at RSPG Cisarua Bogor

These results signify indications of ineffective internal communication in the process of procuring pharmaceutical supplies. Constraints in internal communication are reflected in confusion in the grouping of pharmaceutical supplies from the perspective of budget compilers. In addition, it can be seen in the manual process causing the inability to comprehend the position of the application in real time. Referring to the existing procedure at the hospitals can be the solution, but the fixed procedures have not been obtained. In hospital accreditation, the existence of standard operating procedures has become a necessity. However, the drafting of the regulation is only based on copying the regulations of other hospitals. This deviates from the purpose for which the rules were formed. Hospital regulations are local and customized ensuring differences among hospitals [24]. The variations are based on the adaptation process to the existing ecosystem in the hospitals. Hospitals without standard rules as needed will place the rules as a mere display instead of a work reference. The content of the regulation will also not be fully applied. This creates internal barriers in the hospital's business processes. Internal obstacles in the process of procuring pharmaceutical supplies have been investigated by Saputra in 2019 concluding that unclear coordination is one of the inhibiting factors. This lack of clarity in coordination reflects the absence of standard operating procedures that are fully comprehended by every individual involved in the procurement process of pharmaceutical supplies. Therefore, written procedures become mandatory in a hospital system [25].

5.3 Proposed Efforts to Improve the Procurement Process at RSPG Cisarua Bogor

5.3.1 Proposed improvement of SPO in the process of procuring pharmaceutical supplies

The addition of standard operating procedures in the procurement process must be tackled immediately. In addition to being formed, each element in the procurement process must also understand the procedures and apply them in the work process. Standard procedures are created according to the needs of the organization. Formulating fixed procedures for the procurement process of pharmaceutical supplies becomes the basic reference in the procurement process.

Making standard procedures that are appropriate to the hospital ecosystem will facilitate coordination in activities. Poor coordination will hamper the process of procuring pharmaceutical supplies [25]. Standard fixed procedures for pharmaceutical supplies procurement work are also a formal reference in pharmaceutical supply procurement activities.

Standard operating procedures in each hospital will be different. This depends on the organizational culture, organizational structure, authority between each section and the job description of each individual in the hospital. Improvements to standard operating procedures are customized depending on the needs of the hospital [11, 26].

The need for RSPG Cisarua Bogor as a referral hospital with class A type is effective and efficient hospital governance. Hospital governance refers to written regulations used as a reference and basis for evaluation in hospitals. This evaluation is part of the *continuous improvement* needed during this period [21, 22]. This improvement in operational standards must involve all elements in the hospital related to the policy [5, 16]. Accordingly, the process can ensure that the policies are collectively adhered to, not just a complement to accreditation alone.

5.3.2 Proposed use of planning and procurement indicators in the process of procurement of pharmaceutical supplies

Planning indicators at RSPG Cisarua Bogor have not been calculated by looking at efficiency factors in the process of procuring pharmaceutical supplies. Indicators of the efficiency of procurement of pharmaceutical supplies are not calculated, but data to calculate these indicators exist. The existing data has not been utilized optimally to determine the efficiency of the planning and procurement process at RSPG Cisarua Bogor.

Data that is not processed properly will not be of much use to the organization. If the data is processed properly, it can be important information for organizations to improve themselves. Hospitals as complex business organizations are required to change continuously with the times. Therefore, the information possessed by the hospital must be comprehensive in order to compete well with other hospitals.

The use of procurement indicators has been widely recognized by hospitals in Indonesia. This indicator of the efficiency of pharmaceutical supply procurement includes the process of planning, procurement, storage and distribution. This efficiency indicator is important to be a benchmark for efficiency in the process of planning and procuring pharmaceutical supplies. The results of the indicator can be the basis for internal evaluations performed periodically [20, 27].

The efficiency indicator of drug management in hospital pharmacy popularized by Pudjastuti in 2006 states that this indicator can present valid, sensitive and specific data. The use of planning and procurement efficiency indicators has an important urgency in the context of pharmaceutical supply management.

Indicators that can more sharply describe hospital conditions objectively will ease the difficulties for hospitals to take further steps in their business strategy. This involves the use of pharmaceutical supply management indicators that are more specific to discuss pharmaceutical supply management necessary. By looking at existing indicators, hospitals can formulate steps and strategies at ease. Hospital management is different from one another causing the use of tools that function to sharpen analysis is needed as part of understanding the overall situation and condition of the hospital. In other words, this will not be achieved using common indicators.

5.3.3 ABC VEN proposal as a tool in the procurement process

Understanding the ABC VEN grouping in pharmaceutical supplies can facilitate the setting of pharmaceutical supply procurement priorities. Thus, the utilization of grouping with ABC VEN in procurement systems is helpful for non-pharmaceutical parts to understand the urgency of each item. Grouping with this system is also the basis of communication between pharmaceutical and non-pharmaceutical departments regarding the priority of purchasing pharmaceutical supplies.

The use of ABC VEN as a grouping tool for pharmaceutical supplies helps to identify groups based on importance, cost value and use value [13, 14, 17]. The ABC VEN grouping is widely known among pharmaceutical personnel. The ABC VEN grouping in each hospital is different so that the ABC VEN group is customized. In other words, it cannot be used in other hospitals. Each hospital has a different ABC VEN arrangement on its pharmaceutical supply items.

The ABC VEN guideline, serving as a classification guide in the management of pharmaceutical supplies, will ensure effective communications among hospitals departments. In addition, planning references will enhance the planning reference to determine the priorities of pharmaceutical supply management. Accurate planning will boost hospital performance, especially from the financial aspect. Proper planning will also reduce the chance for drug stock outs.

The shortage of medications is an unavoidable situation within healthcare facilities. The prevention entails precise and meticulous planning of pharmaceutical supplies and it relates to the financial status of the hospital. In this way, maximizing the available budget allocation is required to meet the needs of pharmaceutical supplies in hospitals. Hospitals' failures in planning causes shortage of medication so that urgent procurement becomes necessary. Urgent procurement will certainly increase hospital costs to meet the needs of pharmaceutical supplies. Therefore, the presence of tools in planning and procuring pharmaceutical supplies is essential for effective pharmaceutical supply management.

6 Conclusion

The result findings show that proposed intervention can be carried out in various terms. In the short term, it is urgent to open access to pharmaceutical modules for decision makers and to formulate standard operating procedures containing procurement indicators. In the medium term, the research suggests a development of a hospital information system based on the research results, involving the ABC VEN categorizing and buffer stock as part of the online procurement system. In the long term, it advocates the government regarding the pivotal role of Procurement and Goods/Services Procurement Policy (PPBJ) within the Government Business Entity (BLU) to elevate the effectiveness and efficiency of the pharmaceutical supply procurement process.

7 References

1. Polii S, Posangi J, Manampiring AE. Management of Drug Planning, Procurement, and Control in Hospital Pharmacy Installations. Sam Ratulangi. Sam Ratulangi J Public Heal. 2022;2(2):53.
2. Suyanti T. Drug Management Analysis Of Procurement In Pharmacy Department Of RSUD Gambiran KotaKediri Period 2016. J Fram Indones. 2016;13(2).
3. Riki Y., Lembong, Kalalo M. B. Analysis of Accounting Information System on Procurement of Drug Inventory at GMIM Kalooran Amurang Hospital. LPPM J EcoSosBudKum (Economic, Soc Cult Law). 2022;5(2):593–600.
4. Ningsih A, Fudholi A, Sumarni S. THE RELATIONSHIP OF THE APPLICATION OF ELECTRONIC CATALOGS TO THE EFFICIENCY OF PROCUREMENT AND AVAILABILITY OF DRUGS. J Pharm Manag Serv. 2015;5(4).
5. Ulfah M, Wiedyaningsih C, Endarti D. Evaluation of Drug Management in Planning and Procurement Phase at Muntilan Regional Hospital, Magelang District, 2015 - 2016. Jmpf [Internet]. 2018;8(1):24–31. Available from: https://journal.ugm.ac.id/jmpf/article/view/31883/pdf
6. Rahmiyati AL, Irianto G. Theory and Practice of Hospital Logistics Management. Bandung: Refika Aditama; 2021.
7. Taha NA, Lolo WA, Rundengan G. ANALYSIS OF DRUG LOGISTICS MANAGEMENT AT THE MANADO CITY HEALTH OFFICE PHARMACY INSTALLATION IN 2020. Pharmacon. 2021;10.
8. R.U. Mu'Minin. PROFILE OF ABC ANALYSIS ON DRUG PROCUREMENT MANAGEMENT IN HOSPITALs. Kendari: Mandala Waluya University; 2020.
9. Friska E, Suryoputro A, Kusumastuti W. Analysis of Procurement Process to Ensure Drug Availability at Tugurejo Hospital Semarang. Indones Public Heal Media. 18AD;4.
10. Capritasari R, Kurniawati DR. Analysis of planning and procurement to ensure the availability of drugs in hospitals. Sasambo J Pharm. 2021;2(1):32–6.
11. Health MSF. Managing Access to Medicines and Health Technologies. 2012;
12. Yanti TH, Farida Y. ABC ANALYSIS IN ANTIBIOTIC DRUG PLANNING AT SURAKARTA ORTHOPEDIC HOSPITAL. J Pharm Sci Clin Res. 2016;1:51–7.
13. Dampung V, Maidin A, Mardiana R. APPLICATION OF CONSUMPTION METHOD WITH FORECASTING, EOQ, MMSL AND ABC-VEN ANALYSIS IN PHARMACEUTICAL SUPPLY MANAGEMENT AT PELAMONIA HOSPITAL MAKASSAR. Pharm Media. 2018;14(1):124.
14. Rofiq A, Oetari O, Widodo G. Analysis of Drug Inventory Control Using ABC, VEN and EOQ Methods at Bhayangkara Kediri Hospital. JPSCR J Pharm Sci Clin Res. 2020;5(2):97.
15. Rahmawatie E, Santosa S. DRUG PROCUREMENT PLANNING INFORMATION SYSTEM AT BOYOLALI DISTRICT HEALTH OFFICE. J Pseudocode. 2015;
16. Nesi G, Kristin E. EVALUATION OF DRUG PLANNING AND PROCUREMENT AT THE PHARMACEUTICAL INSTALLATION OF KEFAMENANU HOSPITAL, NORTH CENTRAL TIMOR REGENCY. Indones J Heal Policy. 2018;7(4).
17. Wulandari S, Sugiarto S. Drug Procurement Model with ABC VEN Method at RS X Semarang. Indones J Heal Manag. 2019;7(3):186–90.
18. Falinda R, Andriana F, Sjaaf AC. Evaluation of the selection, planning and procurement stages of drugs at Kembangan Regional Hospital. Syntax Lit Indones Sci J. 7AD;11.
19. Setianto B, Adriansyah AA, Asih AY. Implementation of Lean Management in the

https://doi.org/10.1051/bioconf/20237505012

Pharmacy Unit of Surabaya Islamic Hospital A. Yani. Indones J Heal Manag. 2020;8(2):81–7.

20. Prasetyo J, Octaviani P, Prabandari R. Analysis of Drug Logistics Management Management at the Pharmaceutical Installation of RSUD dr. R. Goeteng Taroenadibrata Purbalingga. 2021.
21. Tirtabayu HE, Susilo W, Hanevi D. Evaluation of Quality Control and Cost Control Policies, Fraud Prevention and Commitment-Based Capitation in the JKN Era in Bengkulu Province Using a Realist Evaluation Approach. Indones Heal Policy. 2020;
22. Oktaviani N, Pamudji G. Evaluation of drug management at the pharmaceutical installation of the NTB Provincial General Hospital in 2017. Indones J Pharm. 2018;15(2):135–47.
23. Sagala J. INVENTORY RISK ANALYSIS AT UKI RSU PHARMACEUTICAL INSTALLATIONS. SIBATIK J Sci J Soc Econ Cult Technol Educ. 2022;1(4):327–34.
24. Regulation of the Minister of Health of the Republic of Indonesia Number 72 of 2016 concerning Pharmaceutical Service Standards in Hospitals.
25. Saputra WA, Puspandari DA, Kurniawan MF. EVALUATION OF DRUG PROCUREMENT WITH E-PURCHASING THROUGH E-CATALOGUE AT GRHASIA MENTAL HOSPITAL YOGYAKARTA SPECIAL REGION IN 2017 – 2018. Indones J Heal Policy. 2019;8(3).
26. Health MS for. The Selection, Procurement, Distribution and Use of Pharmaceuticals. 1997;
27. Karimah C, Arso S., Kusumastuti W. ANALYSIS OF DRUG MANAGEMENT AT THE PROCUREMENT STAGE AT ROEMANI MUHAMMADIYAH HOSPITAL SEMARANG. J Public Health (Bangkok). 2020;8(2).

Practice skills and compliance of private pharmacies with regulations on the prescription drug: a multi-method study in Vietnam

Thuy Nguyen Thi Phuong[1,#,*], *Binh Thanh* Nguyen[1,#], *Thang Do* Xuan[1], *Trung Quang* Vo[2], *Trung Duc* Nguyen[3], *Thuan Nguyen* Van[4], *Susi Ari* Kristina[5]

[1]Hanoi University of Pharmacy, Hanoi 100000, Vietnam
[2]Faculty of Pharmacy, Pham Ngoc Thach University of Medicine, Ho Chi Minh City 700000, Vietnam.
[3]Faculty of Pharmacy, 108 Military Central Hospital, Hanoi 100000, Vietnam.
[4]Faculty of Pharmacy, Military Hospital 175, Ho Chi Minh City 700000, Vietnam.
[5]Faculty of Pharmacy, Universitas Gadjah Mada, Yogyakarta 55281, Indonesia.

Abstract. Professional practice of pharmacists plays a crucial role in the reinforcement of drug retailers' services to achieve optimal health care provision to customers. To evaluate the professional skills and compliance of retail pharmacy staff with selling prescription drugs by surveying patients' knowledge of drugs and role-playing the customer buying antibiotics without a prescription. A cross-sectional study was conducted with two kinds of surveys at 480 drug retail establishments using the cluster sample technique among 12 provinces/cities in Vietnam. Clients were interviewed to assess their knowledge about drugs. Moreover, the method of acting as a client was used in two common scenarios in order to evaluate the implementation of professional regulations and professional practice skills of drug sellers: a child acute respiratory infection (ARI) case and an amoxicillin case without a prescription. The data were presented as frequency and percentage. The basic tests were used to compare the ratios and means between the two groups. The total number of interviewed customers was 2389 while the figure for role-playing was 960 cases. When customers buy medications with a prescription, 100% of those were fully aware of the dosage of the drugs they purchased, which was higher than the scenario of buying without a prescription (93.1%). In role-play scenarios, the rate of drug sellers asking patients to explore information was higher in the ARI children case than in the amoxicillin case. Besides, 100% of customers were consulted on treatment in both cases, and the rate of advising was at a low rate in both scenarios 3.8% in the amoxicillin case compared to 15.4% in the ARI case. Drug sellers did not respond well to requirements in professional practice and were influenced by economic concerns in business.

1 INTRODUCTION

Common factors used to assess a drug retailer's service are the quality of drug supply, drug use counseling, and optimal health care provided to customers by pharmacists. In which, the pharmacist's professional practice plays a crucial role and requires numerous abilities such as obtaining information from the patient, listening to the patient, giving advice, guiding drug use, and suggesting proper treatment or solutions at drug retail establishments. As a result, the National Pharmaceutical Association in Australia launched two processes named "WHAT-STOP-GO" and "CARER" in the operation of selling drugs to customers buying prescription and over-the-counter drugs for pharmacists and pharmacy staff in order to ensure the implementation of drug selling and safe and rational drug use [1]. According to Good Pharmacy Practices (GPP) regulations issued by the Ministry of Health in Vietnam, one of the main activities of a drug retailer, aside from the procurement of medications and quality control, is to sell drugs including the provision of drugs with counseling and guidance on safe and effective drug use for patients [2].

Results of a systematic review conducted by Rosalline Miller in 2011 on the practice of pharmacists in drug retailers in low- and middle-income Asian countries found that the practice of asking, giving advice or consulting skills was still at a low rate in drug sellers during the process of buying and selling drugs to patients or clients [3]. A study conducted at 128 pharmacies (2010) in Penang, Malaysia, using a role-playing setting involving the purchase of medicine to treat colds and flu found that the percentage of drug sellers who provided advice about dosage and warnings were low, reaching 25.0% and 7.7% respectively in the pharmacist in charge, compared to 5.3% and 0.0% in the rest of the pharmacy personnel. Many other important pieces of information and advice were not mentioned by drug sellers [4]. Furthermore, many low- and middle-income countries had violations of legislation prohibiting the sale of prescription drugs without a doctor's prescription, notably antibiotics [3, 5]. Worldwide, the estimated percentage of antibiotic delivery without a prescription at drug shops would be

# These authors are the co-first authors and contributed equally to this work.
* Correspondence:
Email: thuy_ntp@hup.edu.vn

62.1%. The five nations with the highest proportion of antibiotics administered without a prescription in the survey were India, Indonesia, Vietnam, Syria, and Ethiopia [6]. This has been contributing to the worsening of antibiotic resistance, a critical issue that jeopardizes global health security.

Various studies in Vietnam demonstrated that drug sellers continued to have many limitations while dispensing antibiotics without a prescription at drug retailers. A recent study showed that most of the information obtained before deciding to dispense antibiotics was just to ask customers about symptoms and whether they saw a doctor, with rates ranging from 4.5% to 36.0%. Furthermore, drug allergy history was the information that drug sellers questioned at a relatively low percentage, ranging from 0% to 59.4% [6]. Another one conducted in Thanh Hoa province by Nguyen Van Quan using the role-playing method of customers on the case of purchasing antibiotics without a prescription and in insufficient doses found that only 6.5% of drug sellers questioned the prescription drugs, 28.0% of drug sellers had questions about the causes and symptoms of disease, and 34.0% of drug sellers did not raise any questions. The overall percentage of drug suppliers that provide guidance is 68.0%. Notably, just 13.0% of medication retailers advised customers to take full-dose antibiotics [7]. In 2010, Do Thi Quynh Nga and colleagues discovered that 80.0% of antibiotics were sold without a prescription in urban regions compared to 91.0% in rural areas, and awareness of drug vendors and clients regarding antibiotics was particularly inadequate in rural areas [8].

From 2007 to 2013, a number of published studies first revealed the existence of flaws in the professional practice of drug sellers in pharmacies both before and after the implementation of GPP. However, from 2013 to the present, the assessment of drug sellers' compliance in selling prescription pharmaceuticals has received insufficient attention. As a result, this review must be redone, particularly during the period of enhancing GPP implementation. Therefore, the purpose of this study was to assess the professional skills and compliance of retail pharmacy staff with selling prescription drugs using two methods: assessing the patient's knowledge of the purchased drug and role-playing the customer in the situation of buying antibiotics without a prescription.

2 MATERIALS AND METHODS

2.1 Study design

The cross-sectional study was conducted in 12 provinces/cities across the country, with the deliberate sampling approach used in five provinces/cities under the central government: Hanoi, Hai Phong, Da Nang, Ho Chi Minh City, and Can Tho. The other seven provinces/cities are chosen at random from a list of provinces/cities in six economic zones according to Decree 92/2006/ND-CP. Finally, the study selected 480 drug retail establishments using the cluster sample technique, including 220 drug dispensaries (located in rural regions, professionally responsible by pharmacists with a secondary diploma in pharmacy) and 260 drugstores (placed in urban areas, professionally responsible by the pharmacist who owns a Bachelor of Pharmacy). Researchers conducted two kinds of surveys at each retail establishment: interviewing clients buying drugs and playing the role of clients buying drugs.

2.2 Sampling and data collection

The study chose 5 clients who purchased medications from each drug dispensary /drugstore to interview. Customers were interviewed directly after purchasing medications based on a series of questions that included information about the circumstance of purchasing drugs, kinds of purchased drugs, the customer's understanding of the drug, and background information. During the interview, the surveyor used his phone to take images to support the collection of information about medications. The survey consisted of 35 questions, including 22 about general information about purchased drugs, 5 on customers' understanding of purchased medicines, and 8 about information about clients. A customer was considered to know how to take the medicine if he or she knew the dosage per use, the number of times used in a day, the time of taking the medication, how to administrate drug, and typical warnings and precautions.

Parallel to the survey of customers, the study employed the method of acting as a client in two common scenarios in the community in order to evaluate the implementation of professional regulations in selling prescription drugs and professional practice skills or drug sellers. Investigators were chosen and instructed how to play the role of a customer buying drugs. Following that, they acted at the drugstore model with the record by camera and audio to know how to fill out the questionnaires. Following that, each investigator practiced the play role at three different drug retail establishments. The sample size did not contain these drug dispensaries /drugstores. After being fully trained, the investigators conducted surveys of the drug retail establishments in sample size. These drug retail establishments owners were unaware of the investigator's role. The investigator just gave information as a guide to the scenario and accepted any medicine selection and quantity recommendations made by the drug seller. Within 15 minutes of purchasing pharmaceuticals at the drug retail establishment, the investigator recorded information about the drugs, questions, advice, and the content of drug usage instructions for pharmacy customers and stored it on the information collecting sheet. All phone conversations were recorded, to store information and deleted after the questionnaire was completed. The purchased pharmaceutical bag was prominently labeled with information about the trial site and the name of the drug seller. Each drug retail establishment had two investigators who are involved in the following two scenarios:

Scenario 1. A 4-year-old child had an acute respiratory infection (ARI), and the drug seller received preliminary information concerning the child's cough.

https://doi.org/10.1051/bioconf/20237505013

After examining the research and referring to the scenario in the research conducted in Vietnam by the author group of designer Chuc [9], the scenario was constructed. ARI was a common disease that affected many young children in the community in many places with different socioeconomic conditions. The scenario included the following details: 4-year-old child weighing approximately 17kg, coughing for 2 days, cough frequency 2-3 times/hour, no sputum, runny nose with clear nasal fluid but no stuffy nose, no sore throat, no vomiting, slight heat, mild fever of approximately 38 degrees, slight fatigue, no history of illness/allergies, and have not taken any medication, normal breathing rate, no other abnormalities. The typical client in the child ARI case was a female between the ages of 27 and 35.

Scenario 2. Make an offer to purchase a 500mg tablet of amoxicillin. This was one of the most commonly found antibiotic active components in drug retail establishments. The customer bought antibiotics because he or she had symptoms of cough with a sore throat, mild temperature of about 38oC, thick green sputum, no runny nose; had experienced this before and found amoxicillin to be useful; no history of illness or other health problems.

2.3 Data analysis

Collected data was checked, cleaned, and encrypted and some drug-related information was looked up. The data was entered into Epi data 3.1, and SPSS 20.0 software and Excel Microsoft Office 2013 were used to analyze quantitative data. Categorical variables were presented as frequency and percentage. The chi-square test and Fisher's exact test were used to compare the ratios between the two groups. Besides, the Independent T-test and ANOVA test were applied to compare the mean between two groups of subjects if the variable was normally distributed. If the variable was not normally distributed, the Mann-Whitney test and the Kruskal-Wallis test were used.

2.4 Ethics consideration

The research protocol was approved by the Council of the Ministry of Health and the provincial/municipal Department of Health. The data was encrypted, kept secret and was only used for research purposes.

3 RESULTS

There were 11 questionnaires that missed a lot of information for the survey of drug customers. The total number of votes cast for analysis was 2389. The number of surveys completed and assessed for the role of the customer was 480 cases for amoxicillin and 480 cases for pediatric ARI scenarios.

The majority of drug clients surveyed are female (62.9%), with the age range 35 to 55 having the highest prevalence (45.6%). The bulk of clients (51.6%) were laborers or self-employed or had a high school diploma (29.2%). Moreover, 83.8% of customers bought pharmaceuticals two or more times, and the percentage of customers who bought drugs without a prescription was exceptionally high, accounting for 87.9%.

When customers buy medications with a prescription, 100% of those were fully aware of the dosage of the drugs they purchased, which was higher than the scenario of buying without a prescription (93.1%) and there was a significant statistical different ($p<0.05$). Administration awareness had a similar trend with 95.3% of consumers buying pharmaceuticals with a prescription compared to 81.5% without a prescription. Moreover, 79.6% of customers were aware of the side effects of pharmaceuticals purchased. However, customers' knowledge of drug side effects was very poor in both scenarios of buying pharmaceuticals with and without a prescription, with 4.8%.

Table 1. Characteristics of surveyed customers (N=2389)

Gender		***Professional***	
Male	887 (37.1)	Businessman	1233 (51.6)
Female	1502 (62.9)	Office Workers	690 (28.8)
Age group		Retired or Housewife	331 (13.9)
8-34	930 (38.9)	Working in the healthcare field	61 (2.6)
35-55	1089 (45.6)	Other	62 (2.6)
> 55	357 (15.0)	Missing value	12 (0.5)
Missing value	13 (0.5)	***Type of buying drugs***	
Qualification		With a prescription	288 (12.1)
Lower secondary education or below	582 (24.4)	Without a prescription and with prescription drug	601 (25.1)
High School	697 (29.2)	Without a prescription and prescription drug	1500 (62.8)
Associate degree	571 (23.9)	***Frequency of buying drugs at the establishment***	
≥ Bachelor's degree or above	521 (21.8)	First time	372 (15.6)
Missing value	18 (0.7)	Second time or more	2002 (83.8)
		Missing value	15 (0.6)

Table 2. Knowledge of customers about drugs in the survey

Characteristics	With a prescription (n=288)	Without a prescription (n= 2101)	p-value	Total (N=2389)
Dosage	288 (100)	1956 (93.1)	0.017	2255 (93.9)
Administration	285 (95.3)	1713 (81.5)	0.027	1998 (83.3)
Indication	240 (83.3)	1663 (79.1)	0.111	1003 (79.6)
Adverse reactions	21 (7.3)	95 (4.5)	0.179	116 (4.8)

Table 3. Professional practice skills of drug sellers in the Amoxicillin case

Characteristics	*Drugstores (n=260)*	*Drug dispensaries (n=220)*	*P-value*	*Total (N=480)*
Questioning skill				
Have an asking	89 (34.2)	56 (25.5)	0.037	145 (30.2)
Number of questions, mean (SD)	*0.60 (1.17)*	*0.34 (0.73)*	0.018	*0.48 (1.0)*
Min-Max	0-9	0-5	-	0-9
Content of questions				
Have you seen a doctor/ Have you had a prescription?	-	-	-	-
Subjects taking drugs	36 (13.8)	20 (9.1)	0.106	56 (11.7)
Cough (frequency/time/duration)	15 (5.8)	5 (2.3)	0.05	20 (4.2)
Cough nature (sputum/dryness)	5 (1.9)	2 (0.9)	0.35	7 (1.5)
Nose (runny/stuffy nose)	3 (1.2)	-	0.11	3 (0.6)
Throat/pain when swallowing	9 (3.5)	7 (3.2)	0.86	16 (3.3)
Fever	3 (1.2)	-	0.11	3 (0.6)
Risky signs (wheezing, fatigue/weakness, vomiting/chest indrawing)	-	-	-	-
History of allergies	1 (0.4)	-	0.357	1 (0.2)
Medicines/remedies used	1 (0.4)	-	0.357	1 (0.2)
Medical history/related comorbidities	1 (0.4)	-	0.192	1 (0.2)
The type of medicine you want to buy / dosage form for child	47 (18.1)	25 (11.4)	*0.041*	72 (15.0)
The need to take antibiotics	-	-	-	-
Customer's desired number of days of taking medication	-	-	-	-
Reasons to buy medicine	4 (1.53)	4 (1.81)	0.372	7.3 (1.52)
Consulting skills				
Have a consultation	*49 (18.8)*	*46 (20.9)*	0.572	*95 (19.8)*
Content of consultation				
Go to the doctor	-	-	-	-
Written instructions	-	-	-	-
When have to change medicine	*4 (1.6)*	*4 (1.8)*	0.832	*8 (1.7)*
Dosage per time	44 (16.9)	40 (18.2)	0.718	84 (17.5)
Number of times used per day	39 (15.0)	38 (17.3)	0.499	77 (16.0)
When taking medicine	24 (9.2)	22 (10.0)	0.775	46 (9.6)
When taking medicine compared to meals	24 (9.2)	23 (10.4)	0.775	47 (9.8)
Total days of medication	21 (8.1)	18 (8.2)	0.967	39 (8.1)
Effects of drugs	-	-	-	-
Side effects of drugs	-	-	-	-
Should use enough antibiotic treatment time (if any)	*3 (1.2)*	*1 (0.5)*	0.394	*4 (0.9)*
Advice skill				
Have advice	13 *(5.0)*	5 *(2.3)*	0.117	18 *(3.8)*
Content of advice				
Signs that you need to see a doctor	-	-	-	-
Nose and throat hygiene	-	-	-	-
Keep nose and throat warm	*2 (0.8)*	*1 (0.5)*	0.652	*3 (0.6)*
Diet	*2 (0.8)*	-	0.188	*2 (0.4)*
Other advice	*9 (3.6)*	*2 (0.9)*	0.07	*11 (2.3)*
Dispense results				
Only amoxicillin	250 *(96.2)*	207 *(94.1)*		457 (95.2)
Amoxicillin and other drugs	*7 (2.7)*	*3 (1.4)*		10 *(2.1)*
Other antibiotics	3 *(1.2)*	10 (4.5)		13 *(2.7)*

*SD: standard deviation

In the circumstance of asking for buying amoxicillin, only 30.2% of drug sellers asked clients before selling pharmaceuticals; there was a statistically significant difference in this rate between drugstores (34.2%) and drug dispensaries (25.5%) (p=0.037). Furthermore, the average number of questions asked by the drug sellers at the pharmacy was larger (0.60 vs. 0.34), and the difference was statistically significant (p=0.018). When it comes to the content of the question, no drug retail establishment asked for a prescription and only 11.7% of pharma sellers requested information about subjects taking the drug when customers requested to acquire amoxicillin. The most asked information was cough and painful throat when swallowing with 4.2% and 3.3%, respectively. Meanwhile, only 0.6% of information on fever and runny nose/stuffy nose was extracted. With only three sellers asking, information about the patient's history was also underutilized. However, the information that sellers were interested in was the form of drug they want to buy/the form that children can use, which accounts for 15% of the total.

There were no drug sellers persuading customers to see a doctor when they wanted to get amoxicillin for personal use. Regarding the content of drug usage counseling, 80.2% of drug sellers did not provide it and none of them provided written directions for use. When selling amoxicillin, use guidance for customers was relatively limited. The one-time dose and number of times per day with the highest counseled content were only 17.5% and 16.0%, respectively. The frequency per day and the total number of days of taking drugs were lower, at 9.6% and 8.1% in turn. It is noticeable that there was no pharmacist giving information about the drug's effects and side effects. There is no statistically significant difference between the pharmacy and the pharmacy in the counsel content.

Drug sellers advised on warning indicators or non-drug measures in 3.8% of cases. Only 0.6% of them advised keeping the nose and throat warm/avoiding colds, while 0.4% of them advised a balanced diet. However, there was no guidance on when to see a doctor or how to keep your nose and throat clean. Overall, Amoxicillin was offered on-demand by 95.2% of drug sellers, sold with other medications by 2.7%, and replaced by other antibiotics by 2.7%.

Table 4. Professional practice skills of drug sellers in the children ARI case

	Drugstores (n=260)	***Drug dispensaries (n=220)***	***P-value***	***Total (N=480)***
Questioning skill				
Have an asking	253 (97.3)	216 (98.2)	0.524	*469* (97.7)
Number of questions, mean (SD)	4.3 (1.8)	*4.0* (1.8)	0.041	*4.1* (1.8)
Min-Max	0-8	0-8		0-8
Content of questions				
Have you seen a doctor/ Have you had a prescription?	2 *(0.8)*	6 *(1.2)*	0.095	8 *(1.7)*
Subjects taking drugs	235 (90.4)	195 (88.6)	0.532	430 (89.6)
Cough (frequency/time/duration)	180 (69.5)	136 (61.8)	0.08	316 (66.0)
Cough nature (sputum/dryness)	155 (59.6)	88 (40.0)	<0.001	243 (50.6)
Nose (runny/stuffy nose)	159 (61.2)	129 (58.6)	0.57	288 (60.0)
Throat/pain when swallowing	27 (10.4)	9 (4.1)	<0.001	30 (6.2)
Fever	125 (48.1)	94 (42.7)	0.27	219 (45.6)
Risky signs (wheezing, fatigue/weakness, vomiting/chest indrawing)	2 (0.8)	3 (1.4)	0.52	5 (1.0)
History of allergies	4 (1.5)	4 (1.8)	0.811	8 (1.7)
Medicines/remedies used	17 (6.5)	20 (9.1)	0.308	37 (7.7)
Medical history/related comorbidities	1 (0.4)	7 (3.2)	0.017	8 (1.7)
The type of medicine you want to buy / dosage form for child	73 (28.1)	54 (24.5)	0.382	127 (16.5)
The need to take antibiotics	7 (2.7)	9 (4.1)	0.395	16 (3.3)
Customer's desired number of days of taking medication	20 (7.7)	25 (11.4)	0.169	45 (9.4)
Reasons to buy medicine	-	-	-	-
Consulting skills				
Have a consultation	219 (84.2)	188 (85.5)	0.799	407 (84.8)
Content of consultation				
Written instructions	47 (18.1)	41 (18.6)	0.875	*88 (18.3)*
Go to the doctor	3 (1.3)	4 (2.0)	0.548	7 (1.6)
When have to change medicine	-	-	-	-
Dosage per time	180 (69.2)	179 (81.4)	0.002	359 (74.8)
Number of times used per day	200 (76.9)	172 (78.2)	0.742	372 (77.5)
When taking medicine	180 (69.2)	153 (69.5)	0.941	333 (69.4)
When taking medicine compared to meals	113 (43.5)	108 (49.1)	0.218	221 (46.0)
Total days of medication	70 (26.9)	64 (29.1)	0.598	*134 (27.9)*
Effects of drugs	20 (7.7)	8 (3.6)	0.059	28 (5.8)
Side effects of drugs	6 (2.3)	5 (2.3)	0.98	*11 (2.3)*

Should use enough antibiotic treatment time (if any)	9 (8.7)	6 (7.1)	0.687	15 (7.9)
Advice skill				
Have advice	35 (13.5)	39 (17.9)	0.197	74 (15.4)
Content of advice				
Signs that you need to see a doctor	5 (2.1)	3 (1.4)	0.622	8 (1.8)
Nose and throat hygiene	10 (4.1)	12 (5.8)	0.409	22 (4.9)
Keep nose and throat warm	6 (2.5)	7 (3.4)	0.561	13 (2.9)
Diet	14 (5.8)	23 (11.1)	0.057	37 (8.2)
Other advice	46 (19.0)	37 (18.2)	0.823	83 (18.7)
Dispense results				
Total product sold	888	796		1684
Drugs	623 (70.2)	619 (77.8)		1242 *(73.8)*
Dietary supplements	45 (5.1)	37 (4.6)		82 *(4.8)*
Unidentified Drugs	220 (24.7)	140 (17.6)		360 *(21.4)*
Number of products, mean (SD)	*3.42 (1.67)*	*3.62 (1.59)*	*0.333*	*3.51 (1.64)*
Number of establishments selling prescription drugs without a prescription	*189 (72.7)*	*188 (85.5)*	0.001	*377 (78.5)*
Number of establishments selling antibiotics without a prescription	*154 (59.9)*	*160 (71.2)*	0.032	*314 (65.8)*
Number of establishments selling corticoids without a prescription (n=359)	*49 (26.8)*	*73 (41.5)*	0.004	*122 (34.0)*

*SD: standard deviation

In the case of ARI in children, 97.7% of drug suppliers asked consumers before deciding to offer medications. In which, the mean number of questions was 4.1 (SD=1.8). Notably, drug sellers at the drugstores asked more inquiries (4.3 ± 1.8) than at the drug dispensaries (4.0 ± 1.8), and the difference was statistically significant (p=0.041). The percentage of drug sellers inquiring about taking their children to see a doctor was relatively low, accounting for only 1.7 %. 89.6% of drug sellers asked for information about children. Cough features were the most frequently considered (66%), followed by nasal symptoms (60%), and the form of dry cough/sputum (50.6%). The percentage of drug sellers in the drugstores who inquired about the type of dry cough or phlegm cough (59.6%) was greater than in the drug dispensaries (40.0%), and the difference was statistically significant (p<0.001). Fever, on the other hand, was noted by only 45.6% of drug vendors. Only 1.7% of medicine sellers asked for essential information about the patient's history of allergies and comorbidities. The percentage of drugs and measures employed was 7.7%. Furthermore, drug sellers primarily inquired about the requirement for dosage form (syrup or tablet) for children (16.5%), the number of days to take the drug that consumers wanted to purchase (9.4%), and whether the child required antibiotics (3.3%).

There were 407 drug shops providing counsel when selling medications, accounting for 84.8% of the total. Only 18.3% of ARI medicine sellers had written medication instructions. The counseled information with the largest proportion of drug sellers was the dosage of drugs (74.8%) and the number of times used in a day (77.5%), followed by the content of instructions on when to take the drug during the day (69.4%) and when to take the medicine in relation to the meal (46.0%). The least discussed aspect of counseling was drug side effects (2.3%). Notably, 7.9% of medicine vendors recommended that when providing antibiotics to young children, the entire course of treatment must be completed.

A total of 74/480 (15.4%) drug sellers actively provided advice. Diet was the most frequently advised content (8.2%). Furthermore, only 1.6% of drug vendors suggest that children should visit a doctor, while 1.8% of them advised watching for signals that children should see a doctor.

In final, the average number of products consulted and sold was 3.51 ±1.64. On average, sellers sold more products at the drugstore than at the drug dispensaries, but the difference is not statistically significant. Dietary supplements in the form of syrup or oral solution account for 4.8% of the total 1684 goods, while 21.4% of the total medications cannot be identified since they are retailed in vials/opened from blister packs without any information. Notably, sellers at drug dispensaries sold the majority of prescription medications (85.5%), antibiotics (71.2%), and corticosteroids (41.5%). The difference was statistically significant (p<0.05) when compared to the pharmacy.

4 DISCUSSION

The study assessed drug sellers' professional practice skills by playing a role as clients to buy pharmaceuticals without a doctor's prescription. The method of acting as a consumer was applied correctly and delivered valuable data while maintaining objectivity and consistency among drug retail establishments in the examined areas. According to the findings of the study, drug sellers' abilities in the case of purchasing amoxicillin antibiotics and ARI for children were still quite limited.

The first and most important talent in medication sales is questioning ability, which is critical in collecting and assessing the patient's health situation and making proper decisions. According to the findings, 69.8% of drug sellers had no questions before providing the antibiotic amoxicillin to suit the needs of clients. This finding was consistent with findings from Thai and Egyptian studies that used a role-playing scenario in which participants were asked to acquire a prescription medicine [10]. It is probable that this is a common antibiotic that has been randomly asked by many customers to buy, and when the drug sellers believe that it is unnecessary to inquire further about the reason for purchase. In the case of amoxicillin, drug dealers at drugstores performed information extraction activities better (34.2%) than at drug dispensaries (25.5%). Notably, the content asking about the type of drug that customers wanted to buy had the highest rate (15.0%). Moreover, only 1 drug seller mentioned the patient's allergic history before selling amoxicillin, which was lower when compared with similar studies in Saudi Arabia (7.3%) [11], Ethiopia (4.8%) [10].

The circumstance of buying cough medicine for small children is a regular situation at drug retail establishments because people in Vietnam do not have the habit of taking their children to visit a doctor when they have mild respiratory symptoms. In contrast to the case when requesting specific drugs, 97.7% of drug sellers actively asked inquiries to gain more information about the patient before making treatment decisions in this situation. This finding was higher than that of Nguyen TK Chuc [9] and studies in Saudi Arabia [12] và Spain [13]. Besides, 10.4% of drug sellers did not inquire about the child's age, and 63.3% did not ask about the child's weight. This is the most basic information about drug users but they have not been researched by drug sellers before deciding on drug doses for children. However, when compared to studies in Saudi Arabia and Indonesia [11, 14], our study's rate was still higher. Risk factors are crucial markers for drug sellers to determine the degree of common cold respiratory infection with pneumonia, but only 1.7% of drug sellers were interested in exploiting, which was significantly lower than Chuc's prior research did (11%) [9]. This demonstrated that medication sellers were not cautious in their information exploit before selling drugs for customers with common diseases/symptoms.

Eighty-two percent of drug sellers sold amoxicillin with no verbal or written instructions for use. This figure was higher than those found in Ethiopia and Arabia [11, 15]. In the ARI case, 77% of drug sellers instructed drug usage content such as dose and number of times per day. Only 27.9% of drug sellers advised the total number of days spent taking the medicine, and only 5.8% of them gave information about the drug's effect. These rates were all lower than those seen in similar research conducted in Ethiopia [15].

When drug dealers' treatment decisions for ARIs were reviewed, antibiotic misuse (65.2%) with mild respiratory infections and corticosteroid abuse (34.0%) were found. The percentage of drug sellers utilizing corticosteroids for children was significantly greater than the earlier study by Chuc, which found a corticoid rate of only 1% [9]. Such poor treatment choices raise the risk of adverse events and antibiotic resistance. When treating children in drug dispensaries, the percentage of drug dealers selling corticosteroids and antibiotics without a prescription was greater than in drugstores, and the difference was statistically significant. According to previous research findings, drug sellers in rural areas had a limitation of knowledge of antibiotic use than in urban areas [8].

Clients' lack of awareness about the purchased medicines demonstrated a number of weaknesses in the professional practice skills of drug sellers. The percentage of consumers who knew the dose and time of use was higher with prescriptions than the percentage of customers purchasing drugs without a prescription, and the difference was statistically significant. This can be explained by the fact that the consumer who had a prescription was instructed by the doctor to administrate drug properly. Moreover, this was most likely due to the fact that when instructing drug use, drug sellers frequently guided the dose and time of taking the medicine, while the content of adverse effects was rarely addressed.

As a result, it is critical to conduct communication and education for people in order to improve their awareness of safe and effective drug usage. Before making treatment or medication recommendations for both individual drug purchases and symptomatic conditions, drug sellers must exploit adequate information. When selling pharmaceuticals, drug sellers must pay attention to primary patient health, as well as rigorously adhere to legal rules in practice at drug retail establishments in accordance with GPP standards. It is vital to take steps to increase patient knowledge, and accountability and professional ethics of drug sellers, particularly through counseling on safe and effective drug use.

5 CONCLUSION

According to the findings of the research, drug sellers did not respond well to requirements in professional practice, made improper treatment decisions, overuse antibiotics and corticosteroids, and were influenced by economic concerns in business. As a result, the principles of GPP practice had ineffective in improving and overcoming the existence of professional problems in the practice of drug sale; therefore, stronger measures to handle violations were required, as well as strengthening education and improving professional skills for the drug sellers.

ABBREVIATION

GPP: Good Pharmacy Practices; ARI: acute respiratory infection; SD: standard deviation.

ACKNOWLEDGMENTS

This analysis is supported by the participants during the study design and data collection.

 https://doi.org/10.1051/bioconf/20237505013

AUTHORS CONTRIBUTIONS AND DETAILS

All authors critically reviewed, revised the manuscript, and gave the final approval of the version to be published.

FUNDING AND COMPETING INTERESTS

The authors declare that they have no competing interests in this analysis. This research did not receive any specific grant from funding agencies in the public, commercial, or not-for-profit sectors. The authors have no other conflict of interest to declare.

REFERENCES

1. Pharmaceutical Society of Australia, *The Pharmacy Guild, 2013*. 2013.
2. G. P. Mohanta, et al., *Achieving good pharmacy practice in community pharmacies in India.* American journal of health-system pharmacy, 2001. **58**(9): p. 809-810.
3. Rosalind Miller and Catherine Goodman, *Performance of retail pharmacies in low-and middle-income Asian settings: a systematic review.* Health policy and planning, 2016. **31**(7): p. 940-953.
4. Chin Fen Neoh, et al., *Nature and adequacy of information on dispensed medications delivered to patients in community pharmacies: a pilot study from Penang, Malaysia.* Journal of Pharmaceutical Health Services Research, 2011. **2**(1): p. 41-46.
5. Felicity Smith, *The quality of private pharmacy services in low and middle-income countries: a systematic review.* Pharmacy world & science, 2009. **31**(3): p. 351-361.
6. N. T. P. Thúy, et al., *Cung cấp kháng sinh không có đơn tại các cơ sở bán lẻ thuốc trên thế giới: Tổng quan hệ thống và phân tích gộp.* Nghiên cứu dược và thông tin thuốc, 2019: p. 2-10.
7. Quân Nguyễn Văn, *Đánh giá kỹ năng thực hành của nhân viên nhà thuốc đạt nguyên tắc, tiêu chuẩn GPP trên địa bàn tỉnh Thanh Hóa giai đoạn 2012 - 2014.* Tạp chí Dược học, 2015. **2**.
8. Nguyen Thi Kim Chuc, et al., *Antibiotic sales in rural and urban pharmacies in northern Vietnam: an observational study.* BMC Pharmacology and Toxicology, 2014. **15**(1): p. 1-10.
9. Nguyen T. K. Chuc, et al., *Management of childhood acute respiratory infections at private pharmacies in Vietnam.* Annals of Pharmacotherapy, 2001. **35**(10): p. 1283-1288.
10. Abdullah I. Abdelaziz, et al., *Quality of community pharmacy practice in antibiotic self-medication encounters: a simulated patient study in upper Egypt.* Antibiotics, 2019. **8**(2): p. 35.
11. Sinaa Alaqeel and Norah O. Abanmy, *Counselling practices in community pharmacies in Riyadh, Saudi Arabia: a cross-sectional study.* BMC health services research, 2015. **15**(1): p. 1-9.
12. Aref A. Bin Abdulhak, et al., *Non prescribed sale of antibiotics in Riyadh, Saudi Arabia: a cross sectional study.* BMC public health, 2011. **11**(1): p. 1-5.
13. Carl Llor and Josep Maria Cots, *The sale of antibiotics without prescription in pharmacies in Catalonia, Spain.* Clinical Infectious Diseases, 2009. **48**(10): p. 1345-1349.
14. Nazanin Foroutan and Fatemeh Dabaghzadeh, *Evaluating the practice of Iranian community pharmacists regarding oral contraceptive pills using simulated patients.* Pharmacy Practice (Granada), 2016. **14**(4).
15. Wubante Demilew Nigussie and Wondemagegn Mulu Lingerew, *Quality of labeling on dispensed medicines and associated factors among public health care facilities in Bahir dar city, Northwest Ethiopia.* Int. J. Pharma Sci. Res, 2014. **1**: p. 47-55.

Prescriptions Conformity with National Formulary: A Literature Review to Explore the Need for Pharmaceutical Cost Containment

Oskar Skarayadi[1], *Dwi* Endarti[2]*, *Ali* Ghufron Mukti[3] and Satibi[2]

[1]Universitas Jenderal Achmad Yani, Faculty of Pharmacy, Pharmacist Professional Study Program, Cimahi, Indonesia
[2]Universitas Gadjah Mada, Faculty of Pharmacy, Department of Pharmaceutics Yogyakarta, Indonesia
[3]Universtas Gadjah Mada, Faculty of Medicine, Public Health and Nursing, Gadjah Mada Medical Centre Department, Yogyakarta, Indonesia

Abstract. The government has created the National Formulary (Fornas) as a tool for control of quality and cost control. Patients' prescriptions should be based on the national Formulary during the era of national health coverage. If prescriptions complied with the national Formulary, the cost would reduce. This research aimed to determine the factor of non-conformity between prescriptions and the national Formulary measures it and explores the need for pharmaceutical cost containment. This study applied the Preferred Reporting Items for Systematic Reviews and Meta-Analyses (PRISMA) method between 2014 and 2022. The search terms "Kesesuaian Fornas" were utilized on Google Scholar, Pubmed, ScienceDirect, and Scopus. The evaluation was organized by factor and subsequent action. Searches of the databases Google Scholar (2,246), Pubmed (0), ScienceDirect (1), and Scopus (0) yielded a total of 247 results. Following the selection procedure, 34 articles were included in the analysis. The average conformity of prescription with Fornas was 71.02%. Prescription adherence, drug availability, and formulary Conformity with Fornas were the primary factors in the conformity of prescriptions with the National Formulary. Conformity with the national Formulary for prescriptions can be enhanced through three primary actions: increasing prescription adherence, updating the Formulary, and procuring following the electronic catalog.
Keyword: Healthcare Cost, Prescription, National Formulary, Pharmacy, Medicine, Indonesia,

1 BACKGROUND

Health Insurance. Implementing quality control and cost Drug management is essential in the era of National Health Insurance. Implement quality control and cost control. The government has established the National Formulary (Fornas) as a quality control tool and the drug E-catalogue as a price control tool [1]. Fornas is a list of needed and available medications for patients in healthcare facilities. The Formulary is very important because it can improve the quality of health services and the cost-effectiveness of treatment, and it can assist health facilities in understanding the long-term needs and priorities for enhancing the quality and safety of drug use [1],[2]. Formularies can undoubtedly exert a strong influence on prescribing decisions and medication use. When a formulary is utilized effectively, it becomes the critical element of a formulary system, one of the most effective means of ensuring rational drug therapy and controlling drug costs [3].

The conformity of the prescription with Fornas stipulated in a Decree of the Minister of Health of the Republic of Indonesia, where the percentage of the formulary in writing the prescription obtained is considered acceptable if it meets the minimum hospital service standards, requires that the acceptance of the prescription with the formulary be 100 per cent.[4]. The National Formulary (Fornas) implementation process as quality control and cost control in drug administration still has many problems. As a result, the goal of Fornas can not reach the optimum result [1].

Non-conformity between Fornas and prescriptions resulting from non with the formulary causes drug shortages or blanks. On the other hand, there are empty drug stocks, and fulfilling more types of drugs than the standard requires a greater investment. In addition, the service time will increase, prescriptions will be rejected, drug prices will rise, drugs will be unavailable, treatment continuity will remain stagnant, and the total cost of treatment will be high. [5]. There is a significant relationship between the conformity of drug prescriptions with the national Formulary and the difference in drug costs[6]; besides that, it can also cause the total cost of treatment to increase [5]. This study aims to determine whether health institutions can achieve basic service criteria based on Conformity with the National Formulary and explores the need for pharmaceutical cost containment. What distinguishes this research from previous research is that no research has reviewed many articles regarding the relationship between the national Formulary and hospital

* Corresponding email: endarti_apt@ugm.ac.id

prescriptions. This study may provide insights for policymakers and related parties so that every receipt made by hospital doctor will pay attention to write receipts that must conform with the National Formulary.

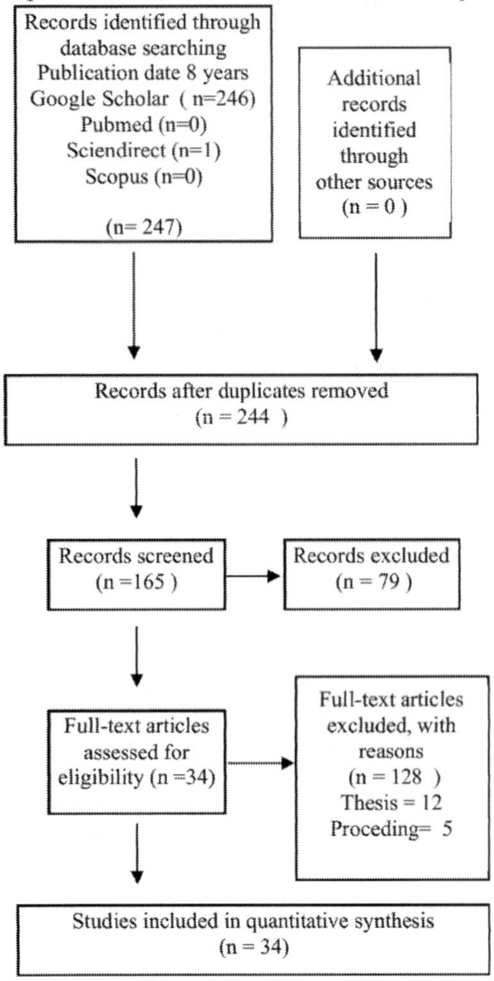

Fig. 1. PRISMA. Flow Diagram

2 METHOD

This research is a literature review to determine the conformity of the prescription with Fornas. Three primary phases were involved in acquiring the data: first, we identified and selected relevant studies on our issue. Second, we evaluate the retrieved research that met all qualifying requirements. Then, for each study, we conducted a review and data extraction.

Relevant research studies were located through an extensive search from 4 electronic databases (Google Scholar, Pubmed, ScienceDirect, and Scopus) on August 2022. We conducted a literature search for published articles in Indonesian and English that reported on prescription with Fornas before recent years (2014 - 2022). the literature review was conducted by entering the keyword "Kesesuaian Fornas." Only articles with well-specified data for the variables of interest were reviewed. Inclusion criteria include a publication detailing the between prescription and national Formulary (Fornas) from health facilities serving BPJS. The selection of studies was guided by PRISMA (Preferred Reporting Items for Systematic Reviews and Meta-Analyses) diagrams. It showed in Figure 1.

The first author extracted data. The data extraction form included details of the study, health Facilities, region of BPJS., the drugs item, factors, , and follow-up. The author's discussion handled disagreements or differences over data extraction. About 247 studies were initially collected and reviewed. After further tracking of references from the retrieved studies, only 34 studies were finally analyzed that met all eligibility criteria.

3 RESULT AND DISCUSSION

According to the literature review, only five health facilities (13,51%) had full prescriptions compliant with the national formulary. The number of facilities is more than 34 because, in two journals, research was done in two health facilities. The average prescription conformity for the national formulary was 71.82%. The review results that contain the Region of BPJS, health facilities, and Conformity of prescriptions with the national Formulay can be seen in Table 1.

This article discusses the factors that lead to prescribing differences from the National Formulary. The review identified ten factors whose specifics are listed in Table 2.

The conformity of prescription with Fornas substantially impacts the service quality at pharmaceutical facilities [32], [37], [43]. Prescriptions for BPJS. Patients use medications not listed in the National Formulary (Fornas) [33] and hospital formulary (FRS), resulting in patient complaints because they must purchase these drugs despite being from a lower socioeconomic status [33]. This occurs because drug prescribing has not completely utilized the Fornas restriction provisions[35]. Non-Fornas drugs predominated in the doctor's prescription pattern, which led to a predominance of these non-Fornas drugs in the subsequent period's medication procurement. Therefore, at the end of the year, when the non-Fornas drugs that were acquired previously are still in the storage facility, the drug inventory balance will be elevated[38]. There are differences between prescription writing and the national Formulary due to patients in emergencies or patients with conditions that prevent them from receiving drugs according to the national Formulary, as well as patient requests to prescribe certain drugs [44].

The availability of medicines in hospital formularies does not meet the National Formulary's standards for drug availability [21], [22], it has impact patients, so patients fail to get therapy [45]. Drug shortages in distributors can lead to doctors writing prescriptions that don't follow the formularies for drug supplies because the names of the drugs recommended have no substitutes that are the same in content as the drugs in the formulary [46] and the inability to find generic drug distributors [11]. The non-conformity of physicians' prescription writing for formularies causes drug stocks to be depleted, inadequate, or excessive [46]. Another problem is that drugs are unavailable because they can only be used in level 2 health facilities [17], and the Primary health care remains to acquire medications from the District Pharmacy Installation, which administers medications using the APBD and APBN. Budgets [47].

Table 1. Facilities for study setting

Reference	Health Facility	Region	Conformity Prescription with Fornas (%)	
[7]	Primary Health Care	1	PHC1	77,01%.
[8]		1	PHC2	73,14%
[9]		1	PHC3	76,58%
[10]		1	PHC4	100%
[11]		1	PHC5	31,96%
[12]		1	PHC6	49,04%
[13]		1	PHC7	80.34 %
[14]		1	PHC8	85,81%
[15]		3	PHC9	70%
[16]		3	PHC10	92,36%
[17]		3	PHC11	50,14%
[17]		3	PHC12	56,85%
[18]		4	PHC13	100%
[19]		4	PHC14	56,29%
[20]		4	PHC15	86,07%
[14]		5	PHC16	73,69%
[21]		5	PHC17	68,04%
[22]		5	PHC18	80±2.1%
[23]		5	PHC19	41,63%
[17]		3	PHC20	61,77%
[24]	Type B Hospital	4	H1	72,18%.
[25]		1	H2	100%
[26]		1	H3	86,1%.
[27]	Type C Hospital	1	H4	61,76%
[28]		1	H5	100%,
[29]		1	H6	Generic drugs 99,58%., Branded drugs 97,21%.
[30]		1	H7	90,37%
[31]		1	H8	99,68%.
[32]		3	H9	71%.
[33]		4	H10	68,35%,
[34]	Type D Hospital	1	H11	57,04%
[35]		1	H12	71,22%.
[36]		1	H13	63%
[37]	Hospital	1	H14	91,70%
[38]		1	H15	93,04%.
[39]		-	H16	52,76%
[40]	Public health Office	5	PH1	90,30%
[17]		3	PH 2	61,59%,

Table 2. The Factor of Prescription with Fornas

Conformity Factors	Number Of Studies	Studies
Prescription adherence	16	[8]–[10], [12], [17], [19], [26], [29]–[35], [37]–[39]
Drug Availability	12	[11]–[17], [21]–[24], [26]
Conformity of the Formulary with the Fornas	13	[14], [18], [23]–[26], [28], [30], [34], [37]–[39], [41]
Doctor's habit	6	[9], [11], [23], [33], [35], [38]
Clinical condition	5	[10], [18], [35], [37], [42]
Procurement	5	[9], [15], [24], [27], [38]
Budget	4	[15], [22], [26], [28]
Doctor knowledge	3	[19], [29], [31]
Human resources	1	[22]
Distribution	1	[22]
Substitution	1	[33]

Doctors prescribe several drugs but are not yet included in Fornas and hospital formulary [14], [18], [28], [41]. The conformity rate of medicines based on Fornas and hospital formularies is still below the standard because hospital formularies do not possess all the Fornas-listed medications [30]. After all, the hospital formulary has not been updated [26], [38].The selection of Hospital formulary drugs that adjust drug items to the national Formulary is still not up to standard [24]. Adaptation of the hospital formulary to the present condition of pharmaceutical services and medications required by patients [34].

Due to the lack of adherence, commitment, and consistency of physicians in writing prescriptions according to the Formulary, there is still a lack of knowledge of the Formulary, and there is no force from hospital leadership toward physicians who write prescriptions outside the Formulary [38].

Depending on the therapy protocol, physicians prescribe medications differently, which leads to inconformity with Fornas [9], [33], [35]. Conditions of non-conformity can also come from the choice of the first-line drug [10]. Other reasons for non-conformity include patients in emergencies, the condition of patients who cannot receive medicines according to the national Formulary, and patient requests to prescribe specific medications [37]. Another reason doctors prescribe drugs that aren't in the Hospital's Formulary is that they assume that the drugs they prescribe are more effective for patients than those in the Hospital's Formulary. This inconformity is especially true for patients with a history of certain diseases only compatible with the drug [38].

The big difference between prescriptions and the national Formulary is that drug procurement policies aren't always followed [9]. For example, the Barito Kuala Regency formulary includes items not yet in the

National Formulary [19]. Another reason is that the suitability of buying drugs with the e-catalogue has not yet reached a common value [24]. Planning for drug needs at Primary health care is still based on how diseases spread, which is how the public health office plans and purchases drugs [27].

Prescription differences with the national Fornas may occur because the DAK (Special Allocation Fund) budget is the only source of funding[15], and the Health Service does not purchase all medicines included in the national Formulary. The limited budget for drug spending[26], [28] means that not all medicines included in the national Formulary are procured by the Health Service [15].

There is a difference between the prescription and the national Formulary due to a prescribing doctor's lack of understanding and socialization regarding the contents of the Hospital formulary and drug names, causing prescription differences with the national formulary[46]. If dissemination of the contents of the national Formulary to doctors is not maximized, then the number of prescriptions following the national Formulary is low [19].

A lack of transportation and drug distribution funding caused low availability and inconformity between prescriptions and national Fornas. Because there is no written recommendation from the health office as a policy requirement for distribution costs, not all Primary health care have budgeted, resulting in inadequate and unequal distribution of pharmaceuticals. Primary health care has always brought the medicines to the district pharmacy facility [22]. There is a shortage in the stock of medicines listed in Fornas because they are not available from distributors; as a result, they have adopted a policy of taking drugs with trade names to fulfill the patient's demand for these medications, thereby enhancing pharmacy services [33]. Inadequate pharmaceutical human resources result in the availability of medications with non-standard Fornas and disease patterns [22].

A suitable method is required to enhance prescription and national formulary conformity. This review explores strategies for improving the conformity of medicines with the National Formulary. It is shown in Table 3.

Table 3. Strategy to improve the prescription Conformity with the National Formulary

Follow Up Action	Number of Studies	Studies
Improving Prescription Adhérence	12	[8], [17], [23], [26], [29]–[32], [37]–[39]
Updating the Formulary	7	[15], [17], [25], [28], [34], [38], [39]
Procurement according to e catalogue	4	[11], [21], [22]
Improving H.R.'s knowledge and skills	3	[22], [25], [38]
Improving I.T	1	[22]
Distribution Cost	1	[22]

4 CONCLUSION

The average prescription conformity with Fornas was 71.02 percent. Adherence to prescriptions, drug availability, and formulary Conformity with Fornas was the primary determinant of prescription conformity with the National Formulary. Conformity with the national Formulary for prescriptions can be improved through three primary actions: increasing prescription adherence, updating the Formulary, and following the electronic catalog when purchasing.

References

1. S. Winda, 'Formularium Nasional (FORNAS) dan e-Catalogue Obat Sebagai Upaya Pencegahan Korupsi dalam Tata Kelola Obat Jaminan Kesehatan Nasional (JKN)', *Integritas*, vol. 4, no. 2, p. 30, 2018, doi: 10.32697/integritas.v4i2.328.
2. Kemenkes RI, 'Keputusan Direktur Jenderal Bina Kefarmasian Dan Alat Kesehatan Nomor Hk.02.03/Iii/1346/2014 Tentang Pedoman Penerapan Formularium Nasional', *J. Chem. Inf. Model.*, vol. 01, no. 01, pp. 1689–1699, 2014.
3. B. J. Mahendrakumar, M. Sowmya, U. M. D, R. Hymavathi, and S. Ramesh, 'Hospital Formulary : An Overview', *Indian J. Pharm. Pract.*, vol. 6, no. 1, pp. 1–5, 2013, [Online]. Available: http://www.ijopp.org/sites/default/files/IJOPP_6_1_2013_2.pdf
4. Kemenkes RI, 'Menteri Kesehatan Republik Indonesia Nomor: 129/Menkes/SK/II/2008', 2008.
5. S. Mahfudoh and T. N. Rohmah, 'Faktor Yang Mempengaruhi Kepatuhan Penulisan Resep Sesuai Formularium', *J. Adm. Kesehat. Indones.*, vol. 3, no. 1, 2015.
6. F. Indrasari, S. Suwarni, and R. Kurniawati, 'Hubungan Kesesuaian Pereesepan Obat', vol. 11, no. 1, pp. 36–41, 2021.
7. Y. Anggriani, R. Rosdiana, and S. Khairani, 'Evaluasi Perencanaan dan Pengadaan Obat di Era Jaminan Kesehatan Nasional (JKN) di Puskesmas Kabupaten Cianjur', vol. 17, no. 02, pp. 425–438, 2021.
8. Imron, Prasetya, Usnawati, and Hidayatullah, 'Evaluasi Kesesuaian Peresepan Dengan Formularium Nasional Dirawat Inap Puskesmas Pesantren Kota Kediri', *Java Heal. J.*, vol. 1, no. 1, pp. 1–171, 2020, [Online]. Available: http://etd.eprints.ums.ac.id/14871/%0Ahttps://doi.org/10.1016/j.cell.2017.12.025%0Ahttp://www.depkes.go.id/resources/download/info-terkini/hasil-riskesdas-2018.pdf%0Ahttp://www.who.int/about/licensing/%0Ahttp://jukeunila.com/wp-content/uploads/2016/12/Dea
9. E. N. Indiarto, F. Herawati, and S. A. Wardani, 'Profil Penggunaan Obat Rasional di Puskesmas Kabupaten Sidoarjo Provinsi Jawa Timur Tahun

2017', *Calyptra*, vol. 9, no. September, 2020.
10. K. Khusna and R. Septiana, 'Kesesuaian Obat Hipoglikemik Oral pada Pasien Diabetes Mellitus Tipe 2 di Fasilitas Kesehatan Tingkat Pertama X di Surakarta', *Indones. J. Pharm. Nat. Prod.*, vol. 2, no. 2, pp. 65–70, 2019, doi: 10.35473/ijpnp.v2i2.261.
11. Mustika, F. Yuliastuti, N. Made, and A. Nila, 'Gambaran Kesesuaian Ketersediaan Obat dengan Formularium Nasional di Puskesmas Muntilan II', vol. 2, no. 1, pp. 1–7, 2022.
12. M. M. A. Saputera, N. Yunika Sari, and F. Jannah, 'Evaluasi Manajemen Pengelolaan Obat Tahap Selection Dan Procurement Di Puskesmas Cempaka Putih Tahun 2018', *J. Insa. Farm. Indones.*, vol. 4, no. 1, pp. 85–94, 2021, doi: 10.36387/jifi.v4i1.686.
13. E. Wulandari and A. Widayati, 'Evaluasi Pengelolaan Obat Di Puskesmas Salaman 1 Kabupaten Magelang Dalam Masa Pandemi COvid-19 Tahun 2020', *JFSP Vol.7, No.2, Novemb. 2021, Hal 95-106 ,pISSN 2549-9068, eISSN 2579-4558 J*, vol. 7, no. 2, pp. 95–106, 2021.
14. Rintanantasari, A. Fudholi, and Satibi, 'Evaluasi Perencanaan dan Pengendalian Obat di Puskesmas Wilayah Kabupaten Tulungagung dan Kota Kupang', vol. 17, no. 3, pp. 296–302, 2021, doi: 10.22146/farmaseutik.v1i1.54063.
15. Azanuddin, M. Tasjiddin, and E. Mustiqawati, 'Analisis Formularium Dinas Kesehatan Kota Baubau', *J. Public Heal. Ind. Nutr.*, vol. 1, no. 1, pp. 68–74, 2021.
16. H. Maspekeh, Satibi, and G. P. Widodo, 'Evaluasi Perencanaan dan Pengadaan Kebutuhan Obat Publik Serta Ketersediaan Obat di Wilayah Kerja Dinas Kesehatan Daerah Kota Tomohon Tahun 2016', *J. Farm. Sains Indones.*, vol. 1, no. 2, pp. 14–25, 2018.
17. L. Prayitno, Y. Yuniar, and T. Rosita, 'Kesesuaian Antara Ketersediaan Antibiotik dan Formularium Nasional Pada Era JKN di Faskes Tingkat Pertama Kota Manado Tahun 2014 - 2017', *Bul. Penelit. Kesehat.*, vol. 48, no. 1, pp. 31–42, 2020, doi: 10.22435/bpk.v48i1.1970.
18. Y. Sukowati, D. R. Laksmitawati, Y. Anggraini, and M. Restina, 'Pengaruh Penerapan Jaminan Kesehatan Nasional Terhadap Profil Obat Pasien Hipertensi di Puskesmas Kecamatan Pulogadung', *Maj. Farm.*, vol. 17, no. 1, p. 97, 2021, doi: 10.22146/farmaseutik.v17i1.55194.
19. Salwati, A. Rahem, and A. A. Prayitno, 'Analisis Hubungan Profil Ketersediaan Obat Terhadap Profil Rasionalitas Peresepan Pada Pasien Rawat Jalan Di Puskesmas Kabupaten Barito Kuala Provinsi Kalimantan Selatan', *Calyptra J. Ilm. Mhs. Univ. Surabaya*, vol. 7, no. 1, pp. 2086–2100, 2018.
20. S. Satibi, E. H. Daulay, G. A. Oviani, K. Erlianti, A. Fudholi, and D. A. Puspandari, 'Performance analysis of pharmacist and influencing factors in the era of National Health Insurance at Puskesmas', *J. Manag. Pharm. Pract.*, vol. 8, no. 1, p. 32, 2018, doi: 10.22146/jmpf.34441.
21. I. M. Bentura, M. Indrisari, and R. Aksa, 'Analisis Kesiapan Obat Dalam Penegakan Diagnosa Wajib BPJS di Puskesmas Busalangga , Kabupaten Rote Ndao-Nusa Tenggara Timur', *J. Pharm. Med. Sci.*, vol. 1, no. 1, pp. 1–5, 2016.
22. I. Carolien, A. Fudholi, and D. Endarti, 'Evaluation Medicine Availability Before and After NHI Implementation at Health Centers in Keerom District, Papua Province', *J. Manaj. dan Pelayanan Farm. (Journal Manag. Pharm. Pract.*, vol. 7, no. 1, p. 31, 2017, doi: 10.22146/jmpf.30294.
23. R. S. Dianingati and S. D. Prasetyo, 'Analisis Kesesuaian Resep untuk Pasien Jaminan Kesehatan Nasional dengan Indikator Peresepan WHO 1993 pada Instalasi Farmasi Rawat Jalan di RSUD Ungaran Periode Januari-Juni 2014', *Maj. Farm.*, vol. 11, no. 3, pp. 362–371, 2015.
24. M. M. A. Saputera, 'Perencanaan Di Era Jaminan Kesehatan Nasional Di Rsud H . Hasan Basery Kandangan', *J. Ilm. Ibnu Sina*, vol. 1, no. September 2016, pp. 248–255, 2016.
25. S. Wibowo, C. Suryawati, and J. Sugiarto, 'Analisis Pengendalian Persediaan Obat-Obatan Instalasi Farmasi RSUD Tugurejo Semarang Selama Pandemi COVID-19', *Manaj. Kesehat. Indones.*, vol. 9, 2021.
26. N. Oktaviani, G. Pamudji, and Y. Kristanto, 'Evaluasi Pengelolaan Obat Di Instalasi Farmasi Rumah Sakit Umum Daerah Provinsi NTB Tahun 2017', vol. 15, no. 2, pp. 135–147, 2018.
27. M. Elisabeth, M. Melia, T. Rame, and W. Lutsina, 'Evaluasi Pengelolaan Obat Dan Bahan Medis Habis Pakai di Puskesmas Oebobo Kota Kupang Pada Tahun 2019', vol. 4, no. 1, pp. 249–254, 2021.
28. Y. M. Indriana, E. S. Darmawan, and A. C. Sjaaf, 'Analisis Pengelolaan Obat di Instalasi Farmasi RSUA Tahun 2020', *J. Kesehat. Masy.*, vol. 11, 2021, [Online]. Available: http://jurnal.unismuhpalu.ac.id/index.php/PJKM/article/view/1512
29. U. Khasanah, G. Mayasari, and I. D. K. Irianto, 'Evaluasi Kesesuaian Peresepan Obat Pada Pasien Rawat Jalan Poli Penyakit Dalam Dengan Formularium Nasional Di Rsud Prambanan', *J. Farmaku (Farmasi Muhammadiyah Kuningan)*, vol. 7, no. 1, pp. 31–38, 2022, doi: 10.55093/jurnalfarmaku.v7i1.283.
30. T. Nugroho, I. Purwidyaningrum, and S. B. Harsono, 'Evaluasi Pengelolaan Obat Dan Strategi Perbaikan Dengan Metode Hanlon Di Instalasi Farmasi Rumah Sakit Angkatan Udara dr . Efram Harsana Madiun', pp. 98–109, 2022.
31. P. R. Veryanti and J. Supriyanto, 'Gambaran Penulisan Obat Generik dan Kesesuaian Peresepan Pada Pasien BPJS Terhadap

Formularium Nasional', *Bali Heal. J.*, vol. 3, no. 2, pp. 87–90, 2019.

32. T. S. Tuloli, N. Rasdianah, and I. P. Gobel, 'Evaluasi Kesesuaian Peresepan Obat Pada Pasien Poliklinik Penyakit Dalam Terhadap Formularium di Rumah Sakit', vol. 4, pp. 251–258, 2022.
33. R. Fitriah, 'Kesesuaian Peresepan Obat Bpjs Berdasarkan Formularium Nasional Dan Formularium Rumah Sakit Di RSD Idaman Banjarbaru', *Media Inf.*, vol. 16, no. 1, pp. 32–37, 2021, doi: 10.37160/bmi.v16i1.397.
34. D. P. Rahmawati, F. Herawati, and S. A. Wardani, 'Evaluasi Penggunaan Obat Di Rumah Sakit Marsudi Waluyo, Kabupaten Malang Tahun 2016', *Calyptra*, vol. 7, no. 2, 2019.
35. K. A. Susanto, M. L. Sitanggang, D. Hassan, and I. Chorida, 'Analysis Of Prescription of National Formularium Medicines Based on The Adjustment of Restriction in Public Hospitals of Sawah Besar 2018', *J. Ilmu Kefarmasian Indones.*, vol. 19, no. 1, pp. 68–72, 2021, [Online]. Available: http://jifi.farmasi.univpancasila.ac.id/index.php/jifi/article/view/1056
36. Y. Wardati, D. A. Deswat, and Z. Muttaqin, 'Pharmacoeconomic Analysis Of Using Cephalosporin Group In Appendicitis Prophylactic In Rumah Sakit Umum Daerah Subang', *J. Ilmu Kefarmasian Indones.*, vol. 16, no. 2, p. 200, 2018, doi: 10.35814/jifi.v16i2.441.
37. W. R. Pratiwi, A. P. Kautsar, and D. Gozali, 'Hubungan Kesesuaian Penulisan Resep dengan Formularium Nasional Terhadap Mutu Pelayanan pada Pasien Jaminan Kesehatan Nasional di Rumah Sakit Umum di Bandung', *Pharm Sci*, vol. 4, no. 1, p. 49, 2017.
38. K. S. Yunarti, 'Analisis Ketersediaan Dan Peresepan Obat Dengan Formularium Di Rumah Sakit X Kabupaten Karanganyer', *J. Bina Cipta Husada*, vol. XVIII, 2022.
39. M. Zulkifli, 'Identifikasi Faktor-Faktor Penyebab Tingginya Saldo Persediaan Obat (Studi Pada RSUD ABC dan RSUD QYZ)', *Account. Bus. Inf. Syst. J.*, vol. 6, no. 3, 2018.
40. E. Taulabi, W. Utami, and A. Rahem, 'Profil Rencana Kebutuan Obat Di Dinas Kesehatan Pada Daerah Tertinggal, Perbatasan dan Kepulauan (Studi Di Kabupaten Belu Nusa Tenggara Timur)', *J. Ilm. Ibnu Sina*, vol. 4, no. 2, pp. 420–431, 2019.
41. A. Patantan, Satibi, and G. P. Widodo, 'Evaluation of drug supply management in drug storage health departemen X In The JKN Era', *CHMK Pharm. Sci. J.*, vol. 2, no. April, pp. 254–259, 2019.
42. R. Yulia, F. Herawati, A. K. Jaelani, and W. Anggraini, 'Evaluasi Penggunaan Antibiotik dan Profil Kuman pada Seksio Sesarea di Rumah Sakit Umum Daerah Kabupaten Pasuruan', *J. Farm. Klin. Indones.*, vol. Vol. 7, no. December, pp. 1–6, 2018.
43. Krisnadewi, 'Evaluasi Standar Pelayanan Minimal Instalasi Farmasi RSUD Waluyo Jati Kraksaan Sebelum dan Sesudah Badan Penyelenggara Jaminan Sosial (BPJS) Kesehatan', *e-Jurnal Pustaka Kesehat.*, vol. 2, no. 2, pp. 192–198, 2014.
44. D. Medisa and S. S. Danu, 'Kesesuaian Resep Dengan Standar Pelayanan Medis dan Formularium pada Pasien Rawat Jalan Jamkesmas', *J. Ilm. Farm.*, vol. 11, no. 1, pp. 20–28, 2015.
45. A. Vermasari, M. Masrul, and H. Yetti, 'Analisis Implementasi Standar Pelayanan Minimal (Spm) Di Instalasi Gawat Darurat (Igd) Rsu Mayjen Ha Thalib Kabupaten Kerinci', *J. Kesehat. Andalas*, vol. 8, no. 2, p. 275, 2019, doi: 10.25077/jka.v8i2.1002.
46. M. Arfania, Ice, and S. Amal, 'Kesesuaian resep terhadap formularium rumah sakit karawang', *J. Pharm. Sci. Med. Res.*, vol. 4, no. 2, pp. 47–54, 2021.
47. T. Lestari, Y. Anggriani, and D. R. Laksmitawati, 'Analisa Kesesuaian Peresepan Obat Pasien Bpjs Kesehatan Dengan Formularium Nasional Di Puskesmas Kabupaten Tangerang Tahun 2016', *J. Farmagazine*, vol. 6, no. 2, p. 57, 2019, doi: 10.47653/farm.v6i2.144.

Return on investment of health intervention among diabetic patient: A systematic review

Kadek Ida Krisnadewi[1,2], *Susi* Ari Kristina[3*], *Dwi* Endarti[3], *Tri* Murti Andayani[4]
[1]Doctoral Program in Pharmaceutical Science, Faculty Pharmacy, Universitas Gadjah Mada, Yogyakarta, Indonesia.
[2]Department of Pharmacy, Faculty of Science and Technology, Universitas Bali Dwipa, Bali, Indonesia.
[3]Department of Pharmaceutics, Universitas Gadjah Mada, Yogyakarta, Indonesia
[4]Department of Pharmacology and Clinical Pharmacy, Universitas Gadjah Mada, Yogyakarta, Indonesia

Abstract. A strategy for reducing healthcare costs of diabetic mellitus (DM) is by reducingbody weight and blood glucose through lifestyle-modification programs such as intensive lifestyle interventions, meal-replacement programs, and physical activity. A return on investment (ROI) analysis needs to be done to see a program's effectiveness, which aims to make this program provide benefits, especially in the long term. In this systematic review, the authors describe ROI, the program, the program's cost, adoption, and implementation. The authors searched the Science Direct, Pubmed, and SAGE databases using the following search terms: 'Return on Investment' AND 'DiabeticMellitus' and Programs". Moreover, hand searching identifies further studies. This review includes eight articles, and most articles on lifestyle modification intervention have great ROI; several articles with public health workers have great ROI. The findings show that long-term and program intensity in a long time horizon plays a major role in outcomes and more significant ROI. Programs with lifestyle intervention fellow health workers or CHWs that helped deliver physical activity, education, and dietary had a good impact in lowering diabetes risk.
Keyword: Diabetic mellituss, life style intervention, program, return on investment.

1 Introduction

The incidence of diabetes mellitus worldwide has increased, which is a global threat [1–3]. The most recent estimate of the number of people with diabetes ranges from 20-79 years of age and is expected to increase to 642 million in 2040, which means that 1 in 10 adults will have diabetes. In addition, in middle and low-income countries, 3 out of 4 adults have diabetes. Diabetes is the cause of 6.7 million deaths worldwide which means that 1 death every 5 seconds. From an economic burden standpoint, diabetes spends 966 billion USD on health costs, and this increase is 315% from the last 15 years [4].

Lifestyle change interventions in improving physical activity and diet can increase the incidence of type 2 diabetes mellitus by 58% within three years; this has been done in the Diabetes Prevention Program (DPP) clinical trial program. This intervention's benefits were also shown for the next 10 to 15 years [5, 6]. The results of the DPP issued a recommendation that overweight or obese adults with at least one risk factor for cardiovascular disease receive behavioral counseling to minimize the severity of complications [3, 7–9].

This intervention also was a strategy to reduce the cost of treating DM, which showed that modifying lifestyle with intensive intervention through a diet program reduces the health cost of DM [10, 11]. Weight loss in DM patients has been shown to provide benefits in terms of health costs, both hospitalization costs and medical expenses in DM patients. Diabetic patients with excess weight incur higher medical costs than diabetic patientswith normal weight [12, 13]. The government's agenda and making a national plan for managing the prevalence of DM, providing a budget for the DM program will be more beneficial in the long term when viewed from the value of the cost of health costs caused by DM disease [40]. The results of the NDPP and the Diabetes Prevention Program Observation Studyconfirm that lifestyle interventions and metformin are equally cost-effective [14].

Evaluation of DM disease prevention programs has beenaffordable, only limited to its implementation and effectiveness in financing and the health of DM sufferers. Economic effectiveness can be assessed by ROI analysis by assessing the benefits of program interventions to obtain a return on investment (ROI) value from the health system perspective [15]. In this systematic review, the authors explain the types of interventions implemented, the costs, and the value of the ROI resulting from the interventions. A review of ROI will be beneficial in making diabetes-related policies. This review will provide priority recommendations that are most effective and profitable long-term so that therapy in diabetes patients will be more cost-effective.

* Corresponding email: susiari_k@ugm.ac.id

2 Methods

2.1 Search Strategy

The authors systematically reviewed the completed ROI values from program interventions aimed at prediabetic patients or diabetics. This systematic review is part of a research study ROI analysis in health programs in Indonesia with a code of ethics number KE/FK/1652/EC/2022. The disbursement database, Pubmed, Science Direct, and SAGE, uses the following search keywords 'Return on Investment' AND 'Diabetic Mellitus' AND 'Intervention'. A hand search was performed to identify further studies. The limit in this particular payment is the last ten years of articles and articles in English.

2.2 Study Selection an Inclusion Criteria

The studies included in this systematic are each ROI study with an intervention design for diabetes patients in the community. These included articles from the past ten years. The authors search for program interventions related to the science and art of promoting and protecting health and well-being to prevent ill health and prolong life through community-organized efforts for diabetes prevention and diabetes in high-income countries. The program is then assessed economically by return on investment analysis. This systematic review is based on items in the Systematic Review and Meta-Analysis (PRISMA) [16]. The authors screen original articles with full text related to interventions in diabetes patients and are in English. Theexclusion criteria in this study did not count the variablereturn on investment, and the article was incomplete.

2.3 Selection Article and Data Extraction

The first investigator (KI) initially screened the article titles. Investigators (KI) independently review the relevance of titles and abstracts of articles that are fit for purpose. The results of the articles were then discussed by (SA) and (DE) by reading the entire manuscript. The investigator (TM) then independently extracted and assessed the quality of the articles. Next, the reviewer (SA) decides on the assessment results and the article's qualit.

3 Result

3.1 Study Selection

An article search was conducted on Pubmed, ScienceDirect, and SAGE, and the article was found (Figure 1). n = 428 journals from PubMed, n = 120 journals from ScienceDirect, and n = 83 from SAGE Journals. Article screening was carried out by reading article titles and abstracts relevant to the inclusion criteria and identifying 89 potentially relevant titles. At Positioning this stage, 89 articles were read thoroughly to assess their suitability with the desired outcome in a systematic review. From this process, articles will be decided according to the objectives of the systematic review, and some articles were excluded for the following reasons, complete articles were not found (n=34), duplicates(n=9), and excluded for other reasons (n=30). In-depth review of the full text was excluded because the results did not meet the inclusion criteria and study protocol. Therefore, eight articles are relevant Place the figure as close as possible after the point where it is first referenced in the text. If there is a large number of figures and tables it might be necessary to place some before their text citation.

3.2 Study Characteristic

Table 1 describes the characteristics. The eight studies were conducted in the UK, USA, and around. The duration of this study ranged from 1 years-10 years. Two perspectives appear in the article: the perspective of the payer and the provider payer perspective is used in articles [15, 17], and the payer perspective is used in articles [15, 16, 18–21]. Studies with a payer perspective are the perspectives of companies such as insurance, private companies, or the government, which will pay for costs related to health services provided to participants. The costs considered in this perspective are direct costs, but the analysis can consider indirect costs such as lost productivity or lost workdays [22]. Lifestyle intervention was a health plan with a healthy diet and physical activity for obese, diabetes, and prediabetes patients [17–19, 23, 24]. Counseling-based programs are carried out in various ways, starting from providing material, medical journals given to the patient, digital intervention, and education face-to-face with patients facilitated by health workers [15, 20, 21]. There was also research on counseling interventions and therapy for DM patients with foot ulcers [20]. A health service provider's perspective is the actual value of the product or health service provided by the health service. These include direct costs, namely professional service fees, hospitalization costs, medical expenses, laboratory examination fees, and consumables [22].

3.3 ROI Specialism Programin Dealing with Diabetes Mellitus

3.3.1 *Counseling health public and health workers' intervention.*

There were three articles on counseling, public health, and health workers' intervention. The value of the cost of intervention from these articles ranges from 115-500 USD. The ROI value range is 7.4-9.01. Figure 2 shows that the highest ROIvalue is 9.01, with a time horizon of 3 years.

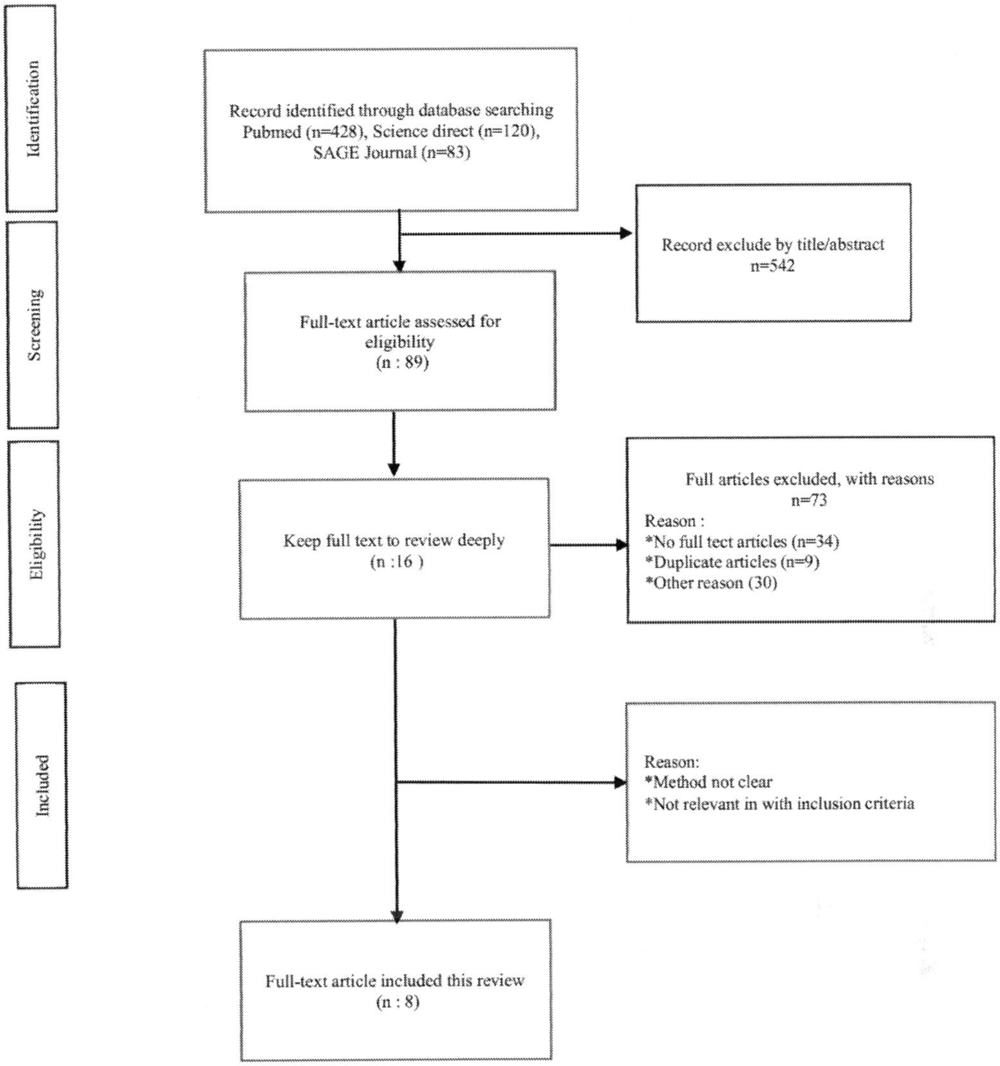

Fig. 1. Prisma Flow Diagram

Table 1. Summary Study Characteristics

Reference	Intervention	Population	ROI	Cost perspective	Discount rate	Time horizon
[24]	Lifestyle intervention	UK	1,42	Provider	3,5%	5 years
[15]	Lifestyle intervention	USA	9,01	Provider and Payer	3%	3 years
[26]	Lifestyle intervention	USA	16,9	Payer	3%	10 year
[18]	Lifestyle intervention	USA	9,64	Payer	None	5 years
[19]	Lifestyle intervention	USA	0,42	Payer	None	3 years
[20]	Counseling health public and health workers intervention	Canada	7,4	Payer	None	1 year
[27]	Counseling and health public and health workers intervention	USA	7,9	Payer	3%	3 year
[17]	Counseling and health public and health workers intervention	California	2,1	Provider and Payer	3%	3 years

. Public Health Programs and Cost of Programs

nce	Intervention	Detail Programs	Cost of Programs/person/USD
	Lifestyle intervention	healthy lifestyle interventions by providing education on healthy diets, physical activity and weight loss targeted at individuals at high risk of DM and DM	284,2
	Lifestyle intervention	Intervention study with 2 groups, control group and intervention group. In the intervention group, healthy lifestyle activities, namely ILI (intensive lifestyle intervention) were carried out. Comparing the cost saving of the two groups	342
	Lifestyle intervention	Weigh and Win (WAW), by Incenta HEALTH LLC is a community-based weight loss program	1487
	Lifestyle intervention	Lifestyle habit intervention through education with an appeal to reduce alcohol, smoking, more excercise and consume more vegetables and fruit. Participants are given a journal that will be checked every month	None
	Lifestyle intervention	Healthy lifestyle interventions for prediabetic patients and seeing the potential for continuity against the risk of DM disease for this group	None
	Counseleling health public and health workers intervention	Diabetic foot care clinical pathway (DFCCP) in diabetic foot wound patients. There are two DM groups, a control group with treatment and an intervention group with treatment	500
	Counseling health public and health workers intervention	Implement National DPP is a partnership of public and private organizations providing intensive training in diet, physical activity, and behavior modification	200-400
	Counseling health public and health workers intervention	Comparing the intervention of DM patients with regular (PCP) therapy vs diabetes intense medical management (DIMMs)	115-164

 https://doi.org/10.1051/bioconf/20237505015

3.3.2 *Counseling health public and health workers' intervention.*

There were five articles with lifestyle interventions. Table 2 shows that cost of lifestyle interventions ranges from 248.2-1487 USD. Several articles do not include the value of the cost of intervention but still display the ROI value. ROI values in lifestyle intervention range from 0.42-16.9. The highest value is 16.9, with an estimated interventionif it is carried out for ten years.

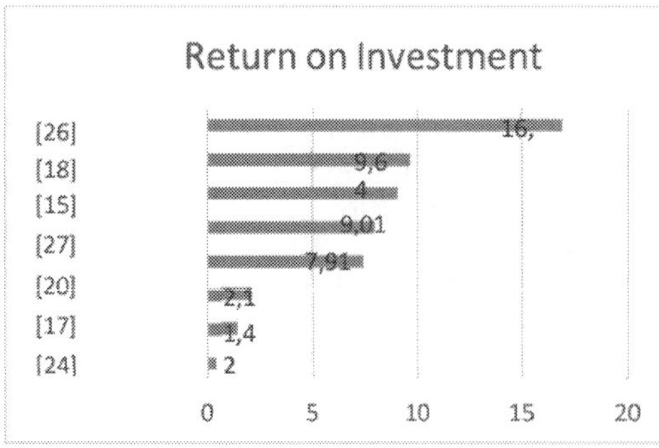

(a)

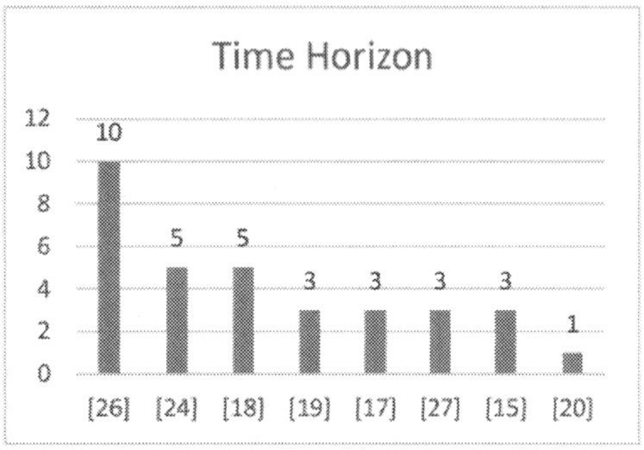

(b)

Fig. 2. The graph gender of (a) Return on Investment and (b) Time Horizon

4 Discussion

ROI can be used in calculating a program's intervention or several interventions if the company has a large-scale scope and can also be one of the measuring indicators to be compared with other interventions [28]. Assessing the ROI ratio is seen from how much the ratio is. ROI is greater than or equal to 1, then the program has a practical value because it has value equal to or more than the cost of implementation. In this case, the ROI value is considered positive. For example, an ROI of 1.8 indicates that for every 1 rupiahinvested in the program, the agency will earn a profit of 1.80 rupiah. If, ROI is less than 1, then the interventionsuffers a loss. In this case, ROI is considered negative. For example, ROI-1.5 shows that for every 1 dollar invested, the agency suffers a loss of 1.50 rupiah. If theROI value is 0.8, then for every 1 dollar invested, 0.8 dollars will be obtained by the agency, while the agencywill lose 0.2 dollars for every 1 dollar spent on the program [29]. Most of the articles do interventions in lifestyle modifications. These lifestyle modifications include physical activity and diet. In the articles, interventions are different; this is also related to the results to be achieved. Several articles measure effectiveness in comparison with a control group. In comparison with the control group, the value of cost saving can also be seen from the comparison with the reduction of the two groups [17, 24]. The article provides an intervention in which participants are asked to complete household tasks, including self-monitoring ofeating and physical activity. The physical activities will be in the form of safety training, stretching, basic strength training, and increasing activity levels witheach predetermined schedule [17–19, 23, 24] There are two interesting findings in a systematic review regarding ROI in interventions in DM patients.

First, even with the most basic economic evaluation, it is clear that most public health interventions include with monitoring and lifestyle intervention are substantial cost savings. It confirms our theory that public health interventions generally offer a sizable ROI. Overall, the results of our systematic review clearly show that public health interventions save costs for health services and the wider economy. Besides, some very rapid: preventive interventions report substantial results within one year. Greater return on investment seen over the ten years horizon. It has significant implications for policymakers, who often work with shorter time horizons (usually 3–5 years). It is in line with previous research that the intervention of physical activity and diet positively affects weight loss and blood sugar [2, 21, 30].

Second, health worker-based interventions provide a relatively high ROI value compared to physical activity and diet alone. Intervention with physical activity and diet provides high effectiveness if given in the long term. This positive result from [15] is driven by the large difference in HbA1c results achieved in the DIMM versus the PCP group, as well as the relatively low cost of pharmacist interventions designed as a limited series of visits to treat patients. Interventions with health workers have many benefits and effectiveness, especially in patient health, and can avoid long-term complications and a better quality of life [31, 32]. Education intervention for diabetic programs in adults showed a good impact on cost saving if given in the long term, at least with 1 year duration [33, 34]. One article is about diabetic foot care, and that showed good ROI. The positive results of providing regular foot wound care to DM patients with foot wounds indicate that cost savings can be obtained if the patient receives regular and long-term care [20].

Cost saving is a benefit variable to measure the successof a program by assessing the cost saving. ROI analysisis obtained by dividing the net benefit value by the intervention cost. In other words, the net benefit here issignificant, obtained from the cost saving minus the

https://doi.org/10.1051/bioconf/20237505015

intervention cost value. The benefit value in the net benefit has the same concept as the net profit value [35]. The current study shows that the impact of an effective program for diabetes in adult patients correlates with cost savings, savings from complications, and also comorbidities [33, 23, 25]. This finding is similar to estimates of cost savings reported in previous studies. For example, 12-month pre- and post-education cost data were examined for diabetes program participants, with $815 direct cost savings through the program and $1504 indirect cost savings through changes in comorbid disease burden [36]. There is a correlation between clinical outcomes and cost-effectiveness in providing usual care and lifestyle interventions for DM patients. Bringing patients, providers, and the healthcare system together and integrating public health workers as liaisons between patients is an excellent approach to effective diabetes management [37]. Other studies that showed cost saving of diabetes intervention showed good member health workers to manage diabetes can reduce comorbid condition, cost, and HbA1C [4, 36, 38].

Studies find it important to consider the long-term impact to capture the full ROI and find positive ROI values [20]. Programs designed for cost-saving can be looked at in the short-term, medium-term, and long-term. To see good and effective service quality requires an affordable intervention program so that it can be carried out in the long term. Identification of good ROI can be seen from how well the profit is given in the long term [39]. The authors recommend that advisory bodies regularly emphasize that public health interventions can offer rapid results, which can increase further in the long term. The limitations of this review are that several articles did not mention the value of the discount rate and cost of intervention. Some articles did not explain in detail the interventions carried out. Very little has been done regarding research conducted by research authors regarding ROI. Many studies on ROI have been conducted in developed countries, but only a few focused on diabetes and public health.

5 Conclusion

The findings show that long-term and program intensity in a long time horizon plays a major role in outcomes and more significant ROI. Programs with fellow lifestyle intervention deliver physical activity, education, and dietary had a good impact in lowering diabetes risk. The authors recommend that advisory bodies regularly emphasize that public health interventions canoffer rapid results, which can increase further in the longterm.

References

1. NCD Risk Factor Collaboration (NCD-RisC). Worldwide trends in diabetes since 1980: a pooled analysis of 751 population-based studies with 4.4 million participants. Lancet, **(27)**,15–30 (2016).

2. MK. Kramer, RG. Miller, LM. Siminerio. Evaluation of a community Diabetes Prevention Program delivered by diabetes educators in the United States: one-year follow up. Diabetes Res Clin Pract. **(3)**, 49-52 (2014).

3. Diabetes Prevention Program Research Group, WC. Knowler, SE. Fowler, RF. Hamman, CA. Christophi, HJ. Hoffman, AT. Brenneman, JO. Brown-Friday, R. Goldberg, E Venditti, and DM Nathan. 10-year follow-up of diabetes incidence and weight loss in the Diabetes Prevention Program Outcomes Study. Lancet.,**(16)**, 77–86 (2009).

4. M. Bansal, M. Shah, B. Reilly, S. Willman, M. Gill , FR. Kaufman. Impact of Reducing Glycated Hemoglobin on Healthcare Costs Among a Population with Uncontrolled Diabetes. Appl Health Econ Health Policy, (6), 75–84, (2018).

5. WC. Knowler, E. Barrett-Connor, SE. Fowler, RF. HammaN, JM. Lachin, EA. Walker, and DM. Nathan. Reduction in the incidence of type 2 diabetes with lifestyle intervention or metformin. N Engl J Med. ,(6), 393–403, (2002).

6. JH. Shubrook, W. Chen, A. Lim. Evidence for the Prevention of Type 2 Diabetes Mellitus. J Am Osteopath Assoc. (11), 30–38, (2018).

7. WH. Herman. The cost-effectiveness of diabetes prevention: results from the Diabetes Prevention Program and the Diabetes Prevention Program Outcomes Study. Clin Diabetes Endocrinol, (12) 1-9, (2015).

8. S. Roberts, E. Barry, D. Craig, M. Airoldi, G. Bevan, T. Greenhalgh. Preventing type 2 diabetes: systematic review of studies of cost-effectiveness of lifestyle programmes and metformin, with and without screening, for pre-diabetes. BMJ Open. (11):017-184, 2017.

9. Taylor R, Valabhji J, Aveyard P, Paul D. Prevention and reversal of Type 2 diabetes: highlights from a symposium at the 2019 Diabetes UK Annual Professional Conference. Diabet Med. (3), 59–65, (2019).

10. LM. Davis, C. Coleman, J. Kiel, J. Rampolla, T. Hutchisen, L. Ford, WS. Andersen, and AH. Mitola, Efficacy of a meal replacement diet plan compared to a food-based diet plan after a period of weight loss and weight maintenance: a randomized controlled trial. Nutr J. Mar (11), 9:11, (2010).

11. X. Zhuo, P. Zhang, HS. Kahn, BH. Bardenheier, R. Li, EW. Gregg. Change in medical spending attributable to diabetes: national data from 1987 to. Diabetes Care 2011. (4), 81–87 (2015).

12. J. Cawley, C. Meyerhoefer, A. Biener, M. Hammer, N. Wintfeld. Savings in Medical Expenditures Associated with Reductions in Body Mass Index Among US Adults with Obesity, by Diabetes Status.

https://doi.org/10.1051/bioconf/20237505015

Pharmacoeconomics. (7), 7–22 (2015.)

13. J. Cawley, C. Meyerhoefer. The medical care costs of obesity: an instrumental variables approach. J Health Econ., (1), 19–30, (2012).

14. WH. Herman, TJ. Hoerger, M. Brandle, K. Hicks, S. Sorensen, P. Zhang, RF. Hamman, RT, Ackermann, MM. Engelgau, and RE. Ratner. The cost-effectiveness of lifestyle modification or metformin in preventing type 2 diabetes in adults with impaired glucose tolerance. Ann Intern Med. (5), 23–32, (2005).

15. JD. Hirsch, M. Bounthavong, A. Arjmand, DR. Ha, CL. Cadiz, A. Zimmerman, H. Ourth, AP. Morreale, SV. Edelman, and MM. Candis. Estimated Cost Effectiveness, Cost Benefit, and Risk Reduction Associated with an Endocrinologist-Pharmacist Diabetes Intense Medical Management "Tune-Up" Clinic. J Manag Care Spec Pharm. (3), 18–26 (2017).

16. MJ. Page, JE. McKenzie, PM. Bossuyt , I. Boutron, TC. Hoffmann, CD. Mulrow, L.Shamseet, JM. Tetzlaff, EA. Akl, SE. Brennann, R.Chou, J. Glanville, JM. Grimshaw, A. Hróbjartsson, MM. Lalu, T. Li, EW. Loder, EM. Wilson, S. McDonald, LA. McGuinness, LA. Stewart, J. Thomas, AC. Tricco, VA, Welch, P. Whiting, D.Moher. The PRISMA 2020 statement: an updated guideline for reporting systematic reviews. Syst Rev. 2021 Mar 29;10(1):89.

17. CM. Horstman, DH. Ryan, LJ. Aronne, CM. Apovian, JP Foreyt, HM. Tuttle, and DA. Williamson. Return on Investment: Medical Savings of an Employer Sponsored Digital Intensive Lifestyle Intervention, Weight Loss. Obesity (Silver Spring). (4):54–61, (2021).

18. ND. White, TL. Lenz, MZ. Skrabal, JJ. Skradski, L. Lipari. Long-Term Outcomes of a Cardiovascular and Diabetes Risk-Reduction Program Initiated by a Self-Insured Employer. Am Health Drug Benefits. (4):77–83, (2018).

19. T. Khan, S. Tsipas, G. Wozniak. Medical Care Expenditures for Individuals with Prediabetes: The Potential Cost Savings in Reducing the Risk of Developing Diabetes. Popul Health Manag. 2017 Oct 1;20(5):389–96.

20. NX. Thanh, K. Dmytruk, P. O'Connell, E. Rogers, D. Fillier, JM. MacRae, C. Thomas, C. Rennie, C. Eitzenberger, C. Newman, B.Match, C. Thompson, J.Nhan and T. Wasylak. Return on investment of the diabetes foot care clinical pathway implementation in Alberta, Canada. Diabetes Res Clin Pract. (165),108-241, (2020).

21. AC. Sylvetsky, SL. Edelstein, G. Walford, EJ. Boyko, ES. Horton, UN. Ibebuogo. A highcarbohydrate, high-fiber, low-fat diet result in weight loss among adults at highriskog type 2 diabetes. The Journal of Nutrition. (11), 2060- 2066, (2016).

22. LA. Sanchez. Applied pharmacoeconomics: evaluation and use of pharmacoeconomic data from the literature. Am J Health Syst Pharm. (16), 38– 40, (1999).

23. BA. Kash, SH. Lin, J.Baek, RL. Ohsfeld. The diabetes management education program in south texas: an economic an clinical impact analysis. Front Public Helath (5), 345, (2017).

24. C. Thomas, S. Sadler, P. Breeze, H. Squires, M. Gillett, A. Brennan. Assessing the potential return on investment of the proposed UK NHS diabetes prevention programme in different population subgroups: an economic evaluation. BMJ Open. (8), 149-153, (2017).

25. FF. Jiao, CSC. Fung, EYF. Wan, AKC. Chan, SM, McGhee, RLP. Kwok. Five-years cost-effective of the multidisciplinary risk assesment and management programme diabetes mellitus (RAMP-DM). American Diabetes Association. (2), 250-257, (2018).

26. TL. Michaud, W. You, KE. Wilson, D. Su, TJ. McGuire, FA. Almeida, AL. Bayer, and PA. Estabrooks. Cost effectiveness and return on investment of a scalable community weight loss intervention. Prev Med. (105), 295– 303, (2017).

27. W. Su, F. Chen, TM. Dall, W. Iacobucci, L. Perreault. Return on Investment for Digital Behavioral Counseling in Patients With Prediabetes and Cardiovascular Disease. Prev Chronic Dis. (13),150-357, (2016).

28. World Health Organization, Programme UND. Non-communicable disease prevention and control: a guidance note for investment cases [Internet]. World Health Organization; 2019 [cited 2023 Jun 4].

29. Agency for Healthcare Research and Quality. Return on Investment Tools." Agency for Healthcare Research and Quality. [Internet]. Agency for Healthcare Research and Quality [cited 2023 Jun 1].

30. MB. Weber, MM. Hennink, KMV. Narayan. Tailoring lifestyle programmes for diabetes prevention for US South Asians. Fam Med Community Health. (2), 29-35, (2020).

31. P. Balagopal, N. Kamalamma, TG. Patel, R. Misra. A community-based participatory diabetes prevention and management intervention in rural India using community health workers. Diabetes Educ. (6), 22–34, (2012).

32. M. Matsushita, A. Muramoto, E. Nomura, Y. Eguchi, A. Kato, Y. Sano, M. Kabayama, M. Arakawa, Y. Oguma, D. Yabe, M.Matsunaga, H.Yatsuya, H. Arima dan K. Tsushita. The Smart Life Stay (SLS) program: effects of a lifestyle intervention program in combination with health tourism and health guidance for type 2 diabetes. Nutr Diabetes. (1), 33-40, (2020).

33. ML. Smith, L. Zhong, S. Lee, SD. Towne, MG. Ory. Effectiveness and economic impact of a diabetes education program among adults with type 2 diabetes in South Texas. BMC Public Health. (21),16-46, (2021).

34. CC. Sweet, CB. Jasik, A. Diebold, A. DuPuis, B. Jendretzke. Cost Savings and Reduced Health Care Utilization Associated with Participation in a Digital Diabetes Prevention Program in an Adult Workforce Population. J Health Econ Outcomes Res. (2), 39–47, (2020).

35. AJ. Vickers, B. Van Calster, EW. Steyerberg. Net benefit approaches to the evaluation of prediction models, molecular markers, and diagnostic tests. BMJ.(3), 52-66 (2016).

36. RM. Turner, Q. Ma, K. Lorig, J. Greenberg, AR. DeVries. Evaluation of a Diabetes SelfManagement Program: Claims Analysis on Comorbid Illnesses, Health Care Utilization, and Cost. J Med Internet Res. (6), 207-211, (2018).

37. SH. Golden, N. Maruthur, N. Mathioudakis, E. Spanakis, D. Rubin, M. Zilbermint and FH. Briggs. The Case for Diabetes Population Health Improvement: Evidence-Based Programming for Population Outcomes in Diabetes. Curr Diab Rep. (7), 51-60, (2017).

38. TP. Gilmer, PJ. O'Connor, WA. Rush, AL. Crain, RR. Whitebird, AM. Hanson, LI.Solberg. Predictors of health care costs in adults with diabetes. Diabetes Care. (1), 59–64, (2005).

39. H. Heun-Johnson, KV. Zuluaga, M. Menchine, S. Starkey, RF. David, S. Seabury. Assessing the Use of Data Systems to Estimate Return-on-Investment of Behavioral Healthcare Interventions: Opportunities and Barriers. J Behav Health Serv Res. (1), 80–94, (2023).

40. P. Soewondo, A. Ferrario, DL Tahapary. Challenges in diabetes management in Indonesia: a literature review. Global Health. 2013 Dec 3;9:63.

Risk of Adverse Drug-Drug Interactions in Heart Failure Patients with Co-morbidity Chronic Kidney Disease Prescribed Polypharmacy

Aldi Suryaman[1], *Mohammad* Bakhriansyah[2*], *Intan* Yustikasari[3,4], *Pandji* Winata Nurikhwan[5], *Dwi* Laksono Adiputro[3,4]

[1]School of Medicine (Clinical Program), Faculty of Medicine, Universitas Lambung Mangkurat, Banjarmasin, Indonesia
[2]Department of Pharmacology, School of Medicine (Academic Program), Faculty of Medicine, Universitas Lambung Mangkurat, Banjarmasin, Indonesia
[3]Department of Cardiology, Faculty of Medicine, Universitas Lambung Mangkurat, Banjarmasin, Indonesia
[4]Department of Cardiology and Vascular Medicine, Ulin Public Hospital, Banjarmasin, Indonesia
[5]Department of Medical Education, School of Medicine (Academic Program), Faculty of Medicine, Universitas Lambung Mangkurat, Banjarmasin, Indonesia

Abstract. Heart failure (HF) is a complex clinical syndrome due to the impairment of myocardial function, valvular or pericardial diseases, or anything interfering blood flow leading to fluid retention. A comorbidity that can accompany HF patients is chronic kidney disease (CKD). This condition requires consumption of several drugs simultaneously (polypharmacy). This study aimed to determine the risk of adverse drug-drug interactions in HF patients with co-morbid CKD prescribed polypharmacy. This is an analytic observational study with a cross-sectional approach in the Cardiac Outpatient Clinic of Ulin Public Hospital, Banjarmasin using medical records from 2020-2021. The interactions were determined using the UpToDate® under the Lexicomp tool. Prevalence Odds Ratio (POR) and 95% Confidence Intervals (95%CI) were determined using Logistic Regression analysis. Of 27 patients, 17 were males (62.96%). Most of the interactions (92.92%) were under category C and the moderate category (82 interactions, 82.83%). For HF patients with CKD, polypharmacy increased the risk of adverse drug-drug interactions by 2.75 times compared to those who were not prescribed polypharmacy, but it was not statistically significant (crude POR 2.75, 95% CI; 0.248-30.512). In conclusion, there is no significant relationship between polypharmacy and the risk of adverse drug-drug interactions in HF patients with CKD.
Keywords: polypharmacy, drug-drug interactions, heart failure, chronic kidney disease, Ulin Public Hospital, Banjarmasin.

1 Introduction

Heart failure (HF) is a complex clinical syndrome as a result of disturbances in myocardial function (systolic and diastolic functions), valve or pericardial diseases, or anything that can interfere blood flow with fluid retention as a result. It is usually manifested as pulmonary congestion, peripheral edema, dyspnoea, and fatigue [1]. As many as 64.3 million people in worldwide is estimated to suffer from HF. In developed countries, the prevalence of HF is generally estimated to be 1% to 2% of the adult population [2]. According to the Sample Registration System (SRS) survey conducted in 2019, heart diseases are the second leading cause of death in Indonesia following stroke [3]. In 2013 the Riset Kesehatan Dasar (Riskesda; Basic Health Research) demonstrated that according to doctors' diagnosis the prevalence of HF in Indonesia is estimated to affect around 229,696-530,068 people (0.13%-0.30%) [4].

HF patients usually have co-morbidities such as chronic kidney disease (CKD). Approximately 70% of the causes of death in patients with CKD are caused by heart diseases. According to Riskesdas in 2018, the prevalence of CKD affected as many as 713,783 people (0.38%) in Indonesia and 11,068 people (0.32%) in the Province of South Kalimantan. The incidence of HF with CKD ranges from 17% to 21% [5].

Drug interactions can be exploited to gain therapeutic advantages and a combination therapy might lead to the optimal treatment. However, these interactions can cause adverse side effects for the patients. Drug interactions commonly occur because of rarely documented and missed observed. Thus the adverse side effects were considered as an increased of disease severity. Around 5-15% of patients are estimated to experience adverse drug interactions that are clinically significant. Risk factors for drug interactions are identified including age, genetic polymorphism, co-morbidities, dosage, number of physicians prescribing drugs, self-medication, and polypharmacy [6].

Polypharmacy can increase the risk of drug-drug interactions. More than 80% of elderly patients take at

* Corresponding email: bakhriansyah@gmail.com

least 1 prescribed drug and more than half of them take 5 or more drugs concomitantly. The higher the number of patients who consume a lot complex drug or regimen, the higher the risk and the more vulnerable to drug-drug interactions [6].

Patients with HF generally have decreased function organs and have co-morbidities that require to take several drugs simultaneously. About 52% of drug categories involved in drug-drug interactions are cardiovascular drugs that lead to an adverse effect on the body ranging from mild to severe levels, such as syncope and the death [6].

There are a limited number of study assessing the risk of adverse drug interaction for patients with cardiovascular diseases patients with co-morbidities. Furthermore, we have not found any studies on the risk of adverse drug-drug interactions specifically for HF patients with comorbidity. Thus, this study aims to assess the risk of adverse drug-drug interaction for HF patients with CKD co-morbid prescribed polypharmacy in Indonesia, especially in South Kalimantan Province. We then performed this study by using secondary data (medical records) at outpatients Cardiology Clinic at Ulin Public Hospital, Banjarmasin, Indonesia.

2 Methods

2.1 Research Design and Data Source (Setting)

This was an observational study with cross sectional approach among HF patients with CKD comorbidity at the Cardiology Outpatient Clinic at Ulin Public Hospital Banjarmasin, Indonesia. Ulin hospital is a referral hospital in the province of South Kalimantan and provinces around. The patients' demographic (sex and age) and medical data (co-morbidities and drug used) were retrospectively collected from their medical records over the period 2020-2021. Only adult patients (>18 years old) were included in this study. Our study protocol had been registered to and approved by the Ethic Committee for Health Research, Medical Faculty, Universitas Lambung Mangkurat, Banjarmasin (No. 568/KEPK-FK ULM/EC/XII/2022).

2.2 Research Variables

Variables of interest in this study were number of drugs prescribed by the cardiologists and the risk of drug-drug interactions. For this study, polypharmacy was determined when a patient was prescribed 5 drugs or more simultaneously [7], otherwise non-polypharmacy.

According to UpToDate® website under Lexicomp tool, drug-drug interaction was classified into Category A (no known interaction); Category B (no further actions needed); Category C (benefits over risks, but therapy should be closely monitored to prevent harm); Category D (considering to modify therapy and accurate actions should be taken to reduce toxicity); and Category X (contraindicated) [8]. In this study, we lumped category A, B, and C as "less risk of adverse drug-drug interaction" and category D and X as "high risk of adverse drug-drug interaction". Meanwhile, the severity of interaction was categorized into **minor** (mild effects, well overcome), **moderate** (moderate effect, may lead to organ damage) and **major** (fatal effect, may lead to the death).

2.3 Statistical Analyses

Demographic and medical data, category of drug-drug interactions, and severity level of the interaction were presented in proportion (%), but number of drugs prescribed was presented in an average number.

In order to estimate the risk of adverse drug-drug interaction among HF patients with CKD prescribed polypharmacy compared to non-polypharmacy, we applied Logistic Regression test. We then presented the data as crude Prevalence Odds Ratios (PORs), 95% Confidence Intervals (95%CIs), and p-value. All the analyses were conducted using a statistical software IBM SPSS version 26. p-value of less than 0.05 was considered statistically significant.

3 Results and Discussion

3.1 Results

We identified 27 HF patients with co-morbid CKD over the period 2020-2021 at the Cardiology Outpatient Clinic, at Ulin Public Hospital, Banjarmasin, Indonesia. Most of the patients were males (17 patients, 63.0%), elderly age (17 patients, 63.0%), and without a history of hypertension (23 patients, 85.2%). In average, 3.7 drugs was prescribed per patient (**Table 1**).

Table 1. Baseline characteristics of heart failure patients with chronic kidney failure

Characteristics	n=27
Sex, n (%)	
● Males	17 (63.0)
● Females	10 (37.0)
Age, n (%)	
● 18-60 years	10 (37.0)
● >60 years	17 (63.0)
Hypertension, n (%)	
● Yes	4 (14.8)
● No	23 (85.2)
Number of drugs prescribed per patient, mean	3.7

For 27 patients, there were 100 drug-drug interactions. Most of the interactions were under category C (92 interactions, 92%), followed by category D and B, i.e., 4 interactions (4%) and 3 interactions (3%), respectively. In contrast, 1 prescription had no known interaction (0%) and no prescription was under category X (0 prescription, 0%) (**Table 2**).

All interactions under category B were classified as moderate. Under category C, interaction between isosorbite dinitrate and candesartan were frequently reported with moderate level of its severity, while under category D, the interaction between warfarin and acetylsalicylic acid was mostly counted with major level of severity (**Appendix**).

According to its severity level, most of drug-drug interactions were moderate (82 interactions, 82.2%), followed by major (11 interactions, 11.1%) and finally minor (6 interactions, 6.1%). Meanwhile, 1 prescription did not have a known interaction (**Table 3**).

As being demonstrated in **Table 4**, HF patients with CKD prescribed polypharmacy were associated with a higher risk of the adverse drug-drug interaction compared to those without polypharmacy by 2.75 times (crude POR 2.750, 95%CI; 0.248-30.512). However, the risk was not statistically significant (p=0.410).

Table 2. Categories of drug-drug interaction

Category	n (%)
• A (no interaction known)	1 (0.0)
• B (no action needed)	3 (3.0)
• C (benefits are outweigh of the risks, but the therapy should be under strict supervision to avoid a potency of harm)	92 (92.0)
• D (considered to modify the therapy and to take the proper action in order to reduce its toxicity)	4 (4.0)
• X (contraindicated, should be avoided)	0 (0.0)
Total	100 (100.0)

Table 3. Level of severity

Category	n (%)
Minor (mild effects, well overcome)	6 (6.1)
Moderate (moderate effect, may lead to organ damage)	82 (82.8)
Major (fatal effect, may lead to the death)	11 (11.1)
Total	99 (100.0)

Table 4. Prevalence odds ratio for the risk of adverse drug-drug interactions for heart failure patients with chronic kidney failure complication prescribed polypharmacy compared to those without polypharmacy

Poly-pharmacy, n (%)	Adverse drug-drug interactions, n (%)		p-value	Crude POR (95%CI)
	High Risk	Less Risk		
Yes	3 (75.0)	79 (82.3)	0.410	2.750 (0.248-30.512)
No	1 (25.0)	17 (17.7)		1

CI= confidence intervals; POR = prevalence odds ratio

3.2 Discussion

In our study most commonly drugs found in drug-drug interaction were furosemide (38 interactions, 38%), isosorbide dinitrate (33 interactions, 33%), and candesartan (22 interaction, 22%). Interaction under category C with the most major severity were ramipril – spironolactone and candesartan – spironolactone. The use of Angiotensin Converting Enzymes-I (ACE-I) or Alpha Receptor Blockers (ARB) drugs in combination with spironolactone may increase serum potassium by ±0.19 mEq/L compared to use single drug of ACE-I or ARB [8]. A combination of isosorbide dinitrate and candesartan has a moderate level of severity. Candesartan is able to increase antihypertension effect of isosorbite dinitrate. This combination may cause organ damage.

Drug-drug interactions under category D consisted of Warfarin – Acetylsalicylic acid, Captopril – Valsartan, and Ramipril – Candesartan. Interaction of Warfarin – Acetylsalicylic Acid has the greatest potential for adverse drug-drug interactions in this study due to its severity level major and fall under category D. Low-dose acetylsalicylate acid and warfarin can be used together in a certain case with close monitoring for increasing signs and symptoms of bleeding. Higher doses of acetylsalicylic acid and other salicylates generally should be avoided. A combination of warfarin and acetylcalycilic acids might be fatal, such as the death. The later drug might increase anti coagulation effect of warfarin.

Captopril–Valsartan and Ramipril–Candesartan interactions include interactions between ACE-I and ARBs drug classes. If the combination should be used, the patients should be monitored very closely for a greater therapeutic response expected for the combination, including monitoring of blood pressure, kidney function, and potassium concentration. For increasing the likelihood of successful initiation. An ACE-I or ARB should be initiated at a lower dose and titrate more slowly if accompanied by chronic kidney failure [9].

A study conducted by Akbar et al concluded that a ratio of potential adverse drug-drug interactions that are not detrimental and detrimental, i.e., 3.6:1 among patient population with general cardiovascular diseases. While the ratio in this research was 2.75:1. It indicates that HF patients with CKD co-morbidity are less likely to experience the adverse interactions than for those general cardiovascular patients [8]. With these risks, caution is needed and accuracy in prescribing drugs, especially in polypharmacy, as well monitoring the patient after being given the drug [6, 10].

The risks of potential adverse drug-drug interactions increase if the number of drugs consumed increase. About 13% of patients taking 2 drugs experienced adverse drug-drug interactions. For those who take 4 drug and >7 drugs the interactions increase to 38% and 82%, respectively [11]. A study by Akbar et al, demonstrated that there was a significant relationship between the number of drugs by >12 and its interactions, even though there was no significant relationship between the number of drugs by 6-12 and drug interactions [9]. Meanwhile, a research conducted by Fatin et al among patients with pneumonia showed that the number of drugs by >6 was associated with 10.1 times of adverse drug-drug interactions [12]. It indicates that a patient's diagnosis and the type of drug also affect adverse drug-drug interactions.

3.3 Potential Clinical Implication

Our findings provide a scientific evidence on the potency of higher risk of adverse drug-drug interaction for HF patients with CKD prescribed polypharmacy. Clinicians should then carefully take into consideration the use of polypharmacy for those patients, especially for elderly. Ultimately, the risks of harm or even fatal incidents affecting patients are being able to avoid.

A review of patient prescriptions for potential adverse drug-drug interactions should be performed by using a proper software and/or clinical pharmacists experienced in drug interactions. In addition, it is necessary to monitor patients' responses and related side effects to find adverse interaction reactions [13].

3.4 Strengths and Limitations

We need to mention several strengths in this study. First, to the best of our knowledge this is the first study assessing the risk of adverse drug-drug interaction for HF patients with CKD prescribed polypharmacy in Indonesia, especially Banjarmasin. Second, Lexicomp® is the most commonly used tool to assess drug interaction. It also classifies drug-drug interaction based on its severity. Finally, this tool has high sensitivity and specificity.

However, this study is a subject to several weakness. First, instead of the actual adverse events, adverse drug-drug interactions were defined theoretically by using a computer tool as a proxy. It might lead to a misclassification bias of the outcome. Second, our concern is the issue of small sample size. Lack of sample size may lead to inability to detect small differences or association between our variables of interest. Third, we did not adjust our model to several potential confounder contributing to the risk of adverse drug-drug interaction such as sex, age, co-morbidities, genetic polymorphism, drug dosage, number of physicians prescribed, and self medication. Finally, the tool Lexicomp® does not provide information on interactions for more than 2 drugs. This tool has low reliability.

4 Conclusion

Our study indicated that HF patients with CKD co-morbid prescribed polypharmacy were associated with a high risk of adverse drug-drug interaction by 2.75 times compared to those without polypharmacy. However, this association was not statistically significant (crude POR 2.75, 95%CI 0.248-30.512).

5 Acknowledgment

We thank all patients contributing in this study.

6 References

1. Perhimpunan Dokter Spesialis Kardiovaskular Indonesia. Pedoman tatalaksana gagal jantung. Perhimpunan Dokter Spesialis Kardiovaskular Indonesia. 2020;2:12-91.
2. Groenewegen AH, Rutten F, Mosterd AW, Hoes A. Epidemiology of heart failure. European Journal of Heart Failure. 2020;22:1343-6.
3. Juliar AR. Manifestasi klinis dan tatalaksana gagal jantung. Jurnal Kedokteran Nanggroe Medika. 2020;3(3):36-45.
4. Riset Kesehatan Dasar (Riskesdas). Badan Penelitian dan Pengembangan Kesehatan Kementerian RI. 2013.
5. Riset Kesehatan Dasar (Riskesdas). Badan Penelitian dan Pengembangan Kesehatan Kementerian RI. 2018.
6. Sulistiyowatiningsih E, Nurul Hidayati S, Febrianti Y. Kajian interaksi obat pada pasien gagal jantung dengan gangguan fungsi ginjal di Instalasi Rawat Inap RSUP DR. Sardjito Yogyakarta periode 2009-2013. Jurnal Ilmiah Farmasi. 2016;12(1):25-31.
7. Khezrian M, McNeil C, Murray A, Myint P. An overview of prevalence, determinants and health outcomes of polypharmacy. Therapeutic Advances in Drug Safety. 2020;11:1-7.
8. Akbar Z, Rehman S, Khan A, Khan A, Atif M, Ahmad N. Potential drug–drug interactions in patients with cardiovascular diseases: findings from a prospective observational study. Journal of Pharmaceutical Policy and Practice. 2021;14(1).
9. Villa-Zapata L, Carhart BS, Horn JR, Hansten PD, Subbian V, Gephart S, et al. Serum potassium changes due to concomitant acei/ARB and Spironolactone Therapy: A systematic review and meta-analysis. American Journal of Health-System Pharmacy. 2021;78(24):2245–55.
10. Herdaningsih S, Muhtadi A, Lestari K, Annisa N. Potensi interaksi obat-obat pada resep polifarmasi: Studi retrospektif pada salah satu apotek di Kota Bandung. Jurnal Farmasi Klinik Indonesia. 2016;5(4):289-91.
11. Magro L, Moretti U, Leone R. Epidemiology and characteristics of adverse drug reactions caused by drug–drug interactions. Expert Opinion on Drug Safety. 2011;11(1):83–94.
12. Fatin M, Pasha Y. Potensi interaksi obat dengan obat pada pasien dewasa dengan pneumonia. Journal of Pharmacopolium. 2021;4(2):98-104.
13. Georgiev K, Hvarchanova N, Georgieva M, Kanazirev B. Potential drug-drug interactions in heart failure patients. International Journal of Pharmacy and Pharmaceutical Sciences. 2019; 37–41.

Appendix

Category	Drug-drug Interaction	Quantity (n)	Level of severity	Effect(s)
A	Warfarin - Atorvastatin	1	-	-
B	Acetylcalicylic acid - $CaCO_3$	1	Minor	$CaCO_3$ decreases serum calicylic consentration
	Levofloxacin - Candesartan	1	Moderate	Candesartan increases aritmogenic effect of levofloxacin. Levofloxacin increases nephrotoxic effect of candesartan
	Warfarin - Spironolacton	1	Moderate	Spironolacton decreases anticoagulant effect of warfarin
C	Isosorbide dinitrate - Candesartan	9	Moderate	Candesartan increases hypotension effect of isosorbide dinitrate
	Candesartan - Furosemide	8	Moderate	Furosemide increases hypotension and nephrotoxic effects of candesartan.
	Bisoprolol - Furosemide	7	Moderate	Furosemide increased hypotension effect of bisoprolol
	Isosorbide dinitrate - Bisoprolol	7	Moderate	Bisoprolol increases hypotensive effect of isosorbide dinitrate
	Isosorbide dinitrate - Furosemide	7	Moderate	Furosemide increases hypotensive effect of isosorbide dinitrate
	Spironolcaton - Furosemide	6	Moderate	Furosemide increases hypotensive effect of spironolacton
	Furosemide - Acetylcalicylic acid	5	Moderate	Acetylcalicylic acid reduces diuretic effect of furosemide. Furosemide increases serum calicylic consentration
	Isosorbide dinitrate - Amlodipine	4	Moderate	Amlodipine increases hypotensive effect of isosorbide dinitrate
	Ramipril - Furosemide	4	Moderate	Furosemide increases hypotensive and nephrotoxic effects of ramipril.
	Spironolacton - Acetylcalicylic acid	4	Minor	Acetylcalicylic acid reduces therapeutic effects of spironolacton
	Ramipril - Spironolacton	4	Major	Spironolacton increases hyperkalemia effect of ramipril
	Isosorbide dinitrate - Ramipril	3	Moderate	Ramipril increases hypotensive effect of isosorbide dinitrate
	Isosorbide dinitrate - Spironolacton	3	Moderate	Spironolacton increases hypotensive effect of isosorbide dinitrate
	Spironolacton - Candesartan	3	Major	Candesartan increases hyperkalemia effect of spironolacton
	Amlodipine - $CaCO_3$	2	Moderate	$CaCO_3$ reduces therapeutic effects of amlodipine
	Clopidogrel - Amlodipine	2	Moderate	Amlodipine reduces therapeutic effects of clopidogrel
	Spironolacton - Atorvastatin	2	Moderate	Atorvastatin increases detrimental/toxic effects of spironolacton, specifically theoretically it increases reduced effect of endogen steroid activity
	Bisoprolol - Digoxin	2	Moderate	Bisoprolol increases bradycardia effect of another drug and vice versa
	Digoxin - Spironolacton	2	Moderate	Spironolacton increases serum digoxin concentration and impair a test for determination of digoxin concentration; false increased or decreased concentration of digoxin
	Clopidogrel - Lansoprazole	1	Major	Lansoprazole reduces serum concentration of active metabolite of clopidogrel
	Allopurinol - Furosemide	1	Moderate	Furisemide increases detrimental/toxic effects and increases serum concentration of allopurinol, specifically furosemide increases oxypurinol concentration, active metabolites of allopurinol
	Allopurinol - Ramipril	1	Major	Ramipril increases potency of allergic reaction or hypersensitivity towards allopurinol
	Allopurinol - Captopril	1	Major	Captopril increases potency of allergic reaction or hypersensitivity towards allopurinol
	Captopril - $CaCO_3$	1	Moderate	$CaCO_3$ decreases serum captopril concentration
	Nifedipine - $CaCO_3$	1	Moderate	$CaCO_3$ decreases therapeutic effect of nifedipine
	Vitamin D - $CaCO_3$	1	Moderate	$CaCO_3$ increases detrimental.toxic effects of vitamin D analog
D	Warfarin -	2	Major	Calicylic increases anticoagulant effect of warfarin

	Acetylcalicylic acid			
	Captopril - Valsartan	1	Moderate	Valsartan increases toxic effect of captopril and serum captopril concentration
	Ramipril - Candesartan	1	Moderate	Candesaran increases toxic effect and serum ramipril concentration of ramipril
X	-	0	-	-

The acceptability of and willingness to pay for a herpes zoster vaccine: A systematic review

Hien Thi Bich Tran[1,*], *Trung* Quang Vo[1], *Tuong* Thi Khanh Tran[2], *Ha* Thuy Nhu Nguyen[1], *Han* Nguyen Ngoc Le[1], *Quan* Ngoc Bui Tran[1], *Hanh* Dung Diep[1], *Quyen* Tai Ung[2], and *Susi* Ari Kristina[3]

[1] Faculty of Pharmacy, Pham Ngoc Thach University of Medicine, Ho Chi Minh City 700000, Vietnam.
[2] Faculty of Medicine, Pham Ngoc Thach University of Medicine, Ho Chi Minh City 700000, Vietnam.
[3] Faculty of Pharmacy, Universitas Gadjah Mada, Yogyakarta 55281, Indonesia.

Abstract Patients, predominantly the elderly, with Herpes Zoster (HZ) not only suffer symptoms of the disease but also bear considerable expenses. This study systematically reviewed the acceptability of and willingness to pay for the HZ vaccine. This review was registered in PROSPERO 2023 (CRD42023403062). We used "acceptance", "willing to pay", and "HZ vaccine" (and variations thereof) as keywords in a systematic search for original English research articles published up to April 7, 2023. The search was conducted over Scopus, PubMed, ScienceDirect, Cochrane, and Google Scholar in accordance with PRISMA 2020 guidelines. The inclusion criteria were as follows: studies (1) that mentioned HZ vaccination, (2) related to acceptability or willingness to pay, and (3) with full texts available and peer-reviewed prior to final publication. Grey literature, letters to editors, commentaries, case reports or series, systematic reviews, meta-analyses, articles of poor quality, and articles with ambiguously defined and measured outcome variables were excluded. The Joanna Briggs Institute (JBI) critical appraisal checklist was used to evaluate the methodological quality of the studies. Finally, the search yielded 24 studies, of which 9 were conducted in Asia, 8 in Europe, and 7 in America. General adults or patients aged 50 or older were often the target populations, for whom treatments were accompanied by healthcare providers' recommendations. The willingness to pay and willingness to accept the vaccine ranged from $8 to $150 and 16.6% to 85.8%, respectively. Compared to the US, Asia and Europe had higher acceptance rates for HZ immunization. The most frequent excuses given for not being vaccinated are side effects, cost, lack of recommendations, anti-vaccination views, ignorance about the HZ vaccine, and the belief that one is not at risk for the disease. National campaigns should be developed to increase public awareness of HZ, and more international research should be conducted to understand the WTA and WTP for HZ immunizations.

Keywords: Acceptability, herpes zoster, systematic review, willingness to pay, vaccination.

1 Introduction

Herpes zoster (HZ), also known as zona or shingles, is caused by the varicella zoster virus (VZV) or human herpesvirus 3. VZV belongs to the *Herpesviridae* family, and its target tissue is the human nervous system. It is not completely eliminated but exists in latent, inactive form in a host's body, especially in the entire nervous system. This persistence can be explained by the similarity of the VZV genome to that of a human (double-stranded DNA). Accordingly, VZV may be reactivated and cause HZ, damaging skin cells, at any time under favorable conditions. Although there is no clear explanation of the reactivation of the virus, scientists have found that cellular-mediated immunity (CMI) is related to the pathogenesis of HZ [1]. More specifically, CMI maintains the latency of VZV in the human body, thereby diminishing CMI function and leading to virus reactivation. Because CMI decreases with age [1], HZ predominantly affects the elderly [2], among whom one of the most common complications is post-herpetic neuralgia (PHN). PHN is a long-lasting pain that can become a chronic condition that ultimately exerts negative effects on the quality of life and daily activities of patients [1] [2]. PHN is a complication of HZ, occurring with a frequency of 5% to 30%, and each year, the global HZ incidence ranges from 3.0 to 5.0 for every 1000 individuals [1].

A study in Thailand showed that the HZ morbidity rate increases gradually with age and the disease spreads

* Corresponding email: hienttb@pnt.edu.vn

across many countries, such as the UK, France, Germany, the US, Canada, Taiwan, and Australia) [3]. The median age of HZ onset is 59.4 years, and 68% of cases are individuals older than 50 years [3]. HZ imposes considerable health care, economic, and societal burdens on countries. HZ patients not only suffer the symptoms of the disease but also bear significant expenses, including the costs of treatments, medical examinations, diagnostic tests, hospitalization, and emergencies [1]. In developed countries, the direct annual costs (e.g., healthcare utilization and medicine costs) of HZ and its complications range from US$2.7 million to US$2.6 billion, while their indirect yearly costs (e.g., loss of productivity and absenteeism) range from US$1.7 to US $241.5 million [4]. To prevent HZ, adults are prescribed and administered a zoster vaccine, which is regarded as effective when it functions as a therapeutic vaccine and induces a more potent immune response to prevent reactivation in an already infected individual with pre-existing immunity to VZV [5]. There are two types of vaccines: a live attenuated vaccine (Zostavax®) and a recombinant subunit vaccine (HZ/su) (Shingrix®) [6].

In the development of a country's vaccination policy, an important index is vaccine acceptance (willingness to accept, WTA), and a high WTA is more likely to result in successful immunization programs [7]. Therefore, studies on WTA are necessary because such investigations reveal patients' assessments of vaccine suitability and vaccination readiness [7]. The belief in and acceptance of a vaccine depends not only on the safety and effectiveness of these preparations but also on health care delivery systems as well as policymakers who develop vaccination requirements [8]. This reality means that a partnership between academic researchers and governments is needed to integrate evidence-informed strategies into vaccination policies and programs, thereby enhancing vaccine WTA [9]. Another important index in this process is willingness to pay (WTP), which reflects how a product is valued by customers. The WTP for vaccines is a monetary indicator that reveals the amount of money that a person is willing to spend to derive the benefits of vaccines [10]. Early scientific investigations of WTP inform policy decisions on sustainable financing mechanisms for vaccination campaigns, particularly in resource-limited countries [11].

Currently, there are relatively few studies on the WTA and WTP for HZ vaccines, and these constructs are mainly treated separately in these studies. For example, Binshan Jiang et al. explored the WTA of HZ vaccines in China and reported that 43.02% of the participants intend to get vaccinated against HZ [12]. Another representative work is that of Eilers et al., who estimated the potential HZ vaccination rate among older adults to be 58.1% [13]. A 2016 study in Italy involving 1001 participants whose mean age was 67 years found that 58% of surveyed individuals support HZ vaccination campaigns and that 73% of them are willing to pay for a vaccine at an ideal cost of €50 [14]. Finally, research in Thailand which was conducted from December 2013 to December 2014 and recruited 118 zoster patients older than 18 years from the Dermatologic Clinic of the Outpatient Department at Siriraj Hospital, showed that HZ vaccines have yet to be included in vaccination policy and that knowledge about HZ among Thai patients remains limited. Although the authors found no data on the WTP for HZ vaccines, they found that the WTP for HZ treatment was THB500 (range: THB50-10,000) or only 4.2% of people's median income per month [15].

As can be seen, some studies have been devoted to the WTA and WTP for HZ vaccination, but only a brief systematic review of these works has been made [7]. A complete and comprehensive systematic review of research on the issues of interest is necessary to determine the intention to acquire vaccination against HZ and promote vaccination among the public. The results of such reviews cannot only contribute to other research and practices but also serve as reference for helping national health systems develop vaccination policies for high-risk groups.

2 Materials and methods

2.1 Research questions

The research questions that guided this work are as follows:

- What is the global level of acceptance of an HZ vaccine?
- What is the WTP for such a vaccine globally?

2.2 Study setting

This systematic review was directed toward studies that administered global surveys.

2.3 Data source and search strategy

Articles were searched from academic thematic databases, including PubMed, ScienceDirect, Scopus, the Cochrane Library, and Google Scholar. The keywords used in the search were "acceptability," "herpes zoster," "willingness to pay," "vaccination," and variations thereof. The search strategies are detailed in **Supplementary File 1**.

2.4 Ethical approval

Ethical approval for this systematic review was not necessary because data were collected from previous studies that were granted such approval. However, the review protocol was registered at the International Prospective Register of Systematic Reviews (registration number: PROSPERO 2023 CRD42023403062).

2.5 Inclusion and exclusion criteria

Studies were included in the analysis if they met the following criteria:

(1) Studies based on PICO elements (population: participants worldwide;

intervention: HZ vaccine; comparison: none; outcome: acceptance and/or WTP; study type: cross-sectional surveys);

(2) Studies on attitudes, reluctance, and/or barriers to HZ vaccine acceptability among a given population;
(3) Papers with the full texts available;
(4) Original research involving cross-sectional surveys and including quantitative, qualitative, or mixed-methods analyses and peer-reviewed research;
(5) Studies published in English.

Studies that were excluded were as follows:

(1) Gray literature, including presented abstracts, letters to editors, commentaries, case reports or series, systematic reviews, and meta-analyses;
(2) Studies focusing on other types of intervention (e.g., surgery, drug administration, radiotherapy);
(3) Poor-quality articles and articles in which the outcome variables were ambiguously defined and measured;
(4) Research with insufficient or no information about the acceptability of or WTP for HZ vaccination.

2.6 Main outcomes

In this systematic review, the primary outcomes of interest were the WTA and WTP for HZ vaccines.

2.7 Data extraction

A quality assurance process was implemented in stages during the critical appraisal of studies. All the articles identified during the database search were downloaded to EndNote version 8 (http://endnote.com/), and duplicate articles were removed from the list. The remaining articles were screened at the title/abstract level, after which a full-text review was conducted by one reviewer. Double screening and record validation was performed by another reviewer. The data extracted included the titles, authors, years of publication, journals, study designs, durations of time search, number of participants, demographic characteristics of participants, WTA (in percentage), and WTP mean ± SD or median.

Two reviewers independently evaluated the full texts and checked for relevance, and discrepancies were resolved by discussion or with reference to a third reviewer if consensus could not be reached. The search method was presented in a PRISMA flow chart showing the included and excluded studies. In the case of missing data and/or additional details, an investigation was carried out by reaching out to the corresponding authors of the reviewed studies. The data was entered into a Microsoft Excel spreadsheet.

2.8 Quality assessment

The risk of bias (quality of research articles) was assessed independently by two researchers, with a third consulted upon a lack of consensus. Two independent reviewers also assessed the quality with which the findings of the articles were reported, and disagreements were resolved through discussion. The assessment tool used in the quality assessment was the Joanna Briggs Institute's (JBI) critical appraisal checklist for analytical cross-sectional studies. The results of the evaluation are presented in **Table 3**.

The above-mentioned checklist was also used to analyze the risk of reporting bias [16]. Eight questions related to the following points were used for this purpose: (1) a clear definition of criteria for inclusion in the sample, (2) detailed descriptions of study subjects and settings, (3) validity and reliability of exposure measurements, (4) objective and standard criteria for measuring the condition, (5) the definition of confounding factors, (6) strategies for dealing with confounding factors, (7) the validity and reliability of outcome measurement, and (8) appropriateness of statistical analyses. The satisfaction of each criterion was denoted by "yes," "no," "unclear," or "not applicable." The risk of bias was considered low, moderate, and high when more than 70%, 50% to 69%, and 0% to 49% of the criteria received a response of "yes" [17].

2.9 Data reporting

The overall process and results of this systematic review were documented according to the flowchart of the Preferred Reporting Items for Systematic Review and Meta-Analysis (PRISMA) [18]. The PRISMA checklist is shown in **Figure 1**.

2.10 Data synthesis

We did not synthesize the results of the data analyses to address the heterogeneity of the cross-sectional survey methods and outcome measures used in the examined studies. Instead, the results were presented using a qualitative synthesis approach that was aimed at identifying methodological and population differences. This systematic review used percentage (%) and US dollars as units of measurement for WTA and WTP, respectively. In cases where countries' currency values reported in studies were not US dollars, units of this parameter were converted on the basis of the exchange rate during the year at which a given study was carried out. The conversion was done on a free online website.

3 Results

The initial database search yielded 17,822 papers, after which title and abstract screening resulted in the exclusion of 17,735 of these. With the removal of duplicates, 44 studies remained. The final sample comprised 24 (0.13%) articles [12-14, 19-39], letters, articles with no full-text version or English version, articles with no data on WTA or WTP for HZ vaccines were excluded from the study (**Figure 1, Table 1**). Among 24 included studies, 21 studies reported the rate of vaccine acceptance, 15 studies were concerned with the refusal or reluctance to receive HZ vaccines, and 4 studies reported the WTP for HZ vaccination. Most of the studies were published between 2021 and 2023, and they were conducted across three continents: Asia, America, and Europe. The target groups were primarily members of the general population aged ≥ 50 years, and the analysis software mainly used was the Statistical Package for the Social Sciences. Acceptability rates among the general population ranged from 16.6% to 85.8%, whereas those among patients ranged from 25.4% to 32%. The willingness to accept HZ vaccines ranged from 55% to 86.9% and was 74.2% among the general population and patients, respectively, upon recommendation by a healthcare provider. The WTA values for HZ vaccines are presented in **Figure 2**. The factors affecting WTA identified in all the studies were divided into four groups: socio-demographic factors, the perception of risk, concerns about vaccines, and others (**Figure 3**) [40]. As previously stated, the JBI critical appraisal checklist was used to evaluate the methodological quality of the studies (**Table 3**). The risk of bias in the 24 studies ranged from moderate to high.

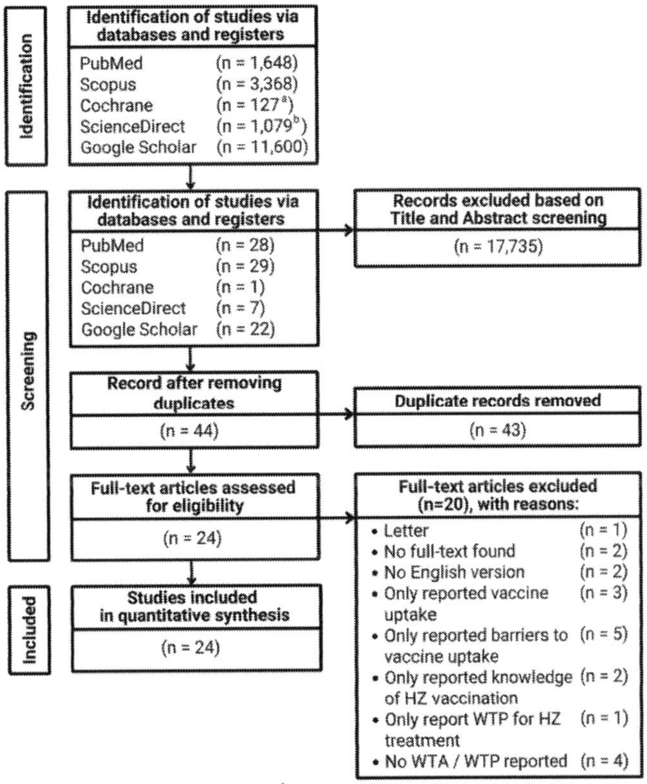

Note: [a]: Only Cochrane reviews; [b]: Only Review articles and Research articles

Fig. 1. PRISMA flowchart (2023).

Table 1. Descriptive summary of the characteristics of the 24 studies in the sample (2023).

Characteristics	WTA		WTP		Refuse/ Hesitancy	
	n	%	n	%	n	%
Publication year	21	100.0	4	100.0	15	100.0
Until 2010	3	14.0	-	-	2	13.0
2011 - 2015	3	14.0	-	-	1	7.0
2016 - 2020	6	29.0	3	75.0	4	27.0
2021 - 2023	9	43.0	1	25.0	8	53.0
Age groups	21	100.0	4	100.0	15	100.0
≥ 50	10	48.0	4	100.0	6	40.0
≥ 60	2	10.0	-	-	2	13.0
≥ 65	2	10.0	-	-	2	13.0
Others	7	32.0	-	-	5	34.0
Target population	21	100.0	4	100.0	15	100.0
General population	12	57.0	2	50.0	7	47.0
Patients	8	38.0	2	50.0	7	47.0
Healthcare provider and General population	1	5.0	-	-	1	6.0
Region	21	100.0	4	100.0	15	100.0
Asia	8	38.1	-	-	7	46.7
Europe	8	38.1	1	25.0	5	33.3
America	5	23.8	3	75.0	3	20.0
Data analysis software	21	100.0	4	100.0	15	100.0
SPSS	9	43	1	25	7	47
R	2	10	-	-	2	13
STATA	2	10	1	25	1	7
SAS	3	14	2	50	1	7
Others	5	23	-	-	4*	26

Note: *: One study does not state its data analysis software

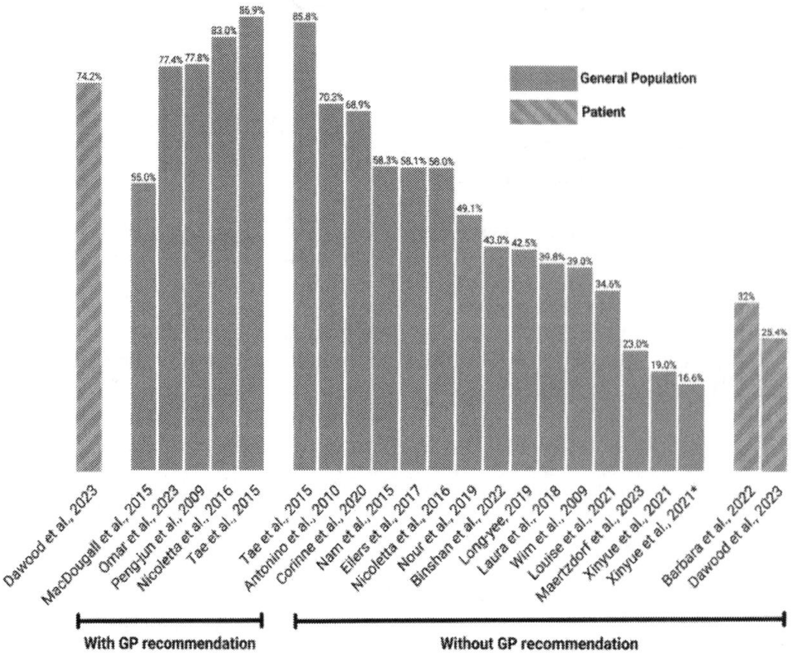

Fig. 2. WTA values for HZ vaccines.

https://doi.org/10.1051/bioconf/20237505017

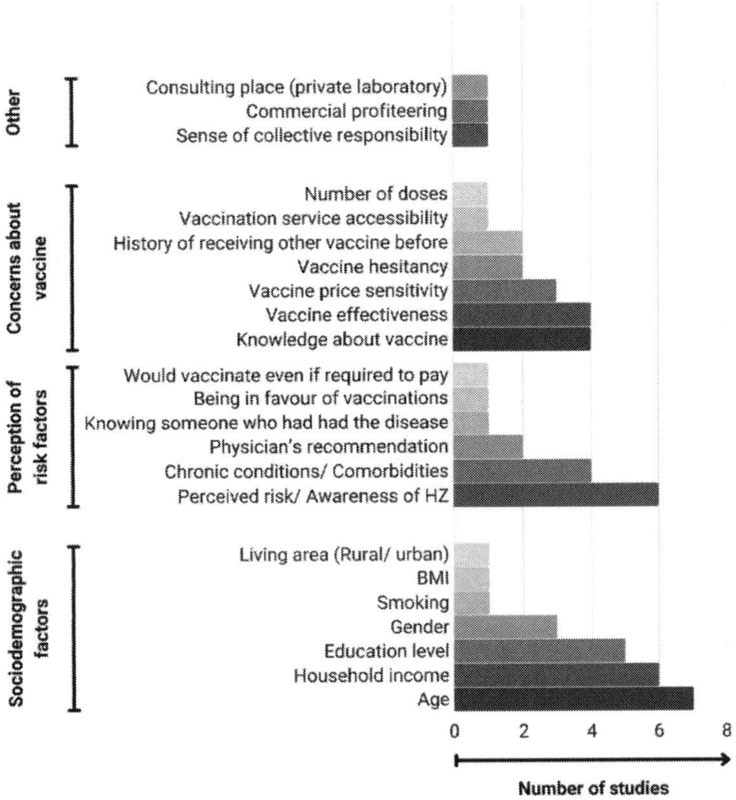

Fig. 3. Factors associated with WTA, as determined using multivariate logistic regression.

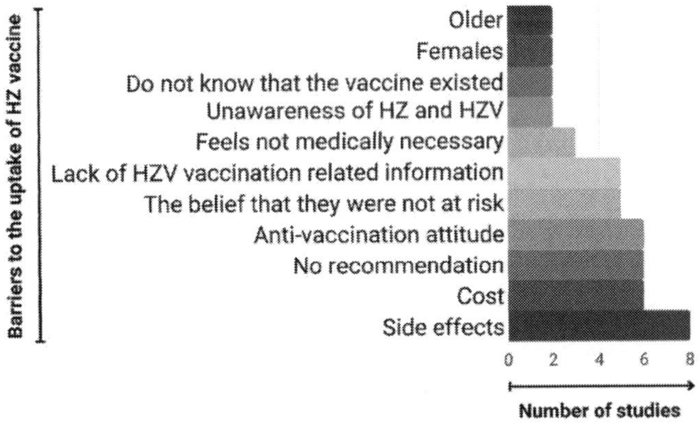

Fig. 4. Barriers to vaccine uptake.

The studies also identified barriers to the uptake of HZ vaccines (**Figure 4**), of which the most popular was side effects (n = 8), followed by cost, the absence of a recommendation, and anti-vaccination attitudes (n = 6). Among the sample, only four mentioned WTP. We found that this parameter oscillates over a wide range of 8 to 150 USD. The participants' preferences for each range of WTP and some of the factors affecting this parameter are presented in **Table 2**.

Table 2. WTP for HZ vaccines and associated factors.

Author, Year	WTP threshold ($USD)	Participant preference (%)	Factors associated with WTP
Brandon J. Patterson et al., 2021	8 - 13 USD*	75%	- Out-of-pocket price - Flu-like symptoms - Injection-site reactions
	140 - 150 USD*	51.20%	
Nour A. Baalbaki et al., 2019	0 USD	31.4%	N/a
	24.99 USD	36%	
	25 - 50 USD	18.7%	
Nicoletta Valente et al., 2016	55.34 USD**	73%	- Higher education - Being in favour of vaccinations in general - General practitioner advice
Osayi E. Akinbosoye et al., 2016	44.06 - 95 USD	N/a	N/a

Note: *: For the second dose; **: Equivalent to €50; N/a: Not available.
€1 = 1.1068 USD (Source: Exchange Rates UK: Euro to US Dollar Spot Exchange Rates for 2016)

Table 3. Risk of bias analysis using the JBI checklist.

Source	Questions								Quality level
	Q1	Q2	Q3	Q4	Q5	Q6	Q7	Q8	
Dawood et al., 2023	1	1	4	4	4	4	1	1	Moderate
Omar et al., 2023	1	1	4	4	4	4	1	1	Moderate
Maertzdorf et al., 2023	1	1	4	4	4	4	4	1	High
Aygin et al., 2022	1	1	4	4	4	4	4	1	High
Nikhita et al., 2022	1	4	4	4	4	4	1	4	High
Barbara et al., 2022	1	1	4	4	4	4	1	4	High
Binshan et al., 2022	1	1	4	4	4	4	1	1	Moderate
Xinyue et al., 2021	1	4	4	4	4	4	1	1	High
Xinyue et al., 2021*	1	4	4	4	4	4	1	1	High
Louise et al., 2021	1	1	4	4	4	4	3	1	High
Brandon et al., 2021	1	1	4	4	4	4	1	1	Moderate
Corinne et al., 2020	1	1	4	4	4	4	1	1	Moderate
Nour et al., 2019	1	1	4	4	4	4	1	1	Moderate
Long-yee et al., 2019	1	1	4	4	4	4	1	1	Moderate
Laura et al., 2018	1	1	4	4	4	4	4	1	High
Eilers et., 2017	1	1	4	4	4	4	1	1	Moderate
Nicoletta et al., 2016	1	1	4	4	4	4	1	1	Moderate
Osayi et al., 2016	1	1	4	4	4	4	4	1	High
Nam et al., 2015	4	1	4	4	4	4	1	1	High
MacDougall et al., 2015	4	1	4	4	4	4	1	4	High
Tae et al., 2015	4	1	4	4	4	4	4	4	High
Antonino et al., 2010	1	1	4	4	4	4	1	4	High
Wim et al., 2009	1	1	4	4	4	4	1	1	Moderate
Peng - jun et al., 2009	4	1	4	4	4	4	4	1	High

Notes: Q: Question; 1: Yes; 2: No; 3: Unclear; 4: Not applicable
Score: The quality assessment score ranged from 0 to 8 based on each question of the Joanna Briggs Institute checklist.

4 Discussion

Evidence from this systematic review indicates that the majority of populations have positive attitudes toward HZ vaccines and would accept their use for prevention. Worldwide, HZ vaccines were regarded as acceptable by 44.9% of the population on average, but 59.6% of individuals expressed a reluctance to get vaccinated. Respondents' attitudes and knowledge about HZ vaccines and the HZ virus were also assessed in all the examined studies. The assessments revealed that the core factors associated with the willingness of individuals over 50 years of age to accept HZ vaccines included age, household income, awareness of HZ and HZ vaccines, vaccine effectiveness, and education level. The side effects of vaccines cause hesitation among people to acquire vaccination, and the high costs of vaccines, the lack of such preparations, and the wave of anti-vaccine sentiments remain a considerable concern in national vaccination programs.

Recommendations from healthcare providers significantly affected vaccination behaviors. A higher vaccine acceptance was observed with the intervention of medical staff, with acceptance rates increasing from

44.9% to 75.7%. However, the number of studies on the influence of caregivers is limited. Therefore, more research should be conducted to determine the willingness of health care staff to introduce HZ vaccines. The systematic review uncovered that the WTP for HZ vaccines worldwide was 56.86 USD, with the cost that people are willing to shoulder peaking at 150 USD. The factors associated with WTP included out-of-pocket expenses, flu-like symptoms, injection site reactions, higher education, being in favor of vaccinations in general, and advice from general practitioners. Respondents' WTP for HZ vaccines was mentioned in only four of the evaluated surveys, and only two of these analyzed related factors. More studies should be carried out to derive additional data for future systematic reviews.

The implications of the review for HZ vaccination policy and practice encompass several aspects:

(1) Vaccine effectiveness: Policy and practice with respect to HZ vaccination may be influenced by the ongoing monitoring of its effectiveness. Research and real-world data analysis can determine how strategies can be adjusted (e.g., providing booster shots or changing vaccine doses).

(2) Immunization programs: The effectiveness, safety, and cost-effectiveness of HZ vaccines can affect national immunization programs. Governments and public health organizations should consider including the vaccine in routine immunization schedules or recommending it for specific high-risk groups, such as immunocompromised individuals or healthcare workers.

(3) Accessibility and coverage: Policy decisions should prioritize ensuring the accessibility and affordability of HZ vaccines. Governments and healthcare systems can explore strategies for increasing vaccine coverage, including providing financial assistance, having vaccination covered by insurance, and integrating it into existing vaccination campaigns.

(4) Education and awareness: Policies and practices should prioritize educating healthcare providers and the general public about HZ vaccines. This can include disseminating accurate information about the benefits, risks, and timing of vaccination, addressing common misconceptions, and promoting vaccine acceptance among eligible individuals.

(5) Research and development: Such efforts can lead to the introduction of new HZ vaccines or improvements to existing ones. Therefore, policy and practice need to remain adaptable to incorporate advancements in vaccine technology, potential changes in dosing schedules, and the development of more effective and durable vaccines.

4.1 Limitations

Several limitations of this systematic review were discovered. The representativeness of the samples in the study was not reliable, only 6 out of 24 studies use a random sampling method. The sample size of most of the studies were limited and there were large differences between studies. The number of studies on willingness to pay for HZ vaccine were extremely meager. Restricting the search to limited databases and English-only peer-reviewed articles is another limitation, as some important articles could have been excluded from the review. This study was purely descriptive statistics, without further research, so its reliability is limited.

5 Conclusion

Our study indicated that the HZ vaccine acceptance rate was higher in Asia and Europe than in the US. Popular reasons for the reluctance to get vaccinated included side effects, cost, the absence of recommendations, anti-vaccination attitudes, the lack of HZ vaccination-related information, and the belief of individuals that they are not at risk of developing the disease. The WTA increased dramatically because of the influence of general practitioners' recommendations, but few studies in the sample mentioned WTP. More explorations should be conducted in different countries to achieve a better overview of WTA and WTP for HZ vaccines, and national programs should be developed to enhance HZ awareness among citizens.

ABBREVIATIONS

HZ: Herpes zoster; VZV: Varicella-Zoster Virus; GP: General practitioner; WTA: Willingness to accept; WTP: Willingness to pay.

CONFLICT OF INTEREST

The authors declare that the research was carried out without potential conflicts of interest.

FUNDING

This study did not receive any external financial funding.

References

1. D. Panatto, N. L. Bragazzi, E. Rizzitelli, P. Bonanni, S. Boccalini, G. Icardi, R. Gasparini, D. Amicizia, Evaluation of the economic burden of Herpes Zoster (HZ) infection: A systematic literature review, Human Vaccines & Immunotherapeutics, 11, 245-62, (2015).
2. J. M. Weinberg, Herpes zoster: epidemiology, natural history, and common complications, Journal of the American academy of dermatology, 57, S130-S5, (2007).
3. B. P. Yawn, D. Gilden, The global epidemiology of herpes zoster, Neurology, 81, 928-30, (2013).
4. P. San Martin, K. Aunhachoke, M. C. F. Batac, K. Lodrono-Lim, C. Kwanthitinan, D. Santoso, T. Fonseka, M. Nguyen, A. Guzman-Holst, Systematic

Literature Review of Herpes Zoster Disease Burden in Southeast Asia, Infectious diseases and therapy, 12, 1553-78, (2023).

5. R. Harbecke, J. I. Cohen, M. N. Oxman, Herpes Zoster Vaccines, J Infect Dis, 224, S429-s42, (2021).
6. A. L. Cunningham, M. J. Levin, Herpes zoster vaccines, The Journal of Infectious Diseases, 218, S127-S33, (2018).
7. O. Damm, J. Witte, W. Greiner, A systematic review of herpes zoster vaccine acceptance, Value in Health, 18, A592, (2015).
8. L. J. F. Rutten, X. Zhu, A. L. Leppin, J. L. Ridgeway, M. D. Swift, J. M. Griffin, J. L. St Sauver, A. Virk, R. M. Jacobson, editors. Evidence-based strategies for clinical organizations to address COVID-19 vaccine hesitancy. Mayo Clinic Proceedings; 2021: Elsevier.
9. K. Attwell, E. Dube, A. Gagneur, S. B. Omer, L. S. Suggs, A. Thomson, Vaccine acceptance: science, policy, and practice in a 'post-fact'world, Vaccine, 37, 677-82, (2019).
10. T. N. Huyen, S. Pumtong, S. Sangroongruangsri, L. Anuratpanich, Acceptability of and willingness to pay for human immunodeficiency virus vaccination: A systematic literature review, Journal of Pharmacy & Pharmacognosy Research, 10, 748-67, (2022).
11. R. K. Alhassan, E. Nketiah-Amponsah, M. Immurana, A. A. Abuosi, Financing COVID-19 vaccination in sub-Saharan Africa: lessons from a nation-wide willingness to pay (WTP) survey in Ghana, BMC Public Health, 22, 1-14, (2022).
12. B. Jiang, Q. Wang, Z. Wang, Y. Xu, T. Yang, W. Yang, M. Jia, L. Feng, Willingness to accept herpes zoster vaccines and the influencing factors in China, BMC Infectious Diseases, 22, 1-9, (2022).
13. R. Eilers, H. De Melker, J. Veldwijk, P. Krabbe, Vaccine preferences and acceptance of older adults, Vaccine, 35, 2823-30, (2017).
14. N. Valente, S. Lupi, A. Stefanati, M. Cova, N. Sulcaj, L. Piccinni, GPs Study Group, G. Gabutti, Evaluation of the acceptability of a vaccine against herpes zoster in the over 50 years old: an Italian observational study, BMJ open, 6, e011539, (2016).
15. O. Chayangsu, S. Jiamton, C. Leeyaphan, N. Prasertworonun, V. Omcharoen, K. Kulthanan, Willingness to pay, quality of life, and knowledge on herpes zoster among Thai patients prior zoster vaccine era, Southeast Asian J Trop Med Public Health, 47, 6, (2016).
16. J. B. Institute, Checklist for Analytical Cross Sectional Studies, (2017).
17. J. B. Institute, The Joanna Briggs Institute critical appraisal tools for use in JBI systematic reviews checklist for analytical cross sectional studies, North Adelaide, Australia The Joanna Briggs Institute, (2017).
18. PRISMA, PRISMA2020 flow diagram, (2020).
19. C. Del Signore, A. Hemmendinger, N. Khanafer, J. Thierry, E. Trépo, G. Martin Gaujard, R. Chapurlat, C. Elias, P. Vanhems, Acceptability and perception of the herpes zoster vaccine in the 65 and over population: A French observational study, Vaccine, 38, 5891-5, (2020).
20. D. Al-Orini, A. A. Alshoshan, A. O. Almutiri, A. A. Almreef, E. S. Alrashidi, A. M. Almutiq, R. Noman, O. Al-Wutayd, Acceptability of Herpes Zoster Vaccination among Patients with Diabetes: A Cross-Sectional Study in Saudi Arabia, Vaccines, 11, (2023).
21. L. Kaplan-Weisman, E. Waltermaurer, C. Crump, Assessing and Improving Zoster Vaccine Uptake in a Homeless Population, Journal of community health, 43, 1019-27, (2018).
22. N. K. Roh, Y. M. Park, H. Kang, G. S. Choi, B. J. Kim, Y. W. Lee, B. L. Lew, W. Y. Sim, Awareness, Knowledge, and Vaccine Acceptability of Herpes Zoster in Korea: A Multicenter Survey of 607 Patients, Annals of dermatology, 27, 531-8, (2015).
23. D. M. MacDougall, B. A. Halperin, D. MacKinnon-Cameron, L. Li, S. A. McNeil, J. M. Langley, S. A. Halperin, The challenge of vaccinating adults: Attitudes and beliefs of the Canadian public and healthcare providers, BMJ Open, 5, (2015).
24. N. A. Baalbaki, J. P. Fava, M. Ng, E. Okorafor, A. Nawaz, W. Chiu, A. Salim, R. Cha, P. E Kilgore, A Community-Based Survey to Assess Knowledge, Attitudes, Beliefs and Practices Regarding Herpes Zoster in an Urban Setting, Infectious diseases and therapy, 8, 687-94, (2019).
25. A. Parlato, V. R. Spica, M. Ciccozzi, F. Farchi, F. Gallè, V. Di Onofrio, E. Franco, G. Liguori, Compliance with herpes zoster vaccination in young and adult individuals in two regions of Italy, BMC Public Health, 10, (2010).
26. O. S. Alhothali, A. S. Alhothali, A. A. Hanif, M. F. Bondagji, A Cross-Sectional Study of the Knowledge, Practice, and Attitude Towards Herpes Zoster Vaccination Among the General Population in the Western Region …, Cureus, (2023).
27. W. Opstelten, G. A. van Essen, E. Hak, Determinants of non-compliance with herpes zoster vaccination in the community-dwelling elderly, Vaccine, 27, 192-6, (2009).
28. A. Bayraktar-Ekincioglu, E. Kara, M. Bahap, Does information by pharmacists convince the public to get vaccinated for pneumococcal disease and herpes zoster?, Irish Journal of Medical …, (2022).
29. O. E. Akinbosoye, M. S. Taitel, J. Grana, C. Macpherson, Factors associated with Zostavax abandonment, American Journal of Pharmacy Benefits, 8, 84-9, (2016).
30. X. Lu, J. Lu, L. Zhang, K. Mei, B. Guan, Y. Lu, Gap between willingness and behavior in the vaccination against influenza, pneumonia, and herpes zoster among Chinese aged 50–69 years, Expert Review of Vaccines, 20, 1147-52, (2021).
31. P.-j. Lu, G. L. Euler, A. O. Jumaan, R. Harpaz, Herpes zoster vaccination among adults aged 60 years or older in the United States, 2007: Uptake of the first new vaccine to target seniors, Vaccine, 27, 882-7, (2009).
32. N. Kalra, T. Kalra, S. Mishra, S. Basu, N. Bhatnagar, Hesitancy for Adult Vaccines Among Healthcare Providers and their Family Members in Delhi, India: A Cross-Sectional Study, Dialogues Health, 1, 100044, (2022).

33. B. P. Yawn, D. D. Merrill, S. Martinez, E. Callen, J. Cotton, D. Williams, N. Y Loskutova, Knowledge and Attitudes Concerning Herpes Zoster among People with COPD: An Interventional Survey Study, Vaccines, 10, (2022).
34. X. Lu, J. Lu, F. Zhang, A. L. Wagner, L. Zhang, K. Mei, B. Guan, Y. Lu, Low willingness to vaccinate against herpes zoster in a Chinese metropolis, Hum Vaccin Immunother, 17, 4163-70, (2021).
35. L. A. B. Nicholls, A. J. Gallant, N. Cogan, S. Rasmussen, D. Young, L. Williams, Older adults' vaccine hesitancy: Psychosocial factors associated with influenza, pneumococcal, and shingles vaccine uptake, Vaccine, 39, 3520-7, (2021).
36. L. Y. Cheng, A pilot study to assess the awareness of herpes zoster and the attitudes towards herpes zoster vaccination among Chinese patients attending a government general out-patient clinic in Hong Kong, Hong Kong Practitioner, 41, (2019).
37. B. J. Patterson, K. Myers, A. Stewart, B. Mange, E. M. Hillson, C. Poulos, Preferences for herpes zoster vaccination among adults aged 50 years and older in the United States: results from a discrete choice experiment, Expert review of vaccines, 20, 729-41, (2021).
38. T. U. Yang, H. J. Cheong, J. Y. Song, J. Y. Noh, W. J. Kim, Survey on public awareness, attitudes, and barriers for herpes zoster vaccination in South Korea, Hum Vaccin Immunother, 11, 719-26, (2015).
39. K. M. Maertzdorf, M. L. Rietman, M. S. Lambooij, Willingness to get vaccinated against influenza, pneumococcal disease, pertussis, and herpes zoster A pre-COVID-19 exploration among the older adult …, Vaccine, (2023).
40. T. Babatope, V. Ilyenkova, D. Marais, COVID-19 vaccine hesitancy: a systematic review of barriers to the uptake of COVID-19 vaccine among adults in Nigeria, Bull Natl Res Cent, 47, 45, (2023).

The Economic Evaluation of HPV Vaccination as a Prevention of Non-Cervical Cancer Using Modelling Approaches: A Systematic Review

Nur Rochmah Hidayati[1], *Dwi* Endarti[1*], *Didik* Setiawan[2]

[1]Faculty of Pharmacy, University of Gadjah Mada, Yogyakarta, Indonesia
[2]Faculty of Pharmacy, University of Muhammadiyah Purwokerto, Purwokerto, Indonesia

Abstract. The cost-effectiveness of human papillomavirus (HPV) vaccine for non-cervical cancer including anogenital cancers and head and neck cancers has been evaluated in many countries. This is important for the decision-makers in health policy as they have a limited budget for an enormous option of health technology. The objective of this study is to systematically review the health economic studies in order to evaluate the cost-effectiveness of HPV vaccination for preventing non-cervical cancer including anogenital cancer and head and neck cancer. Approximately 496 articles were obtained from three databases (ProQuest: 430, Scopus: 40 and Pubmed: 29). The screening and selection of the literature was performed according to the Preferred Reporting Items for Systematic Reviews and Meta-Analyses (PRISMA) statement. The final articles were assessed by using the Consolidated Health Economics Evaluation Reporting Standards (CHEERS) checklist. Based on the review of fifteen articles, most articles concluded that HPV vaccination was cost-effective compared with current standard care. One study showed that HPV vaccination may be cost-effective if only the price is lowered. HPV vaccination is considerably a cost-effective solution to prevent non-cervical cancer based on the studies conducted in many countries particularly the ones with high income.

Keywords: *cost-effective, human papillomavirus, vaccine, non-cervical cancer, modeling*

1 Introduction

Human papillomavirus (HPV) infection is now a well-established cause of cervical cancer, anogenital cancers (anus, vulva, vagina and penis) and head and neck cancers. In the global, it is estimated as 8% of the cancer burden, or approximately 690,000 cancer cases annually [1]. Although anogenital cancers are much less frequent compared to cervical cancer, their association with HPV makes them potentially preventable and subject to similar preventative strategies as those for cervical cancer [2]. Research has demonstrated that HPV vaccines are safe and effective in reducing HPV-related infections, genital warts, and pre-cancer [3]. Some countries have started to use HPV vaccination as a prevention of other HPV-related cancers including anogenital cancer and head and neck cancer. One effective prevention and control strategy is the use of HPV vaccines designed to prevent HPV infection and HPV-related cancers.

It is believed that the HPV vaccine plays an important role in preventing cervical cancer in women. However, HPV vaccination is not just an issue for women since men also have risks of getting HPV infection. Moreover, high HPV vaccine coverage in men has significant benefits for women by reducing the risk of cervical cancer [4]. In 2019, four programs started as gender-neutral in Dominica, Niue, Saint Kitts and Nevis, and Saint Lucia) and 10 more expanded vaccinations to males in Belgium, Belize, Chile, Denmark, Germany, Guyana, Ireland, Luxembourg, United Kingdom, and Uruguay [5].

HPV Vaccination Program in Indonesia has been introduced (partially in several provinces) since 2016 and is expected to be fully implemented in 2023 covering all regions and targeting girls only for grade 5-6 elementary school.

This information is notably important for the decision-makers in health policy as they have a limited budget for an enormous option of health technology. Therefore, the objective of this study is to systematically review the health economic studies in order to evaluate the cost-effectiveness of HPV vaccination for non-cervical cancer including anogenital cancers and head and neck cancers.

*Corresponding email: endarti_apt@ugm.ac.id

2 Method

The screening and the selection of the articles were performed according to the Preferred Reporting Items for Systematic Reviews and Meta-Analyses (PRISMA) statement [6] **(Fig. 1)**.

2.1 Data Source and Search Strategy

A systematic search was developed in three main databases (Pubmed, Scopus and ProQuest) that was performed on 28-30 July 2023 using some keywords in order to reach between the comprehensive and specific search results as explained below:

Pubmed (filters: free full text, last 10 years, written in English)
("cost-effectiveness"[Title/Abstract] OR "cost utility"[Title/Abstract] OR "cost benefit"[Title/Abstract] OR "economic evaluation"[Title/Abstract]) AND ("human papillomavirus"[Title/Abstract] OR "HPV"[Title/Abstract]) AND ("vaccination"[Title/Abstract] OR "vaccine"[Title/Abstract] OR "immunization"[Title/Abstract]) AND ("non cervical cancer"[Title/Abstract] OR "anogenital cancer"[Title/Abstract] OR "head and neck cancer"[Title/Abstract] OR "penile cancer"[Title/Abstract] OR "anal cancer"[Title/Abstract] OR "vaginal cancer"[Title/Abstract] OR "vulvar cancer"[Title/Abstract] OR "oropharyngeal cancer"[Title/Abstract] OR "oral cavity cancer"[Title/Abstract] OR "nasopharyngeal cancer"[Title/Abstract])

Scopus (filters: articles, written in English, last 10 years)
(cost-effectiveness OR cost-utility OR cost-benefit OR "economic evaluation") AND ("human papillomavirus" OR hpv) AND (vaccine OR vaccination OR immunization) AND (non-cervical AND cancer OR "anogenital cancer" OR "head and neck cancer" OR "penile cancer" OR "anal cancer" OR "vulvar cancer" OR "vaginal cancer" OR "oropharyngeal cancer" OR "oral cavity cancer" OR "nasopharyngeal cancer")

ProQuest (filters: scholarly journal, article, in the last 10 years, English, not SR)
(cost-effectiveness OR cost-utility OR cost-benefit OR "economic evaluation") AND ("human papillomavirus" OR hpv) AND (vaccine OR vaccination OR immunization) AND (non-cervical AND cancer OR "anogenital cancer" OR "head and neck cancer" OR "penile cancer" OR "anal cancer" OR "vulvar cancer" OR "vaginal cancer" OR "oropharyngeal cancer" OR "oral cavity cancer" OR "nasopharyngeal cancer")

All references were imported into CSV file and transferred to Microsoft Excel for duplicate removal and screening by title and abstract.

2.2 Eligibility Criteria

The inclusion criteria in this systematic review were health economic studies on HPV vaccination for non-cervical cancer including anogenital cancer and head and neck cancers since 2017 and articles were written in English.

The study selection process consisted of two main stages. Initially, titles and abstracts of all references were screened to identify potentially relevant articles. Subsequently, the full texts of these identified articles were retrieved and evaluated further. During this process, any duplicate articles were removed, as well as those that were not relevant to the topic under investigation. The inclusion criteria of this review were full economic evaluations, available in full-text, published in English, presenting data on Quality-Adjusted Life Years (QALY) and Incremental Cost-Effectiveness Ratio (ICER). Additionally, the articles had to be original research and openly accessed. Several types of articles were excluded from the review, such as review articles, reports, letters, comments and book chapters. Since this systematic review is specific to the economic evaluation of non-cervical cancer, the articles that only discuss cervical cancer and precancerous would be excluded. Furthermore, the economic evaluation that did not use a modeling approach was also excluded.

2.3 Study Quality Assessment

The quality assessment of the included studies was conducted by the reviewer using the Consolidated Health Economics Evaluation Reporting Standards (CHEERS) checklist [7].

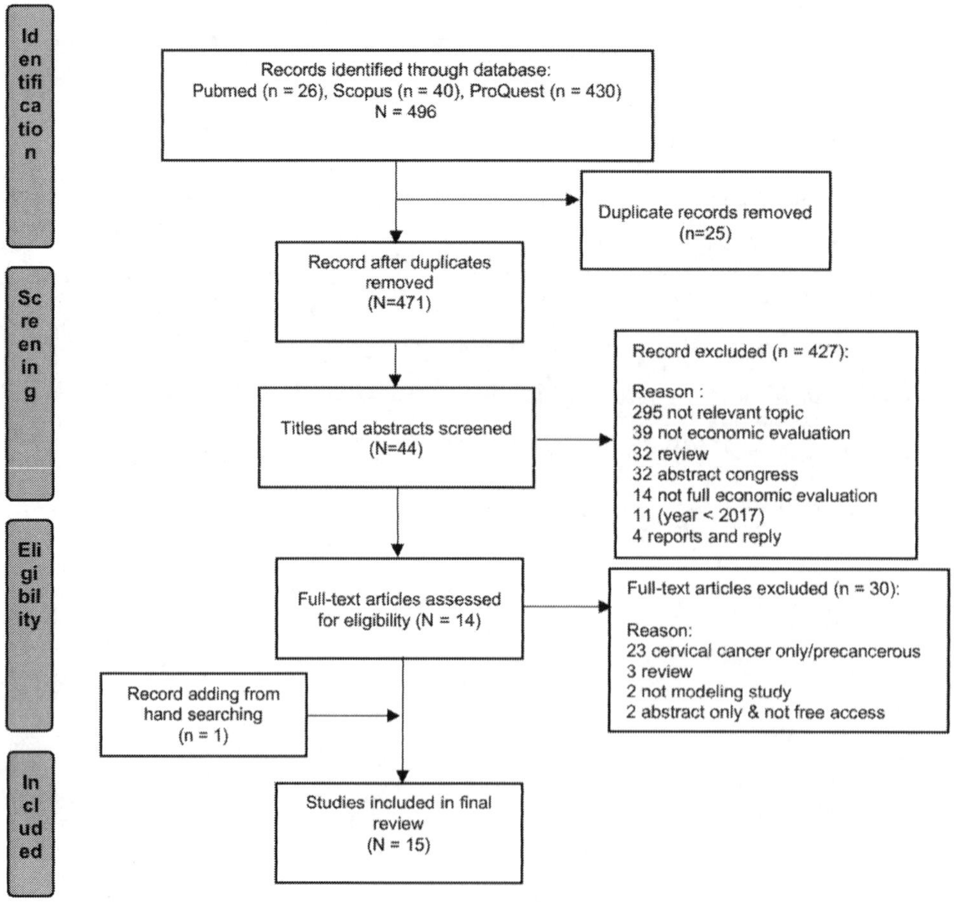

Fig. 1. PRISMA Flowchart

2.4 Data Extraction

The reviewer (NRH) independently extracted information related to health economic analysis from each selected study. The extracted information included the first authors' names and publication years, the country where the study was conducted, the type of health economic study, the study's objective, the research methodology (model) used, the perspective adopted in the analysis, the time horizon considered, details about the HPV vaccination methods (such as vaccine price, age target, and coverage), the discount rate used, clinical outcomes, the calculated ICER, the study's conclusion, and any sensitivity analyses results.

2.5 Data Analysis Method

To ensure whether the intervention of each study is cost-effective or not, the Incremental Cost-Effectiveness Ratio (ICER) should be compared with the cost-effectiveness threshold (CET) of each country. All costs, ICER and threshold were converted to US dollar in 2022 value by taking inflation into account [8] [9].

3 Result

From the search, approximately 496 articles were obtained from three databases (ProQuest: 430, Scopus: 40 and Pubmed: 29). Finally, we found 15 articles that are eligible to be reviewed from Pubmed, Scopus and ProQuest databases.

3.1 Study Characteristics

In this review, 15 included studies were from 11 countries: The United States (n=3), China/Hongkong (n=3), Japan (n=1), Spain (n=1), Sweden (n=1), the Netherlands (n=1), France (n=1), Belgium (n=1), Italy (n=1), Singapore (n=1), Australia (n=1) (Table 1).

A cost- utility analysis (CUA), adopting Quality Adjusted Life Years (QALYs) as the main study outcome, was performed by 6 different studies. The other 2 studies performed cost effectiveness analysis (CEA) that used clinical parameters as study outcome, 7 studies combining CEA and CUA simultaneously.

Almost all studies using HPV vaccination as a prevention not only to one specific disease/cancer but to multiple kinds of cancer attributable to HPV, both cervival cancer and non-cervical cancer such as anogenital cancer (penile cancer, anal cancer, vulvar cancer vaginal cancer) and head and neck cancer (oropharyngeal cancer) [10]–[22]. Only two studies from China and Australia that specifically assessed cost-effectiveness of HPV vaccination to prevent anal cancer, particularly in men who have sex with men (MSM)[23], [24].

Most of the studies tried to expand the HPV vaccination program to both girls and boys. There are 9 studies compared gender-neutral vaccination versus girl only vaccination [10], [17]–[19], [21], [22], [25] [20] [15], two studies assessed the cost-effectiveness HPV vaccination program for specific high risk group e.i. men who have sex with men (MSM) [24] [23]. Other two studies proposed the extended HPV vaccination for "mid adults" (females aged 12-26 years and males aged 12-21 years)[12], [13]. One study compared the cost-effectiveness of routine HPV vaccination, catch up vaccination and no vaccination in girls and women [16]. The last study compared routine vaccination with catch up versus without catch up [10].

Table 1. Study Characteristics of the health economic studies of HPV Vaccination for Non-cervical cancer

No.	Authors, years of publication	Country	Type of Study	Type of Cancer	Study Objective
1	Li et al., 2023 [24]	China	CEA	Anal cancer	to evaluate the effectiveness and cost-effectiveness of different HPV vaccination strategies among MSM in China
2	Cheung et al., 2023[14]	Hong Kong	CUA	Anal cancer, cervical cancer, CIN 1, CIN 2/3, head and neck cancer, penile cancer, vaginal cancer, VaIN, vulvar cancer	to estimate the potential health and economic impact of a routine gender-neutral vaccination (GNV) approach compared with the current female-only vaccination (FOV) strategy
3	Wahab et al., 2023 [20]	Singapore	CUA	Anal cancer, penile cancer, oropharyngeal cancer, cervical cancer	to assess if including adolescent boys in Singapore's school-based HPV vaccination programme is cost-effective from the healthcare perspective
4	Linertová R. et al., 2022 [17]	Spain	CUA	Head and neck and penile cancers	to determine the cost-effectiveness of extending HPV vaccination in Spain to include boys, given that the adolescent female population is already being vaccinated
5	Choi et al., 2022 [15]	US	CEA, CUA	Oropharingeal Cancer	to evaluate the impact of increased HPV vaccination coverage on HPV-associated OPC incidence and costs
6	Simoen et al., 2021 [19]	Belgium	CEA, CUA	Cervical cancer, vaginal cancer, vulvar cancer, anal cancer, penile	to assess whether expanding these programs to gender-neutral vaccination (GNV) with the 9vHPV vaccine is a cost-effective

https://doi.org/10.1051/bioconf/20237505018

No.	Authors, years of publication	Country	Type of Study	Type of Cancer	Study Objective
				cancer, head and neck cancer	strategy in Belgium
7	Cody P. et al., 2021 [16]	Japan	CEA	Intraepithelial neoplasia, cervical, vulvar, vaginal, nal, oral cavity, oropharyngeal cancer	to assess the health impact and cost effectiveness of routine and catch-up vaccination of girls and women aged 11–26 years with a 4-valent (4vHPV) or 9-valent HPV (9vHPV) vaccine in Japan compared with no vaccination
8	Majed et al., 2021 [25]	France	CEA, CUA	cervical, vulvar, vaginal, anal, head and neck cancer	to assess the public health impact and cost-effectiveness of a 9-valent GNV compared with girls-only vaccination program (GOV).
9	Cheung et al., 2021 [10]	Hong Kong	CUA	cervical precancers, cervical cancer, vaginal precancers, vaginal/vulvar cancer, and anal cancer.	to assess the public health impact and cost-effectiveness of implementing routine 9vHPV vaccination (12-year-olds) with or without catch-up 9vHPV vaccination (13–18-year-olds) in HK.
10	Simons et al., 2020 [21]	The Netherlands	CEA, CUA	anal cancer, penile cancer, oropharyngeal cancer	to assess the cost-effectiveness of a gender-neutral HPV vaccination program in the Netherlands.
11	Chesson et al., 2020 [13]	US	CEA, CUA	cancers in females (cervical, vaginal, vulvar, anal, and oropharyngeal); and cancers in males (anal, oropharyngeal, and penile).	to assess incremental costs and benefits of a human papillomavirus (HPV) vaccination program expanded to include "mid-adults" (adults aged 27 through 45 years) in the United States.
12	Chesson et al., 2018 [26]	US	CEA, CUA	anogenital cancers (cervical, vaginal, vulvar, anal, and/or penile), oropharyngeal cancer, cervical intraepithelial neoplasia (CIN)	to assess the health impact and cost-effectiveness of harmonizing female and male vaccination recommendations by increasing the upper recommended catch-up age of HPV vaccination for males from age 21 to age 26 years.
13	Wolff et al., 2018 [22]	Sweden	CUA	HPV-related cancers (cervical, genital, anal and oropharyngeal cancer), and cervical intraepithelial neoplasia (CIN).	to assess cost-effectiveness of expanding the Swedish HPV-vaccination program to include preadolescent boys, by comparing health-effects and costs of HPV-related disease, with a sex-neutral vaccination program versus only vaccinating girls
14	Mennini FS. et al., 2017 [18]	Italy	CUA	CIN 1,2,3, cervival vancer, VaIN2, VaIN3,vaginal cancer,vulvar cancer, penile cancer, anal cancer, head and	to provide realistic estimates of the epidemiological and economic impact of the implementation of the 9-valent HPV vaccine program for both girls and boys in Italy compared to the current

https://doi.org/10.1051/bioconf/20237505018

No.	Authors, years of publication	Country	Type of Study	Type of Cancer	Study Objective
				neck cancer, genital warts, recurrent respiratory papillomatosis	clinical practice using a 4-valent (HPV 6/11/16/18) or bivalent (HPV 16/18) vaccine for girls only
15	Zhang et al., 2017 [23]	Australia	CEA, CUA	Anal cancer	to investigate the effectiveness and cost-effectiveness of a targeted human papillomavirus (HPV) vaccination program for young (15–26) men who have sex with men (MSM).

Abbr. CEA: cost-effectiveness analysis, CUA: cost utility analysis, HPV: human papillomavirus, MSM: men who have sex with men, GNV: gender-neutral vaccination, FOV: female-only vaccination, OPC: oropharyngeal cancer, US: United States,CIN: cervical intraepithelial neoplasia, VaIN: vaginal intraepithelial neoplasia

3.2 Study Design

Since the clinical outcomes of HPV vaccination, such as reduction in cancer incidence and mortality are difficult to obtain from clinical trials, a mathematical model is commonly used in the cost-effectiveness analysis of HPV vaccination. All studies implemented modeling (Table 2). Many of the studies implemented dynamic transmission models (5 studies). There are studies using mathematical models, dynamic compartmental model, dynamic population-based and Markov model.

The perspective chosen for economic evaluation is a crucial factor that significantly impacts both the data utilized in the analysis and the resulting conclusions drawn from the studies. Different studies have employed various perspectives, such as the health care system (eight studies) and health care payer (five studies). However, in two studies, the perspective of their health economic analysis was not explicitly mentioned. In addition, none of the studies utilized the societal perspective in their evaluations.

To achieve a comprehensive understanding of the cost-effectiveness of HPV vaccination, it is essential to employ a long time horizon since the full impact of the vaccination on cancer incidence and mortality reduction will only become evident after several decades of implementation. The majority of studies have utilized a lifetime horizon, typically spanning 100 years, to capture the long-term effects. However, there are also instances where studies employed shorter time horizons, such as 10 years or 85 years. Interestingly, one study took a different approach by using a specific time frame in years, covering the period between 2017 and 2036.

In health economic studies, especially when using modeling and considering long time horizons, the discount rate becomes an important consideration. Most studies have utilized a discount rate of 3%. However, there are also studies that have employed different discount rates, such as 1.5%, 2%, 4%, and 5%.

Regarding the type of HPV vaccine evaluated, the majority of studies focused on the 9-valent (nonavalent) HPV vaccine. Only one study separately examined the 2-valent (bivalent) and 4-valent (quadrivalent) HPV vaccines. Additionally, some studies compared the 4-valent vaccine to the 9-valent vaccine, and others compared the 2-valent vaccine to the 9-valent vaccine. Moreover, there was a study that compared three different HPV vaccines: the bivalent, quadrivalent, and nonavalent versions.

Vaccine coverage varied from 24.9-90% in many countries based on the vaccination program. Vaccine price is also varied in many countries. Most studies targeted HPV vaccination for boys and girls from age 9-14 years old. There are four studies that targeted from age 12 to 45 years especially study that aimed to extend the upper age limit of HPV vaccination.

Table 2. Methodological Aspects and the Base-case Paramaters of Health Economic Studies of HPV vaccination

:s, years of lication	Method	Perspective	Time Horizon	Vaccine Type	Vaccine Price (in 2022 USD)	Age Target (years old)	Coverage
, 2023 [24]	Markov model	NS	10 years	bivalent, quadrivalent and nonavalent	NS	27-45	50%
et al., 2023	Dynamic transmission model	healthcare payer	100 years	nonavalent	USD 203.19	12	70%
et al., 2023	Papillomavirus Rapid Interface for Modelling and Economics (PRIME)	healthcare	lifetime	bivalent and nonavalent	bivalent : USD 100.47 nonavalent : USD 307.13	13	80%
á et al., 7]	a dynamic population-based model with a discrete-time Markov approach	Health care system	lifetime/until 99 years	nonavalent	USD 63.38	12	80% for girl 70.2% fo boys
t al., 2022	Decision analytic Markov model	Healthcare	Until age 85	NS	NS	9+	80%
et al., 2021	Transmission dynamic model	Payer	100 years	bivalent and nonavalent	bivalent : USD 68.69 nonavalent : USD 154.84	12	90, 50, 50%
et al., 2021	mathematical model	Healthcare System	100 years	quadrivalent and nonavalent	quadrivalent : USD 124.83 nonavalent : USD 235.24	12-16 years old	70%
et al.,]	transmission dynamic model	Payer	100 years	nonavalent	USD 163.23	11-14 years old	60%
et al., 2021	transmission dynamic model	healthcare payer	100 years	bivalent and nonavalent	bivalent : USD 125.92 nonavalent : USD 204.9	12 (routine vaccination), 13-18 (catch up)	70% (routin vaccination), catch u (30%)
et al., 2020	static markov model	healthcare-payer	Lifetime	bivalent	USD70 – USD147	12	30% for boys

https://doi.org/10.1051/bioconf/20237505018

:s, years of lication	Method	Perspective	Time Horizon	Vaccine Type	Vaccine Price (in 2022 USD)	Age Target (years old)	Coverage
ı et al., 2020	simplified dynamic mathematical transmission model	healthcare	100 years	nonavalent	US$171.93 (public cost) and US$259.08 (private sector cost)	27- 45	29.5% (femal 12 y, 24.9% (male 1: years0
ı et al., 2018	deterministic, dynamic, population-based model	Healthcare System	100 years	nonavalent	US$141.45 (public cost) and US$236.56 (private sector cost)	22-26	29.5% (femal 12 y, 24.9% (male 1: years0
:t al., 2018	dynamic compartmental model	Healthcare System	100 years	NS	NS	10	80%
i et al., 2017	dynamic transmission model	Healthcare system	100 years	nonavalent	USD 179.35	12	71.10%
:t al., 2017	compartmental mathematic model	NS	2017-2036	quadrivalent	NS	15-26	80%

ation; NS: not stated

3.3 Study Outcome

For all CUA studies, the health outcome was Quality Adjusted Life Years (QALY), while for CEA studies the outcome can be clinical, e.i. the number of HPV-related cancer averted, number of death caused by HPV-related cancer averted. The main information provided by health economic studies is the ICER which explains the cost-effectiveness of a new intervention in comparison with a gold standard, previous recommendation or existing intervention (Table 3).

For high income countries like Australia, European Countries (The Netherlands, Belgium, Italy, Spain and Swedia), and Singapore, the cost-effectiveness threshold has been officially set in each country. Surprisingly, based on the study by Chesson in 2018 and 2020 United States which is the most high-income country has not established an official CET [12] [13]. However, the other study from US used willingness to pay (WTP) as less than USD169,615 as a cost-effectiveness threshold [15]. Moreover, other countries that have not established the threshold used 1-3 times Gross Domestic Product (GDP) per capita as recommended by the World Health Organization (WHO) [28]. Both France and Hong Kong used one-time GDP as a cost-effectiveness threshold [25] [14].

Whereas, CE Threshold can affect the cost-effectiveness of HPV vaccination in a country. Based on the review of fifteen articles, most articles concluded that HPV vaccination was cost-effective compared with current standard care. One study conducted in Spain showed that HPV vaccination may be cost-effective if only the price is lowered [17]. A study by Choi stated that HPV vaccination was only cost-effective for males [15]. Studies conducted by Chesson, in 2018 to expand HPV vaccination for males from age 21 to age 26 years and in 2020 to include adults aged 27-45 years resulted that HPV vaccination is much less cost-effective than the comparison strategy of routine vaccination for adolescents at age ages 11 to 12 years and catch-up vaccination for women through age 26 years and men through age 21 years [12], [13].

Table 3. Clinical and economic outcomes of health economic studies of HPC vaccine

Authors, years of publication	Health Outcome	ICER (in 2022 USD)	CET (in 2022 USD)	Conclusion
Li et al., 2023 [24]	ICER and the prevented infections and deaths	- When the annual vaccination rate increased by 30%, the ICER: USD33,521.75/QALY. - When the vaccine price decreased by 60%, the ICER was reduced to USD 7,344.44 /QALY	USD 37,260	HPV vaccination can effectively reduce the prevalence and mortality of related diseases among MSM in China, especially quadrivalent vaccines for anogenital warts and nine-valent vaccines for anal cancer. MSM aged 27-45 years were the optimal group for vaccination. Annual vaccination and appropriate adjustment of vaccine price are necessary to further improve the cost-effectiveness.
Cheung et al., 2023 [14]	QALY& ICER	USD 36,305/QALY	USD 55,847	These results highlight the potential value of a routine GNV program with the 9vHPV vaccine among 12-year-olds in Hong Kong to reduce the public health and economic burden of HPV-related diseases.
Wahab et al., 2023 [20]	ICER, number of HPV-related cancers averted	USD 15,525/ QALY	USD 36,757	Gender-neutral vaccination program using the bivalent HPV vaccine may be cost-effective in Singapore
Linertová et al., 2022 [17]	QALYs	USD 43,149/QALY,	USD 31,690	Gender-neutral vaccination in Spain, using the 9-valent vaccine, offers more benefits than any other modeled strategy. In comparison with the girls-only vaccination it would not cost-effective in the conservative base case. However, a

Authors, years of publication	Health Outcome	ICER (in 2022 USD)	CET (in 2022 USD)	Conclusion
				decrease of the vaccine price or the inclusion of vaccine effectiveness against additional health outcomes would make it a cost-effective option.
Choi et al., 2022 [15]	ICER, ICER/QALY	Increasing HPV vaccination coverage to 80% was cost-effective for males with ICER of USD98,309/QALY gained. For females, increasing the vaccination coverage result an ICER of USD364,930, which was not cost-effective	USD169,615 (based on WTP)	HPV vaccination rates would likely provide a cost-effective way to reduce the oropharyngeal cancer incidence, particularly among males.
Simoen et al., 2021 [19]	ICER.	Both regional and national catch-up GNV strategies were projected to reduce cumulative HPV-related disease costs and were estimated to be cost-effective compared with FOV with incremental cost-effectiveness ratios of USD 10,067, USD5,219, and USD 7,650 per QALY in the three programs, respectively.	USD 41,208	A gender-neutral vaccination (GNV) strategy with the 9vHPV vaccine can reduce the burden of HPVrelated disease and is cost-effective compared with female-only vaccination (FOV) for both regional vaccination programs and the national catch-up program in Belgium.
Cody P. et al., 2021 [16]	QALY, ICER	- ICER no vaccination vs quadrivalent: USD 10,145/QALY, - ICER no vaccination vs quadrivalent, nonavalent + Catch-up: USD 15,962/QALY	USD 52,010/ QALY	A vaccination program with a 9-valent vaccine targeting 12 to 16 year-old girls together with a temporary catchup program will avert significant numbers of cases of HPV-related diseases among both men and women. Furthermore, such a program was the most cost effective among the vaccination strategies we considered, with an ICER well below a threshold of USD 52,010/QALY
Majed et al., 2021 [25]	disease cases averted and QALY	Introducing male vaccination in France with the nonavalent results in an ICER: - Base case: USD 34,661/QALY - All HPV-related diseases were included compared with girls-only vaccination: USD21,253/QALY	USD 47,408 (1xGDP of France)	In France, GNV has a significant impact in terms of public health benefits and may be considered cost-effective compared with GOV at low and high coverage rates

Authors, years of publication	Health Outcome	ICER (in 2022 USD)	CET (in 2022 USD)	Conclusion
Cheung et al., 2021 [10]	ICER/ QALY	- routine + catch-up nonavalent versus routine nonavalent alone: USD4,404/QALY - routine nonavalent alone versus screening only were USD 3758/ QALY	USD 56,254 (1xGDP of Hong Kong)	the current Hong Kong vaccination strategy can be considered cost-effective and will provide maximum health benefit. The results support addition of the routine 9vHPV vaccine with or without catchup 9vHPV vaccination to the regional vaccination program in Hong Kong.
Simons et al., 2020 [27]	ICER/ QALY, ICER/LY	USD 10,232 per QALY gained	USD 27,994	Vaccination of boys, additional to girls, will prevent a relevant number of cancers in both boys and girls. Based on the current Dutch situation vaccination of HPV in boys is likely cost-effective.
Chesson et al., 2020 [13]	ICER/ QALY	Expanding the 9vHPV program to include mid-adults was estimated to cost USD 701,551 per additional QALY gained when including adults through age 30 years, and USD 779,991 per additional QALY gained when including adults through age 45 years	NS	Mid-adult vaccination is much less cost-effective than the comparison strategy of routine vaccination for all adolescents at ages 11 to 12 years and catch-up vaccination for women through age 26 years and men through age 21 years.
Chesson et al., 2018 [12]	ICER/ QALY	USD278,900/ QALY	NS	The cost-effectiveness ratios are not so favorable as to make a strong economic case for recommending expanding male vaccination, yet are not so unfavorable as to preclude consideration of expanding male vaccination
Wolff et al., 2018 [22]	QALY	The cost per gained QALY was estimated to USD 55,989	USD 69,986	Introducing a sex-neutral HPV-vaccination program would be good value for money also in Sweden where there this 80% coverage in the current HPV-vaccination program for preadolescent girls. The cost-effectiveness of a sex-neutral program is highly dependent on the price of the vaccine, the lower the price the more favourable it is to also vaccinate boys
Mennini et al., 2017 [18]	QALY	USD 15,637/QALY	USD 37,363–USD 59,781/ QALYs	The switch to the nine-valent vaccine in Italy can further reduce the burden associated to cervical cancer and HPV-related diseases and is highly cost-efective
Zhang et al., 2017 [23]	ICER	USD 5,824 per QALY	USD 42,898	A boys program that achieved coverage of about 84% will result in a 90% reduction in HPV. A targeted program for young MSM

Authors, years of publication	Health Outcome	ICER (in 2022 USD)	CET (in 2022 USD)	Conclusion
				is cost-effective if timely implemented.

Abbreviation; NS: not stated, WTP: willingness to pay

3.4 Sensitivity Analysis

The sensitivity analysis method that used in fifteen articles were varied (Table 4). Most studies used one-way/univariate sensitivity analysis to address uncertainty and assess the robustness of the results in the health economic study [14] [17] [15] [19] [16] [25] [10] [22] [18]. One study used probabilistic sensitivity analysis only [24]. One study combined one-way and multi-way sensitivity analysis [12], two studies used both deterministic and probabilistic sensitivity analysis [13], [20], [21]. Interestingly, there was a study used different type of sensitivity analysis named Latin model output [23]. LHS is one of popular sampling Hypercube sampling (LHS) with a Monte Carlo Simulation Method to estimate the uncertainties in the method for Monte Carlo Simulation that the researcher can divide the input space into equal intervals and select a single random value for each interval. LHS can enhance the accuracy and efficiency of the simulation by ensuring uniform coverage of the input space and preventing random number clustering [29]. Parameters that affect the most to the ICER such as discount rate [16] [17], [25] [18], vaccine price [12], [13], [21], [24], duration of vaccine protection [18] [25], vaccination coverage [7] [19] [10], utility of disease [16] [20], treatment cost.

4 DISCUSSION

This systematic review was undertaken to address the shortcomings of a previous study conducted by Abidi et al. in 2020 [30]. The earlier review primarily focused on economic evaluations of HPV vaccination in the context of cervical cancer prevention. In contrast, the current review aims to fill the gap by specifically examining the cost-effectiveness of HPV vaccination for preventing non-cervical cancers, including anogenital cancers such as anal, penile, vulvar, vaginal, and head and neck cancers like oropharyngeal cancer.

The majority of the studies reviewed were carried out in high-income countries, where specific cost-effectiveness thresholds have been established. By comparing the Incremental Cost-Effectiveness Ratio (ICER) with these thresholds, it is possible to determine whether HPV vaccination is considered cost-effective for preventing non-cervical cancer. Based on the review, the majority of studies indicated that HPV vaccination is indeed cost-effective. However, in one study from Spain, certain adjustments may be required, such as reducing the vaccine price, to achieve cost-effectiveness for HPV vaccination in preventing non-cervical cancer. Another study aimed to reach 80% vaccination coverage only cost-effective for males, not for females. And the last two studies recommend that HPV vaccination is only cost-effective for routine vaccination for girls and boys ages 11-12 years and catch-up vaccination for women through age 26 years and men through age 21 years, not more than that.

All the health economic studies included in this review adopted a modelling approach to estimate the long-term costs and outcomes associated with the use of HPV vaccines, focusing on the progression of cancer. The majority of these studies utilized a dynamic model, which offers a more comprehensive depiction of how infectious diseases, like HPV infection, spread throughout a population, considering factors such as herd immunity. Dynamic models require more extensive and complex data, including information like sexual contact matrices and force of infection, which might be more readily available in developed countries. On the other hand, some studies opted for a static model, commonly known as a Markov model. This approach was deemed sufficient for describing the natural history of non-cervical cancer in a population, particularly when vaccination coverage is potentially high. The static model allows for a simpler representation of the disease progression in such cases [31].

Sensitivity analysis has a critical role in health economic studies. The most popular type is one-way/univariate sensitivity analysis. There were some studies that combined two types of sensitivity analysis to assess the robustness of the result of the study.

The systematic review presented here encountered certain limitations. One notable limitation was the lack of available studies from low-middle-income countries. This is concerning as these countries tend to have higher HPV infection incidence and face resource constraints, necessitating cost-effective strategies. To address this gap and develop more effective vaccination approaches, future studies should focus on non-cervical diseases in low-middle-income countries as potential outcomes that can influence the effectiveness of HPV vaccination. By exploring these aspects, researchers can better tailor vaccination strategies to meet the specific needs and challenges of different regions, including those with limited budgets.

Table 4. Sensitivity analysis of health economic studies of HPV vaccination

Authors, years of publication	Sensitivity Analysis	Sensitive Parameter
Li et al., 2023 [24]	Probabilistic sensitivity analysis	vaccine rate and price
Cheung et al., 2023 [14]	One-way sensitivity analysis	Not stated
Wahab et al., 2023 [20]	One-way and probabilistic sensitivity analysis	utility weights assigned to cervical and anal cancer survivors
Linertová et al., 2022 [17]	One-way sensitivity analysis	Coverage (87% in girls and 70% in boys) and discount rate
Choi et al., 2022 [15]	One-way sensitivity analyses: complication rates from cancer treatment, annual costs of local/regional and distant OPC, efficacy of HPV vaccine, and utility score and mortality rate from distant OPC complication. Probabilistic sensitivity analysis	complication rates from local/ regional OPC treatment, cancer treatment cost for males and vaccine efficacy for females
Simoen et al., 2021 [19]	Deterministic one-way sensitivity analyses were conducted to assess the sensitivity of ICER values to variables that have been shown to be impactful to cost effectiveness	vaccine label, vaccine coverage rate (VCR)
Cody P. et al., 2021 [16]	One-way sensitivity analyses to test sensitivity of outcomes to uncertainty in epidemiology, vaccination coverage, health utilities, discount rates, and cost inputs	discount rate, disease utility
Majed et al., 2021 [25]	Deterministic one-way sensitivity analyses	Duration of protection, discount rate
Cheung et al., 2021 [10]	Deterministic one-way sensitivity analyses	vaccination coverage and discount rate
Simons et al., 2020 [27]	One-way/univariate deterministic sensitivity analysis (DSA) and probabilistic sensitivity analysis (PSA)	variation in vaccine price, herd immunity from females and vaccine efficacy
Chesson et al., 2020 [13]	one-way and multi-way probabilistic sensitivity analysis	current and historical vaccination coverage, vaccine price, and the impact of HPV diseases on quality of life
Chesson et al., 2018 [12]	one-way and probabilistic sensitivity analysis	Vaccine price
Wolff et al., 2018 [22]	Deterministic sensitivity analyses were performed to investigate how varying the input parameters affected the results.	Vaccine coverage (higher than 90%)
Mennini et al., 2017 [18]	One-way sensitivity and scenario analyses were conducted and the cost-effectiveness results are displayed in Tornado diagrams	discount rate, duration of protection
Zhang et al., 2017 [23]	Latin Hypercube sampling with a Monte Carlo Simulation Method	Not stated

5 CONCLUSION

HPV vaccination is considerably as a cost-effective solution to prevent non-cervical cancer based on the studies conducted in many countries particularly the ones with high income. It is recommended to conduct an economic evaluation to estimate the cost-effectiveness HPV vaccination as a prevention of non-cervical cancers in low and middle income country like Indonesia. It is suggested to begin the study by expanding the HPV vaccination target to boys or adult women through ages 26 years.

DECLARATION OF COMPETING INTEREST

The author declares that she has no competing interests or personal relationships that could have appeared to influence the work reported in this study.

ACKNOWLEDGMENTS

This study was a part of thesis' requirement in Master in Pharmacy Management Program, Faculty of Pharmacy at Universitas Gadjah Mada.

REFERENCES

1. C. De Martel, D. Georges, F. Bray, J. Ferlay, and G. M. Clifford, "Global burden of cancer attributable to infections in 2018: a worldwide incidence analysis," *Lancet Glob. Health*, vol. 8, no. 2, pp. e180–e190, Feb. 2020, doi: 10.1016/S2214-109X(19)30488-7.
2. L. Bruni, A. Saura-Lázaro, A. Montoliu, M. Brotons, L. Alemany, M. S. Diallo, O. Z. Afsar, D. S. LaMontagne, L. Mosina, M. Contreras, M. Velandia-González, R. Pastore, M. Gacic-Dobo, and P. Bloem, "HPV vaccination introduction worldwide and WHO and UNICEF estimates of national HPV immunization coverage 2010–2019," *Prev. Med.*, vol. 144, p. 106399, Mar. 2021, doi: 10.1016/j.ypmed.2020.106399.
3. G. K. Shapiro, "HPV Vaccination: An Underused Strategy for the Prevention of Cancer," *Curr. Oncol.*, vol. 29, no. 5, pp. 3780–3792, May 2022, doi: 10.3390/curroncol29050303.
4. K. Zou, Y. Huang, and Z. Li, "Prevention and treatment of human papillomavirus in men benefits both men and women," *Front. Cell. Infect. Microbiol.*, vol. 12, p. 1077651, Nov. 2022, doi: 10.3389/fcimb.2022.1077651.
5. L. Bruni, A. Saura-Lázaro, A. Montoliu, M. Brotons, L. Alemany, M. S. Diallo, O. Z. Afsar, D. S. LaMontagne, L. Mosina, M. Contreras, M. Velandia-González, R. Pastore, M. Gacic-Dobo, and P. Bloem, "HPV vaccination introduction worldwide and WHO and UNICEF estimates of national HPV immunization coverage 2010–2019," *Prev. Med.*, vol. 144, p. 106399, Mar. 2021, doi: 10.1016/j.ypmed.2020.106399.
6. M. J. Page, J. E. McKenzie, P. M. Bossuyt, I. Boutron, T. C. Hoffmann, C. D. Mulrow, L. Shamseer, J. M. Tetzlaff, E. A. Akl, S. E. Brennan, R. Chou, J. Glanville, J. M. Grimshaw, A. Hrobjartsson, M. M. Lalu, T. Li, E. W. Loder, E. Mayo-Wilson, S. McDonald, L. A. McGuinness, L. A. Stewart, J. Thomas, A. C. Tricco, V. A. Welch, P. Whiting, and D. Moher, "The PRISMA 2020 statement: an updated guideline for reporting systematic reviews," *BMJ*, p. n71, Mar. 2021, doi: 10.1136/bmj.n71.
7. D. Husereau, M. Drummond, S. Petrou, C. Carswell, D. Moher, D. Greenberg, F. Augustovski, A. H. Briggs, J. Mauskopf, E. Loder, and on behalf of the CHEERS Task Force, "Consolidated Health Economic Evaluation Reporting Standards (CHEERS) statement," *BMJ*, vol. 346, no. mar25 1, pp. f1049–f1049, Mar. 2013, doi: 10.1136/bmj.f1049.
8. "Konversi Mata Uang - Investing.com," Investing.com Indonesia. Accessed: Sep. 05, 2023. [Online]. Available: https://id.investing.com/currency-converter/
9. "Inflation Calculator | Find US Dollar's Value From 1913-2023." Accessed: Sep. 05, 2023. [Online]. Available: https://www.usinflationcalculator.com/
10. T. H. Cheung, S. S. Y. Cheng, D. C. Hsu, Q. W.-L. Wong, A. Pavelyev, A. Walia, K. Saxena, and V. S. Prabhu, "The impact and cost-effectiveness of 9-valent human papillomavirus vaccine in adolescent females in Hong Kong," *Cost Eff. Resour. Alloc.*, vol. 19, no. 1, p. 75, Dec. 2021, doi: 10.1186/s12962-021-00328-x.
11. M. Jit, M. Brisson, J.-F. Laprise, and Y. H. Choi, "Comparison of two dose and three dose human papillomavirus vaccine schedules: cost effectiveness analysis based on transmission model," *BMJ*, vol. 350, no. jan06 14, pp. g7584–g7584, Jan. 2015, doi: 10.1136/bmj.g7584.
12. H. W. Chesson, E. Meites, D. U. Ekwueme, M. Saraiya, and L. E. Markowitz, "Cost-effectiveness of nonavalent HPV vaccination among males aged 22 through 26 years in the United States," *Vaccine*, vol. 36, no. 29, pp. 4362–4368, Jul. 2018, doi: 10.1016/j.vaccine.2018.04.071.
13. H. W. Chesson, E. Meites, D. U. Ekwueme, M. Saraiya, and L. E. Markowitz, "Cost-effectiveness of HPV vaccination for adults through age 45 years in the United States: Estimates from a simplified transmission model," *Vaccine*, vol. 38, no. 50, pp. 8032–8039, Nov. 2020, doi: 10.1016/j.vaccine.2020.10.019.
14. T. H. Cheung, S. S. Y. Cheng, D. Hsu, Q. Wing-Lei Wong, A. Pavelyev, I. Sukarom, and K. Saxena, "Health impact and cost-effectiveness of

implementing gender-neutral vaccination with the 9-valent HPV vaccine in Hong Kong," *Hum. Vaccines Immunother.*, vol. 19, no. 2, p. 2184605, 2023, doi: 10.1080/21645515.2023.2184605.
15. S. E. Choi, A. Choudhary, J. Huang, S. Sonis, A. R. Giuliano, and A. Villa, "Increasing HPV vaccination coverage to prevent oropharyngeal cancer: A cost-effectiveness analysis," *Tumour Virus Res.*, vol. 13, p. 200234, Jun. 2022, doi: 10.1016/j.tvr.2021.200234.
16. P. Cody, K. Tobe, M. Abe, and E. H. Elbasha, "Public health impact and cost effectiveness of routine and catch-up vaccination of girls and women with a nine-valent HPV vaccine in Japan: a model-based study," *BMC Infect. Dis.*, vol. 21, no. 1, p. 11, Jan. 2021, doi: 10.1186/s12879-020-05632-0.
17. R. Linertová, C. Guirado-Fuentes, J. Mar-Medina, and C. Teljeur, "Cost-effectiveness and epidemiological impact of gender-neutral HPV vaccination in Spain," *Hum. Vaccines Immunother.*, vol. 18, no. 6, p. 2127983, Nov. 2022, doi: 10.1080/21645515.2022.2127983.
18. F. S. Mennini, P. Bonanni, F. Bianic, C. de Waure, G. Baio, G. Plazzotta, M. Uhart, A. Rinaldi, and N. Largeron, "Cost-effectiveness analysis of the nine-valent HPV vaccine in Italy," *Cost Eff. Resour. Alloc. CE*, vol. 15, p. 11, 2017, doi: 10.1186/s12962-017-0073-8.
19. S. Simoens, A. Bento-Abreu, B. Merckx, S. Joubert, S. Vermeersch, A. Pavelyev, S. Varga, and E. Morais, "Health Impact and Cost-Effectiveness of Implementing Gender-Neutral Vaccination With the 9-Valent Human Papillomavirus Vaccine in Belgium," *Front. Pharmacol.*, vol. 12, p. 628434, Apr. 2021, doi: 10.3389/fphar.2021.628434.
20. M. T. Wahab, R. K. J. Tan, A. R. Cook, and K. Prem, "Impact of including boys in the national school-based human papillomavirus vaccination programme in Singapore: A modelling-based cost-effectiveness analysis," *Vaccine*, vol. 41, no. 12, pp. 1934–1942, Mar. 2023, doi: 10.1016/j.vaccine.2023.02.025.
21. J. J. M. Simons, N. Vida, T. A. Westra, and M. J. Postma, "Cost-effectiveness analysis of a gender-neutral human papillomavirus vaccination program in the Netherlands," *Vaccine*, vol. 38, no. 30, pp. 4687–4694, Jun. 2020, doi: 10.1016/j.vaccine.2020.05.031.
22. E. Wolff, K. M. Elfström, H. Haugen Cange, S. Larsson, H. Englund, P. Sparén, and A. Roth, "Cost-effectiveness of sex-neutral HPV-vaccination in Sweden, accounting for herd-immunity and sexual behaviour," *Vaccine*, vol. 36, no. 34, pp. 5160–5165, Aug. 2018, doi: 10.1016/j.vaccine.2018.07.018.
23. L. Zhang, D. G. Regan, J. J. Ong, M. Gambhir, E. P. F. Chow, H. Zou, M. Law, J. Hocking, and C. K. Fairley, "Targeted human papillomavirus vaccination for young men who have sex with men in Australia yields significant population benefits and is cost-effective," *Vaccine*, vol. 35, no. 37, pp. 4923–4929, Sep. 2017, doi: 10.1016/j.vaccine.2017.07.078.
24. Y. Li, Y.-F. Lin, X. Wu, X. Zhou, T. Tian, Z. Guo, L. Fu, L. Yang, Z. Lu, S. Fan, Y. Lu, W. Ke, and H. Zou, "Effectiveness and cost-effectiveness of human papillomavirus vaccination strategies among men who have sex with men in China: a modeling study," *Front. Immunol.*, vol. 14, p. 1197191, Jun. 2023, doi: 10.3389/fimmu.2023.1197191.
25. L. Majed, X. Bresse, N. El Mouaddin, A. Schmidt, V. J. Daniels, A. Pavelyev, L. Levy-Bachelot, and E. Elbasha, "Public health impact and cost-effectiveness of a nine-valent gender-neutral HPV vaccination program in France," *Vaccine*, vol. 39, no. 2, pp. 438–446, Jan. 2021, doi: 10.1016/j.vaccine.2020.10.089.
26. H. W. Chesson, E. Meites, D. U. Ekwueme, M. Saraiya, and L. E. Markowitz, "Updated medical care cost estimates for HPV-associated cancers: implications for cost-effectiveness analyses of HPV vaccination in the United States," *Hum. Vaccines Immunother.*, vol. 15, no. 7–8, pp. 1942–1948, 2019, doi: 10.1080/21645515.2019.1603562.
27. J. J. M. Simons, N. Vida, T. A. Westra, and M. J. Postma, "Cost-effectiveness analysis of a gender-neutral human papillomavirus vaccination program in the Netherlands," *Vaccine*, vol. 38, no. 30, pp. 4687–4694, Jun. 2020, doi: 10.1016/j.vaccine.2020.05.031.
28. Indonesian HTA Commitee, *Health Technology Assessment (HTA) Guideline.pdf*. Ministry of Health Republic of Indonesia, 2017.
29. "How do you choose the best sampling method for Monte Carlo simulation in OR?" Accessed: Sep. 05, 2023. [Online]. Available: https://www.linkedin.com/advice/3/how-do-you-choose-best-sampling-method-monte
30. S. Abidi, S. Labani, A. Singh, S. Asthana, and P. Ajmera, "Economic evaluation of human papillomavirus vaccination in the Global South: a systematic review," *Int. J. Public Health*, vol. 65, no. 7, pp. 1097–1111, Sep. 2020, doi: 10.1007/s00038-020-01431-1.
31. D. Setiawan, M. P. Oktora, R. Hutubessy, A. Riewpaiboon, and M. J. Postma, "The health-economic studies of HPV vaccination in Southeast Asian countries: a systematic review," *Expert Rev. Vaccines*, vol. 16, no. 9, pp. 933–943, Sep. 2017, doi: 10.1080/14760584.2017.1357472.

The Role of E-Prescription in Reducing Medication Error in The Prescribing Stage at The Puskesmas Mojoanyar, East Java

Izhhar Firas Widianto[1], *Akhmad* Kharis Nugroh[1], and *Chairun* Wiedyaningsih[1*]

[1]Department of Pharmaceutics, Faculty of Pharmacy, Universitas Gadjah Mada

Abstract. Electronic prescription or e-prescription has been considered as a solution to overcome errors in the early stage of medicines use. However, the electronic prescribing is still not widely implemented in Indonesia. The purpose of this study was to identify the role of electronic prescription in minimizing the incidence of medication errors at the prescribing stage. A cross-sectional study was conducted to compare errors between electronic and hand-written prescriptions in primary health care Mojoanyar. Mojokerto, East Java. Prescriptions were collected retrospectively using purposive and quota sampling based on predetermined criteria. A checklist form was used to collect and review prescriptions during the period February - March 2021 (hand-written prescriptions) and December 2022 - January 2023 (e-prescriptions). Errors in prescribing were identified in aspects related to the prescription writing process (incompleteness information; illegibility writing) and aspects related to drug selection decisions (drug interaction, drug dosage form, drug dosing and therapeutic duplication). The Chi-square test was used for testing relationships between categorical variables as appropriate. The data were presented in frequency and percentage using descriptive statistics. A total of 2570 prescriptions from 656 patients were reviewed for errors in which 1275 (328 patients) hand-written and 1295 (328 patients) electronic prescriptions. The results showed that incompleteness information was found to be significantly higher in hand-written compared to electronic prescriptions ($p < 0.05$). Illegible writing was only found in 92 (7.22%) hand-written prescriptions. The risk of drug interactions and other errors in drug selection decisions were not reduced by electronic prescribing. Electronic prescription was able to reduce prescribing errors, especially in the writing process, while for treatment decision it was necessary to add feature to the electronic support system.
Keywords: errors, electronic prescribing; hand-written, primary health care

1 Introduction

Prescription is an important aspect that needs attention by health professionals, because there are still many cases of errors in drug dispensing caused by errors in reviewing prescriptions. Study classification according to error types in a systematic review of medication errors in Southeast Asian countries reported that prescribing errors occurred 7%-35.4% [1]. In Indonesia, cases of medication errors during the prescribing phase were still relatively high [2].

Along with the development of technology, many health institutions have changed the method of writing prescriptions from hand-written to electronic (e-prescribing) systems, Computerized Physician Order Entry (CPOE). Hellström *et.al* 2009 [3] stated that electronic prescribing is easier to use and can improve patient safety. However, at this time, electronic prescription has only been used as a substitute for manual paper, and it has not been utilized optimally. Most of the electronic prescription that have been implemented, are still not optimally supported by the features in CPOE.

There are still very few research reports on identify of prescribing errors in Indonesia. The Mojoanyar sub-district was chosen because the primary health care in that area had just implemented electronic prescribing, therefore study was needed to compare errors between hand-written and e- prescribing. The results of this study are expected to add to the existing literature and can be very useful in developing policies and frameworks for the prevention of prescription-related medication errors.

2 Methods

2.1 Aim and objectives

The purpose of the study was to identify the impact of electronic prescription on reducing medication errors compared with handwritten prescriptions.

2.2 Study design and setting

A retrospective cross-sectional evaluation to assess errors in primary health care, Mojoanyar. Mojokerto, East Java settings was conducted. A check list sheet was used to collect the data and review to the type of prescribing errors during the period of February - March 2021 (hand-written prescriptions) and December 2022 -

* Corresponding author: chairun_wied@ugm.ac.id

January 2023 (e-prescriptions). The sampling method was purposive and quota sampling based on predetermined criteria.

2.3 Study sample

The study sample consisted of electronic and hand-written prescriptions. The study used the following inclusion criteria: (1) outpatient prescriptions, (2) contains at least one drug for oral administration (3) more than one drug prescribed for systemic use (4) contains prescription drugs to be compounded. The sample size required for this study was calculated based on the sample calculation formula for the hypothesis test of the difference in the proportions of the two populations (p1 was the proportion of electronic prescriptions and p2 was the proportion of non-electronic prescriptions). Determination of p1 and p2 taken from research conducted by Gandhi *et al.,* 2005 [4].

2.4 Prescribing errors evaluation and statistical analysis

Prescribing errors in both handwritten and electronic prescriptions were evaluated in terms of: aspects related to the prescription writing process and aspects related to drug selection decision. The prescription writing process was evaluated based on the completeness and legibility of the writing regarding the information being prescribed. The completeness information included: how to administer the drug orally, how to use special preparations/dosage forms, frequency of use of drugs/dose regimens and dosage strengths. Information on how to administer the drug orally referred to the accuracy of how to use the drug, including before, with or after meals. While the variable regarding specific use was information for drugs with special dosage forms where the information on how to use them needs to be clarified, for example sprayed, dissolved, etc. Legibility evaluation was carried out by reading the written prescription whether it was easy or difficult to read and understand by one of the selected puskesmas staff and the main researcher. Evaluation of inappropriate drug selection decision included drug dosage form, drug dosing, therapeutic duplication and drug interaction. Inaccuracy in the selection of drug dosage forms if an error occurs in determining the dosage form, including in the selection of the type of drug dosage form to be compounded. Drug dosing was inappropriate dosage based on the recommended dosage from the medicine leaflet or MIMS. Symptomatic drugs or symptom relievers were not included in the category of inappropriate drug dosing. Therapeutic duplication was the practice of prescribing multiple medications for the same indication or purpose without a clear distinction of when one agent should be administered over another. The drug interaction was when there were two or more drugs affect each drug. To find out whether or not this interaction exists used the sites drugs.com and Medscape. The interaction findings based on R/ then were classified based on MIMS Indonesia online (https://www.mims.com/indonesia), Medscape (https://reference.medscape.com/drug-interactionchecker) and drug.com (https://www.drugs.com/drug_interactions.html). For testing relationships between categorical variables, analysis statistic with tables 2x2 was used. A Chi-squared test for categorical variables was used as appropriate. A p value of less than 0.05 was considered as significant. Descriptive statistics were used to illustrate results.

2.5 Ethical considerations

The study was carried out after obtaining an ethically certificate with the number KE/FK/0064/EC/2023 from the Medical and Health Research Ethics Committee (MHREC), Faculty of Medicine, Public Health and Nursing Universitas Gadjah Mada-DR Sardjito General Hospital, Indonesia.

3 Results

A total of 2570 prescriptions from 656 patients were reviewed for prescribing errors in which 1275 (328 patients) hand-written and 1295 (328 patients) electronic prescriptions. The results show that the total number of errors in hand-written and electronic prescriptions based on number of R/ was 662 (51.92%) & 495 (38.22%) respectively, while results based on number of patients were 370 R/ (112.8%) prescription errors for hand-written prescriptions and 278 R/ (84,8%)

Table 1. Prescribing errors found in hand-written (category A) and electronic prescriptions (category B)

Variable	Category A		Category B	
Based on number of R/	**N=1275**		**N=1295**	
	frequency (n)	%	frequency (n)	%
incompleteness information	408	61.6	329	66.5
illegibility writing	92	13.9	0	0.0
drug interaction	144	21.7	125	25.2
drug dosage form	11	1.7	26	5.3
drug dosing	3	0.5	6	1.2
therapeutic duplication	4	0.6	9	1.8
Total errors	662	51.92	495	38.22
Based on number of patients	**N= 328**		**N= 328**	
	frequency (n)	%	frequency (n)	%
incompleteness information	200	40.4	178	64.0
illegibility writing	92	18.6	0	0.0
drug interaction	60	12.1	59	21.2
drug dosage form	11	2.2	26	9.4
drug dosing	3	0.6	6	2.2
therapeutic duplication	4	0.8	9	3.2
Total errors	370	112.8	278	84.8

for electronic prescriptions. In addition, it was also found that there were drug interactions occurring in 125 (25.2%) based on R/ or 59 (21.2%) based on patients on electronic and 144 (21.7%) based on R/ or 60 (12.1%) based on patients on hand-written prescriptions. Table 1 shows the prescribing errors found in hand-written and electronic prescriptions. The problem of incompleteness information in prescription writing was still the highest in relation to prescribing errors, both in hand-written and electronic prescriptions, where these errors account for around half of all errors. The high error rate of incompleteness information was due to the fact that the evaluation focuses on the completeness of the information on the drug use dosage regimens (see Table 2). Reading what information in prescriptions was clearly no difficulty when compared to reading information in handwritten form, as can be seen in Table 1, the number of illegibility in e-prescriptions and hand-written prescriptions was 0 (0.0%) & 92 (13.9%) respectively.

Further evaluation regarding the incompleteness of information was: how to administer the drug orally, how to use special preparations/dosage forms, frequency of use of drugs/dose regimens and dosage strengths. The results of further evaluation regarding the incompleteness information are presented in Table 2. The variable administration described whether the physicians mentioned how to use the drug orally in writing the prescription. Physicians generally did not provide this information, even though it's actually easy to write in Latin abbreviations, for example *pc* (*post coenam*), which means after meal. Even though the dosage strength variable in electronic prescribing was always completely written, incomplete information was still found in electronic prescription for administration variables, specific uses, and dosage regimens.

Table 2. Distribution of incompleteness of information found in hand-written (category A) and electronic prescriptions (category B)

Variable	Category A		Category B	
Based on number of R/	**N=1275**		**N=1295**	
	frequency (n)	(%)	frequency (n)	(%)
administration	349	85.6	291	88.4
specific use	45	11.0	22	6.7
dosage regimen	9	2.2	16	4.9
dosage strength	5	1.2	0	0.00
Total incompleteness	408	32.00	329	25.41
Based on number of patients	**N= 328**		**N= 328**	
	frequency (n)	(%)	frequency (n)	(%)
administration	210	78.0	172	81.9
specific use	45	16.7	22	10.5
dosage regimen	9	3.4	16	7.6
dosage strength	5	1.9	0	0.0
Total incompleteness	269	82.0	210	64.0

Based on the drug selection decision, the drug interactions were still found for both electronic and handwritten prescriptions. Therefore, further evaluation in terms of level drug interactions based on the number of R/ was carried out and and classified in several levels, including: minor, moderate and major. The moderate level of drug interactions was the most frequent interaction category (Table 3).

Chi-square analysis using a 2x2 contingency table was performed to describe whether there was a difference in the proportion of incompleteness informa tion between handwritten and electronic prescriptions either based on the number of R/ or based on the number of patients. The results of the analysis showed that there was a significant difference ($p<0.05$) between handwritten prescriptions and electronic prescriptions regarding the incompleteness of the information. The Chi square was also performed to analyse the difference proportion of drug interaction between handwritten and electronic prescriptions either based on the number of R/s or based on the number of patients. The results showed that there was no significant difference with regard to the findings of drug interactions in handwritten and electronic prescriptions (Table 4).

Table 3. Evaluation of the level of drug interactions from handwritten (category A) and electronic prescriptions (category B) based on the number of R/

Variable	Category A N=144		Category B N=125	
number of drug interactions based on R/	**n**	**%**	**n**	**%**
Level 1 – Minor	33	22.92	28	22.4
Level 2 – Moderate	84	58.33	78	62.4
Level 3 -Major	27	18.75	19	15.2

Table 4. Difference in proportion of incompleteness information and drug interaction between handwritten (category A) and electronic prescriptions (category B)

Variable	A	B	p	OR	95% CI	
Evaluation based on number of R/						
Incompleteness information						
Not complete	408	329	0.0002	1.38	L	1.16367
Complete	867	966			U	1.64065
Drug Interaction						
Interaction occurred	144	125	0.1741	1.19	L	0.9252
No interaction	1131	1170			U	1.5350
Evaluation based on number of patients						
Incompleteness information						
Not complete	200	178	0.0000	2.44	L	1.8244
Complete	128	278			U	3.2642
Drug Interaction						
Interaction occurred	60	59	0.9193	1.02	L	0.6862
No interaction	268	269			U	1.5185

* A = Category A; B = Category B ;
L = Lower 95% CI ; U = Upper 95% CI

4 Discussion

Therapeutic management is a multistep process, and errors may occur at any step, from prescribing to administering the medication. Studies have shown that

https://doi.org/10.1051/bioconf/20237505019

drug prescription errors are the most frequent [5-7]. This study was carried out on the prescription errors in hand written and electronic prescriptions of outpatients in a primary health care, East java. Results of the present study revealed that the most common prescribing error was incompleteness prescription writing, both in hand-written and electronic prescriptions. Although previous studies have provided data related to incompleteness information in prescribing, not all studies can be compared since the methods and subjects evaluated are different. The current study is close to the one obtained by al-Madadha *et al* 2014 [8] after the implementation of the electronic system. Al- Madadha *et al* 2014 [8] study also has evaluated incomplete information in terms of drug use. The most common serious errors were dosage form not mentioned (11.6%) and drug strength not mentioned (6.2%)[8], while in the current study, the most common error was not mentioning how to use the drug (before, with or after meals). As expected, the effect of the electronic system on prescribing errors was due to the elimination of unclear handwriting errors that is consistent with the reported data from Hitti *et al* 2017 [9]. It was detected in this present study that as many as 1275prescriptions were prescribed to 328 patients, 13.9% of prescriptions found difficult to read. Prescription illegibility can lead to an increase in the risk of medication errors [10]. With the implementation of an electronic writing system, it will be able to erase prescription errors, especially data about patient characteristics

Drug interactions are a major cause of morbidity and a major source of treatment ineffectiveness. The results of the current study show that drug interactions were still found in electronic prescribing. Moderate drug interactions were the most frequent category of interactions. Selecting drug from a database that has been integrated into prescribing is much easier than writing prescription conventionally. This can assist physicians in deciding to look for alternative drugs through the database if the prescribed drugs interact. However, errors related to drug interactions did not decrease significantly with the presence of electronic prescribing at the primary health care where this study was conducted. This is because the computerized system at the current study setting, the Mojoanyar primary health care, Mojokerto, East Java had not been integrated with the drug interaction database. In electronic prescribing, special attention is needed for systems that facilitate the automatic detection of drug interactions. Several software's have been widely used to detect possible drug interactions in prescribing [11, 12]. Therefore, it is highly recommended to add a feature to electronic prescribing that can automatically give an alert signal when a physician makes error in selecting a drug. This feature will greatly assist physicians in making treatment decisions, so there is no need to check repeatedly when making drug choices. This feature can be in the form of an application or a link that can be directly integrated into the prescribing program

5 Limitation of the study

The limitation deals with the methodological approach to evaluation, as in this study was carried out retrospectively, and sampling was carried out purposively. Therefore, it is still limited to generalize to the population. Although several interviews were conducted to understand the problem of errors that occurred, the interviews were not conducted in depth with qualitative methods and only checklists were used to assess aspects of prescribing errors.

6 Conclusion

The electronic prescribing was a useful tool for reducing the number of prescribing errors in primary health care, Mojoanyar. Mojokerto, East Java settings, although, more advanced system with decision support features may be needed for more impact on the mistakes in making decisions.

Acknowledgment: The authors thank and appreciate the primary health care Mojoanyar. Mojokerto, East Java for the research permit.
Conflict interests: The authors declare no competing or potential conflicts of interest concerning the research and publication of this article

References

1. Salmasi S, Khan TM, Hong YH, Ming LC, Wong TW. Medication errors in the Southeast Asian countries: A systematic review. PLoS One. 2015;10(9):1–19.
2. Damiti S, Thalib S, Kamba V, Ysrafil Y, Hartati H. Analysis of prescribing error incidence in out-patient prescription at community health center of Tilamuta Boalemo Regency. Sci Midwifery. 2022;10(5):4357–63.
3. Hellström L, Waern K, Montelius E, Strand B, Rydberg T, Petersson G. Physicians attitudes towards ePrescribing -evaluation of a swedish full-scale implementation. BMC Med Inform Decis Mak. 2009;9(1):1–10.
4. Gandhi TK, Weingart SN, Seger AC, Borus J, Burdick E, Poon EG, et al. Outpatient prescribing errors and the impact of computerized prescribing. J Gen Intern Med. 2005;20(9):837–41.
5. Lewis, P.J., Dornan, T., Taylor D et al. Prevalence, Incidence and Nature of Prescribing Errors in Hospital Inpatients. Drug Saf. 2009;32:379–89.
6. Garfield S, Reynolds M, Dermont L, Franklin BD. Measuring the severity of prescribing errors: A systematic review. Drug Saf. 2013;36(12):1151–7.
7. Mahomedradja RF, Schinkel M, Sigaloff KCE, Reumerman MO, Otten RHJ, Tichelaar J van AM. Factors influencing in-hospital prescribing errors: A systematic review. Br J Clin Pharmacol. 2023;89(6):1724–35.

 https://doi.org/10.1051/bioconf/20237505019

8. al- Madadha RA., Al-farajat MA, Alyazjeen AA, Masarweh H, Alshqairat T, Haddadin SS, et al. Prescriptions errors: a comparison of handwritten and computerized prescriptions at Royal Medical Services, Jordan. Zagazig J Pharm Sci. 2014;23(2):44–8.
9. Hitti E, Tamim H, Bakhti R, Zebian D, Mufarrij A. Impact of internally developed electronic prescription on prescribing errors at discharge from the Emergency Department. West J Emerg Med. 2017;18(5):943–50.
10. Albarrak AI, Al Rashidi EA, Fatani RK, Al Ageel SI, Mohammed R. Assessment of legibility and completeness of handwritten and electronic prescriptions. Saudi Pharm J [Internet]. 2014;22(6):522–7. Available from: http://dx.doi.org/10.1016/j.jsps.2014.02.013
11. Kulkarni V, Bora SS, Sirisha S, Saji M, Sundaran S. A study on drug–drug interactions through prescription analysis in a South Indian teaching hospital. Ther Adv Drug Saf. 2013;4(4):141–6.
12. Rogero-Blanco E, Del-Cura-González I, Aza-Pascual-Salcedo M, García de Blas González F, Terrón-Rodas C, Chimeno-Sánchez S, et al. Drug interactions detected by a computer-assisted prescription system in primary care patients in Spain: MULTIPAP study. Eur J Gen Pract. 2021;27(1):90–6.

The Social Support of Extended Family as the protective factor of Stunting among Migrant Labour Families in Magetan, East Java

Hadi Sucipto[1], *Nurhadi* Nurhadi[2], *Supriyati* Supriyati[3*]

[1]Magetan Health Office, Jl. Imam Bonjol no 4 Magetan East Java 63361 Indonesia
[2]Faculty of Social and Political Science, Jl. Socio Humaniora Bulaksumur Yogyakarta, 55281 Indonesia
[3]Department of Health Behavior, Environment, and Social Medicine, Jl. Sekip Utara Yogyakarta 55281 Indonesia

Abstract. Indonesian government targeted the stunting prevalence less than 14% by 2024. There are complex social determinants of stunting, including socio-economic background of the family, culture, parenting as well as health literacy. Migrant labour family as the vulnerable population was facing with those social determinants of stunting. This qualitative study was aimed to explore the social determinant of stunting among migrant labour family in Magetan District, East Java Province, Indonesia. A total of 15 informants who selected purposively were participated in this study. Data was collected through in-depth interviews and unstructured observations. Data analysis performed by applied Open Code 4.03 Software. Moreover, triangulation, member checking, and peer debriefing were the strategies of trustworthiness. The migrant labours in Magetan were young families with low and medium educational background. They were exposed to the parenting problems such as toddlers' eating behaviours, lack of nutrition intake, family and social bonding, as well as growth monitoring system. Meanwhile, the family social support helped them a lot to dial with these problems. This study suggests the important to improve and empower family's capacity in parenting issues. Keywords: Social determinant of health, family empowerment, migrant worker, parenting, stunting

1 Background

Indonesia has the highest stunting prevalence among southeast Asia countries. Stunting remains the public health issues in Indonesia. Indonesia's Nutrition Status Survey shows that the Indonesian stunting prevalence in 2022 is 21.6% [1,2]. Moreover, the Indonesian Government has targeted to reduce the stunting prevalence in to 14% in 2024 [3]. Stunting prevention and control become the national priority program and invite various stakeholders to participate in the program.

Studies showed that there were complex social determinants of stunting including child characteristics, inadequate water and supply, parental smoking, family characteristics (i.e. parent education level, parent economic background, parent occupation), parenting, food insecurity, and low caregiver education [2-5]. Low level of parent education and income contributes to the stunting incidence [6]. Moreover, Win et al. [7] explained that children of working mothers had 4.5 times increased odds of stunting.

Migrant labours families were vulnerable population. Study found that children of migrant labours families had low immunization, undernutrition and other health problems [8]. Magetan District, East Java Province had high number of migrant labour in East Java and also had high prevalence of stunting (17.2%). Furthermore, the local government had launched *Desmigratif* (*Desa Migrant Produktif* – The productive migrant village) by developing community parenting among migrant labour families [9]. This study was aimed to explore the social determinant of stunting among migrant labour families in Magetan District, East Java, Indonesia.

2 Method

This qualitative study carried out from April to June 2023 using case study approach [10, 11]. This study was focus on the social determinants of stunting among migrant labour family with children under five years old who experience stunting in Magetan District, East Java Province Indonesia.

A number of 15 people were recruited purposively as informants. They were significant person of the children under five years old who experiencing stunting from the migrant labour families (father, mother, caregivers or relatives), village midwives, *Posyandu* (health integrated post) cadres, nutritionist of the primary health care, member of the community development board, as well as local leaders. Data were collected through in-depth interviews by using pre tested interview guidelines. Besides, unstructured observation had been done for completing this study. In-

* Corresponding author: supriyati@ugm.ac.id

depth interviews had performed in person with each informant and the interviews duration was about 50 minutes for each informant. Moreover, the unstructured observations had been conducted to observe the informants' activities as the triangulation strategy.

The first in-depth interview went to the significant person of toddler with stunting to explore the health problems related stunting, the social determinants of stunting including parenting practice, and to explore the way of coping mechanism among them. We met the family who one of the family members became migrant labour in Asia countries. Those in-depth interviews carried out in their house. Afterward, we conducted the in-depth interviews with P*osyandu* cadres, village midwives, the nutritionist of the primary health care, local leaders, and the member of the community development board. Unstructured observation conducted in the *Posyandu* and the informant house to observe the parenting practice of the toddler families with stunting and the environment of migrant labour family to support the in-depth interviews results.

Data was analysed qualitatively by applied thematic analysis and used the Open Code 4.03 Software. Furthermore, member checking, triangulation and peer debriefing applied for trustworthiness. This study was reviewed and approved by the Medical and Health Research Ethic Committee of the Faculty of Medicine, Public Health and Nursing Universitas Gadjah Mada (REF No.: KE-FK-0523-EC-2023).

3 Results

3.1 Informant characteristics

There were two groups of informants. First group was the significant person of the toddlers with stunting. Secondly, related stakeholders of the stunting prevention and the migrant workers in Magetan. Table 1 described the first group informants' characteristics.

Table 1. Informants' characteristics of the family member

Age	From 28 to 40 years old
Education background	From Elementary School to Senior High School
Occupational	Informal workers such as carpenter, farmer, or housewife
Number of children	1-3 people
Live with	Nuclear family or extended family
Toddlers with stunting	1-2 children
Migrant labours	4 women and 3 men
Countries of destination	Taiwan (4), Malaysia (2), and Singapore (1)
Type of work	Household workers (4), shopkeeper (1), wood factory worker (1), electronic factory worker (1)
Period of work in the destination countries	From 3 months to 17 years

Table 1 showed that informants came from the low socio-economic status, and they were spent many years as the migrant labours. There was a different type of work between men and women. Women tended to be household worker. As opposite, men tended to be factory worker or shopkeeper. Besides, this study showed that the father who lived with their toddlers with stunting played a doble task as informal workers such as being carpenters or farmers) and also taken care of their toddler. Meanwhile, women who lived with their toddlers with stunting more focus on caring for their children (as a housewife), and only one woman that actively in online store.

Taiwan was the most favourites destination country for migrant labour from Magetan. As young families, they had 1 - 3 children. Unfortunately, this study showed that one or two toddlers was diagnosed as toddler with stunting. The stunting toddlers were boys and girls.

The second group of informants were stakeholders, who had medium and high socio-economic background. There were frequently interacts with the toddlers with stunting and their families. Their ages were from 32 to 42 years old.

3.2 Health problems related to stunting

Toddlers with stunting among migrant labour were faced with complex problems related health. The problems came through all phases and all levels of influence (individual, family, and community or social system), as Figure 1.

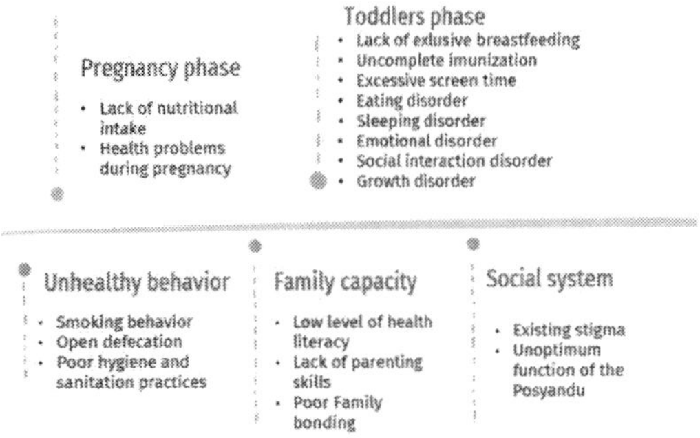

Fig. 1. Health problems related to stunting

Figure 1 showed that the migrant labour family had low level of health literacy, lack of parenting skills and had poor bonding especially between parent and their children. Interestingly, there were difference pattern between family with women migrant labours and family with men migrant labours. Fortunately, families with men as migrant labour had a better family bonding compare to the families with women migrant labour.

https://doi.org/10.1051/bioconf/20237505020

The excessive screen time lead the other problems such as eating disorder, sleep disorder, social interaction disorder as well as emotional disorder. Furthermore, the migrant family also lack of parenting skills and had low level of health literacy. Some of their fathers also smoker. In addition, there were exposed to the stigma related to stunting. They were ashamed if their children diagnosed as toddler with stunting.

> *"...my child suffered from diarrhea for up to 20 days. People said it was okay. That was part of the growth phase and these would made my child more agile in walking (ngentheng-enthengi) (T, 40 years old, woman, her husband was a migrant labour in Malaysia)*

> *"...(they) had threatened not to come to the posyandu, when we told them that their child was suffering stunting..." (L, 38 years old, Posyandu Cadre)*

These was a big dilemma. In the one hand, if the *Posyandu* cadre did not convey the truth about the stunting status, so their family would not immediately conduct the prompt treatment. On the other hand, if Posyandu cadre tell it, they would no longer come in to Posyandu and the childrens' growth could not be monitored.

3.3 Parenting practices among migrant labour

This study showed that the parenting practices of migrant labour families was the results of various social factors interaction. The social factors include the desire to be a happy family in the one hand, and the limited abilities on the other hand. A happy family was described as a family that was able to meet the needs of its family members, and money was considered the most potential resource to meet the various needs of the family members. Economic issues was the main reason of people to become a migrant labour. They were facing with a lot of problems related to economic so they were decided to become migrant labour.

> *"...e...we had a lot of responsibilities related to the economic burden... and.... It was seem like there was no other option (we could take..). It was not easy to earn money here..." (J, 40 years old, man, his wife is migrant labour in Taiwan)*

> *"...poor family... they were categorized as the poor family. They were also received the cash transfer program from the government..but... it did not their economic problems, so they become migrant labour..." (F, 42 years old, Community leader)*

They became the migrant labours as the best solution on the economic problems, since they observed, their neighbour has managed to earn a lot of money in the short time by becoming migrant labour. Furthermore, the migrant labour tried hard to "buy their family happiness" by sending a lot of money and fulfilled many kinds of family needs. They allow their children to have many toys and smartphone, as Figure 2.

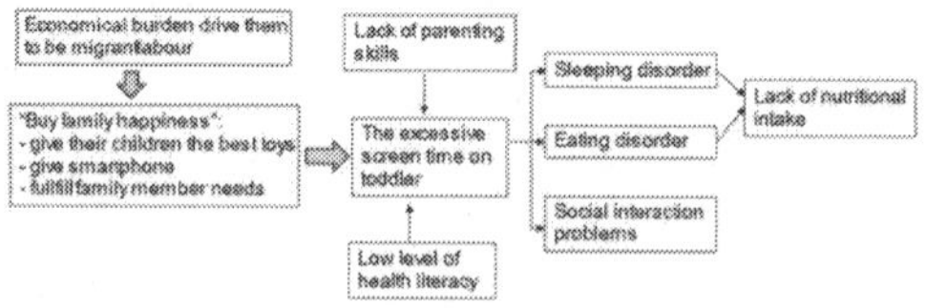

Fig. 2. Buy family happiness

The migrant worker had a communication barrier with their children. Eventhough they bought a smartphones for their children but still there was a communication barrier.

> *'...She asks for a toy... when she asks for a toy she calls her mother... unfortunately, if she does not ask then she does not want to be called... so... when she has a desire then she call her mother ... but she doesn't have a desire he doesn't want to be called (T, 38 years old, man, his wife is migrant worker in Taiwan)*

In fact, the migrant worker had a heavy burden. They spend a lot of time to get money. Long working hours and or very limited opportunities to go home and meet with their family were the example of migrant worker sacrifices, as quotes below. It was their way to buy the family happiness.

> *"...(my wife as migrant labour) work start from 05.00 am until 11.00 pm or even 12.00 pm.."* (J, 40 years old, man, his wife was a migrant labour at Taiwan)

> *"...(my husband) work at wood factory, and he has not back home yet since the first year. Now was year fourth, and our children have not meet her father yet in person... " (D, 34 years old, toddler mother, her husband work as migrant labour at Malaysia)*

The limited abilities of the family member were the other problem of the toddler with stunting family. Low levels of health literacy and the lack or parenting skills lead them make unhealthy decision such as giving smartphone to their toddler as a playmate. As a result, their toddlers spend their days and nights on smartphones. They did not aware on the risk of the excessive screen time among children. Their parents were happy with it because their toddler looked calm and happy. Unfortunately, other problems arise such as sleep disorders, eating disorders, and lack of social interaction.

> *"... it was very difficult to get my child to sleep. Until very late at night, he played his smart phone. Usually, he sleeps at 12.00 pm or 01.00 am. He does not sleep all the day.." (D, 34 year old, woman, his husband was a migrant labour at Malaysia)*

> *"...it was only 3 spoons in two days.. it was quite difficult.. (to make their children eat properly), ..yet.. sometimes she only eat snack.. (J, 40 years old. Man. his wife is migrant labour in Taiwan)*

> *".. no..never...he does not like to eat egg or chicken.. for chicken, he only eat the souce.." (D, 34 year old, woman, his husband was a migrant labour at Malaysia)*

According to the data analysis, stunting among migrant labour families was a complex public health issue.

4 Discussion

This study showed that there were complex problems related to the social determinants on stunting in toddlers from the migrant labour family. Achieving the family happiness was the greatest desire of the family. Each family member, especially parent definitively would like to achieve it. Being migrant labour was the alternative solution of the vulnerable population (people who came from the low socio-economic background and did not have enough resources to fulfil their basic needs). Afterward, they would "buy the family happiness" with their money. Unfortunately, the low level of health literacy and the lack of parenting skills lead them take unhealthy decision such as giving their toddler smartphones that caused various health problems arose.

Parent of toddlers with stunting in this study had low and medium education background who live in rural area. Mother with low education and living in rural area were risk factors of stunting [4-6, 12]. They were also had low level of health literacy include eating behavior for children under five years old. A comprehensive intervention should be done to improve nutritional status during pregnancy, exclusive breastfeeding phase, complementary feeding and improving healthy eating behavior among children under five years old [12-14]. The impact of the poor nutritional status during pregnancy and the first five years of life is not limited to the children growth phase, but also potentially improve adult morbidity and mortality [2,13].

The intervention should engage community and other stakeholders and develop public policies that enable social action. The role of the government as the regulator is needed [15]. The migrant labour community was the vulnerable group whereas potentially experience to health inequality and inequity. This study showed that the policy intervention for the vulnerable population should be not limited to the economic sector, but also covered public health issues such as healthy behavior and parenting. Parenting and family issues lead to the wellbeing achievement of the community [16-17]. The implementation of *Desmigratif* which has been developing community parenting among migrant labour families must strengthen family function as a whole. Family social capital helps people to have healthy behavior [18]. Family is the important actor in public health [19].

This study indicates that family happiness was the highest desire of family, and all family members were encouraged to make it happen. They were decided to become migrant worker to increase their economic capacity in order to able fulfil the family member needs. Unfortunately, the increasing economic capacity which was imbalance with the implementation of the comprehensive family function became new threat of the family happiness. Family dimension such as family cohesion and communication were crucial for the family happiness [16-19]. Moreover, the excessive screen time of children under five years old among migrant labour family was disrupt the family cohesion and communication. The awareness of the parenting skills among vulnerable group was challenging for further study.

This paper helps public health experts when developing public health intervention on stunting issue, especially for the vulnerable group. Unfortunately, this study did not explore yet the *Desmigratif* program from the program developer perspective. This is the weakness of this study, despite the interesting findings.

5 Conclusion

This study indicates that "buy family happiness" was the highest desire of the migrant labour family, and it was causing the complex problems related to stunting. Thus, public health intervention for stunting must be covered the comprehensive healthy behaviours, parenting skills and also empowered the family function.

We acknowledge the local government and the health officer of Magetan District for their hand in this study.

References

1. Badan Kebijakan Pembangunan Kesehatan. Buku Saku Hasil Survei Status Gizi Indonesia 2022. Kementerian Kesehatan. (2022)
2. Miranda AV, Sirmareza T, Nugraha RR, et al. Towards stunting eradication in Indonesia: Time to invest in community health workers. Public Health Chall. **2**, e108. https://doi.org/10.1002/puh2.108 (2023)
3. Herawati DMD, Sunjaya DK. Implementation outcomes of National Convergence Action Policy to Accelerate Stunting Prevention and Reduction at the Local Level in Indonesia: A Qualitative Studi. Int J Environ Res Public Health. **19**, 20:13591. doi: 10.3390/ijerph192013591. PMID: 36294173; PMCID: PMC9602846. (2022)
4. Yani DI, rahayuwati L, Sari CMW, Komariah M, Fauziah SR. Family Household Characteristics and Stunting: An Update Scoping Reviewe. Nutrients. **15**, 1: 233. Doi. 10.3390/nu15010233. PMID: 36615889; PMCID: PMC9824547. (2023)
5. Beal T, Tumilowicz A, Sutrisna A, Izwardy D, Neufeld LM. A review of child stunting determinants in Indonesia. Matern Child Nutr. **14**, 4: e12618. Doi: 10.1111/mcn.12617. Epub 2018 May 17. PMID: 29770565; PMCID: PMC6175423. (2018)
6. Ulfah, IF, Nugroho AB. Menilik Tantangan Pembangunan Kesehatan di Indonesia: Faktor Penyebab Stunting di Kabupaten Jember. J.Sos.Polit. **6**, 201-2013. doi:10.22219/sospol.v6i2.12899 (2020)
7. Win H., Shafique S., Mizan S., Wallenborn J., Probst-hensh N, Fink G. Association between mother's work status and child stunting in urban slums: a cross-sectional assessment of 346 child-mother dyads in Dhaka, Bangladesh. Archives of Public Health. **80**, 192.

https://doi.org/10.1186/s13690-022-00948-6. (2022)

8. Roshania RP, Giri R, Cunningham SA, Young MF, Webb-Girard A, Das A, Mala GS, Srikantiah S, Mahapatra T, Ramakrishnan U. Early life migration and undernutrition among circular migrant children: An observational study in the brick kilns of Bihar, India. J Glob Health. **12,** 04008. doi: 10.7189/jogh.12.04008. PMID: 35136599; PMCID: PMC8818295. (2022).
9. Nuraeni, Y., Pembangunan Desa Migran Produktif (Desmigratif) Dengan Pendekatan Perencanaan Secara Holistik, Temanik, Integratif, dan Spasial. *J. Ketenagakerjaan* 16: 29–47. doi:10.47198/naker.v16i1.90. (2021)
10. Yin RK., Case Study Research Design and Method. 5th ed. Thousand Oaks. CA:Sage. (2014)
11. Utarini, A., Tak Kenal Maka Tak Sayang Penelitian Kualitatif Dalam Pelayanan Kesehatan. Gadjah Mada Univ. Press (2021)
12. Suratri MAL, Putro G, Rachmat B, Nurhayati, Ristrini, Pracoyo NE, Yulianto A, Suryatma A, Samsudin M, Raharni. Risk Factors for Stunting among Children under Five Years in the Province of East Nusa Tenggara (NTT), Indonesia. Int J Environ Res Public Health. **20**, 2:1640. doi: 10.3390/ijerph20021640. PMID: 36674391; PMCID: PMC9865789. (2023)
13. De Sanctis V, Soliman A, Alaaraj N, Ahmed S, Alyafei F, Hamed N. Early and Long-term Consequences of Nutritional Stunting: From Childhood to Adulthood. Acta Biomed. **92**, 1:e2021168. doi: 10.23750/abm.v92i1.11346. PMID: 33682846; PMCID: PMC7975963. (2021)
14. Juarez M, Dionicio C, Sacuj N, Lopez W, Miller AC, Rohloff P. Community-Based Interventions to Reduce Child Stunting in Rural Guatemala: A Quality Improvement Model. Int J Environ Res Public Health. **18**, 2:773. doi: 10.3390/ijerph18020773. PMID: 33477580; PMCID: PMC7831302. (2021)
15. Bhutta ZA, Akseer N, Keats EC, Vaivada T, Baker S, Horton SE, Katz J, Menon P, Piwoz E, Shekar M, Victora C, Black R. How countries can reduce child stunting at scale: lessons from exemplar countries. Am J Clin Nutr. **112**(Suppl 2):894S-904S. doi: 10.1093/ajcn/nqaa153. PMID: 32692800; PMCID: PMC7487427. (2020)
16. Thomas PA, Liu H, Umberson D. Family Relationships and Well-Being. Innov Aging. **3**, igx025. doi: 10.1093/geroni/igx025. Epub 2017 Nov 11. PMID: 29795792; PMCID: PMC5954612. (2017)
17. Izzo F, Baiocco R, Pistella J. Children's and Adolescents' Happiness and Family Functioning: A Systematic Literature Review. Int J Environ Res Public Health. **19**, 24:16593. doi: 10.3390/ijerph192416593. PMID: 36554474; PMCID: PMC9778774. (2022)
18. Supriyati S., Kun H., Adianto RAW., Wahyuni A., Sebastian KH., Aditya EN., Sofyan MS., Family social capital in the public response to COVID-19 in Indonesia. Journal of Community Empowerment for Health. **4**, 3. p.196-202. https://doi.org/10.22146/jcoemph.64567. https://jurnal.ugm.ac.id/jcoemph/article/view/64567/32765 (2021)
19. Barnes MD., Hanson CL, Novilla LB, Magnusson BM, Crandall AC, and Bradford G., Family-Centered Health Promotion: Perspectives for Engaging Families and Achieving Better Health Outcomes. INQUIRY: The Journal of Health Care Organization, Provision, and Financing. **57**, https://doi.org/10.1177/0046958020923537 (2020)

Urban-Rural Difference in Adherence Treatment of Hypertensive Patients In South Sumatra Indonesia

Yopi Rikmasari[1,2], *Tri Murti Andayani*[3*], *Susi Ari Kristina*[4], and *Dwi Endarti*[4]
[1] Doctoral Program in Pharmacy, Faculty of Pharmacy, Universitas Gadjah Mada, Yogyakarta, Indonesia.
[2] Faculty of Pharmacy, Sekolah Tinggi Ilmu Farmasi Bhakti Pertiwi, Palembang, Indonesia
[3] Department of Pharmacology and Clinical Pharmacy, Faculty of Pharmacy, Universitas Gadjah Mada, Yogyakarta, Indonesia
[4] Department of Pharmaceutic, Faculty of Pharmacy, Universitas Gadjah Mada, Yogyakarta, Indonesia

Abstract. Treatment adherence is essential for controlling blood pressure and preventing complications. The availability of information regarding factors related to adherence is needed to design appropriate interventions. However, this information is still limited. This study aims to evaluate medication adherence and identify the main factors associated with hypertension patients living in urban and rural areas. A cross-sectional study was conducted in ten primary healthcare facilities in South Sumatra Province, consisting of 5 urban and 5 rural areas, totaling 458 hypertensive patients. Medication adherence was measured using the MGLS questionnaire, social support with the MSPSS, and level of knowledge with HK-LS. Data were analyzed using descriptive statistics, bivariate tests using chi-square, independent T-test, and mann-whitney tests, and then continued multivariate logistic regression analysis. Adherence to hypertension medication in urban (38.2%) and rural (23.6%) areas differed significantly (p=0.000). Medication adherence in urban areas was related to work (OR=4.787, p=0.000), social support (OR=5.054, p=0.000), and level of knowledge (OR=6.558, p=0.000). In rural areas, medication adherence is associated with social support (OR=4.696, p=0.000), knowledge level (OR=12.555, p=0.022), high/middle school education (OR =3.290, p=0.000), bachelor or above education (OR=12.871, p=0.000). Most patients are non-adherent to treatment hypertension in urban and rural areas. The factors most related to medication adherence in urban areas were employment status, social support, and knowledge about hypertension, while education, social support, and knowledge were the factors most related to adherence in rural areas. Interventions must be designed based on the information obtained to improve medication adherence.
Keywords: hypertension, adherence, urban, rural

1 Introduction

The global prevalence of hypertension is high and continues to increase [1]. At 30-70 years old, hypertension sufferers are estimated to have doubled between 1990 and 2019. The prevalence of hypertension in high-income countries has decreased, but in low-middle-income countries, it has increased [2]. In Indonesia, the ministry of health reported that the national hypertension prevalence had increased in 2013 by 25.8% to 34,1% in 2018 [3]. Hypertension is the leading cause of premature death globally, accounting for 10.4 million deaths per year, and is a cause of disability-adjusted life-years (DALYs) worldwide. Suboptimal blood pressure is a risk factor often associated with cardiovascular and cerebrovascular disease and is the leading cause of kidney disease [4,5].

This complication can be prevented by controlling blood pressure. Blood pressure control is influenced by individual factors and system/provider factors [6,7]. The main factors affecting blood pressure control in patients receiving drugs are prescribing adequate amounts and doses of antihypertensive drugs and medication adherence [8]. Compliance with treatment is a process in which patients take medication according to prescription [9]. A systematic review and meta-analysis reported that the percentage of hypertensive patients who adhered to treatment was 45.2% and significantly affected uncontrolled blood pressure [10]. Social/economic, health system, condition-related, therapeutic, and patient factors are five dimensions that interact with each other in influencing medication adherence. The patient's ability to follow treatment is often influenced by more than one obstacle from influential factors, so interventions to improve adherence need to be targeted at how to overcome these obstacles. Therefore, health professionals must be able to identify systematically to assess the barriers that exist in patients [11].

Community in urban and rural areas have different demographic, economic, and social-environmental characteristics, so it is possible for variations in medication adherence. Studies in China show that hypertension patients' knowledge level in rural areas is worse than in urban areas, with the main influencing factors being gender and age [12]. However, studies in Brazil show the opposite; adherence rates are better in

*Corresponding author email: trimurtia@ugm.ac.id

rural areas than in urban areas [13]. A study at Columbia showed no difference in the level of adherence to treatment of hypertension patients in urban and rural areas [14].

In Indonesia, the level of adherence to the treatment of hypertensive patients varies. Some research data in primary care facilities show different results, including 33% adherence to treatment with factors that affect employment status, distance, level of knowledge, motivation to seek treatment, and family support [15]. In other healthcare facilities, a compliance rate of 36.1% was reported, with factors influencing knowledge, motivation, staff support, and family support [16]. Another study reported a compliance rate of 36.58% with an influential factor, namely education [16]. Similar studies reported a compliance rate of 42.4% with influential factors, namely co-morbidities, older age, and higher education, which support adherence to hypertension treatment [17]. There have been many studies to find out the level of compliance. However, studies comparing compliance levels in urban and rural areas and identifying related factors still need to be completed.

Based on the description above, we conducted a study to evaluate medication adherence and identify the main factors associated with medication adherence in hypertensive patients in Indonesia's urban and rural areas. When non-compliance is known, interventions should focus on improving and maintaining long-term compliance using different approaches based on the obstacles encountered [8].

2 Methods

2.1 Study design

A cross-sectional study on hypertensive patients was conducted in ten primary healthcare facilities, consisting of 5 places in urban areas (Palembang City) and 5 in rural areas (Oku Selatan district) in South Sumatra Province from November 2021 to February 2022 using convenience sampling. Data were obtained from medical records and questionnaires. This research has received approval from the ethical committee of the Faculty of Medicine, Gadjah Mada University, with the number KE/FK/0833/EC/2021.

2.2 Sample

The research subjects were hypertensive outpatients enrolled in ten primary healthcare facilities. Patients in this study met the inclusion criteria: adult patients aged $\geq$ 18 years and willing to participate in the study, diagnosed with essential hypertension and had undergone treatment for at least 3 months, patients taking antihypertensive drugs, and able to communicate in Indonesian. Exclusion criteria: patients with complicated diseases (heart disease, stroke, and kidney failure), pregnant and patients with motor disorders that limit daily activities. The number of samples was determined according to the sample calculation in logistic regression; the sample size per predictor was 15-30 subjects [18]. In this study, there were 13 predictors, so the total sample size was at least 15 subjects per predictor. The final sample size was at least 195 for each region, so the total sample was at least 390. In this study, there were 458 research subjects.

2.3 Study instrument

Socioeconomic data were obtained from questionnaires covering age, gender, education, employment status, monthly income, marital status, health insurance, and distance from home to primary healthcare facilities. Therapeutic data and clinical conditions were obtained from the patient's medical records. Investigators documented the amount of antihypertensive, duration of illness, presence of comorbidities, and blood pressure.

In this study, three questionnaires were used, namely the Multidimensional Scale of Perceived Social Support (MSPSS), Hypertension Knowledge-Level Scale (HK-LS), and Morisky Green Levine Adherence Scale (MGLS). The MSPSS questionnaire used to assess social support was first developed by Zimet et al. (1988), having a Cronbach alpha coefficient of 0.85 – 0.9. In addition, the MSPSS has been used in hypertensive patients with a Cronbach alpha coefficient of 0.96, indicating a reliable questionnaire. The MSPSS questionnaire consists of 12 items with 3 subscales: family, friends, and other influential individuals. Patients were asked to respond from 1 (strongly disagree) to 7 (strongly agree). After adding up, the scores will be in the range of 12 – 84. The bigger the score, the better social support. The cut-off point method is used to determine the category of social support [19].

The HK-LS questionnaire, used to measure the level of knowledge developed by Erkoc et al., (2012), has a Cronbach alpha coefficient of 0.82 and has been tested for validity and reliability for hypertensive patients in Indonesia. Internal consistency validity (Cronbach α 0.758), test re-test reliability (Spearman rank correlation 0.890), and discriminative validity (Mann Whitney $p < 0.05$) show that the HK-LS questionnaire is valid and reliable [20,21]. The HK-LS questionnaire consists of 22 questions, with the answer options being true, false, or do not know. The correct answer was given a score of 1, and the wrong or did not know given a score of 0. A total score of 22 was obtained if the respondent answered all questions correctly. The level of knowledge is declared low if the score is $\leq$ 17 and high if the score is between 18 – 22.

The level of adherence was measured using the MGLS questionnaire consisting of 4 items that have been tested for validity and reliability in Indonesia with internal consistency reliability with Cronbach's α = 0.651, test-retest reliability with Spearman's rank correlation = 0.425 and convergent validity with r = 0.58 [22]. Each answer "yes" is given a score of 1, and "no" is given a score of 0. The assessment results are totaled, and it is concluded that a score of 4 indicates low compliance, 1-3 indicates moderate compliance, and a score of 0 is high compliance. In this study, the level of adherence was divided into 2 categories: patient adherence to taking medication for patients with high compliance (score 4) and non-adherence for low-moderate adherence (score 1-3). The Cronbach alpha coefficients for MSPS, HK-LS, and MGLS were 0.702 - 0.858, 0.713 - 0.862, and MGLS 0.656, respectively in this study.

2.4 Data collection

Data were collected by 10 pharmacists in charge of the pharmacy room at each primary healthcare facilities. Pharmacists as research assistants were given 2-4 hours of training focusing on research objectives, instruments used, and data collection methods. Patients presenting for outpatient treatment at primary healthcare facilities were identified and invited to participate in the study by research assistants. Patients who met the inclusion criteria signed prior informed consent to participate, allowing the investigators to access the patient's medical records. The research assistant explained the purpose of the study and provided standard instructions before the patient filled out the questionnaire. The research assistant was on hand to fill out the questionnaire with the participants to answer questions and clarify issues if needed.

2.5 Data analysis

Descriptive statistics summarize socioeconomic, therapeutic, and clinical characteristics based on urban and rural area criteria. Continuous variables are reported using the mean with a standard deviation. The bivariate test uses independent T-test or Mann-Whitney for numerical variables and uses chi-square for categorical variables. Logistic regression backward method was used to identify predictors of compliance. All variables related to medication adherence in the bivariate analysis had a $p \leq 0.25$ value in the final model. Significance is set at p-value <0.05. All statistical tests use SPSS version 26.

3 Results

3.1 Socioeconomic Characteristics, disease knowledge and social support

A total of 500 patients participated in this study, 250 patients each in urban and rural areas. Participants who met the inclusion criteria and filled out the complete questionnaire were 225 people in urban areas and 233 people in rural areas, so the final total participants obtained were 458 people (**Table 1**). A total of 10 patients received <3 months of treatment and 32 incomplete questionnaires.

Table 1. Socioeconomic characteristics, disease knowledge and social support of hypertension patients in urban and rural

Variables		Urban (n=225)		Rural (n=233)	
		n	%	n	%
Age, years	≥ 60	106	47.1	100	42.9
	45 – 59	100	44.4	107	45.9
	≥ 18 – 45	19	8.4	26	11.2
	Mean (SD)	58.27±9.259		56.26±8.670	
Gender	Female	144	64.0	139	59.7
	Male	81	36.0	94	40.3
Education	Primary and below	66	29.3	95	40.8
	High/middle school	112	49.8	97	41.6
	Bachelor and above	47	20.9	41	17.6
Working status	Retired/not working	84	37.3	42	18.0
	Working	141	62.7	191	82.0
Monthly income (IDR)	< 1.000.000	70	31.1	64	27.5
	1.000.000 - < 3.000.000	82	36.4	120	51.5
	3.000.000 – 5.000.000	53	23.6	49	21,0
	>5.000.000	20	8.9	0	0
Marital status	Divorce/single	34	15.1	34	14.6
	Married	191	84.9	199	85.4
Medical insurance	No insurance	22	9.8	10	4.3
	Indonesian National Social Health Insurance	203	90.2	223	95.7
Distance	>5 km	22	9.8	35	15
	1-5 km	106	47.1	131	56.2
	< 1 km	97	43.1	67	28.8
Disease knowledge	Low (≤ 17)	110	48.9	159	68.2
	High (>18-22)	115	51.1	74	31.8
	Mean (SD)	16.40±3.462		15.53±3.154	
Social support	Low (≤ 62)	102	45.3	145	62.2
	High (63-84)	123	54.7	88	378
	Mean (SD)	64.79±12.708		58.48±8.725	

Abbreviations: IDR, Indonesia Rupiah

3.2 Therapeutic and clinical outcomes

Table 2 shows therapeutic and clinical outcomes include number of antihypertensive drugs, duration of illness, comorbid and blood pressure.

3.3 Adherence to medication

There are differences in adherence levels in urban and rural areas (p=0.002). The level of adherence is better in urban than rural areas (**Table 3**).

Adherence with treating hypertension patients among healthcare facilities in each region varies. The highest adherence rate was in urban areas, namely hypertensive patients in urban primary healthcare facilities 2 (43.5%) and the lowest in urban primary healthcare facilities 1 (28.6%), while in rural areas the highest in rural primary healthcare facilities 2 (49.2%) and the lowest in rural primary healthcare facilities 5 (2%) as shown in **Figure 1**.

Table 2. Therapeutic and clinical characteristics of hypertensive patients in urban and rural areas

Variables		Urban (n=225)		Rural (n=233)	
		n	%	n	%
Number of antihypertensive drugs	2	30	13.3	31	13.3
	1	195	86.7	202	86.7
Duration of illness, years	≥5	76	33.8	66	28.3
	1 - < 5	131	58.2	117	50.2
	< 1	18	8.0	50	21.5
Comorbid	With comorbids	66	29.3	55	23.6
	No comorbids	159	70.7	178	76.4
Blood pressure	Not controlled	166	73.8	170	73.0
	Controlled	59	26.2	63	27.0
	Mean SBP	150.03±18.78		147.47±18.28	
	Mean DBP	89.49±10.08		90.19±10.05	

Abbreviations: SBP = Systolic Blood Pressure
DBP = Diastolic Blood Pressure

Table 3. Patient adherence in urban and rural areas

Areas	Adherence (n=141)		Non Adherence (n=317)		*p*	(95% C.I.)
	n	%	n	%		
Urban	82	36.4	143	63.6	0.002	1.947 (1.293 – 2.932)
Rural	53	22.7	180	77.3		

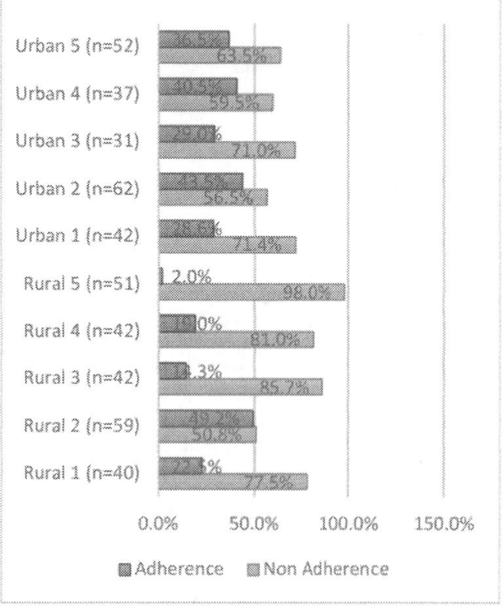

Fig. 1. Level of adherence in each health center in urban and rural areas

Patients in urban areas do not adhere to treatment because they forget to take their medication. In contrast, if they feel healthier in rural areas, they sometimes stop taking medication. This complete information is in **Table 4.**

3.4 Factors related to adherence in urban and rural

Table 5 summarises the bivariate compliance test with socioeconomic factors, clinical characteristics and therapy in urban and rural areas. Selection is made to determine the variables included in the multivariate analysis. Variables with a p-value <0.250 were continued into multivariate analysis to determine the factors most related to medication adherence. In urban areas, the variables of working status (p=0.000), monthly income (p=0.000), social support (p=0.000) and level of knowledge (p=0.000) are included in the multivariate analysis. In rural areas, the variables were education (p=0.000), working status (p=0,109), monthly income (p=0.010), distance from home to health care facilities (p=0.002), social support (p=0.000), level knowledge (p=0.000) and duration of illness (0.185) were included in the multivariate analysis.

3.5 Logistic regression test results

Table 6 summarizes the logistic regression analysis results using the backward method. In urban areas, the final model related to compliance is working status, social support and level of knowledge. In contrast, in rural areas, the final model obtained is social support and level of education.

4 Discussion

Most were female in urban (64.0%) and rural (59.7%) areas. In urban areas, education is dominated by high/middle school (49.8%) and rural (41.6%). Most of the patients were working, but patients do not work/retire in urban areas than in rural areas. The highest income per month is in the range of 1,000,000 - < 3,000,000 IDR amounting to 36.4% in urban areas and 51.5% in rural areas; there is no income > 5.000.000 IDR in rural areas. Most of them are married, with a percentage of 84.9% in urban areas and 85.4% in rural areas. Most of them are Indonesian National Social Health Insurance participants, but the percentage of non-Indonesian National Social Health Insurance patients is higher in rural areas (95.7). The distance from home to primary health care facilities is mostly 1-5 Km, 47.1% in urban and 56.2% in rural areas. However, data also shows distances > 5 Km are more in rural areas than urban areas, and distance < 1 Km is more in urban than rural areas. Social support in urban areas is better than in rural areas, where social support in urban areas is high (54.7%), while in rural areas, it is dominated by low social support (62.2%). The average social support score in urban areas is at a high rating, while in rural areas, it is low. The knowledge about hypertension is low in both

Table 4. The percentage of respondents' answers to each item of the MGLS questionnaire

No	Questions	Urban (n=225) Yes	%	No	%	Rural (n=233) Yes	%	No	%
1	Have you ever forgotten to take your medication?	95	42.2	130	57.8	124	53.2	109	46.8
2	Are you sometimes negligent in taking medicine?	84	37.3	141	62.7	124	53.2	109	46.8
3	When you feel better, do you sometimes stop taking your medication?	87	38.7	138	61.3	149	63.9	84	36.1
4	Sometimes if you feel uncomfortable/worse while taking medicine, do you stop taking it?	86	38.2	139	61.8	148	63.5	85	36.5

Table 5. Patient characteristics related to medication adherence in urban and rural areas on bivariate tests

Variables		Urban Non-adherence (n=154) n	%	Urban Adherence (n=71) % n	%	p	Rural Non-adherence (n=180) n	%	Rural Adherence (n=53) % n	%	p
Age, years	Mean (SD)	58.28±8.987		58.24±9.771		0.978c	56.66±8.518		54.91±9.122		0.322a
Gender	Female	91	63.2	53	36.8	0.995b	104	74.8	35	25.2	0.359b
	Male	52	64.2	29	35.8		76	80.9	18	19.1	
Education	Primary and below	39	59.1	27	40.9	0.645b	87	91.6	8	8.4	0.000b*
	High/middle school	74	66.1	38	33.9		75	77.3	22	22.7	
	Bachelor and above	30	63.8	17	36.2		18	43.9	23	56.1	
Working status	Retired/not working	72	85.7	12	14.3	0.000b*	28	66.7	14	33.3	0.109b*
	Working	71	50.4	70	49.6		152	79.6	39	20.4	
Monthly income (IDR)	< 1.000.000	54	77.1	16	22.9	0.000b*	53	82.8	11	17.2	0.010b*
	1.000.000 - < 3.000.000	47	57.3	35	42.7		97	80.8	23	19.2	
	3.000.000 – 5.000.000	37	69.8	16	30.2		30	61.2	19	38.8	
	>5.000.000	5	25.0	15	75.0		0	0	0	0	
Marital status	Divorce/single	19	55.9	15	44.1	0.415b	25	73.5	9	26.5	0.735b
	Married	124	64.9	67	35.1		155	77.9	44	22.1	
Medical insurance	No insurance	15	68.2	7	31.8	0.809b	7	70	3	30	0.699b
	Indonesian National Social Health Insurance	128	63.1	75	36.9		173	77.6	50	22.4	
Distance	>5 km	15	68.2	7	31.8	0.759b	21	60.0	14	40.0	0.002b*
	1-5 km	69	65.1	37	34.9		112	85.5	19	14.5	
	< 1 km	59	60.8	38	39.2		47	70.1	20	29.9	
Disease knowledge	Mean (SD)	15.24±3.565		18.43±2.085		0.000a*	14.77±2.973		18.11±2.276		0.000a*
Social support	Mean (SD)	61.62±12.394		70.32±11.343		0.000a*	57.27±8.778		62.60±7.209		0.000a*
Duration of illness, years	≥5	46	60.5	30	39.5	0.630b	49	74.2	17	25.8	0.185b*
	1 - < 5	84	64.1	47	35.9		96	82.1	21	17.9	
	< 1	13	72.2	5	27.8		35	70.0	15	30.0	
Comorbid	With comorbids	45	68.2	21	31.8	0.437b	41	74.5	14	25.5	0.716b
	No comorbids	98	61.6	61	38.4		139	78.1	39	21.9	
Number of antihypertensive drugs	2	16	53.3	14	46.7	0.296b	26	83.9	5	16.1	0.475b
	1	127	65.1	68	34.9		154	76.2	48	23.8	

[a]mann-whitney, [b]Chi square, [c]Independen T-test, * Shows a variable entered in a multivariate test

Table 6. Results of logistic regression analysis of medication adherence in urban and rural areas

Urban				Rural		
Variables	n (=225)	OR (95% C.I.)	*p*	n (=233)	OR (95% C.I.)	*p*
Working status						
Retired/not working	84	1				
Working	141	4.787 (2.149 – 10.663)	0.000			
Social support						
Low	102	1		145	1	
High	123	5.054 (2.370 – 10.775)	0.000	88	4.696 (2.045-10.783)	0.000
Diseases knowledge						
Low	110	1		159		
High	115	6.558 (3.098 – 13.880)	0.000	74	12.555 (5.429-29.034)	0.022
Education						
Primary and below				95	1	
High/middle school				97	3.290 (1.186 – 9.130)	0.000
Bachelor and above				41	12.871 (3.986 – 41.562)	0.000

areas, namely 48.9% in urban areas and 68.2% in rural areas. The average score in both areas is at a low level.

Most patients used a single antihypertensive drug in urban and rural areas (86.7%). Duration of illness was between 1 - < 5 years, 58.2% in urban and 50.2% in rural areas. However, the percentage of patients with a duration of illness < 1 year was higher in rural than urban areas. Most patients had no comorbidities, 70.7% in urban and 76.4% in rural areas, with uncontrolled blood pressure at 73.8% in urban and 73.0% in rural areas.

There is a differences between urban and rural areas with the level of medication adherence (p=0.002). Patients in urban areas are 1.947 times more likely to adhere to medication than hypertensive patients in rural areas to comply. The level of adherence is better in urban areas than in rural areas. This finding is consistent with the results of previous studies in China [12]. The prevalence of medication adherence in this study was low in both urban and rural areas and is consistent with previous findings [10,12,23,24]. Differences in socioeconomic, therapeutic and clinical characteristics may cause differences in treatment adherence. Regional, economic and cultural differences between rural and urban areas can lead to differences in adherence to treatment of hypertension sufferers between rural and urban residents. Knowing and understanding village-urban differences in treatment adherence and their influencing factors is an essential step in determining targeted strategies to increase adherence of people with hypertension in different places of residence [25,26]. Patients in urban areas do not adhere to treatment because they forget to take their medication. In contrast, if they feel healthier in rural areas, they sometimes stop taking medication. Previous studies have also reported forgetting to take medication as a cause of non-adherence in addition to multiple daily doses, financial constraints and side effect [27]. Reasons for non-compliance including forgetfulness, drugs side effects, shortage of drugs, poly pharmacy and the asymptomatic nature of hypertension [28].

In urban areas, the variables of working status, monthly income, social support and level of knowledge are included in the multivariate analysis. In rural areas, the variables were education, working status, monthly income, distance from home to health care facilities, social support, level knowledge and duration of illness were included in the multivariate analysis.

Logistic regression test results show in urban areas, patients who work are 4.787 times more likely to have high adherence than patients who are not working/retired to comply. Patients with high social support are 5.054 times more likely to have high medication adherence than patients with low social support, and patients with a high level of knowledge are 6.558 times more likely to have high medication adherence than patients with a low level of knowledge. In contrast to rural areas, patients with middle/high school education are 3.290 times more likely to have high adherence than those with primary or below education to comply, patients with bachelor or above education are 12.871 times more likely to have high adherence than patients with primary and below education for hypertensive patients with good social support is 4.696 times more likely to have high medication adherence than patients with poor social support to comply. Patients with a high level of knowledge are 12.555 times more likely to have high medication adherence than patients with a low level of knowledge to comply.

Social support and level of knowledge are factors that are significantly related to compliance both in urban and rural areas. Knowledge is most related to medication adherence in urban, whereas education and knowledge are most related to rural areas. Previous studies have confirmed that treatment adherence positively increases with increased social support for hypertensive patients [29]. Patient knowledge about hypertension is a determinant of good adherence in hypertensive patients [30]. Differences in education levels may be the reason for different adherence. Urban residents tend to be more educated than rural residents and have more knowledge about hypertension [25]. This study found that working patients tended to be more compliant with treatment, unlike previous research [31]. One study reported that in addition to social support, age, household income, duration of diagnosis, the number of antihypertensive tablets taken in each dose, and the frequency of taking medication every day, were related to medication adherence [23]. Several factors in this study followed previous studies that knowledge, beliefs and attitudes, mental-personality traits, culture and lifestyle, access to health services, individual internal incentives, family support, and support from health care providers influence adherence to hypertensive patients [32].

Studies of medication prevalence and adherence help explain the determinants of adherence models. Static variables may not change with intervention, but behavioural and dynamic variables are very likely to receive the intervention [11]. In this study, both in urban and rural areas, adherence was influenced by social support and level of knowledge, which are dynamic variables, so interventions can be carried out to increase patient adherence to treatment.

Interventions to improve medication adherence in hypertensive patients can be carried out at the doctor, patient, drug therapy, and healthcare system levels [33]. Collaborating between doctors, nurses, and pharmacists, self-management with a simple system, using reminders, having group sessions, instruction combined with motivational strategies, health system support for monitoring, and financially supporting collaboration between healthcare providers, nurses and pharmacists) can help patient adherence [34]. The use of technology can have a positive impact on supporting hypertension self-management, particularly in medication adherence [35].

The limitations of the first study were the small number of samples and the sampling method, which was not randomized. Secondly, compliance, social support and knowledge used a self-reported questionnaire which could lead to memory bias—a small of the possible influencing factors. Future research needs to be conducted in a larger random sample and involves other variables that affect medication adherence.

5 Conclusion

Most patients are non-adherent to treatment in urban and rural areas. Hypertensive patients in urban areas have better adherence than rural patients in this study. The factors most related to medication adherence in urban areas were employment status, social support, and knowledge about hypertension. Education, social support, and knowledge were the factors most related to adherence in rural areas. Our study provides insights into medication adherence in hypertensive patients and related factors for urban and rural Indonesian populations. Stakeholders and health professionals can consider our findings to develop health programs. Social support and level of knowledge are dynamic behavioural factors, so intervention strategies involving family and health worker support, as well as providing education using appropriate methods, should be considered in this population.

Acknowledgement

We want to thank Sekolah Tinggi Ilmu Farmasi Bhakti Pertiwi Palembang Indonesia for the financial support given for this research, and all pharmacist for the support for this research.

Conflict Interests

The authors declare no competing or potential conflicts of interest concerning the research and publication of this article

References

1. MH. Forouzanfar, P. Liu, GA. Roth, SNgM. Biryukov, L. Marczak, Alexander.L, K.Estep, KH. Abate, TF. Akinyemiju, R.Ali, N. Alvis-Guzman, P.Azzopardi, A. Banerjee, T.Barnighausen, A. Basu, T. Bekele, DA. Bernett, S. Biadgilign, F. Catala-Lopez, VL. Feigin, JC. Fernandes, F. Fischer, AA.Gebru, P.Gona, R. Gupta, GJ.Hankey, JB.Jonas, SE. Judd, Y. Khang, Khosravi A, YJ.Kim, RW. Kimokoti, Y. Kokubo, D. Kolte, A. Lopez, PA. Lotufo, R. Malekzadeh, YA. Melaku, GA. Mensah, A. Misganaw, AH. Mokdad, AE. Moran, H. Nawaz, B. Neal, FN. Ngalesoni, T. Ohkubo, F. Pourmalek, A. Rafay, RK. Rai, D. Rojas-Rueda, UK. Sampson, IS. Santos, M. Sawhney, AE.Schutte, SG. Sepanlou, GT. Shifa, I. Shiue, BA. Tedla, AG. Thrift, M. Tonelli, T. Truelsen, N. Tsilimparis, KN. Ukwaja, OA. Uthman, T. Vasankari, N. Venketasubramanian, VV. Vlassov, T. Vos, R. Westerman, LL.Yan, Y. Yano, N. Yonemoto, MEL. Zaki, CJL. Murray. Global burden of hypertension and systolic blood pressure of at least 110 to 115 mmHg, 1990-2015. Journal of the American Medical Association. (**317**), 165–182, (2017).
2. WHO. World health statistics 2022 (Monitoring health of the SDGs). (2022).
3. Kemenkes. Laporan Nasional Riset Kesehatan Dasar. Kementrian Kesehatan Republik Indonesia. Jakarta. (2019).
4. E. Gakidou. Global, regional, and national comparative risk assessment of 84 behavioural, environmental and occupational, and metabolic risks or clusters of risks, 1990-2016: A systematic analysis for the Global Burden of Disease Study 2016. Lancet. (**390**), 1345–1422, (2017).
5. CY. Hsu, CE. McCulloch, J. Darbinian, AS. Go, C. Iribarren. Elevated blood pressure and risk of end-stage renal disease in subjects without baseline kidney disease. Archives of Internal Medicine. (**165**), 923–928 (2005).
6. RM. Carey, P. Muntner, HB. Bosworth, PK. Whelton. Prevention and Control of Hypertension: Journal of the American College of Cardiology Health Promotion Series. (**72**),1278–1293, (2018).
7. T. Unger, C. Borghi, F. Charchar, NA. Khan, NR. Poulter, D. Prabhakaran, A. Ramirez, M. Schlaich, GS. Stergiou, M. Tomaszewki, RD. Wainford, B.

Williams, AE. Schutte. 2020 International Society of Hypertension Global Hypertension Practice Guidelines. Hypertension. (**75**), 1334–1357, (2020).

8. B. Williams, G. Mancia, W. Spiering, EA. Rosei, M. Azizi, M. Burnier, DL. Clement, A. Coca, G. de-Simone, A. Domoniczak, T. Kahan, F. Mahfoud, J. Redon, L. Ruilope, A. Zanchetti, M. Kerins, SE. Kjeldsen, R. Kreutz, S. Laurent, GYH. Lip, R. McManus, K. Narkiewicz, F. Ruschitzka, RE. Schmieder, E. Shyakhto, C. Tsioufis, V. Aboyans, I. Desormais. Practice guidelines for the management of arterial hypertension of the European society of cardiology and the European society of hypertension ESC/ESH task force for the management of arterial hypertension. European Heart Journal. (**36**), 2284–2309, (2018).
9. B. Vrijens, S. De Geest, DA. Hughes, K. Przemyslaw, J. Demonceau, T. Ruppar, F. Dobbels, E. Fargher, V. Morrison, P. Lewek, M. Matyjaszczyck, C. Mshelia, W. Cline, JK. Aronson, J. Urquhart. A new taxonomy for describing and defining adherence to medications. Br J Clin Pharmacol. (**73**), 691–705, (2012).
10. TM. Abegaz, A. Shehab, EA. Gebreyohannes, AS. Bhagavathula, AA. Elnour. Nonadherence to antihipertensive drugs a systematic review and meta analysis. Medicine (Baltimore). (**96**), e5641, (2017).
11. WHO. Adherence to long-term therapies : Evidence for Action. World Health Organization. (2003).
12. J. Pan, H. Yu, B. Hu, Q. Li. Urban-Rural Difference in Treatment Adherence of Chinese Hypertensive Patients. Patient Prefer Adherence. (**16**), 2125–2133, (2022).
13. P. Magnabosco, EC. Teraoka, EM. De Oliveira, EA. Felipe, D. Freitas, LM. Marchi-Alves. Comparative analysis of non-adherence to medication treatment for systemic arterial hypertension in urban and rural populations. Rev Lat Am Enfermagem. (**23**), 20–27, (2015).
14. C. Arbuckle, D. Tomaszewski, BD. Aronson, L. Brown, J. Schommer, D. Morisky, E. Linstead. Evaluating factors impacting medication adherence among rural, urban, and suburban populations. J Rural Health. (**34**), 339–346, (2018).
15. QP. Rasajati, BB. Raharjo, DNA. Ningrum . Faktor-faktor yang berhubungan dengan kepatuhan pengobatan pada penderita hipertensi di wilayah kerja puskesmas kedungmundu Kota Semarang. Unnes J Public Health. (**4**), 16–23, (2015).
16. GW. Pratama, NLP. Ariastuti. Faktor – faktor yang mempengaruhi kepatuhan pengobatan hipertensi pada lansia binaan Puskesmas Klungkung 1. E-Jurnal Med Udayana. (**5**), 1–13, (2016).
17. Y. Rikmasari, A. Rendowati, A. Putri. Factors influencing adherence to using antihypertensive drugs: Cross Sectional Study at the Puskesmas Social Palembang. J Penelitian Sains. (**22**), 87–94. (2020).
18. PB. Palmer, DG. O'Connell. Regression analysis for Prediction : Understanding the Process. Cardiopulm Phys Ther J. (**20**), 23–26, (2009).
19. Y. Gurmu, D. Gela, F. Aga. Factors associated with self-care practice among adult diabetes patients in West Shoa Zone, Oromia Regional State, Ethiopia 11 Medical and Health Sciences 1117 Public Health and Health Services. BMC Health Serv Res. (**18**), 4–11, (2018).
20. SB. Erkoc, B. Isikli, S. Metintas, C. Kalyoncu. Hypertension Knowledge-Level Scale (HK-LS) a study on development, validity and reliability. Int J Environ Res Public Heal. (**9**), 1018–1029, (2012).
21. I. Ernawati, SS. Fandinata, SN. Permatasari. Translation and validation of the indonesian version of the hypertension knowledge-level scale. Open Access Maced J Med Sci. (**8**), 630–637. 2020
22. SA. Kristina, LR. Putri, DA. Riani, Z. Ikawati, D. Endarti. Validity of self-reported measure of medication adherence among diabetic patients in indonesia. International Research Journal of Pharmacy. (**10**), 144–148, (2019).
23. C. Ma. A cross-sectional survey of medication adherence and associated factors for rural patients with hypertension. Applied Nursing Research. (**31**), 94–99, (2016).
24. O. Amaral, C. Chaves, J. Duarte, E. Coutinho, P. Nelas, O. Preto. Treatment adherence in hypertensive patients – a cross-sectional study. Procedia - Soc Behav Sci. (**171**), 1288–1295, (2015).
25. DF. Teshome, KB. Bekele, YA. Habitu, AA. Gelagay. Medication adherence and its associated factors among hypertensive patients attending the Debre Tabor General Hospital, Northwest Ethiopia. Integr Blood Press Control. (**10**), 1–7, (2017).
26. A. Hussein, MS. Awad, HEM. Mahmoud. Patient adherence to antihypertensive medications in upper Egypt: a cross-sectional study. Egyptian Heart Journal. (**72**), 1-8, (2020).
27. NM. Adidja, VN. Agbor, JA. Aminde, CA. Ngwasiri, KB. Ngu, LN Aminde. Non-adherence to antihypertensive pharmacotherapy in Buea, Cameroon: A cross-sectional community-based study. BMC Cardiovasc Disord. (**18**),1–9, (2018).
28. AM. Al-mehza, FA. Al-muhailije, MM. Khalfan, AA. Al-yahya. Drug compliance among hypertensive patients; an Area Based Study. Eur J Gen Med. (6), 6–10, (2009).
29. GB. Turan, M. Aksoy, B. Çiftçi. Effect of social support on the treatment adherence of hypertension patients. J Vasc Nurs. (37), 46–51, (2019).
30. B. Jankowska-Polańska, I. Uchmanowicz, K. Dudek, G. Mazur. Relationship between patients' knowledge and medication adherence among patients with hypertension. Patient Prefer Adherence. (**10**), 2437–2447, (2016).
31. GKY. Lee, HHX. Wang, KQL. Liu, Y. Cheung, DE Morisky, MCS. Wong. Determinants of medication adherence to antihypertensive medications among a Chinese population using

Morisky Medication Adherence Scale. PLoS One. (**8**), e62775, (2013).
32. M. Ashoorkhani, R. Majdzadeh, J. Gholami, H. Eftekhar, A. Bozorgi. Understanding non-adherence to treatment in hypertension : A qualitative study. IJCBNM. (**92**), 314–323, (2018).
33. M. Burnier, BM. Egan. Adherence in hypertension: A review of prevalence, risk Factors, impact, and management. Circ Res. (**124**), 1124–114, (2019).
34. DH. Smith, MO. Keeffe-rosetti, AA. Owen-smith, C. Rand, J. Tom, S. Vupputuri, R. Laws, A. Waterbury, DD. Hankerson-Dyson, C. Yonehara, A. wiliams, J. Schneider, JF. Dickerson, WM. Volmer. Improving adherence to cardiovascular therapies : an economic evaluation of a randomized pragmatic trial. Value Health. (**19**), 176–184, (2019).
35. CH. Still, S. Margevicius, C. Harwell, MC. Huang, L. Martin, PB. Dang, JT. Wright. A community and technology-based approach for hypertension self-management (Coachman) to improve blood pressure control in african americans: Results from a pilot study. Patient Preference Adherence. (**14**), 2301–2313, (2020).

Workers' Compliance with Covid-19 Prevention and Control Protocols in X Mining Company

Rizky Noorleta Putri[1], *Vena* Jaladara[1,2*], and *Supriyati* Supriyati[1,2]

[1]Department of Health Behavior, Environment and Social Medicine, Faculty of Medicine, Public Health and Nursing, University of Gadjah Mada (UGM) in Yogyakarta, Indonesia
[2]Center of Health Behavior and Promotion, Faculty of Medicine, Public Health and Nursing, University of Gadjah Mada (UGM) in Yogyakarta, Indonesia

Abstract. The mining sector presents various risks that can contribute to the spread of Covid-19, such as confined work areas and high worker mobility. The Indonesian government has made several efforts to prevent the spread of Covid-19, including the implementation of health protocols as mitigation measures in workplaces. Employee compliance is a crucial factor in the successful implementation of health protocols in the workplace. The purpose of this study is to analyze the determinant factors of employee compliance in implementing Covid-19 health protocols at Company X. A cross-sectional study was conducted from July to August 2022 with a self-administered questionnaire. The sample of the study was drawn from mining workers in Company X as research subjects. The finding reveals that out of 185 respondents, 91.4% exhibited high compliance in implementing health protocols in the workplace. There is a relationship between knowledge, attitude, company support, vaccination history, and the level of employee compliance in implementing health protocols in the working environment of Company X. This study revealed that company support was the most dominant factor influencing worker compliance. Therefore, good company support is necessary for ensuring employees' safety and health compliance behavior.
Keywords: Covid-19; mining; compliance; safety; prevention

1 Background

The mining industry is one of the sectors with the highest potential hazards. One of the current health risks in the mining sector is the transmission of Covid-19 among workers. Approximately 4,000 mine workers from 18 countries had been infected with Covid-19 from the beginning of the outbreak until June 2020 [1]. The cumulative number of confirmed Covid-19 cases among mine workers in Indonesian state-owned enterprises (BUMN) and their subsidiaries until September 2020 was 1,219 people [2].

The Ministry of Energy and Mineral Resources (ESDM) issued Circular Number 02.E/04/DJB/2020 addressed to the Director General of Minerals and Coal regarding the prevention and handling of Covid-19 in the mineral and coal mining industries. Apart from the implementation of infection prevention and control programs in the workplace, one factor that determines the success of implementation is the involvement of workers. Worker involvement can be seen through their compliance in implementing health protocols in their daily lives, both at the workplace and outside work area [3]. Assessing compliance behavior in the implementation of health protocols is crucial in preventing the spread of Covid-19.

Even if an organization has a good infection prevention program, low compliance becomes a major obstacle to the success of implementing Covid-19 health protocols. The non-compliance of the public with health protocols has led to high infection rates of Covid-19 in Indonesia [4]. Several previous studies have mentioned that compliance with Covid-19 health protocols is influenced by knowledge, attitudes, motivation, and the availability of facilities. Knowledge about Covid-19 directly influences individuals' attitudes toward Covid-19, thereby increasing compliance with health protocol implementation [5,6,7]. Other studies have also revealed that the level of worker compliance is influenced by perceived management or company support [8,9]. Company support can create a positive work climate, enabling workers to feel safe, and indirectly encouraging them to adopt safety and health behaviors in the workplace, including improving compliance [3].

Based on previous research on compliance with health protocols and preliminary studies from both sources mentioned, the author is interested in investigating the employee compliance in implementing Covid-19 prevention and control protocols.

2 Methods

2.1 Design and Participants

This is quantitative cross-sectional research. This study is conducted to examine a specific population or sample and will be statistically analyzed [10]. The population in

* Corresponding email: vena.jaladara@mail.ugm.ac.id

this study is all employees at Company X, totaling 1,000 people. The sample size is determined using the simple random sampling technique, which can be calculated using the Slovin's formula:

$$n = \frac{N}{N\,x\,e^2 + 1}$$

$$n = \frac{1.00}{((1.000\,x\,(0{,}05)^2)+1)} = \frac{1.000}{3{,}5} = 286 \text{ subjects}$$

The sample size taken is 286 individuals. In this study, the author added 10% more samples than the calculated total sample size to account for potential dropouts. The samples were selected based on inclusion criteria of being over 18 years old, willing to participate in the study, willing to complete the questionnaire, and exclusion criteria of not completing the questionnaire and refusing to participate in the study.

2.2 Ethical Consideration

The research protocol of this study has been approved by the Ethics Committee of the Faculty of Medicine, Public Health, and Nursing at Gadjah Mada University (KE/FK/0681/2022). All participants read the informed consent form and confirmed their interest in participating before starting the survey. The confidentiality of participant identities is maintained throughout the research by keeping participant information confidential.

2.3 Research Tools

The questionnaire on knowledge related to Covid-19 consists of 14 statements that are modified from information about Covid-19 and health protocols in Covid-19 prevention form [11,12,13]. The scoring of the questionnaire is based on the Guttman Scale, with true or false answer choices. Favorable statements with the "True" option are given a score of "1," and the "False" answer is given a score of "0," while for unfavorable statements, the "True" option is given a score of "0" and the "False" option is given a score of "1.".

The questionnaire on attitudes towards Covid-19 consists of 8 statements that are modified from [5]. The statements cover cognitive (thoughts or ideas), affective (emotions/feelings), and conative (observable behavior) aspects. The measurement of the questionnaire uses a Likert Scale with 4 answer alternatives: Strongly Disagree, Disagree, Agree, and Strongly Agree. Favorable statements with the "Strongly Agree (SA)" option are given a value of 4, "Agree (A)" is given a value of 3, "Disagree (D)" is given a value of 2, and "Strongly Disagree (SD)" is given a value of 1. Conversely, unfavorable statements with the "Strongly Agree (SA)" option are given a value of 1, "Agree (A)" is given a value of 2, "Disagree (D)" is given a value of 3, and "Strongly Disagree (SD)" is given a value of 4.

The questionnaire on perceived company support consists of 10 questions that are modified from the study [3], designed to assess the influence of organizational climate on infectious disease prevention in the workplace. The questions cover information, communication, availability of facilities and infrastructure, and management values, with all statements being favorable. The assessment of the questionnaire uses a Likert Scale with 4 answer alternatives: (1) Strongly Disagree, (2) Disagree, (3) Agree, (4) Strongly Agree.

The questionnaire on compliance with health protocols consists of 15 statements that are modified from the study [5]. The statements in the questionnaire cover workers' behavior in implementing Covid-19 prevention and control measures, especially in the workplace. The scoring of the questionnaire uses a Likert Scale with 4 answer alternatives: (1) Never, (2) Sometimes, (3) Often, and (4) Always.

2.4 Data Collection

To understand the level of knowledge about Covid-19 and health protocols in the workplace, attitudes towards Covid-19, perceived company support by employees, as well as employee compliance in implementing health protocols in the workplace, a descriptive analysis was conducted based on the scoring of questionnaires collected from respondents. Documentation and observations were conducted regarding the condition of the work area, facilities and infrastructure for the prevention and control of Covid-19 in workplace, as well as the employees' activities in implementing health protocol in the workplace.

2.5 Statistical Analysis

2.5.1 Validity Test

Validity testing of the knowledge, attitude, company support, and compliance questionnaires was conducted using Pearson correlation with a correlation coefficient (r) table value of 0.3610 (two-tailed significance level α=0.05). The validity test of the knowledge questionnaire yielded correlation coefficient (r) values ranging from 0.363 to 0.695 (r observed > r table), indicating that all the questions were valid and could be used as a research tool. For the attitude questionnaire, the calculated correlation coefficients (r) ranged from 0.535 to 0.751 (r observed > r table), indicating that all the questions regarding attitudes towards Covid-19 were valid and could be used as a research tool. The company support questionnaire obtained correlation coefficients (r) ranging from 0.819 to 0.917 (r observed > r table), indicating that all the questions in the questionnaire were valid and could be used as a research tool. Regarding the compliance questionnaire, the calculated correlation coefficients (r) ranged from 0.381 to 0.801 (r observed > r table), indicating that all the questions in the compliance questionnaire were valid and could be used as a research tool.

2.5.2 Reliability Test

Reliability testing of the research instruments for the variables of knowledge about Covid-19, attitudes towards Covid-19 and health protocols, perceived company support, and compliance were measured using Cronbach's Alpha reliability coefficient (α) based on Table 1. The instruments are considered reliable if the calculated value is greater than 0.60. The results of the reliability tests for each variable are presented in the following table:

Table 1. Reliability Test Results

Variable	α	Description
Knowledge About Covid-19	0.752	Reliable
Attitudes towards Covid-19 and Health Protocols	0.766	Reliable
Perceived Company Support	0.971	Reliable
Workers' Compliance	0.887	Reliable

3 Results

This study originally required a sample of 318 respondents, but only 196 samples were collected. The sociodemographic data of the respondents including age, gender, duration of employment in the company, income level, educational level, history of Covid-19 infection, and history of Covid-19 vaccination. After data cleaning, a total of 185 respondents were obtained with the following characteristics:

Table 2. Respondent Characteristics

Respondent Characteristics	n (%)
Age	
≥ 30 Years Old	143 (77.30)
<30 Years Old	42 (22.70)
Gender	
Male	151 (81.62)
Female	34 (18.38)
Duration of Employment in the company	
≥10 Years	61 (32.97)
<10 Years	124 (67.03)
Educational Level	
Senior High School	86 (46.49)
University	99 (53.51)
Income Level (Indonesian Rupiah)	
3.000.000-5.000.000	16 (8.65)
5.000.000-10.000.000	46 (24.86)
10.000.000-15.000.000	37 (20.00)
>15.000.000	86 (46.49)
History of Covid-19 Infection	
Have	142 (76.76)
Have Not	43 (23.24)
History of Covid-19 Vaccination	
2 Doses	29 (15.68)
3 Doses	156 (84.32)

Table 2 shows that majority of respondents are over 30 years old, most of them are male with less than 10 years of work experience. All workers have graduated from high school and college with varying income levels. As many as 78% of respondents had been infected with Covid-19 and all of them had received at least 2 doses of Covid-19 vaccine.

Table 3. Distribution of Research Variable Frequency

Variable	Category	Frequency (n)	Percentage (%)
Compliance Level	High	169	91.4
	Low	16	8.6
Knowledge Level	Good	167	90.3
	Bad	18	9.7
Employee's Attitude	Positive	177	95.7
	Negative	8	4.3
Perceived Company Support	Good	182	98.4
	Bad	3	1.6
	TOTAL	185	100

Table 3 shows that 91.4% of respondents have high compliance in implementing the Covid-19 prevention and control in the workplace. As many as 90.3% of respondents have good knowledge about Covid-19 and health programs. Most of the respondents (95.7%) had a positive attitude towards Covid-19 and health protocols, and as many as 98.4% of respondents perceived a good company's support.

Table 4. Bivariate Analysis' Results

Variable	Compliance in Implementing Health Protocols: Obedient n (%)	Not Obedient n (%)	Total	95% CI	*p-value*
Level of Knowledge					
High Knowledge	156 (93.4)	11 (6.6)	167 (100)	0.078-0.360	**0.002**
Poor Knowledge	13 (72.2)	5 (27.8)	18 (100)		
Attitudes towards Covid-19 and Health Protocols					
Positive	164 (92.7)	13 (7.3)	177 (100)	0.072-0.355	**0.003**
Negative	3 (37.5)	5 (62.5)	8 (100)		
Perceived Company Support					
Good	168 (92.3)	14 (7.7)	182 (100)	0.121-0.398	**<0.001**
Bad	1 (33.3)	2 (66.7)	3 (100)		

Based on Table 4, it can be concluded that the majority of respondents who have a high level of knowledge about Covid-19 and comply with health protocols also have a high level of knowledge (93.4%). The analysis results indicate a relationship between the level of knowledge about Covid-19 and health protocols and the level of compliance in implementing health protocols in the workplace (p=0.002). The majority of respondents (92.7%%) with good compliance had a positive attitude towards Covid-19 and health protocols. The analysis results showed a relationship between attitudes towards Covid-19 and health protocols and the level of compliance in implementing health protocols in the workplace (p=0.003). The majority of respondents

(92.3%) perceived good company support and were compliant in implementing health protocols in the workplace. The analysis results indicate a relationship between perceived company support and the level of compliance in implementing health protocols among employees (p<0.001).

Table 5. Bivariate Analysis' Results from Sociodemographic Characteristic

Variable	Compliance in Implementing Health Protocols		Total	95% CI	*p-value*
	Obedient n (%)	Not Obedient n (%)			
Age					
≥ 30 years old	130 (90,9)	13 (9,1)	143 (100)	-0,177 – 0,120	0,695
< 30 years old	39 (92,9)	3 (7,1)	42 100)		
Gender					
Men	136 (90,1)	15 (9,9)	151 (100)	-,0241 – 0,053	**0,192**
Women	33 (97,1)	1 (2,9)	34 (100)		
Work Experience					
≥ 10 years	56 (91,8)	5 (8,2)	61 (100)	-0,137 – 0,159	0,879
< 10 years	113 (91,1)	11 (8,9)	124 (100)		
Educational Level					
University	9 (92,9)	7 (7,1)	99 (100)	-0,089 – 0,207	0,413
High School	77 (89,5)	9 (10,5)	86 (100)		
Income Level (IDR)					
> 15 mils.	77 (89,5)	9 (10,5)	86 (100)		
10 – 15 mils.	35 (94,6)	2 (5,4)	37 (100)	-0,206- 0,090	0,423
5 – 10 mils.	41 (89,1)	5 (10,9)	46 (100)		
3 – 5 mils.	16 (100)	0	16 (100)		
History of Covid-19 Infection					
Have	129 (90,8)	13 (9,2)	142 (100)	-0,180 – 0,116	0, 658
Have Not	40 (93,0)	3 (7,0)	43 (100)		
History of Covid-19 Vaccination					
3 Doses (*booster*)	146 (93,6)	10 (6,4)	156 (100)	0,037- 0,324	**0,012**
2 Doses	23 (79,3)	6 (20,7)	29 (100)		

Based on Table 5, it can be concluded that most respondents in the age group of 30 years old and above have high compliance (90.9%). As many as 90.1% of respondents were male and had high level of compliance. Workers with a higher level of education are more compliant with the implementation of health protocols in the workplace. Most workers (91.1%) with less than 10 years of working experience adhere to the implementation of the health program, with varying levels of income. A total of 142 respondents had been infected with Covid-19, and 129 of them had high adherence (90.8%) toward health protocols. All workers have received the Covid-19 vaccine from both the government and the company. A total of 146 respondents who had received 3 doses of Covid-19 vaccine (93.6%) adhered to the implementation of the health protocol in the workplace. Of all these sociodemographic characteristics, only the Covid-19 Vaccine History variable had an association on Worker Compliance in implementing health protocols (p=0.012).

Table 6. Multivariate Multiple Logistic Regression Analysis' Results

Variable	z	*p-value*	OR	95% CI
Knowledge Level	1,478	0,028	4,386	1,176-16,354
Attitudes towards Covid-19 and Health Protocols	2,011	0,016	7,473	1,4,63-38,170
Perceived Company Support	3,213	0,015	24,865	1,889-327,355

In Table 6, it can be concluded that the final model obtained includes independent variables that have a statistically significant influence, namely knowledge level, attitude, and company support (p-value<0.05). The formed model is considered appropriate as it satisfies the model's significance, indicated by the omnibus test with a p-value<0.001. The obtained coefficient of determination (R-squared) value is 0.194, meaning that the independent variables included in the model can explain 19.4% of the variance in compliance, while the remaining 80.6% is influenced by other factors. The variable that has the most dominant relationship with employee compliance is company support (z = 3.213). From the model, it can be observed that employees with good company support are 24 times more likely to comply with health protocols in the workplace compared to employees with insufficient company support.

4 Discussion

4.1 The Relationship Between Knowledge Regarding Covid-19 in The Workplace and Employee Compliance in Implementing Health Protocols in Workplace

The results of this study indicate a significant relationship between knowledge of COVID-19 and employee compliance in implementing COVID-19 health protocols. This suggests an improvement in compliance with an increase in knowledge. This finding is supported by the bivariate analysis results, which showed that knowledge has a p-value<0.05, indicating a significant association with compliance. The data analysis of the 185 respondents in this study reveals that

https://doi.org/10.1051/bioconf/20237505022

the majority of them have a high level of knowledge, with an accuracy rate of 90.3%. The research observations indicate that the company has provided education and training regarding COVID-19 to the employees. The company has also made promotive efforts by displaying posters and banners in various areas in the workplace. The questionnaire responses have further confirmed that the employees of Company X possess good knowledge about COVID-19.

The education and training provided primarily cover common clinical symptoms experienced by patients, the risk of transmission and severity of COVID-19, as well as the implementation of health protocols as preventive and control measures both at the workplace and at home. The success of these promotive efforts is reflected in the questionnaire responses of the respondents, with the majority answering correctly to the statements in the distributed questionnaire. However, there are certain aspects where many respondents lack knowledge, such as information regarding the use of antibiotics as a treatment for COVID-19, the effectiveness of using a face shield without a mask, and the possibility of transmission without exhibiting fever symptoms.

A high level of knowledge means that respondents understand the dangers and the process of COVID-19 transmission, which is then followed by compliance in implementing health protocols. This study is in line with some theories, which state that knowledge levels influence the level of compliance in implementing health protocols. The higher an individual's knowledge level, the higher their tendency to comply with health protocol implementation [14,15]. Similarly, previous studies have indicated a relationship between knowledge levels and compliance [16,17]. Both studies suggest that non-compliance is due to a lack of knowledge among workers. This study aligns with the findings of another research that good knowledge, along with positive attitudes of workers and company commitment, leads to an increased compliance with COVID-19 prevention protocols [18]. In conclusion, this study shows that the level of knowledge of employees at Company X regarding COVID-19 and health protocols significantly influences their compliance in preventing and controlling COVID-19 in the workplace.

4.2 The Relationship Between Attitudes Towards Covid-19 in The Workplace and Employee Compliance in Implementing Health Protocols in The Workplace

The results of this study indicate a significant relationship between workers' attitude towards Covid-19 and their compliance in implementing Covid-19 prevention and control protocols. This is supported by the statistical test results, which showed a p-value of 0.002 ($p<0.05$) for the relationship between attitudes and compliance. These findings demonstrate that individuals with positive attitudes are more likely to exhibit compliant behavior in implementing health protocols and following regulations for the prevention and control of Covid-19 in the workplace. This is consistent with some theories which suggests that attitudes are related to individuals' intentions and readiness to behave in accordance with stimuli in performing certain actions, in this case, the implementation of Covid-19 prevention and control measures [19,20].

The study conducted reveals that there are still some respondents who exhibited negative attitudes, which can be interpreted as a sign of boredom and lack of interest in Covid-19 and the implementation of health protocols. Even though the compliance level of workers is relatively high, the observation results of this study showed that some workers rarely wore masks in their daily activities. This might be because of some workers experiencing breathing difficulty because they were working in underground mines. This finding aligns with previous study which suggests that a decrease in compliance behavior may be due to feelings of boredom resulting from limited activities that have been restricted since the start of the pandemic [21]. Based on the observations, respondents have adapted to the presence of Covid-19 in their daily lives, leading to a reduction in anxiety as well.

This study is consistent with previous research which found a relationship between attitudes and the level of compliance in implementing Covid-19 health protocols [22]. Based on the theories and previous research mentioned above, it can be concluded that in this study, an individual's attitude has been proven to have a correlation with compliance in implementing prevention and control protocols for Covid-19 in the workplace. Similarly, in another study [23], it was found that attitudes are associated with both compliance and non-compliance, and compliance should be approached by analyzing the constructive and destructive aspects of individual attitudes. Positive attitudes will lead to constructive behavior, which in this study refers to compliance behavior. Therefore, it can be concluded that the positive and constructive attitudes of employees at Company X are associated with compliance in the prevention and control of Covid-19 in the workplace.

4.3 The Relationship Between Perceived Company Support Among Employees in The Workplace and Worker Compliance in Implementing Health Protocols at Workplace

The results of this study have shown that there is a significant relationship between Perceived Company Support in the workplace and Workers' Compliance in the Implementation of Covid-19 Prevention and Control Protocols. This is supported by the statistical test results indicating that knowledge about compliance has a p-value of <0.001. This finding proves that individuals with high compliance were also influenced by the support from the company to implement and follow regulations for Covid-19 prevention and control in the workplace.

One of the supports that can be provided to workers is the availability of adequate facilities and infrastructure for Covid-19 prevention and control in the workplace. A previous study stated that the availability of prevention facilities and infrastructure in the workplace affects workers' compliance in implementing health protocols in the workplace [24[. The availability of facilities and infrastructure in the workplace is the responsibility of the company owners. Therefore, if these needs are met, workers will feel the positive management support. This company has been following the government regulations for Covid-19 prevention and control since the beginning of the pandemic. They also have a special budget to support Covid-19 prevention and control in the workplace to their maximum capacity. The observation results in the study also showed that Company X prioritizes the safety and health of workers by implementing lockdown policies and shifting the focus from production processes to the recovery and health of workers as the top priority when multiple positive cases were found in the workplace. The initial case of Covid-19 caused many workers to be unable to meet their targets, causing a decline in the mining process and output. This is one of the main reasons this company prioritizes the wellbeing of their employees. Workers who have a healthy work environment can increase their work productivity and job satisfaction [25].

The observation results showed that the availability of facilities and infrastructure in efforts to prevent Covid-19 has been fulfilled by 98%. Researchers observed that temperature checking was no longer available in the work area, but it was only carried out by the health workers at the company's main entrance gate as a screening method before employees begin their roster shift. This was done because temperature checking using the contactless thermometer was not effective in early detection of Covid-19 cases anymore because not all those infected with Covid-19 show symptoms of fever [26]. In addition, the company has a PCR inspections schedule that is routinely carried out on workers which is more effective in detecting Covid-19 cases.

Good management support can influence the psychological aspects of workers. When the psychological aspects of workers are well fulfilled, it becomes one of the triggering factors for improving their behavior [27]. This study is consistent with another research which proves that workers will find it easier to implement or apply Covid-19 prevention and control programs if they are supported by the company [28]. Based on the theories and previous research mentioned above, it can be concluded that company support in this study for ensuring compliance in the implementation of Covid-19 prevention and control protocols in the workplace is very high. Similarly, previous research emphasizes the need for company support to ensure effective implementation of health protocols [29]. High compliance among workers in company X in adhering to these rules must be initiated by company support with a shared commitment between management and workers to provide tangible support. In conclusion, this study shows that the perceived company support by workers in company X has an impact on the level of compliance in the implementation of Covid-19 prevention and control protocols in the workplace.

4.4 The Relationship Between the History of Covid-19 Vaccination in The Workplace and The Compliance of Workers in Implementing Health Protocols in The Workplace

The analysis results in this study indicate a significant relationship between the history of Covid-19 vaccination and the compliance of workers in implementing Covid-19 prevention and control protocols. This suggests an increase in the variable of knowledge with compliance. This is supported by the bivariate test results, which show that knowledge regarding compliance has a p-value of 0.012 ($p<0.05$). Most workers have received the Covid-19 vaccine up to the third dose, while the remaining workers have only received two doses.

The observation findings indicate that the workers of the company have been compliant in implementing Health Protocols in the Workplace by receiving vaccinations and undergoing screenings before starting work in their respective environment. Company X mandates that workers need to receive Covid-19 vaccine, as recommended by the government, with a minimum of two doses. Based on the observations made by the researchers, workers received the Covid-19 vaccine from their respective areas through government programs usually conducted at health centers, hospitals, and other public facilities. Vaccination is one of the preventive measures taken by company X considering the high mobility of the workers.

One of the policy efforts made by Company X is social distancing in the workplace, provision of self-isolation rooms, provision of vaccination for employees, and implementation of health protocols in the workplace. These policies are made as preventive and mitigation efforts against Covid-19, tailored to the characteristics of the mining Company X. This demonstrates that individuals who receive vaccinations will certainly implement or follow regulations for the prevention and control of Covid-19 in the work environment. This is inconsistent with some theories that suggest that the Covid-19 vaccine may lead to a decrease or relaxation in the adherence to health protocols because it is assumed to be safe from Covid-19 [21,30]. The study is in line with some research that shows that even though individuals have been vaccinated, they are not completely free from the transmission of Covid-19. Instead, it is one of the preventive measures and helps in suppressing the spread of Covid-19 [31].

This study is consistent with research stating that the more people receive vaccinations, the higher the compliance level with protocol implementation tends to be, as it is considered a preventive behavior [32]. Similarly, previous research indicates that vaccines can reduce the risk of spreading Covid-19 infections, but other preventive measures should still be followed. Communication and education regarding the importance of vaccination also need to be enhanced, as vaccination is the most effective way to prevent infectious diseases [33,34]. Previous research suggests that a lack of education about vaccination can lead to misinformation about vaccines, causing individuals to hesitate in getting vaccinated. This hesitation can significantly affect individuals' compliance with vaccination, which is one of the preventive measures in workplace health protocols [35]. In conclusion, this study indicates that the Covid-19 vaccination history can influence their compliance with prevention and control measures for Covid-19 in the workplace.

4.5 The Relationship Between Knowledge, Attitude, Perceived Company Support, and The History of Covid-19 Vaccination in The Workplace with The Compliance of Workers in Implementing Health Protocols in The Workplace

The results of this study indicate that there is a correlation between Company Support in the workplace and Workers' Compliance in implementing Covid-19 prevention and control protocols. This is supported by the results of the multivariate analysis, which show that knowledge, attitude, and perceived company support have a 19.4% influence on compliance levels. Company support has the most dominant influence on workers' compliance behavior, followed by individual attitudes and knowledge. Workers who receive good company support tend to have high levels of compliance, and those with positive attitudes are more likely to adhere to health protocols. This is consistent with a study suggesting that Workers' Compliance in implementing Covid-19 prevention and control protocols should be supported by positive attitudes and adequate support to ensure effective implementation [8].

This study is consistent with the research conducted by [36], which demonstrates that workers who have high compliance also have high attitudes and knowledge, with the suggestion of company support through gradual supervision by management. This study also aligns with research [37] that identifies the factors influencing compliance with health protocols, such as supervision from management or supervisors in the workplace, and with the support of a company, compliance can be smoothly implemented. Similar to the research conducted by [38], it is not only the influence of individual characteristics on compliance with Personal Protective Equipment (PPE) usage but also the influence of perceived vulnerability, perceived benefits, training, and supervision, which are all signals for compliance with protocols.

A good working climate is created by company management to support and protect their employees at work [3]. Workers would find it easier to implement the Covid-19 prevention and control programs if they were supported by the company. Clear communication regarding health protocols as well as providing training in preventing Covid-19 can help workers better understand the intent and purpose of implementing these health protocols {28]. It can be concluded from previous research related to perceived company support in the workplace and worker compliance, that high support and good communication is needed so that workers can successfully implementing health protocols for prevention and control in the working environment.

Factors that affect employee compliance with health and safety measures at work are related to the work environment, such as the theory by [39] which states that good compliance and safety awareness in employees of health and safety policies and their implications can improve worker welfare. Likewise with compliance in health behavior based on social norms. In the context of the work environment, it can be used to analyze how social norms in the workplace can affect employee health behavior. Behavior research based on social norms according to [40] that employee health compliance tends to come from the completeness of the facilities provided by the company which will then be followed by the good living habits of the workers.

From the results of the theory and research above, it can be concluded that creating a system and culture that encourages employee compliance in good health behavior at work is an important step for the company. If this is successfully adopted in the work environment, it can improve employee welfare, reduce absenteeism, and increase productivity. To implement these steps, it is important for companies to involve employees in the decision-making process and implementing health programs can help build a positive health culture that adheres to health behavior in the workplace. It can be concluded that in this study it proves that perceived company support at Company X influence the Employee Compliance in the Implementation of the Covid-19 Prevention and Controls.

5 Conclusion and Recommendation

The level of knowledge, attitudes towards Covid-19 and health protocols, as well as perceived company support, influence workers' compliance with Covid-19 prevention and control protocols at PT. X by 19.4%, while the remaining 80.6% is attributed to other factors. Workers with good company support are 24 times more likely to comply with health protocols in the workplace compared to those who perceive less company support.

It is expected that the company will continue its obligation in implementing health protocols in the workplace because it is an adaptive health behavior that can protect both workers and the company. Health protocols in the workplace not only serve to prevent Covid-19 but also address various other health issues that may disrupt work activities.

References

1. Leahy, S. Mines are Hostspots for Spread of Covid-19, study finds. (2020) https://www.theguardian.com/environment/2020/j

un/05/mines-coronavirus-hotspots-report-us-canada

2. Wareza, M. 1.219 Positif Covid-19 di BUMN Tambang, Freeport Terbanyak (2020)
3. Hubert, P., Abdel Hadi, S., Mojzisch, A., & Häusser, J. A. The effects of organizational climate on adherence to guidelines for COVID-19 prevention. Social Science and Medicine, 292. (2022) https://doi.org/10.1016/j.socscimed.2021.114622
4. Tanto, & Handayani. Literature Review: Determinan Kepatuhan Terhadap Protokol Kesehatan COVID-19 di Indonesia. 2(11), 127–136. (2022) https://doi.org/https://doi.org/10.33221/jikm.v11i02.1274
5. Wijayanti, P. Y., & Yuantari, M. G. C. Kepatuhan Penerapan Protokol Kesehatan pada Pekerja di Tempat Ibadah Selama Pandemi Covid-19. Jurnal Penelitian Dan Pengembangan Kesehatan Masyarakat Indonesia, 2(2), 129–140. (2021) https://doi.org/10.15294/jppkmi.v2i2.52426
6. Tanto, & Handayani. Literature Review: Determinan Kepatuhan Terhadap Protokol Kesehatan COVID-19 di Indonesia. 2(11), 127–136. (2022) https://doi.org/https://doi.org/10.33221/jikm.v11i02.1274
7. A Angin, S. R., & Astutik, E. KNOWLEDGE, ATTITUDE, AND PERCEPTION OF PEOPLE IN COMPLIANCE WITH THE COVID-19 HEALTH PROTOCOL. Jurnal Berkala Epidemiologi, 10(1), 103–110. (2022)
8. Sangadji, N. W., & Aningsih. HUBUNGAN ANTARA DUKUNGAN MANAJEMEN, PENGETAHUAN DAN SIKAP TERHADAP KEPATUHAN KARYAWAN PADA PROTOKOL PENCEGAHAN COVID 19 DI PT. ASURANSI SINARMAS JAKARTA.pdf. Jurnal Ilmu Kesehatan Masyarakat, (2021), 38–44. https://doi.org/10.19184/ikesma/v090.27115
9. Nguyen, H. N., & Tran, M. D. The effect of perceived organizational support on employee work engagement: An empirical study in Vietnam. Journal of Asian Finance, Economic and Business, 8(6), 0415–0426. (2021). https://doi.org/10.13106/jafeb.2021.vol8.no6.0415
10. Sugiyono. Metode Penelitian Kuantitatif, Kualitatif dan R&D. Bandung: Alfabeta.CV (2013)
11. World Health Organization. Coronavirus Disease (Covid-19) (2020)
12. Centers for Disease Control and Prevention. Hierarchy of Controls. (2015).
13. Kementerian Kesehatan RI. Peta Sebaran Covid-19. (2020)
14. Rincón Uribe, F. A., Godinho, R. C. de S., Machado, M. A. S., Oliveira, K. R. da S. G., Neira Espejo, C. A., de Sousa, N. C. V., de Sousa, L. L., Barbalho, M. V. M., Piani, P. P. F., & Pedroso, J. da S. (2021). Health knowledge, health behaviors and attitudes during pandemic emergencies: A systematic review. PloS One, 16(9), e0256731. https://doi.org/10.1371/journal.pone.0256731
15. Sangadji, N. W., & Aningsih. HUBUNGAN ANTARA DUKUNGAN MANAJEMEN, PENGETAHUAN DAN SIKAP TERHADAP KEPATUHAN KARYAWAN PADA PROTOKOL PENCEGAHAN COVID 19 DI PT. ASURANSI SINARMAS JAKARTA.pdf. Jurnal Ilmu Kesehatan Masyarakat, 38–44. (2021) https://doi.org/10.19184/ikesma/v090.27115
16. Farida Yeni. Analisis Hubungan Pengetahuan Terhadap Kepatuhan Terapi Pada Pasien Hipertensi di Puskesmas Pucang Sawit Surakarta. Vol 6, No 3 (2021)
17. Wahyuni. GAMBARAN KARAKTERISTIK DAN PENGETAHUAN COVID-19 PADA REMAJA DI DAERAH SIDOREJO SUKOHARJO. doi: https://doi.org/10.36419/jki.v13i2.624 (2021)
18. Ahmad Baijuri. Faktor-Faktor yang berhubungan dengan kepatuhan terhadap protokol pencegahan Covid-19 pada pegawai Minimarket di Wilayah Kota Jakarta Tahun 2021. (2021)
19. Ajzen, IcekAttitudes, personality, and behavior. Milton Keynes: Open University Press dan Chicago, IL: Dorsey Press. (1991)
20. Fenitra, R. M., Praptapa, A., Suyono, E., Kusuma, P. D. I., & Usman, I. Factors Influencing Preventive Intention Behavior Towards COVID-19 in Indonesia. The Journal of Behavioral Science (TJBS), 16(1), 14–27. (2021)
21. Andersson, O., Campos-Mercade, P., Meier, A. N., & Wengström, E. Anticipation of COVID-19 vaccines reduces willingness to socially distance. Journal of Health Economics, 80. (2021) https://doi.org/10.1016/j.jhealeco.2021.102530
22. Angin, S. R., & Astutik, E. KNOWLEDGE, ATTITUDE, AND PERCEPTION OF PEOPLE IN COMPLIANCE WITH THE COVID-19 HEALTH PROTOCOL. Jurnal Berkala Epidemiologi, 10(1), 103–110. (2022)
23. Passini. The Obedience-disobedience Dynamic and the Role of Responsibility. Journal of Community & Applied Social Psychology 20(1):1-14. (2009) doi: 10.1002/casp.1000
24. Nuriati, Y., Heryana, A., Mustikawati, I. S., & Sangadji, N. W. (2021). Persepsi Karyawan terhadap Ketersediaan Fasilitas dan Sarana Penanganan Covid-19 di Tempat Kerja Berhubungan dengan Kepatuhan Protokol Kesehatan. Jurnal Kesehatan Masyarakt, 9(4), 566–575
25. Voordt, T. and Jensen, P. The Impact of Healthy Workplaces on Employee Satisfaction, Productivity, and Cost. Journal of Corporate Real Estate, 25(1):29-49. (2023) doi: 10.1108/JCRE-03-2021-0012

26. Wright, W. and MacKowiak, P. Why Temperature Screening for Coronavirus Disease 2019 with Noncontact Infrared Thermometers Does Not Work. Open Forum Infectious Disease, 8(1):4-6. (2021). doi: 10.1093/ofid/ofaa603
27. Sun Li. Perceived Organizational Support: A Literature Review. International Journal of Human Resource Studies 9(3):155. doi: 10.5296/ijhrs.v9i3.15102
28. Donaghy, J. Organisational support improves adherence to infection prevention and control guidelines. Evidence-Based Nursing, 25(1), 10. (2022) https://doi.org/10.1136/ebnurs-2020-103305
29. Kementerian Kesehatan RI. 2020. Peta Sebaran Covid-19. https://covid19.go.id/peta-sebaran-covid19.
30. Wright, L., Steptoe, A., Mak, H. W., & Fancourt, D. Do people reduce compliance with COVID-19 guidelines following vaccination? A longitudinal analysis of matched UK adults. Journal of Epidemiology and Community Health, 76(2), 109–115. (2022) https://doi.org/10.1136/jech-2021-217179
31. Zheng, C., Shao, W., Chen, X., Zhang, B., Wang, G., & Zhang, W. Real-world effectiveness of COVID-19 vaccines: a literature review and meta-analysis. International Journal of Infectious Diseases, 114, 252–260. (2022) https://doi.org/10.1016/j.ijid.2021.11.009
32. Sara Rubinelli. Knowledge of COVID-19 and Its Relationship with Preventive Behaviors and Vaccination among Adults in Northern Thailand's Community. Int J Environ Res Public Health. 2022 Feb; 19(3): 1521. (2022) doi: 10.3390/ijerph19031521
33. Branko. Compliance with Preventive Measures and COVID-19 Vaccination Intention among Medical and Other Healthcare Students. Int. J. Environ. Res. Public Health 2022, 19, 11656. (2022) https://doi.org/10.3390/ijerph191811656
34. Kin Israel. Impact of Covid-19 Vaccination on the Risk Developing Long-Covid and on Existing long-Covid Symptoms: A Systematic Review. eClinicalMedicine. (2022) Doi: https://doi.org/10.1016/j.eclinm.2022.101624
35. Michael Gisondi. Misinformation About COVID-19 Vaccines on Social Media: Rapid Review. PubMed Central. (2022) Doi: 10.2196/37367
36. Adriansyah, A. A., Suyitno, S., & Sa'adah, N. Kepatuhan Penggunaan Alat Pelindung Diri (Apd) Masker Ditinjau Dari Pengetahuan Dan Sikap Pekerja. Ikesma, 17(1), 39. (2021) https://doi.org/10.19184/ikesma.v17i1.22452
37. Sari. Illness Representation dan Kepatuhan Masyarakat Pada Anjuran Pemerintah. Fakultas Psikologi Universitas Semarang (2021)
38. Muk Thar. Psychological health during the coronavirus disease 2019 pandemic outbreak. Int J Soc Psychiatry. 2020 Aug;66(5):512-516. (2020) doi: 10.1177/0020764020925835
39. Smith, J. The Role of Compliance in Workplace Health and Safety. Journal of Occupational Health Psychology. (2018)
40. Tracy Kolbe-Alexander. The relationship between workplace environment and employee health behaviors in a South African workforce. J Occup Environ Med. (2019)

AUTHOR INDEX